Methods in Enzymology

Volume 214
CAROTENOIDS
Part B
Metabolism, Genetics, and Biosynthesis

METHODS IN ENZYMOLOGY

EDITORS-IN-CHIEF

John N. Abelson   Melvin I. Simon

DIVISION OF BIOLOGY
CALIFORNIA INSTITUTE OF TECHNOLOGY
PASADENA, CALIFORNIA

FOUNDING EDITORS

Sidney P. Colowick and Nathan O. Kaplan

Methods in Enzymology
Volume 214

Carotenoids

Part B
Metabolism, Genetics, and Biosynthesis

EDITED BY

Lester Packer

DEPARTMENT OF MOLECULAR AND CELL BIOLOGY
UNIVERSITY OF CALIFORNIA, BERKELEY
BERKELEY, CALIFORNIA

Editorial Advisory Board

Louise Canfield
John E. Hearst
Norman Krinsky

Maria A. Livrea
James A. Olson
George A. Wolf

ACADEMIC PRESS, INC.
Harcourt Brace Jovanovich Publishers
San Diego   New York   Boston
London   Sydney   Tokyo   Toronto

This book is printed on acid-free paper. ∞

Copyright © 1993 by ACADEMIC PRESS, INC.
All Rights Reserved.
No part of this publication may be reproduced or transmitted in any form or by any means, electronic or mechanical, including photocopy, recording, or any information storage and retrieval system, without permission in writing from the publisher.

Academic Press, Inc.
1250 Sixth Avenue, San Diego, California 92101-4311

United Kingdom Edition published by
Academic Press Limited
24–28 Oval Road, London NW1 7DX

Library of Congress Catalog Number: 54-9110

International Standard Book Number: 0-12-182115-3

PRINTED IN THE UNITED STATES OF AMERICA
93 94 95 96 97 98   EB   9 8 7 6 5 4 3 2 1

Table of Contents

CONTRIBUTORS TO VOLUME 214 ix

PREFACE xiii

VOLUMES IN SERIES. xv

Section I. Metabolism

A. Absorption and Transport

1.	Carotenoid Absorption in Humans	PHYLLIS E. BOWEN, SOHRAB MOBARHAN, AND J. CECIL SMITH, JR.	3
2.	Evaluation of Carotenoid Intake	MARC S. MICOZZI	17
3.	Uptake and Cleavage of β-Carotene by Cultures of Rat Small Intestinal Cells and Human Lung Fibroblasts	GIORGIO SCITA, GREGORY W. APONTE, AND GEORGE WOLF	21
4.	Association of Carotenoids with Human Plasma Lipoproteins	BEVERLY A. CLEVIDENCE AND JOHN G. BIERI	33
5.	Translocation of Carotenoid Droplets in Goldfish Xanthophores	T. T. TCHEN AND JOHN D. TAYLOR	47
6.	Solubilization, Cellular Uptake, and Activity of β-Carotene and Other Carotenoids as Inhibitors of Neoplastic Transformation in Cultured Cells	ROBERT V. COONEY, T. JOSEPH KAPPOCK, IV, AO PUNG, AND JOHN S. BERTRAM	55
7.	Differentiation between Central and Excentric Cleavage of β-Carotene	GUANGWEN TANG AND NORMAN I. KRINSKY	69

B. Tissue Distribution

8.	Carotenoid–Protein Complexes	M. R. LAKSHMAN AND CHITUA OKOH	74
9.	Analysis of Carotenoids in Human Plasma and Tissues	ROBERT S. PARKER	86
10.	Assessing Variability in Quantitation of Carotenoids in Human Plasma: Variance Component Model	ANNA R. GIULIANO, MONIKA B. MATZNER, AND LOUISE M. CANFIELD	94

11. Analysis of Carotenoids in Human and Animal Tissues	HAROLD H. SCHMITZ, CHRISTOPHER L. POOR, ERIC T. GUGGER, AND JOHN W. ERDMAN, JR.	102
12. Plasma Carotenoid Levels in Anorexia Nervosa and in Obese Patients	CHERYL L. ROCK AND MARIAN E. SWENDSEID	116
13. Psoralen Photosensitization and Plasma and Cutaneous β-Carotene Concentrations in Hairless Mice	WENDY S. WHITE, HAROLD B. FAULKNER, AND DAPHNE A. ROE	124

C. Transformation and Metabolic Regulation

14. Human Metabolism of Carotenoid Analogs and Apocarotenoids	SIHUI ZENG, HAROLD C. FURR, AND JAMES A. OLSON	137
15. Metabolism of Carotenoids and *in Vivo* Racemization of (3*S*,3'*S*)-Astaxanthin in the Crustacean *Penaeus*	KATHARINA SCHIEDT, STEFAN BISCHOF, AND ERNST GLINZ	148
16. Assay for Carotenoid 15,15'-Dioxygenase in Homogenates of Rat Intestinal Mucosal Scrapings and Application to Normal and Vitamin A-Deficient Rats	LAURENCE VILLARD-MACKINTOSH AND CHRISTOPHER J. BATES	168
17. Control of Carotenoid Synthesis by Light	KLAUS HUMBECK AND KARIN KRUPINSKA	175
18. Functions of Carotenoids in Photosynthesis	RICHARD J. COGDELL AND ALASTAIR T. GARDINER	185
19. Retinoic Acid Synthesis from β-Carotene *in Vitro*	JOSEPH L. NAPOLI	193
20. β-Carotene Modulation of Delayed Light Emission from Aggregated Chlorophyll	SHAN YUAN YANG	202
21. Relation of Cis–Trans Isomers of Carotenoids to Developmental Processes	H. J. NELIS, P. SORGELOOS, AND A. P. DE LEENHEER	208
22. *In Vitro* Biological Methods for Determination of Carotenoid Activity	JOEL L. SCHWARTZ	226
23. Enzymatic Conversion of all-*trans*-β-Carotene to Retinal	M. R. LAKSHMAN AND CHITUA OKOH	256
24. Methods for Investigating Photoregulated Carotenogenesis	MIKIRO TADA	269
25. Photoinduction of Carotenoid Biosynthesis	JAVIER ÁVALOS, EDUARDO R. BEJARANO, AND E. CERDÁ-OLMEDO	283

Section II. Molecular and Cell Biology

A. Genetics

26. Evolutionary Conservation and Structural Similarities of Carotenoid Biosynthesis Gene Products from Photosynthetic and Nonphotosynthetic Organisms — GREGORY A. ARMSTRONG, BHUPINDER S. HUNDLE, AND JOHN E. HEARST — 297

27. Cloning of Carotenoid Biosynthetic Genes from Maize — BRENT BUCKNER AND DONALD S. ROBERTSON — 311

28. Protection by Cloned Carotenoid Genes Expressed in *Escherichia coli* against Phototoxic Molecules Activated by Near-Ultraviolet Light — R. W. TUVESON AND G. SANDMANN — 323

B. Biosynthesis

29. Biosynthesis of Carotenoids: An Overview — T. W. GOODWIN — 330

30. Carotenoid Analysis in Mutants from *Escherichia coli* Transformed with Carotenogenic Gene Cluster and *Scenedesmus obliquus* Mutant C-6D — GERHARD SANDMANN — 341

31. Effects of δ-Aminolevulinic Acid and 3-Amino-1,2,4-Triazole on Carotenoid Accumulation — GEORGES T. DODDS, SYAMALA S. ASHTAKALA, AND STEPHEN W. LAMOUREUX — 348

32. Plant Phytoene Synthase Complex: Component Enzymes, Immunology, and Biogenesis — BILAL CAMARA — 352

33. Solubilization and Purification Procedures for Phytoene Desaturase from *Phycomyces blakesleeanus* — PAUL D. FRASER AND PETER M. BRAMLEY — 365

34. Functional Analysis and Purification of Enzymes for Carotenoid Biosynthesis Expressed in Photosynthetic Bacteria — GLENN E. BARTLEY, ANETTE KUMLE, PETER BEYER, AND PABLO A. SCOLNIK — 374

35. β-Carotene Synthesis in *Rhodotorula* — HEBE MARTELLI AND IRACEMA M. DA SILVA — 386

36. Characteristics of Membrane-Associated Carotenoid-Binding Proteins in Cyanobacteria and Prochlorophytes — K. J. REDDY, GEORGE S. BULLERJAHN, AND LOUIS A. SHERMAN — 390

37. Xanthosomes: Supramolecular Assemblies of Xanthophyll–Chlorophyll *a/c* Protein Complexes — TETZUYA KATOH, AYUMI TANAKA, AND MAMORU MIMURO — 402

38. Photoregulated Carotenoid Biosynthetic Genes of GIORGIO MORELLI,
     Neurospora crassa   MARY ANNE NELSON,
                                                     PAOLO BALLARIO, AND
                                                     GIUSEPPE MACINO   412

AUTHOR INDEX 425

SUBJECT INDEX 443

Contributors to Volume 214

Article numbers are in parentheses following the names of contributors.
Affiliations listed are current.

GREGORY W. APONTE (3), *Department of Nutritional Sciences, University of California, Berkeley, Berkeley, California 94720*

GREGORY A. ARMSTRONG (26), *Department of Plant Genetics, Institute for Plant Sciences, Swiss Federal Institute of Technology (ETH), CH-8092 Zürich, Switzerland*

SYAMALA S. ASHTAKALA (31), *Department of Biology, Concordia University, Montreal, Quebec, Canada H3G 1M8*

JAVIER ÁVALOS (25), *Departamento de Genética, Universidad de Sevilla, E-41012 Seville, Spain*

PAOLO BALLARIO (38), *Dipartimento di Genetica e Biologia Molecolare, Centro di Studio per gli Acidi Nucleici, Università di Roma "La Sapienza," 00185 Rome, Italy*

GLENN E. BARTLEY (34), *DuPont Central Research and Development, Wilmington, Delaware 19880*

CHRISTOPHER J. BATES (16), *Dunn Nutritional Laboratory, Medical Research Council, Cambridge CB4 1XJ, England*

EDUARDO R. BEJARANO (25), *Departamento de Biología Celular y Genética, Universidad de Malaga, E-29071 Málaga, Spain*

JOHN S. BERTRAM (6), *Molecular Oncology Program, Cancer Research Center of Hawaii, University of Hawaii, Honolulu, Hawaii 96813*

PETER BEYER (34), *Institut für Biologie II, Albert Ludwig Universität, D-7800 Freiburg, Germany*

JOHN G. BIERI (4), *National Institute of Diabetes and Digestive and Kidney Diseases, U.S. Department of Health and Human Services, National Institutes of Health, Bethesda, Maryland 20892*

STEFAN BISCHOF (15), *Department of Vitamin and Nutrition Research, F. Hoffmann-La Roche, Ltd., CH-4002 Basel, Switzerland*

PHYLLIS E. BOWEN (1), *Department of Nutrition and Medical Dietetics, University of Illinois at Chicago, Chicago, Illinois 60612*

PETER M. BRAMLEY (33), *Department of Biochemistry, Royal Holloway, University of London, Surrey TW20 0EX, England*

BRENT BUCKNER (27), *Division of Science, Northeast Missouri State University, Kirksville, Missouri 63501*

GEORGE S. BULLERJAHN (36), *Department of Biological Sciences, Center for Photochemical Sciences, Bowling Green State University, Bowling Green, Ohio 43403*

BILAL CAMARA (32), *Institut de Biologie Moléculaire des Plantes, Centre National de la Recherche Scientifique, F-67084 Strasbourg, France*

LOUISE M. CANFIELD (10), *Department of Biochemistry, University of Arizona, Tuscon, Arizona 85721*

E. CERDÁ-OLMEDO (25), *Departamento de Genética, Universidad de Sevilla, E-41012 Seville, Spain*

BEVERLY A. CLEVIDENCE (4), *Lipid Nutrition Laboratory, Beltsville Human Nutrition Research Center, U.S. Department of Agriculture, Agricultural Research Service, Beltsville, Maryland 20705*

RICHARD J. COGDELL (18), *Department of Botany, University of Glasgow, Glasgow G12 8QQ, Scotland*

ROBERT V. COONEY (6), *Molecular Oncology Program, Cancer Research Center of Hawaii, University of Hawaii, Honolulu, Hawaii 96813*

IRACEMA M. DA SILVA (35), *Departamento de Engenharia Bioquímica, Universidade Federal do Rio de Janeiro, Rio de Janeiro 21941, Brazil*

A. P. DE LEENHEER (21), *Laboratoria voor Medische Biochemie en voor Klinische Analyse, Universiteit Gent, B-9000 Gent, Belgium*

GEORGES T. DODDS (31), *Department of Agricultural Engineering, Macdonald College of McGill University, Ste. Anne-de-Bellevue, Quebec, Canada H9X 3V9*

JOHN W. ERDMAN, JR. (11), *Division of Nutritional Sciences, University of Illinois at Urbana-Champaign, Urbana, Illinois 61801*

HAROLD B. FAULKNER (13), *Faculty of Medicine, University of Toronto, Toronto, Ontario, Canada M5S 1A8*

PAUL D. FRASER (33), *Department of Biochemistry, Royal Holloway, University of London, Surrey TW20 0EX, England*

HAROLD C. FURR (14), *Department of Nutritional Sciences, University of Connecticut, Storrs, Connecticut 06269*

ALASTAIR T. GARDINER (18), *Department of Botany, University of Glasgow, Glasgow G12 8QQ, Scotland*

ANNA R. GIULIANO (10), *Department of Family and Community Medicine, University of Arizona, Tuscon, Arizona 85716*

ERNST GLINZ (15), *Department of Pharmaceutical Research, F. Hoffmann-La Roche, Ltd., CH-4002 Basel, Switzerland*

T. W. GOODWIN (29), *Department of Biochemistry, University of Liverpool, Liverpool, England*

ERIC T. GUGGER (11), *Department of Food Science, University of Illinois at Urbana-Champaign, Urbana, Illinois 61801*

JOHN E. HEARST (26), *Department of Chemistry, University of California, Berkeley, Berkeley, California 94720*

KLAUS HUMBECK (17), *Institut für Allgemeine Botanik, D-2000 Hamburg 52, Germany*

BHUPINDER S. HUNDLE (26), *Department of Neurology, School of Medicine, University of California, San Francisco, San Francisco, California 94110*

T. JOSEPH KAPPOCK, IV (6), *Molecular Oncology Program, Cancer Research Center, University of Hawaii, Honolulu, Hawaii 96813*

TETZUYA KATOH (37), *Department of Botany, Faculty of Science, Kyoto University, Kyoto 606-01, Japan*

NORMAN I. KRINSKY (7), *Department of Biochemistry, Tufts University School of Medicine, Boston, Massachusetts 02111*

KARIN KRUPINSKA (17), *Institut für Allgemeine Botanik, D-2000 Hamburg 52, Germany*

ANETTE KUMLE (34), *Institut für Biologie II, Albert Ludwig Universität, D-7800 Freiburg, Germany*

M. R. LAKSHMAN (8, 23), *Department of Medicine, George Washington University, Washington, D.C. 20037*

STEPHEN W. LAMOUREUX (31), *Pharmaceutical Division, Boehringer Manheim, Laval, Quebec, Canada H7V 4A2*

SOHRAB MOBARHAN (1), *Department of Medicine/Gastroenterology, Loyola University Medical Center, Maywood, Illinois 60153*

GIUSEPPE MACINO (38), *Dipartimento di Biopatologia Umana, Policlinico Umberto I, Università di Roma "La Sapienza," 00164 Rome, Italy*

HEBE MARTELLI (35), *Departmento de Engenharia Bioquímica, Universidade Federal do Rio de Janeiro, Rio de Janeiro 21941, Brazil*

MONIKA B. MATZNER (10), *Arizona Cancer Center, University of Arizona, Tuscon, Arizona 85724*

MARC S. MICOZZI (2), *National Museum of Health and Medicine, Armed Forces Institute of Pathology, Washington, D.C. 20306*

MAMORU MIMURO (37), *National Institute for Basic Biology, Okazaki, Aichi 444, Japan*

GIORGIO MORELLI (38), *Unità di Nutrizione Sperimentale, Istituto Nazionale della Nutrizione, 00178 Rome, Italy*

JOSEPH L. NAPOLI (19), *Department of Biochemistry, School of Medicine and Bio-*

medical Sciences, State University of New York at Buffalo, Buffalo, New York 14214

H. J. NELIS (21), *Laboratoria voor Medische Biochemie en voor Klinische Analyse, Universiteit Gent, B-9000 Gent, Belgium*

MARY ANNE NELSON (38), *Department of Biology, University of New Mexico, Albuquerque, New Mexico 87131*

CHITUA OKOH (8, 23), *Department of Medicine, George Washington University, Washington, D.C. 20037*

JAMES A. OLSON (14), *Department of Biochemistry and Biophysics, Iowa State University, Ames, Iowa 50011*

ROBERT S. PARKER (9), *Division of Nutritional Sciences, Cornell University, Ithaca, New York 14853*

CHRISTOPHER L. POOR (11), *Division of Nutritional Sciences, University of Illinois at Urbana-Champaign, Urbana, Illinois 61801*

AO PUNG (6), *Molecular Oncology Program, Cancer Research Center, University of Hawaii, Honolulu, Hawaii 96813*

K. J. REDDY (36), *Department of Biological Sciences, State University of New York at Binghamton, Binghamton, New York 13902*

DONALD S. ROBERTSON (27), *Department of Zoology and Genetics, Iowa State University, Ames, Iowa 50011*

CHERYL L. ROCK (12), *Program in Human Nutrition, The University of Michigan, Ann Arbor, Michigan 48109*

DAPHNE A. ROE (13), *Division of Nutritional Sciences, Cornell University, Ithaca, New York 14853*

GERHARD SANDMANN (28, 30), *Lehrstuhl für Physiologie und Biochemie der Pflanzen, Universität Konstanz, D-7750 Konstanz, Germany*

KATHARINA SCHIEDT (15), *Department of Vitamin and Nutrition Research, F. Hoffmann-La Roche, Ltd., CH-4002 Basel, Switzerland*

HAROLD H. SCHMITZ (11), *Department of Food Science, North Carolina State University, Raleigh, North Carolina 27695*

JOEL L. SCHWARTZ (22), *Department of Oral Pathology, Harvard School of Dental Medicine, Boston, Massachusetts 02115*

GIORGIO SCITA (3), *Department of Nutritional Sciences, University of California, Berkeley, Berkeley, California 94720*

PABLO A. SCOLNIK (34), *DuPont Central Research and Development Department, Wilmington, Delaware 19880*

LOUIS A. SHERMAN (36), *Department of Biological Sciences, Purdue University, West Lafayette, Indiana 47907*

J. CECIL SMITH, JR. (1), *Vitamin and Mineral Nutrition Laboratory, Beltsville Human Nutrition Research Center, U.S. Department of Agriculture, Agriculture Research Service, Beltsville, Maryland 20705*

P. SORGELOOS (21), *Laboratory of Aquaculture and Artemia Reference Center, University of Gent, B-9000 Gent, Belgium*

MARIAN E. SWENDSEID (12), *School of Public Health, University of California, Los Angeles, Los Angeles, California 90024*

MIKIRO TADA (24), *Department of Biofunction and Genetic Resources, Faculty of Agriculture, Okayama University, Okayama 700, Japan*

AYUMI TANAKA (37), *Department of Botany, Faculty of Science, Kyoto University, Kyoto 606-01, Japan*

GUANGWEN TANG (7), *Gastrointestinal Nutrition Laboratory, U.S. Department of Agriculture, Human Nutrition Research Center on Aging, Tufts University, Boston, Massachusetts 02111*

JOHN D. TAYLOR (5), *Department of Biological Sciences, Wayne State University, Detroit, Michigan 48202*

T. T. TCHEN (5), *Department of Chemistry, Wayne State University, Detroit, Michigan 48202*

R. W. TUVESON (28), *Department of Microbiology, University of Illinois at Urbana-Champaign, Urbana, Illinois 61801*

LAURENCE VILLARD-MACKINTOSH (16), Department of Public Health and Primary Care, University of Oxford, Oxford, OX2 6HE, England

WENDY S. WHITE (13), Department of Food Science and Human Nutrition, Iowa State University, Ames, Iowa 50011

GEORGE WOLF (3), Department of Nutritional Sciences, University of California, Berkeley, Berkeley, California 94720

SHAN YUAN YANG (20), Department of Photosynthesis, Shanghai Institute of Plant Physiology, 200032 Shanghai, China

SIHUI ZENG (14), Department of Biochemistry and Biophysics, Iowa State University, Ames, Iowa 50011

Preface

Carotenoids are unique pigments synthesized in photosynthetic higher plants and photosynthetic microorganisms which serve essential functions in the protection against singlet oxygen-generated damage by photosensitized reactions. In the plant they participate in light-harvesting reactions and may act as a protective covering for certain higher plant species. In animals, carotenoids have attracted considerable interest, beginning over fifty years ago when it was discovered that they are precursors for vitamin A and its derivatives. Mammaliam species do not synthesize carotenoids or vitamin A. These essential substances are therefore derived solely from plant carotenoids, which are absorbed and stored in tissues and subsequently metabolized.

From a biomedical and biotechnological viewpoint, the carotenoids have also received considerable attention recently. In numerous epidemiological studies, the dietary intake of foods rich in carotenoids, but not of preformed vitamin A, has been associated with a reduced risk of some types of cancer. Chemical studies have shown that the carotenoids are very powerful quenchers of singlet oxygen and other reactive oxygen species.

Carotenoids, which are considered to be safe and stable, are currently used primarily as colorants in food products such as margarine. However, there is increasing interest in their possible role in maintaining health particularly in regard to lowering the risk of cancer. Many multivitamin supplements now contain β-carotene.

In addition to being of widespread biomedical interest, the carotenoids and their biosynthetic pathways are also being considered as targets for precise inhibitors that would act as herbicides.

Studies of the genetics and molecular biology of carotenoids in microbial photosynthetic systems, or nonphotosynthetic organisms such as *Escherichia coli,* are leading to the identification of the gene sequence homologies and differences between microbial and higher plant systems. This area of genetics and molecular and cell biology is undergoing rapid development. It is suspected that carotenoids, or perhaps their metabolic products, may act as regulators of gene expression.

In addition to research on their natural occurrence, structure, and biological activity, the distribution of carotenoids in animal and plant tissues has also been receiving considerable attention.

These developments have led to an explosion of activity in this research area. This volume and its companion, Volume 213 of *Methods in Enzymology,* present a comprehensive and state-of-the-art compilation of the molecular and cellular methodologies needed for pursuing research with carotenoids.

LESTER PACKER

METHODS IN ENZYMOLOGY

VOLUME I. Preparation and Assay of Enzymes
Edited by SIDNEY P. COLOWICK AND NATHAN O. KAPLAN

VOLUME II. Preparation and Assay of Enzymes
Edited by SIDNEY P. COLOWICK AND NATHAN O. KAPLAN

VOLUME III. Preparation and Assay of Substrates
Edited by SIDNEY P. COLOWICK AND NATHAN O. KAPLAN

VOLUME IV. Special Techniques for the Enzymologist
Edited by SIDNEY P. COLOWICK AND NATHAN O. KAPLAN

VOLUME V. Preparation and Assay of Enzymes
Edited by SIDNEY P. COLOWICK AND NATHAN O. KAPLAN

VOLUME VI. Preparation and Assay of Enzymes (*Continued*)
Preparation and Assay of Substrates
Special Techniques
Edited by SIDNEY P. COLOWICK AND NATHAN O. KAPLAN

VOLUME VII. Cumulative Subject Index
Edited by SIDNEY P. COLOWICK AND NATHAN O. KAPLAN

VOLUME VIII. Complex Carbohydrates
Edited by ELIZABETH F. NEUFELD AND VICTOR GINSBURG

VOLUME IX. Carbohydrate Metabolism
Edited by WILLIS A. WOOD

VOLUME X. Oxidation and Phosphorylation
Edited by RONALD W. ESTABROOK AND MAYNARD E. PULLMAN

VOLUME XI. Enzyme Structure
Edited by C. H. W. HIRS

VOLUME XII. Nucleic Acids (Parts A and B)
Edited by LAWRENCE GROSSMAN AND KIVIE MOLDAVE

VOLUME XIII. Citric Acid Cycle
Edited by J. M. LOWENSTEIN

VOLUME XIV. Lipids
Edited by J. M. LOWENSTEIN

VOLUME XV. Steroids and Terpenoids
Edited by RAYMOND B. CLAYTON

VOLUME XVI. Fast Reactions
Edited by KENNETH KUSTIN

VOLUME XVII. Metabolism of Amino Acids and Amines (Parts A and B)
Edited by HERBERT TABOR AND CELIA WHITE TABOR

VOLUME XVIII. Vitamins and Coenzymes (Parts A, B, and C)
Edited by DONALD B. MCCORMICK AND LEMUEL D. WRIGHT

VOLUME XIX. Proteolytic Enzymes
Edited by GERTRUDE E. PERLMANN AND LASZLO LORAND

VOLUME XX. Nucleic Acids and Protein Synthesis (Part C)
Edited by KIVIE MOLDAVE AND LAWRENCE GROSSMAN

VOLUME XXI. Nucleic Acids (Part D)
Edited by LAWRENCE GROSSMAN AND KIVIE MOLDAVE

VOLUME XXII. Enzyme Purification and Related Techniques
Edited by WILLIAM B. JAKOBY

VOLUME XXIII. Photosynthesis (Part A)
Edited by ANTHONY SAN PIETRO

VOLUME XXIV. Photosynthesis and Nitrogen Fixation (Part B)
Edited by ANTHONY SAN PIETRO

VOLUME XXV. Enzyme Structure (Part B)
Edited by C. H. W. HIRS AND SERGE N. TIMASHEFF

VOLUME XXVI. Enzyme Structure (Part C)
Edited by C. H. W. HIRS AND SERGE N. TIMASHEFF

VOLUME XXVII. Enzyme Structure (Part D)
Edited by C. H. W. HIRS AND SERGE N. TIMASHEFF

VOLUME XXVIII. Complex Carbohydrates (Part B)
Edited by VICTOR GINSBURG

VOLUME XXIX. Nucleic Acids and Protein Synthesis (Part E)
Edited by LAWRENCE GROSSMAN AND KIVIE MOLDAVE

VOLUME XXX. Nucleic Acids and Protein Synthesis (Part F)
Edited by KIVIE MOLDAVE AND LAWRENCE GROSSMAN

VOLUME XXXI. Biomembranes (Part A)
Edited by SIDNEY FLEISCHER AND LESTER PACKER

VOLUME XXXII. Biomembranes (Part B)
Edited by SIDNEY FLEISCHER AND LESTER PACKER

VOLUME XXXIII. Cumulative Subject Index Volumes I–XXX
Edited by MARTHA G. DENNIS AND EDWARD A. DENNIS

VOLUME XXXIV. Affinity Techniques (Enzyme Purification: Part B)
Edited by WILLIAM B. JAKOBY AND MEIR WILCHEK

VOLUME XXXV. Lipids (Part B)
Edited by JOHN M. LOWENSTEIN

VOLUME XXXVI. Hormone Action (Part A: Steroid Hormones)
Edited by BERT W. O'MALLEY AND JOEL G. HARDMAN

VOLUME XXXVII. Hormone Action (Part B: Peptide Hormones)
Edited by BERT W. O'MALLEY AND JOEL G. HARDMAN

VOLUME XXXVIII. Hormone Action (Part C: Cyclic Nucleotides)
Edited by JOEL G. HARDMAN AND BERT W. O'MALLEY

VOLUME XXXIX. Hormone Action (Part D: Isolated Cells, Tissues, and Organ Systems)
Edited by JOEL G. HARDMAN AND BERT W. O'MALLEY

VOLUME XL. Hormone Action (Part E: Nuclear Structure and Function)
Edited by BERT W. O'MALLEY AND JOEL G. HARDMAN

VOLUME XLI. Carbohydrate Metabolism (Part B)
Edited by W. A. WOOD

VOLUME XLII. Carbohydrate Metabolism (Part C)
Edited by W. A. WOOD

VOLUME XLIII. Antibiotics
Edited by JOHN H. HASH

VOLUME XLIV. Immobilized Enzymes
Edited by KLAUS MOSBACH

VOLUME XLV. Proteolytic Enzymes (Part B)
Edited by LASZLO LORAND

VOLUME XLVI. Affinity Labeling
Edited by WILLIAM B. JAKOBY AND MEIR WILCHEK

VOLUME XLVII. Enzyme Structure (Part E)
Edited by C. H. W. HIRS AND SERGE N. TIMASHEFF

VOLUME XLVIII. Enzyme Structure (Part F)
Edited by C. H. W. HIRS AND SERGE N. TIMASHEFF

VOLUME XLIX. Enzyme Structure (Part G)
Edited by C. H. W. HIRS AND SERGE N. TIMASHEFF

VOLUME L. Complex Carbohydrates (Part C)
Edited by VICTOR GINSBURG

VOLUME LI. Purine and Pyrimidine Nucleotide Metabolism
Edited by PATRICIA A. HOFFEE AND MARY ELLEN JONES

VOLUME LII. Biomembranes (Part C: Biological Oxidations)
Edited by SIDNEY FLEISCHER AND LESTER PACKER

VOLUME LIII. Biomembranes (Part D: Biological Oxidations)
Edited by SIDNEY FLEISCHER AND LESTER PACKER

VOLUME LIV. Biomembranes (Part E: Biological Oxidations)
Edited by SIDNEY FLEISCHER AND LESTER PACKER

VOLUME LV. Biomembranes (Part F: Bioenergetics)
Edited by SIDNEY FLEISCHER AND LESTER PACKER

VOLUME LVI. Biomembranes (Part G: Bioenergetics)
Edited by SIDNEY FLEISCHER AND LESTER PACKER

VOLUME LVII. Bioluminescence and Chemiluminescence
Edited by MARLENE A. DELUCA

VOLUME LVIII. Cell Culture
Edited by WILLIAM B. JAKOBY AND IRA PASTAN

VOLUME LIX. Nucleic Acids and Protein Synthesis (Part G)
Edited by KIVIE MOLDAVE AND LAWRENCE GROSSMAN

VOLUME LX. Nucleic Acids and Protein Synthesis (Part H)
Edited by KIVIE MOLDAVE AND LAWRENCE GROSSMAN

VOLUME 61. Enzyme Structure (Part H)
Edited by C. H. W. HIRS AND SERGE N. TIMASHEFF

VOLUME 62. Vitamins and Coenzymes (Part D)
Edited by DONALD B. MCCORMICK AND LEMUEL D. WRIGHT

VOLUME 63. Enzyme Kinetics and Mechanism (Part A: Initial Rate and Inhibitor Methods)
Edited by DANIEL L. PURICH

VOLUME 64. Enzyme Kinetics and Mechanism (Part B: Isotopic Probes and Complex Enzyme Systems)
Edited by DANIEL L. PURICH

VOLUME 65. Nucleic Acids (Part I)
Edited by LAWRENCE GROSSMAN AND KIVIE MOLDAVE

VOLUME 66. Vitamins and Coenzymes (Part E)
Edited by DONALD B. MCCORMICK AND LEMUEL D. WRIGHT

VOLUME 67. Vitamins and Coenzymes (Part F)
Edited by DONALD B. MCCORMICK AND LEMUEL D. WRIGHT

VOLUME 68. Recombinant DNA
Edited by RAY WU

VOLUME 69. Photosynthesis and Nitrogen Fixation (Part C)
Edited by ANTHONY SAN PIETRO

VOLUME 70. Immunochemical Techniques (Part A)
Edited by HELEN VAN VUNAKIS AND JOHN J. LANGONE

VOLUME 71. Lipids (Part C)
Edited by JOHN M. LOWENSTEIN

VOLUME 72. Lipids (Part D)
Edited by JOHN M. LOWENSTEIN

VOLUME 73. Immunochemical Techniques (Part B)
Edited by JOHN J. LANGONE AND HELEN VAN VUNAKIS

VOLUME 74. Immunochemical Techniques (Part C)
Edited by JOHN J. LANGONE AND HELEN VAN VUNAKIS

VOLUME 75. Cumulative Subject Index Volumes XXXI, XXXII, XXXIV–LX
Edited by EDWARD A. DENNIS AND MARTHA G. DENNIS

VOLUME 76. Hemoglobins
Edited by ERALDO ANTONINI, LUIGI ROSSI-BERNARDI, AND EMILIA CHIANCONE

VOLUME 77. Detoxication and Drug Metabolism
Edited by WILLIAM B. JAKOBY

VOLUME 78. Interferons (Part A)
Edited by SIDNEY PESTKA

VOLUME 79. Interferons (Part B)
Edited by SIDNEY PESTKA

VOLUME 80. Proteolytic Enzymes (Part C)
Edited by LASZLO LORAND

VOLUME 81. Biomembranes (Part H: Visual Pigments and Purple Membranes, I)
Edited by LESTER PACKER

VOLUME 82. Structural and Contractile Proteins (Part A: Extracellular Matrix)
Edited by LEON W. CUNNINGHAM AND DIXIE W. FREDERIKSEN

VOLUME 83. Complex Carbohydrates (Part D)
Edited by VICTOR GINSBURG

VOLUME 84. Immunochemical Techniques (Part D: Selected Immunoassays)
Edited by JOHN J. LANGONE AND HELEN VAN VUNAKIS

VOLUME 85. Structural and Contractile Proteins (Part B: The Contractile Apparatus and the Cytoskeleton)
Edited by DIXIE W. FREDERIKSEN AND LEON W. CUNNINGHAM

VOLUME 86. Prostaglandins and Arachidonate Metabolites
Edited by WILLIAM E. M. LANDS AND WILLIAM L. SMITH

VOLUME 87. Enzyme Kinetics and Mechanism (Part C: Intermediates, Stereochemistry, and Rate Studies)
Edited by DANIEL L. PURICH

VOLUME 88. Biomembranes (Part I: Visual Pigments and Purple Membranes, II)
Edited by LESTER PACKER

VOLUME 89. Carbohydrate Metabolism (Part D)
Edited by WILLIS A. WOOD

VOLUME 90. Carbohydrate Metabolism (Part E)
Edited by WILLIS A. WOOD

VOLUME 91. Enzyme Structure (Part I)
Edited by C. H. W. HIRS AND SERGE N. TIMASHEFF

VOLUME 92. Immunochemical Techniques (Part E: Monoclonal Antibodies and General Immunoassay Methods)
Edited by JOHN J. LANGONE AND HELEN VAN VUNAKIS

VOLUME 93. Immunochemical Techniques (Part F: Conventional Antibodies, Fc Receptors, and Cytotoxicity)
Edited by JOHN J. LANGONE AND HELEN VAN VUNAKIS

VOLUME 94. Polyamines
Edited by HERBERT TABOR AND CELIA WHITE TABOR

VOLUME 95. Cumulative Subject Index Volumes 61–74, 76–80
Edited by EDWARD A. DENNIS AND MARTHA G. DENNIS

VOLUME 96. Biomembranes [Part J: Membrane Biogenesis: Assembly and Targeting (General Methods; Eukaryotes)]
Edited by SIDNEY FLEISCHER AND BECCA FLEISCHER

VOLUME 97. Biomembranes [Part K: Membrane Biogenesis: Assembly and Targeting (Prokaryotes, Mitochondria, and Chloroplasts)]
Edited by SIDNEY FLEISCHER AND BECCA FLEISCHER

VOLUME 98. Biomembranes (Part L: Membrane Biogenesis: Processing and Recycling)
Edited by SIDNEY FLEISCHER AND BECCA FLEISCHER

VOLUME 99. Hormone Action (Part F: Protein Kinases)
Edited by JACKIE D. CORBIN AND JOEL G. HARDMAN

VOLUME 100. Recombinant DNA (Part B)
Edited by RAY WU, LAWRENCE GROSSMAN, AND KIVIE MOLDAVE

VOLUME 101. Recombinant DNA (Part C)
Edited by RAY WU, LAWRENCE GROSSMAN, AND KIVIE MOLDAVE

VOLUME 102. Hormone Action (Part G: Calmodulin and Calcium-Binding Proteins)
Edited by ANTHONY R. MEANS AND BERT W. O'MALLEY

VOLUME 103. Hormone Action (Part H: Neuroendocrine Peptides)
Edited by P. MICHAEL CONN

VOLUME 104. Enzyme Purification and Related Techniques (Part C)
Edited by WILLIAM B. JAKOBY

VOLUME 105. Oxygen Radicals in Biological Systems
Edited by LESTER PACKER

VOLUME 106. Posttranslational Modifications (Part A)
Edited by FINN WOLD AND KIVIE MOLDAVE

VOLUME 107. Posttranslational Modifications (Part B)
Edited by FINN WOLD AND KIVIE MOLDAVE

VOLUME 108. Immunochemical Techniques (Part G: Separation and Characterization of Lymphoid Cells)
Edited by GIOVANNI DI SABATO, JOHN J. LANGONE, AND HELEN VAN VUNAKIS

VOLUME 109. Hormone Action (Part I: Peptide Hormones)
Edited by LUTZ BIRNBAUMER AND BERT W. O'MALLEY

VOLUME 110. Steroids and Isoprenoids (Part A)
Edited by JOHN H. LAW AND HANS C. RILLING

VOLUME 111. Steroids and Isoprenoids (Part B)
Edited by JOHN H. LAW AND HANS C. RILLING

VOLUME 112. Drug and Enzyme Targeting (Part A)
Edited by KENNETH J. WIDDER AND RALPH GREEN

VOLUME 113. Glutamate, Glutamine, Glutathione, and Related Compounds
Edited by ALTON MEISTER

VOLUME 114. Diffraction Methods for Biological Macromolecules (Part A)
Edited by HAROLD W. WYCKOFF, C. H. W. HIRS, AND SERGE N. TIMASHEFF

VOLUME 115. Diffraction Methods for Biological Macromolecules (Part B)
Edited by HAROLD W. WYCKOFF, C. H. W. HIRS, AND SERGE N. TIMASHEFF

VOLUME 116. Immunochemical Techniques (Part H: Effectors and Mediators of Lymphoid Cell Functions)
Edited by GIOVANNI DI SABATO, JOHN J. LANGONE, AND HELEN VAN VUNAKIS

VOLUME 117. Enzyme Structure (Part J)
Edited by C. H. W. HIRS AND SERGE N. TIMASHEFF

VOLUME 118. Plant Molecular Biology
Edited by ARTHUR WEISSBACH AND HERBERT WEISSBACH

VOLUME 119. Interferons (Part C)
Edited by SIDNEY PESTKA

VOLUME 120. Cumulative Subject Index Volumes 81–94, 96–101

VOLUME 121. Immunochemical Techniques (Part I: Hybridoma Technology and Monoclonal Antibodies)
Edited by JOHN J. LANGONE AND HELEN VAN VUNAKIS

VOLUME 122. Vitamins and Coenzymes (Part G)
Edited by FRANK CHYTIL AND DONALD B. MCCORMICK

VOLUME 123. Vitamins and Coenzymes (Part H)
Edited by FRANK CHYTIL AND DONALD B. MCCORMICK

VOLUME 124. Hormone Action (Part J: Neuroendocrine Peptides)
Edited by P. MICHAEL CONN

VOLUME 125. Biomembranes (Part M: Transport in Bacteria, Mitochondria, and Chloroplasts: General Approaches and Transport Systems)
Edited by SIDNEY FLEISCHER AND BECCA FLEISCHER

VOLUME 126. Biomembranes (Part N: Transport in Bacteria, Mitochondria, and Chloroplasts: Protonmotive Force)
Edited by SIDNEY FLEISCHER AND BECCA FLEISCHER

VOLUME 127. Biomembranes (Part O: Protons and Water: Structure and Translocation)
Edited by LESTER PACKER

VOLUME 128. Plasma Lipoproteins (Part A: Preparation, Structure, and Molecular Biology)
Edited by JERE P. SEGREST AND JOHN J. ALBERS

VOLUME 129. Plasma Lipoproteins (Part B: Characterization, Cell Biology, and Metabolism)
Edited by JOHN J. ALBERS AND JERE P. SEGREST

VOLUME 130. Enzyme Structure (Part K)
Edited by C. H. W. HIRS AND SERGE N. TIMASHEFF

VOLUME 131. Enzyme Structure (Part L)
Edited by C. H. W. HIRS AND SERGE N. TIMASHEFF

VOLUME 132. Immunochemical Techniques (Part J: Phagocytosis and Cell-Mediated Cytotoxicity)
Edited by GIOVANNI DI SABATO AND JOHANNES EVERSE

VOLUME 133. Bioluminescence and Chemiluminescence (Part B)
Edited by MARLENE DELUCA AND WILLIAM D. MCELROY

VOLUME 134. Structural and Contractile Proteins (Part C: The Contractile Apparatus and the Cytoskeleton)
Edited by RICHARD B. VALLEE

VOLUME 135. Immobilized Enzymes and Cells (Part B)
Edited by KLAUS MOSBACH

VOLUME 136. Immobilized Enzymes and Cells (Part C)
Edited by KLAUS MOSBACH

VOLUME 137. Immobilized Enzymes and Cells (Part D)
Edited by KLAUS MOSBACH

VOLUME 138. Complex Carbohydrates (Part E)
Edited by VICTOR GINSBURG

VOLUME 139. Cellular Regulators (Part A: Calcium- and Calmodulin Binding Proteins
Edited by ANTHONY R. MEANS AND P. MICHAEL CONN

VOLUME 140. Cumulative Subject Index Volumes 102–119, 121–134

VOLUME 141. Cellular Regulators (Part B: Calcium and Lipids)
Edited by P. MICHAEL CONN AND ANTHONY R. MEANS

VOLUME 142. Metabolism of Aromatic Amino Acids and Amines
Edited by SEYMOUR KAUFMAN

VOLUME 143. Sulfur and Sulfur Amino Acids
Edited by WILLIAM B. JAKOBY AND OWEN GRIFFITH

VOLUME 144. Structural and Contractile Proteins (Part D: Extracellular Matrix)
Edited by LEON W. CUNNINGHAM

VOLUME 145. Structural and Contractile Proteins (Part E: Extracellular Matrix)
Edited by LEON W. CUNNINGHAM

VOLUME 146. Peptide Growth Factors (Part A)
Edited by DAVID BARNES AND DAVID A. SIRBASKU

VOLUME 147. Peptide Growth Factors (Part B)
Edited by DAVID BARNES AND DAVID A. SIRBASKU

VOLUME 148. Plant Cell Membranes
Edited by LESTER PACKER AND ROLAND DOUCE

VOLUME 149. Drug and Enzyme Targeting (Part B)
Edited by RALPH GREEN AND KENNETH J. WIDDER

VOLUME 150. Immunochemical Techniques (Part K: *In Vitro* Models of B and T Cell Functions and Lymphoid Cell Receptors)
Edited by GIOVANNI DI SABATO

VOLUME 151. Molecular Genetics of Mammalian Cells
Edited by MICHAEL M. GOTTESMAN

VOLUME 152. Guide to Molecular Cloning Techniques
Edited by SHELBY L. BERGER AND ALAN R. KIMMEL

VOLUME 153. Recombinant DNA (Part D)
Edited by RAY WU AND LAWRENCE GROSSMAN

VOLUME 154. Recombinant DNA (Part E)
Edited by RAY WU AND LAWRENCE GROSSMAN

VOLUME 155. Recombinant DNA (Part F)
Edited by RAY WU

VOLUME 156. Biomembranes (Part P: ATP-Driven Pumps and Related Transport: The Na,K-Pump)
Edited by SIDNEY FLEISCHER AND BECCA FLEISCHER

VOLUME 157. Biomembranes (Part Q: ATP-Driven Pumps and Related Transport: Calcium, Proton, and Potassium Pumps)
Edited by SIDNEY FLEISCHER AND BECCA FLEISCHER

VOLUME 158. Metalloproteins (Part A)
Edited by JAMES F. RIORDAN AND BERT L. VALLEE

VOLUME 159. Initiation and Termination of Cyclic Nucleotide Action
Edited by JACKIE D. CORBIN AND ROGER A. JOHNSON

VOLUME 160. Biomass (Part A: Cellulose and Hemicellulose)
Edited by WILLIS A. WOOD AND SCOTT T. KELLOGG

VOLUME 161. Biomass (Part B: Lignin, Pectin, and Chitin)
Edited by WILLIS A. WOOD AND SCOTT T. KELLOGG

VOLUME 162. Immunochemical Techniques (Part L: Chemotaxis and Inflammation)
Edited by GIOVANNI DI SABATO

VOLUME 163. Immunochemical Techniques (Part M: Chemotaxis and Inflammation)
Edited by GIOVANNI DI SABATO

VOLUME 164. Ribosomes
Edited by HARRY F. NOLLER, JR. AND KIVIE MOLDAVE

VOLUME 165. Microbial Toxins: Tools for Enzymology
Edited by SIDNEY HARSHMAN

VOLUME 166. Branched-Chain Amino Acids
Edited by ROBERT HARRIS AND JOHN R. SOKATCH

VOLUME 167. Cyanobacteria
Edited by LESTER PACKER AND ALEXANDER N. GLAZER

VOLUME 168. Hormone Action (Part K: Neuroendocrine Peptides)
Edited by P. MICHAEL CONN

VOLUME 169. Platelets: Receptors, Adhesion, Secretion (Part A)
Edited by JACEK HAWIGER

VOLUME 170. Nucleosomes
Edited by PAUL M. WASSARMAN AND ROGER D. KORNBERG

VOLUME 171. Biomembranes (Part R: Transport Theory: Cells and Model Membranes)
Edited by SIDNEY FLEISCHER AND BECCA FLEISCHER

VOLUME 172. Biomembranes (Part S: Membrane Isolation and Characterization)
Edited by SIDNEY FLEISCHER AND BECCA FLEISCHER

VOLUME 173. Biomembranes [Part T: Cellular and Subcellular Transport: Eukaryotic (Nonepithelial) Cells]
Edited by SIDNEY FLEISCHER AND BECCA FLEISCHER

VOLUME 174. Biomembranes [Part U: Cellular and Subcellular Transport: Eukaryotic (Nonepithelial) Cells]
Edited by SIDNEY FLEISCHER AND BECCA FLEISCHER

VOLUME 175. Cumulative Subject Index Volumes 135–139, 141–167

VOLUME 176. Nuclear Magnetic Resonance (Part A: Spectral Techniques and Dynamics)
Edited by NORMAN J. OPPENHEIMER AND THOMAS L. JAMES

VOLUME 177. Nuclear Magnetic Resonance (Part B: Structure and Mechanism)
Edited by NORMAN N. OPPENHEIMER AND THOMAS L. JAMES

VOLUME 178. Antibodies, Antigens, and Molecular Mimicry
Edited by JOHN J. LANGONE

VOLUME 179. Complex Carbohydrates (Part F)
Edited by VICTOR GINSBURG

VOLUME 180. RNA Processing (Part A: General Methods)
Edited by JAMES E. DAHLBERG AND JOHN N. ABELSON

VOLUME 181. RNA Processing (Part B: Specific Methods)
Edited by JAMES E. DAHLBERG AND JOHN N. ABELSON

VOLUME 182. Guide to Protein Purification
Edited by MURRAY P. DEUTSCHER

VOLUME 183. Molecular Evolution: Computer Analysis of Protein and Nucleic Acid Sequences
Edited by RUSSELL F. DOOLITTLE

VOLUME 184. Avidin-Biotin Technology
Edited by MEIR WILCHEK AND EDWARD A. BAYER

VOLUME 185. Gene Expression Technology
Edited by DAVID V. GOEDDEL

VOLUME 186. Oxygen Radicals in Biological Systems (Part B: Oxygen Radicals and Antioxidants)
Edited by LESTER PACKER AND ALEXANDER N. GLAZER

VOLUME 187. Arachidonate Related Lipid Mediators
Edited by ROBERT C. MURPHY AND FRANK A. FITZPATRICK

VOLUME 188. Hydrocarbons and Methylotrophy
Edited by MARY E. LIDSTROM

VOLUME 189. Retinoids (Part A: Molecular and Metabolic Aspects)
Edited by LESTER PACKER

VOLUME 190. Retinoids (Part B: Cell Differentiation and Clinical Applications)
Edited by LESTER PACKER

VOLUME 191. Biomembranes (Part V: Cellular and Subcellular Transport: Epithelial Cells)
Edited by SIDNEY FLEISCHER AND BECCA FLEISCHER

VOLUME 192. Biomembranes (Part W: Cellular and Subcellular Transport: Epithelial Cells)
Edited by SIDNEY FLEISCHER AND BECCA FLEISCHER

VOLUME 193. Mass Spectrometry
Edited by JAMES A. MCCLOSKEY

VOLUME 194. Guide to Yeast Genetics and Molecular Biology
Edited by CHRISTINE GUTHRIE AND GERALD R. FINK

VOLUME 195. Adenylyl Cyclase, G Proteins, and Guanylyl Cyclase
Edited by ROGER A. JOHNSON AND JACKIE D. CORBIN

VOLUME 196. Molecular Motors and the Cytoskeleton
Edited by RICHARD B. VALLEE

VOLUME 197. Phospholipases
Edited by EDWARD A. DENNIS

VOLUME 198. Peptide Growth Factors (Part C)
Edited by DAVID BARNES, J. P. MATHER, AND GORDON H. SATO

VOLUME 199. Cumulative Subject Index Volumes 168–174, 176–194 (in preparation)

VOLUME 200. Protein Phosphorylation (Part A: Protein Kinases: Assays Purification, Antibodies, Functional Analysis, Cloning, and Expression)
Edited by TONY HUNTER AND BARTHOLOMEW M. SEFTON

VOLUME 201. Protein Phosphorylation (Part B: Analysis of Protein Phosphorylation, Protein Kinase Inhibitors, and Protein Phosphatases)
Edited by TONY HUNTER AND BARTHOLOMEW M. SEFTON

VOLUME 202. Molecular Design and Modeling: Concepts and Applications (Part A: Proteins, Peptides, and Enzymes)
Edited by JOHN J. LANGONE

VOLUME 203. Molecular Design and Modeling: Concepts and Applications (Part B: Antibodies and Antigens, Nucleic Acids, Polysaccharides, and Drugs)
Edited by JOHN J. LANGONE

VOLUME 204. Bacterial Genetic Systems
Edited by JEFFREY H. MILLER

VOLUME 205. Metallobiochemistry (Part B: Metallothionein and Related Molecules)
Edited by JAMES F. RIORDAN AND BERT L. VALLEE

VOLUME 206. Cytochrome P450
Edited by MICHAEL R. WATERMAN AND ERIC F. JOHNSON

VOLUME 207. Ion Channels
Edited by BERNARDO RUDY AND LINDA E. IVERSON

VOLUME 208. Protein–DNA Interactions
Edited by ROBERT T. SAUER

VOLUME 209. Phospholipid Biosynthesis
Edited by EDWARD A. DENNIS AND DENNIS E. VANCE

VOLUME 210. Numerical Computer Methods
Edited by LUDWIG BRAND AND MICHAEL L. JOHNSON

VOLUME 211. DNA Structures (Part A: Synthesis and Physical Analysis of DNA)
Edited by DAVID M. J. LILLEY AND JAMES E. DAHLBERG

VOLUME 212. DNA Structures (Part B: Chemical and Electrophoretic Analysis of DNA)
Edited by DAVID M. J. LILLEY AND JAMES E. DAHLBERG

VOLUME 213. Carotenoids (Part A: Chemistry, Separation, Quantitation, and Antioxidation)
Edited by LESTER PACKER

VOLUME 214. Carotenoids (Part B: Metabolism, Genetics, and Biosynthesis)
Edited by LESTER PACKER

VOLUME 215. Platelets: Receptors, Adhesion, Secretion (Part B)
Edited by JACEK J. HAWIGER

VOLUME 216. Recombinant DNA (Part G)
Edited by RAY WU

VOLUME 217. Recombinant DNA (Part H)
Edited by RAY WU

VOLUME 218. Recombinant DNA (Part I) (in preparation)
Edited by RAY WU

VOLUME 219. Reconstitution of Intracellular Transport
Edited by JAMES E. ROTHMAN

VOLUME 220. Membrane Fusion Techniques (Part A) (in preparation)
Edited by NEJAT DÜZGÜNES

VOLUME 221. Membrane Fusion Techniques (Part B) (in preparation)
Edited by NEJAT DÜZGÜNES

VOLUME 222. Proteolytic Enzymes in Coagulation, Fibrinolysis, and Complement Activation (Part A: Mammalian Blood Coagulation Factors and Inhibitors) (in preparation)
Edited by LASZLO LORAND AND KENNETH G. MANN

VOLUME 223. Proteolytic Enzymes in Coagulation, Fibrinolysis, and Complement Activation (Part B: Complement Activation, Fibrinolysis, and Nonmammalian Blood Coagulation Factors) (in preparation)
Edited by LASZLO LORAND AND KENNETH G. MANN

VOLUME 224. Molecular Evolution: Producing the Biochemical Data (in preparation)
Edited by ELIZABETH A. ZIMMER, THOMAS J. WHITE, REBECCA L. CANN, AND ALLAN C. WILSON

VOLUME 225. Guide to Techniques in Mouse Development (in preparation)
Edited by PAUL M. WASSARMAN AND MELVIN L. DEPAMPHILIS

Section I

Metabolism

A. Absorption and Transport
Articles 1 through 7

B. Tissue Distribution
Articles 8 through 13

C. Transformation and Metabolic Regulation
Articles 14 through 25

[1] Carotenoid Absorption in Humans

By PHYLLIS E. BOWEN, SOHRAB MOBARHAN, and J. CECIL SMITH, JR.

Introduction

A number of methods have been employed to investigate the absorption of carotenoids in humans. The most common has been the acute or chronic dosing of a carotenoid such as β-carotene and the measurement of the plasma response.[1-4] Area under the curve, peak response, time to peak, and slope can yield important information when comparing different carotenoids, formulations, and foods; however, this technique does not provide the absolute quantity or percent absorbed. Metabolic balance techniques have also been employed.[5,6] The gastrointestinal lavage technique, wherein the entire gastrointestinal tract of the subject is cleansed of ingesta prior to consumption of the compound or food component whose absorption is being measured, has been reported for assessing the absorption of calcium[7] but more recently was applied for measuring carotenoid absorption in humans.[8] Metabolic balances can provide estimations of the apparent disappearance of various carotenoids from the gut and therefore a crude estimation of the amount absorbed under various experimental conditions, assuming no degradation or microbial synthesis within the gut. Simultaneous measurement of various tissue responses[9,10] provides valuable information in conjunction with these techniques. The availability of stable isotopes of various carotenoids holds great promise to provide more definitive data in the characterization and quantitation of carotenoid ab-

---

[1] E. D. Brown, M. S. Micozzi, N. E. Craft, J. G. Bieri, G. Beecher, B. K. Edwards, A. Rose, P. R. Taylor, and J. C. Smith, *Am. J. Clin. Nutr.* **49**, 1258 (1989).
[2] C. T. Henderson, S. Mobarhan, P. Bowen, M. Stacewicz-Sapontzakis, P. Langenberg, R. Kioni, D. Lucchesi, and S. Sugarman, *J. Am. Coll. Nutr.* **8**, 625 (1988).
[3] N. V. Dimitrov, C. Meyer, D. E. Ullrey, W. Chenoweth, A. Michelakis, W. Malone, C. Boone, and G. Fink, *Am. J. Clin. Nutr.* **48**, 298 (1988).
[4] G. Maiani, S. Mobarhan, M. Ceccanti, L. Ranaldi, S. Gettner, P. Bowen, H. Friedman, A. A. DeLorenzo, and A. Ferro-Luzzi, *Eur. J. Clin. Nutr.* **43**, 749 (1989).
[5] O. A. Roels, M. Trout, and R. Dujacquier, *J. Nutr.* **65**, 115 (1958).
[6] C. N. Rao and B. S. N. Rao, *Am. J. Clin. Nutr.* **23**, 105 (1970).
[7] G. W. Bo-Linn, G. R. Davis, D. J. Buddrus, S. G. Morawski, C. Santa Ana, and J. S. Fordtran, *J. Clin. Invest.* **73**, 640 (1984).
[8] A. Shiau, S. Mobarhan, M. Stacewicz-Sapontzakis, P. Bowen, C. Ford, Y. Liao, and R. Benya, *J. Am. Coll. Nutr.* **9**, 533 (1990).
[9] B. A. Clevidence and J. G. Bieri, this volume [4].
[10] R. S. Parker, this volume [9].

sorption in humans. Finally, animal models which have been shown to closely resemble human absorption dynamics are needed to elucidate the mechanism(s) of carotenoid absorption and the regulation of conversion of provitamin A carotenoids to vitamin A. Potential models include monkeys,[11] ferrets,[12,13] and young calves.[14] Intestinal cell models may also be helpful.[15]

Plasma/Sera Response Technique

The relative absorption of carotenoids involves measuring the changes in their respective plasma concentrations following ingestion. The carotenoid may be in the form of a purified compound or ingested as a food item (i.e., fruit, vegetable, or juice). In simplest terms, healthy human subjects ingest quantitated amounts of carotenoids (food or purified), and the changes in plasma or serum concentration at various time intervals following ingestion is determined. Single or multiple daily doses may be given. Relative differences in absorption are determined by comparing the plasma carotenoid concentrations at various postingestion intervals with baseline values.

Experimental Design

Acute Single Ingestion. For an acute single ingestion, a crossover design is recommended so that each treatment will be tested by two or more groups of subjects depending on the number of treatments. The treatment could include a purified carotenoid, a food item selected for its relatively high concentration of a specific carotenoid, or a placebo containing the same vehicle as that used to "carry" the purified compound. The relatively slow rate of appearance and time required to reach maximum concentration (24–48 hr) necessitate that study periods be no shorter than 7 days, excluding the equilibrium period.[1] Figure 1 outlines, in generic form, a crossover design for an acute single ingestion experiment. Two treatment periods of 14 days each are separated by a 42-day "washout" period. The purpose of the washout period, when the subjects consume self-selected

---

[11] N. I. Krinsky, M. M. Mathews-Roth, S. Welankiwar, P. K. Sehgal, N. C. Lausen, and M. Russet, *J. Nutr.* **120,** 81 (1990).

[12] J. D. Ribaya-Mercado, S. C. Holmgren, J. G. Fox, and R. M. Russell, *J. Nutr.* **119,** 665 (1989).

[13] E. T. Gugger, T. L. Bierer, T. M. Henze, W. S. White, and J. W. Erdman, Jr., *J. Nutr.* **122,** 115 (1992).

[14] C. L. Poor, T. L. Bierer, N. R. Merchen, and G. C. Fahey, *FASEB J.* **5,** A1322 (1991).

[15] T. C. Quick and D. E. Ong, *Biochemistry* **29,** 1116 (1990).

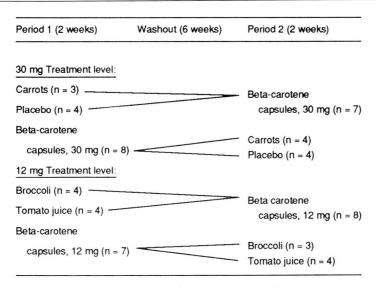

FIG. 1. Crossover experimental design for a study to determine the plasma carotenoid response to an acute single ingestion of a high-carotenoid vegetable or juice, purified β-carotene, or placebo. Numbers of subjects are given in parentheses. See text for procedural details. [From E. D. Brown, M. S. Micozzi, N. E. Craft, J. G. Bieri, G. Beecher, B. K. Edwards, A. Rose, P. R. Taylor, and J. C. Smith, *Am. J. Clin. Nutr.* **49,** 1258 (1989). Reproduced with permission from the American Society for Clinical Nutrition.]

diets, is to allow the plasma carotenoids of interest to return to baseline concentrations. A washout period of 42 days has been sufficient to allow α- and β-carotene, lutein/zeaxanthin, cryptoxanthin, lycopene, and total carotenoids to return to near baseline values.[1]

During the treatment periods, the subjects consume controlled low-carotenoid diets for 3 days before being dosed and for the remaining 11 days after dosing. The treatment, vegetable or purified carotenoid in a capsule, is given on day 3 with breakfast. The subjects are divided into two strata based on their baseline fasting plasma total carotenoid levels prior to consuming the controlled low-carotenoid diets and then randomly assigned to two major groups to receive a carotenoid dose of either 12 or 30 mg. These two groups are further subdivided into three subgroups (placebo, vegetable, or pure carotenoid) for the crossover design. The serving size of vegetable or juice is calculated from previous intralaboratory high-performance liquid chromatography (HPLC) analysis of the actual batch to be used, prepared as to be eaten, that is, cooked or raw. If analytic capabilities are limited, the carotenoid content can be estimated using databases of food carotenoids; however, because carotenoid levels vary by

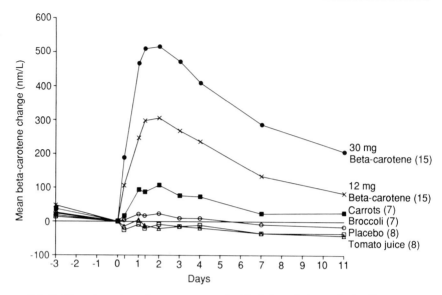

FIG. 2. Mean β-carotene plasma response of subjects in a study employing the crossover design outlined in Fig. 1. Numbers of subjects per treatment are given in parentheses. Note the greater response with the purified β-carotene (30 to 12 mg) compared to a serving of carrots containing 29 mg of β-carotene. [From E. D. Brown, M. S. Micozzi, N. E. Craft, J. G. Bieri, G. Beecher, B. K. Edwards, A. Rose, P. R. Taylor, and J. C. Smith, *Am. J. Clin. Nutr.* **49**, 1258 (1989). Reproduced with permission from the American Society for Clinical Nutrition.]

cultivar, storage, and preparation, inferences which can be drawn by comparisons are limited.

Figure 2 shows the mean β-carotene sera/plasma response of subjects in a study using the single ingestion crossover design as outlined in Fig. 1.[1] The serving size of the vegetable was based on actual HPLC analysis to provide approximately 12 or 30 mg of the primary carotenoid(s) in each vegetable as consumed (β-carotene in carrots, lutein plus β-carotene in broccoli, and lycopene in tomato juice).

Chronic Multiple Daily Ingestion. For the experiment involving chronic multiple daily ingestion, following the equilibrium period, each subject consumes one of the treatments at meals in conjunction with the basal low-carotenoid diet. The treatments may include a single fruit, vegetable, purified carotenoid(s), or a placebo. The length of investigation necessary for maximum plasma carotenoid concentrations to be achieved, excluding the prestudy period, is approximately 6 weeks. A poststudy followup of 4 weeks is optimal and allows monitoring of the plasma carotenoids.[16]

Subject Selection. Criteria to consider in the selection of subjects are gender, age, smoking status, medications including vitamin and mineral supplements, weight for height, plasma cholesterol, triglyceride, and lipoprotein levels because all of these attributes are known to affect plasma levels of the carotenoids.[4,17-21] If healthy subjects are to be used for the assessment of normal absorption the following tests should be performed on fasting blood: tests for liver function, thyroid function, as well as electrolytes and glucose; blood counts including differential leukocytes and platelets. Urinalysis should include glucose, electrolytes, bilirubin, and blood cells. Premenopausal women should have a test for pregnancy.

A number of conditions are known to affect absorption, metabolism, or clearance of carotenoids and include (1) defects and diseases of malabsorption, (2) sickle cell anemia, (3) alcoholism, (4) diseases that are associated with the generation of free radicals or peroxidation products (hemopoietic protoporpheria, surgery, trauma, heart attack, etc.), and (5) the use of mineral oil. Subjects should have no history of chronic diseases as assessed by a physician. Likewise, the subjects should be willing, psychologically stable, and available for whole days for acute studies. The relative ease of phlebotomy for a particular subject is of critical importance in these studies requiring multiple blood draws.

Finally, if subjects will be divided into groups for treatment, prestudy fasting plasma carotenoid levels should be used to assign individuals to groups so that each group will consist of equivalent numbers with high (>50 percentile) or low (<50 percentile) concentrations of total carotenoids. For one study, initial serum β-carotene levels were the greatest predictors of subsequent chronic increased levels in an intervention trial,[21] presumably because individuals vary greatly in their ability to absorb carotenoids.[1-3] Prospective subjects who have very high plasma carotenoid

---

[16] J. C. Smith, M. S. Micozzi, E. D. Brown, B. K. Edwards, J. G. Bieri, P. R. Taylor, F. Khachik, and G. R. Beecher, *FASEB J.* **5**, A4078 (1991).

[17] M. Stacewicz-Sapontzakis, P. E. Bowen, J. W. Kikendall, and M. Burgess, *J. Micronutr. Anal.* **3**, 27 (1987).

[18] S. B. Sugarman, S. Mobarhan, P. E. Bowen, M. Stacewicz-Sapontzakis, P. Langenberg, C. Henderson, W. S. Stryker, L. A. Kaplan, E. A. Stain, M. J. Stampfer, A. Sober, and W. C. Willett, *Am. J. Epidemiol.* **127**, 283 (1988).

[19] C. L. Rock and M. E. Swendseid, this volume [12].

[20] L. K. Bjornson, H. J. Kaydan, E. Miller, and A. N. Moshell, *J. Lipid Res.* **17**, 343 (1976).

[21] J. M. Sullivan, P. E. Bowen, J. W. Kikendall, P. Langenberg, M. Sapuntzakis, M. Burgess, and S. Moody, *FASEB J.* **5**, A1073 (1991).

levels and a history of β-carotene supplementation or vegetable juice consumption should be excluded since their large tissue stores may increase the variability of carotenoid levels over time and thus change the time course of the parameters to be studied.

Prestudy Equilibrium Period. The previous diet will affect baseline values, so a prestudy controlled diet must be fed; 4 or more days are recommended. Bowen[22] found that even after 4 days of a low-carotenoid diet, one of nine subjects had serum α- and β-carotene levels 20% higher than baseline. Prestudy diet records indicated that she had consumed approximately one and one-half cups of carrots on the day before starting the low-carotenoid diet. Therefore subjects should be instructed to avoid consuming large portions of carotenoid-containing foods before starting the equilibrium period.

Basal Diet. Ideally the basal diet is formulated from commonly ingested foods and should contain minimal quantities of carotenoids (i.e., <1 mg/day). To obtain such a diet low in specific carotenoids, the inclusion of fruits and vegetables should be minimized, and the remaining ones should contain negligible quantities of the carotenoids to be investigated. More complete data on the carotenoid content of foods are forthcoming and will be helpful in formulating basal diets.[23] Once the diet is planned, actual chemical analysis of the carotenoids in the basal diet is desirable, although it may not be feasible owing to laboratory limitations. The basal diet should be adequate in essential vitamins and minerals; therefore, it may be necessary to provide specific supplements, particularly vitamin C and folate. However, the most practical approach may be to provide a daily supplement to all subjects that assures the Recommended Dietary Allowance for all essential nutrients.[24] Particular attention should be paid to the level of the fat-soluble vitamins supplied in the basal diet since competition and synergism with various carotenoids may be operating in absorption, metabolism, and clearance of these substances. The levels provided should be clearly stated in scientific reports.

Because the fat content of the diet may limit the efficiency of absorption if low, it is recommended that it approximate that of the typical diet of the relevant population. The amount of polyunsaturated fatty acids should also be controlled (e.g., 6.75 g/1000 kcal for typical diet in the United States, P:S ratio 0.5) since they have been noted to decrease micelle

---

[22] P. E. Bowen, unpublished data, University of Illinois at Chicago Health Sciences Center, 1989.
[23] J. M. Holden and A. R. Mangels, *J. Am. Diet Assoc.,* in press (1993).
[24] National Research Council, "Recommended Dietary Allowances," 10th Ed. National Academy Press, Washington, D.C., 1989.

inclusion of carotenoids,[25] increase the load of oxidative products, and increase the requirement for tocopherol.[24] Likewise, dietary fiber should also be controlled since reduced β-carotene utilization in animals[26] and impaired β-carotene response in humans[27] have been reported. In the United States, typical diets have been reported to contain 7–10 g of dietary fiber per 1000 kcal.[28] Energy intake should be adjusted to maintain body weight (± 1.0 kg) throughout the study period.

The 1 to 7-day cycle menus can be used depending on the length of the study, but nutrients, noted above, should be within 10% of basal diet specifications for each day in the cycle in order to avoid daily variability. On absorption study days, the same basal diet and the same times for consumption must be used throughout the study. Therefore, the test carotenoid-containing foods should be easily incorporated into the basal diet, and a comparable placebo food(s) providing equivalent energy, fat, dietary fiber, and fat-soluble vitamins must be planned. The food items (fruits and vegetables) providing the carotenoid(s) to be studied should be purchased in a single batch in amounts sufficient to last the entire study (this limits fruits and vegetables to frozen or canned supplies). Ideally, the serving size should be calculated to provide the desired quantity of carotenoid based on chemical analysis of the actual lot to be used. The treatment foods should be cooked according to instructions on the package from the supplier (e.g., precooked, flash-frozen foods are boiled in serving bags).

Blood Collection. At the end of the equilibrium period, blood samples are obtained using appropriate syringes/Vacutainers. The blood is centrifuged at 2000 g for 20 min at room temperature. Serum is preferred to plasma, since anticoagulants have been demonstrated to result in incomplete recovery of specific and total carotenoids[29] as well as α-tocopherol and retinol. Although fasting blood is recommended, Brown *et al.*[30] reported no difference between the concentrations of seven carotenoid fractions in blood from the same individuals who were fasted overnight compared to blood samples obtained up to 4 hr after breakfast. Protect all samples from photooxidation by carrying out the procedure in subdued or yellow lighting. Plasma or serum samples are more vulnerable to photoox-

---

[25] A. R. Giuliano, M. B. Matgner, and L. M. Canfield, this volume [10].
[26] J. W. Erdman, Jr., G. C. Fahey, Jr., and C. B. White, *J. Nutr.* **116**, 2415 (1986).
[27] C. L. Rock, personal communication, School of Public Health, University of Michigan, Ann Arbor, MI, 1991.
[28] G. Block and E. Lanza, *J. Natl. Cancer Inst.* **79**, 83 (1987).
[29] J. C. Smith, Jr., S. Lewis, J. Holbrook, K. Seidel, and A. Rose, *Clin. Chem.* **33**, 814 (1987).
[30] E. D. Brown, A. Rose, N. Craft, K. E. Seidel, and J. C. Smith, Jr., *Clin. Chem.* **35**, 310 (1989).

idation than whole blood. Whole blood samples may be left at room temperature, protected from light, for up to 4 hr without the loss of the major carotenoids in serum. Likewise, no significant differences were found in the concentrations of seven carotenoid fractions or total carotenoids in plasma samples that were frozen ($-70°$) immediately after separation as compared with replicate samples maintained in the dark at room temperature.[31] Hemolysed samples should be avoided since hemolysis may cause accelerated oxidation of the carotenoids depending on subsequent handling conditions.

Serum samples should be vigorously mixed before aliquoting to avoid the layering of carotenoids. Under even gentle centrifugation conditions carotenoids tend to be more concentrated near the red blood cell interface. For storage, the samples are aliquoted into screw-capped Cryotubes (Thomas Scientific, Swedesboro, NJ), which are filled to capacity to minimize oxidation by air. Some investigators fill the headspace with nitrogen, although one study reported no beneficial effect under the conditions of their study.[31] The samples should be protected from light and stored at $-70°$ or lower and not thawed until analyzed.[31] To prevent dehydration in the freezer, the tubes, held in an upright position in appropriate holders, may be sealed along with ice cubes in plastic bags.

When multiple samples taken within a single day are required, the 21-gauge needle of a butterfly infusion set is inserted into a peripheral vein located on the back of the hand or on the forearm of the subject and firmly taped in place. Three milliliters of a heparin solution [1000 units (U)/ml] is added to 30 ml of sterile normal saline to produce a 100 unit/ml dilution. After each blood sample is collected, 3 ml of this solution is injected to inhibit clot formation in the vein. Immediately before the next timed blood sampling, this heparin is removed by syringe. A vein can be kept open, with care, for 12-hr periods. Care should be taken to remove all the heparin solution before a blood sample is obtained because of the possibility of dilution. Dilution can be assessed by measuring plasma levels of another carotenoid which is not under investigation. For example, if β-carotene absorption is being tested, the lutein or lutein/zeaxanthin peak can be measured because its concentration should remain constant over the day of sampling or be depressed to a slight degree over a 24-hr period.

Serum Analysis. High-performance liquid chromatography (HPLC) is the method of choice for the analysis of individual carotenoids. Numerous variations are now available. Early techniques identified and quantitated three to five carotenoids. Recently, HPLC techniques have been developed

---

[31] N. E. Craft, E. D. Brown, and J. C. Smith, Jr., *Clin. Chem.* **34,** 44 (1988).

that have reported the identification and quantitation of more than a dozen carotenoids.[32,33] The importance of using an internal standard and highly purified standards must be stressed. There is now available from the National Institute of Science and Technology (Gaithersburg, MD 20899) a certified reference material (SRM) for β-carotene in human plasma (SRM 968). A more recent SRM (968A) includes a certified value for plasma β-carotene and "informational" values for α-carotene, lycopene, cryptoxanthin, lutein, and zeaxanthin. Because of the expense and the limited quantities available, the SRMs should be used to establish accurately the concentrations of individual carotenoids in a single large (50–100 ml) plasma "pool." This plasma pool whose carotenoid concentration has been established by cross-referencing to the SRMs can be aliquoted into small vials of 1–2 ml (stored at $-70°$) to be used as daily control sera. Thus, the preparation procedures and instrumentation can be monitored daily for accuracy. Precision (within ±5%) is required to interpret absorption curve information. Samples from a single subject should be assessed at one time within the same HPLC assay to minimize error.

Estimation of Carotenoid Absorption by Balance Techniques

Metabolic balance has been a traditional method for the study of substances that are not subject to biological modification in the gastrointestinal tract. This technique should be used with caution with regard to carotenoids since gastrointestinal modification of these compounds is not well defined. However, the technique may be the only practical approach for assessing absorption of carotenoids in a mixed diet consumed over long periods of time.

The basic of the metabolic balance method is the careful measurement of all intake (input) and excretion (output) of the carotenoid(s) of interest over a specific period of time. In the case of carotenoids, no urinary excretion of either free or conjugated carotenoids has been found, and although some carotenoids are lost with exfoliation from skin, loss via the skin is thought to be negligible and has not been quantitated. Therefore, balance studies involve the estimation of total carotenoid (or the carotenoid of interest) intake for a period of days called the balance period (usually 5–8 days) and the collection and analysis of all fecal output for carotenoid(s) for the same time period. Output is subtracted from input and divided by the number of days of the balance period to obtain the quantity of carotenoid apparently absorbed in the body per day. Equilibration

---

[32] F. Khachik, G. R. Beecher, M. B. Goli, and W. R. Lusby, *Pure Appl. Chem.* **63**, 71 (1991).
[33] N. I. Krinsky, M. D. Russet, G. J. Handelman, and M. D. Snodderly, *J. Nutr.* **120**, (1990).

periods, basal diets, subject selection, and collection of blood are similar to those discussed under Plasma/Sera Response Technique; therefore, only additional concerns and methodology will be discussed.

Experimental Design

A crossover design similar to Fig. 1 is recommended for balance studies except the periods must be longer. A low-carotenoid basal diet plus pure carotenoids, whole food, or an entire diet containing carotenoids may be used. Because the balance technique requires the subtraction of a large number (total output) from a larger number (total input) to obtain a relatively small number (apparent absorption), it is important to conduct such studies with great precision under highly controlled conditions. Therefore, the exact quantity of carotenoid consumed and the exact quantity excreted must be determined. Present food table estimations of intake are inadequate except for planning the diet to be used or when a no-carotenoid basal diet is involved.

The balance technique produces a systematic error on the side of estimating a greater absorption than actually occurs since the extent of bacterial degradation or other destruction of the carotenoids in the gastrointestinal tract is unknown. Thus, the technique is best used to compare the apparent absorption of carotenoids under various conditions. Only approximate estimates of the amount or percent absorbed can be obtained using this technique.

Brown et al.[1] found that a 42-day washout period was sufficient between treatments for plasma carotenoids to return to baseline; Mobarhan et al.[34] found that a 2-week carotenoid-free diet reduced serum carotenoids by 50%. However, serum β-carotene remained 400 and 650% above baseline values 4 to 6 weeks after subjects received 15 or 120 mg of β-carotene for 21 days, respectively. Therefore, it may be necessary to include washout periods of up to 2 or more months depending on the dose and the duration of the experimental period. Experimental periods must be of sufficient duration to attain equilibrium and accommodate a balance period at equilibrium. The balance period should be at least 5 days and preferably longer to assure the collection of sufficient fecal output to be characteristic of usual output. The length of time to reach equilibrium is dependent on the dose of the carotenoid. For example, many subjects who ingested a 15-mg β-carotene supplement daily as a part of an intervention trial had higher serum β-carotene concentrations at 18 months compared to their

---

[34] S. Mobarhan, P. Bowen, B. Anderson, M. Evans, M. Stacewicz-Sapuntzakis, S. Sugarman, P. Simms, D. Lucchese, and H. Friedman, *Nutr. Cancer* **14**, 195 (1990).

6-month levels,[22] whereas women taking a diet of six servings of fruits and vegetables per day experienced a 2- to 3-fold increase in serum levels of five major carotenoids at 3 weeks but experienced no further increase in serum carotenoids at 4 weeks.[22] For shorter studies, investigators should plan the lowest doses of carotenoids that are still consistent with obtaining observable effects.

Fecal Preparation and Analysis. Nonabsorbable markers such as brilliant blue (Sigma, St. Louis, MO) are consumed with the first meal in the balance period and with the first meal excluded from the balance period to mark the stool output that is associated. Another approach is to consume a nonabsorbable plant sterol during the balance period and correct output for the percent sterol lost in the collection as a means of assuring a complete collection.[35] Each defecation is collected in a separate preweighed plastic container with lid and labeled with subject number, time, and date. Whole stools are immediately frozen until compositing. We have noted no migration of carotenoids into the plastic when low doses of carotenoids are given, but this could be a problem with higher doses.

For compositing, containers are lined up by subject, date, and time and the presence of markers. Samples to be included in the composite are weighed and thawed and placed in a preweighed 4 liter stainless steel blender. The blender is weighed again when the entire 5–8 days of stools has been added. Avoid rubber spatulas because they absorb carotenoids. A 1% by weight solution of pyrogallol in water is added slowly and blended until the fecal material is the consistency of thick pea soup, and the weight of the container is again recorded. Fifty-gram samples (50 g) are poured into preweighed containers and lyophilized until constant weight is achieved. The containers are immediately capped and sealed with tape, then stored frozen until analyzed for carotenoid content. Care must be taken to prevent the oxidation of carotenoids as fecal material is more vulnerable than plasma, probably because it contains fewer natural antioxidants.

Dried samples are pulverized in a blender to a fine powder. Tomato skins and other carotenoid-containing components that resist pulverization can cause problems in obtaining consistent samples. All operations must be carried out under subdued or yellow light. Twenty milligrams of pulverized fecal sample is weighed and placed in a test tube to which 0.5 ml of 10% KOH (w/v) in water and 0.5 ml of 2% (w/v) pyrogallol in ethanol is added. Duplicate samples are used. Each tube is mixed thoroughly for 30 sec, then placed in a 65° water bath for 60 min. Two milliliters of hexane is added to each tube and mixed thoroughly for 1 min followed by centrifu-

---

[35] B. R. Kottko and M. T. R. Subbiatr, *J. Lab. Clin. Med.* **80**, 530 (1972).

gation at 2000 g for 5 min at room temperature. The hexane supernatant, now a yellow color, is transferred to a separate tube. The hexane extraction is repeated 3–4 times until no yellow color is visible in the hexane. One milliliter of distilled water is added to the hexane and mixed thoroughly for 1 min and again centrifuged for 5 min. The hexane layer is removed to a new tube, and the hexane wash is repeated on the water remaining. The tubes containing the hexane solutions are evaporated to dryness in a Speed-Vac Concentrator (Savant Instruments, Inc., Hicksville, NY) centrifuge, usually about 30 min. At this point, the samples are ready to be solubilized with mobile phase or other solvent for HPLC analysis.

Estimation of Apparent Absorption of β-Carotene by Total Gastrointestinal Lavage

The following method has been used in several studies for evaluation of the absorption of calories, carbohydrates, and calcium,[36-39] and it may prove applicable for carotenoids. It begins with a preparatory washout in which the entire alimentary tract is cleansed of all food and fecal material by a special lavage solution that does not cause significant loss of electrolytes and does not affect the absorption of calcium, D-xylose, glucose, water, and various electrolytes.[34,35] After the first lavage, the subject consumes a meal with or without the test carotenoid. All fecal output is collected from this point. After 24 hr a second lavage solution is given, and all effluent is collected and combined for carotenoid analysis. The completeness of collection is evaluated by the recovery of polyethylene glycol (PEG).

Experimental Design

A crossover design with a washout period is the design of choice for this type of study. Subjects serve as their own control because each subject has characteristic absorption of β-carotene. The effect of fat and energy intake on the absorption of single doses of β-carotene can be determined.[8] Each test should be separated by at least a 3-week washout period, and placebos for β-carotene are assessed for each meal type. Subjects should be given a

[36] G. W. Bo-Linn, G. R. Davis, D. J. Buddrus, S. G. Morawski, C. Santa Ana, and J. S. Fordtran, *J. Clin. Invest.* **73,** 640 (1984).
[37] G. W. Bo-Linn, C. A. Santa Ana, S. G. Morawski, and J. S. Fordtran, *N. Engl. J. Med.* **317,** 1413 (1982).
[38] M. S. Sheikh, C. A. Santa Ana, M. J. Nicar, L. R. Schiller, and J. S. Fordtran, *N. Engl. J. Med.* **317,** 532 (1987).
[39] G. R. Davis, C. A. Santa Ana, S. G. Morawski, and J. S. Fordtran, *Gastroenterology* **78,** 991 (1980).

list of carotenoid-containing foods that they should avoid 3 to 4 days prior to each test.

They are also asked to fast overnight for 12 hr.

On day 1, subjects come at 8 AM and drink 1 gallon of Colyte (Reed & Carnick, Inc., Piscataway, NJ) (contains, in g/liter, 60 g PEG, 1.46 g NaCl, 0.745 g KCl, 1.68 g Na_2Co_3, and 5.68 g Na_2SO_4) which they drink over 1.5 hr. After subjects drink the Colyte, the last clear rectal effluent is collected (usually 1–2 hr later and stored at $-70°$ until analyzed). The low-carotenoid basal meal (high fat or low fat) is given with the test dose of carotenoid (15 mg of β-carotene), and this is called 0 hr. For the next 24 hr, subjects may only consume noncaloric fluids, for example, water and diet soft drinks, no caffeine. It is important to encourage the consumption of large amounts of these fluids. Subjects are allowed to return home after 8 hr since serum response curves are also generated during this time, but they are given containers to collect all stool until the following morning. On day 2, another gallon of Colyte is consumed at 8 AM, and all rectal effluents are collected until there is no more Colyte remaining in the colon ($\sim$ 3 hr later). All stool and effluent produced after 0 hr of day 1 is pooled. The subjects eat low-carontenoid meals of their own choosing for the next 38 hr and return home with containers to collect stool for an additional 24 hr, which they are asked to keep on dry ice.

Preparation of Effluent for Analysis. The entire collection of effluent is thawed within 2 weeks and placed in a 4 liter blender along with 10 ml of 2-octanol (Eastman Kodak Co., Rochester, NY) per 1500 ml of effluent to prevent foaming. The mixture is blended for 10 min and then poured into a large (4 liter) beaker or stainless steel pan placed on top of a magnetic stirrer. The mixture is continuously stirred by means of a magnetic stir bar. Since particles containing β-carotene or other carotenoids may float to the top, preventing homogeneous sampling, the upper portion of the mixture is removed and filtered. The residue is then dried (lyophilized), weighed, and analyzed separately for carotenoid content, while the filtrate is poured back into the remaining mixture. Duplicate aliquots of 50 ml are then removed for lyophilization after addition of 1 ml of 5 mg/ml butylated hydroxytoluene (BHT) in ethanol to each aliquot to prevent oxidation. We have noted losses of 23–26% in sample β-carotene when BHT is not added.

After 2–3 days of freeze-drying, the samples come to constant weight. The dried effluent is then placed in a small blender and pulverized until homogeneous. The extraction procedure is identical to that described earlier (see Estimation of Carotenoid Retention by Balance Techniques, section on fecal preparation and analysis). The concentration of carotenoids in the total effluent is calculated by averaging the amount of carotenoids in each 50-ml aliquot and then multiplying this number by the total volume of effluent divided by 50 ml. The total carotenoid content of the filtered residue is then added to yield the total carotenoid content of the effluent.

Assessment of Technique for Estimating Apparent Absorption. A series of experiments were performed on young men to ascertain the limitations of the technique. Subjects were given placebo or β-carotene with various types of meals as described under experimental design. When placebo was given without a meal (24 hr fast), β-carotene was not detected in the effluent of any of the subsequent stools. When a 15-mg dose of β-carotene was given without a meal, 12.7 ± 0.4 mg of β-carotene was retrieved in the stool ($n = 3$). This indicates that only 2.3 ± 0.4 mg (or 13–18%) of the original dose ingested was apparently absorbed. When the β-carotene was given with a 1000 kcal, 40% fat diet, 7.9 ± 1.9 mg of β-carotene was retrieved in the first 24-hr washout. Stools collected after 24 hr (including collections assayed as long as 72 hr after the ingested dose) contained only trace amounts of β-carotene (between 0 and 4% of the 15-mg dose) and other carotenoids. This was probably due to the self-selected, low-carotenoid diets consumed by the young men after the first 24-hr period.

The Colyte was successful in cleaning the gastrointestinal tract on the first washout. Only 5–10% of the PEG was retained after 4 hr (PEG analysis is performed by a turbidometric assay[40]). The effluent was also assayed by bomb calorimetry and Kjeldahl analysis. There were minimal amounts of energy and nitrogen in the stool, which varied slightly depending on the diet provided. PEG was found to contain an average of 5.24 kcal/g. There was also a slight weight loss (2 pounds) in the young men over the 24-hr tolerance test period. Healthy older men (60–80 years) were also able to tolerate the lavage procedure on repeat occasions.

The 24-hr period allows time for the entire meal to be absorbed. All the β-carotene available for retrieval also appears to be accounted for within 24 hr, since subsequent stools contain negligible amounts of the test carotenoid (β-carotene).

The potential advantage of the gastrointestinal lavage technique is the ability to carefully control conditions allowing a more precise estimation of the quantities of carotenoids retained in the gastrointestinal tract, and thus available for absorption, over a specified time period. Such precision is required when assessing apparent retention of graded doses of a single purified carotenoid, nutrient composition of the diet, or time course of disappearance of the carotenoid from the gastrointestinal tract in comparison to plasma response. The technique also holds promise in conjunction with the use of specific stable isotopes of individual carotenoids to assess

[40] S. Hyden, *Kingl Lantbrukshoosk Ann.* **22**, 139 (1955).

isotope dilution from endogenous carotenoids in an effort to determine true absorption. However, it is recognized that further evaluation and confirmation are necessary and that the procedure is best carried out in a clinical setting.

[2] Evaluation of Carotenoid Intake

By MARC S. MICOZZI

Human Dietary Assessment of Carotenoid Intake

People eat foods, not nutrients, and foods contain many nutrients simultaneously. It is important to recognize that the total diet must be analyzed in order to determine the dietary intake of individual nutrients, such as carotenoids. Collecting dietary information on one or a few nutrients may be acceptable under selected circumstances. However, for epidemiological research, to control the effects of confounding factors, or to analyze effect modification, it is usually unwise and a false economy to restrict data collection to a limited list of one or a few nutrients. Studies that do not attempt to assess the entire diet may result in false conclusions regarding nutrient intake.

Generally, the goal of dietary assessment by various methods is to measure the "individual's usual diet." Like so many measurement techniques, dietary assessment methods are suited to obtaining a view of "normal" dietary patterns. Acute, unusual dietary events or exposures often will be unobserved or unmeasured. Since most dietary studies attempt to determine the association of habitual dietary patterns with other variables, the usual dietary assessment methods are generally suitable. An additional goal of nutritional assessment methods is to derive data on nutrient intake that are suitable for assessment of individual experience, not just group intake.

The validity of all food intake data in terms of nutrient intake is based on the availability and accuracy of food composition data. It is important to have accurate and complete data on food composition to be able to determine the nutrient intake from the foods consumed. Methods for nutrient composition analysis are critical for this step and are described for carotenoids elsewhere in this volume.

Finally, the effects of individual nutrient supplementation on evaluation of carotenoid intake must be taken into account. The methods described here are for estimating dietary intake excluding vitamin supplements. Since individuals and populations vary in the extent to which they consume vitamin supplements, it is important to be able to evaluate dietary intake methods independently, as well as vitamin supplementation. Several specific methods exist for dietary assessment of nutrient intake.

Daily Dietary Recall

The respondent is asked to report his or her complete dietary intake of all foods and beverages consumed over the past 24 hr. Portion size must be accurately obtained, preferably through the use of three-dimensional models and/or picture books. The database must be sufficiently detailed to permit coding of a great variety of cuts of meat, food preparation methods, etc. This method requires a highly trained interviewer and 20–30 min to perform. It does not provide valid estimates of the usual diet of individuals.

Diet Records

In the diet record method, the diet is recorded at the time it is consumed. The diet record may be completed for 1 or 2 days. The accuracy of recorded values may decline after the first 2 days. Consecutive days also provide less information about usual diet than do the same number of nonconsecutive days. Therefore, multiple-day diet records may be obtained in the form of 1 or 2 day units distributed over longer periods of time.

Because the diet record is kept by the individual respondent, respondents must be literate and familiar with the concept of careful, conscientious record-keeping. The effect of carefully recording food intake may cause a modification in normal eating patterns.

Food-Frequency Questionnaires

The respondent is requested to indicate the usual frequency of consumption of each of a list of foods or, sometimes, food groups. Questionnaires may be quantified so that nutrients such as carotenoids are estimated in addition to simple food frequencies. A weighting factor such as nutrient quantity in a "standard portion" is multiplied by the frequency of consumption to obtain a nutrient score or estimate of the quantitative nutrient content of the diet.

Food lists are often targeted toward the assessment of a single nutrient, such as carotenoids, or only a few nutrients of interest. Such lists include only those foods designed to capture the intake of that nutrient. Such questionnaires may be self-administered among individuals or large population samples. If information is not obtained on portion size, the method is termed a food-frequency assessment.

Diet History Questionnaires

If information is obtained on portion size, the method is termed a diet history. This method involves the reporting of overall patterns of dietary intake over extended periods of time, often over the past year. A 1-year period is generally long enough to allow patterns to encompass normal seasonal variations and smooth out short-term aberrations. Unlike a 24-hr dietary recall, the cognitive process involved does not require individual memory of specific instances or events, but rather involves the recognition and reporting only of patterns of dietary behavior.

The "advantage" of obtaining portion size in the diet history, versus the food-frequency questionnaire, may be relative as there is controversy about the need for portion size information in epidemiologic studies. For vitamin A, for example, there has been little or no advantage to obtaining portion size with respect to ranking or relative risk of cancer.

Errors of Dietary Assessment

There are errors inherent in all methods of human dietary assessment, such as imprecision and misclassification. Imprecision results in the absolute value of the nutrient intake estimate differing from the true nutrient intake of the individual to a significant degree. Imprecision is undesirable if one wishes to counsel individuals regarding their dietary intake, for example. It may be less serious in a study of the relations of dietary nutrient intake to other factors, unless misclassification is also present. Misclassification results in individuals not being categorized correctly along the distribution of nutrient intake. A dietary method may be imprecise but still satisfactory in terms of classifying individuals correctly.

When investigators take errors into account in estimating confidence limits, the obvious uncertainty in dietary assessment may be discouraging. However, a clear understanding of the limitations of human dietary assessment is the first step toward overcoming them.

Correlation between Dietary Carotenoid Intake and Blood Carotenoid Levels

Carotenoid intake may also be evaluated on the basis of resulting levels of carotenoids in body fluids and tissues, or as reflected in the coloration of skin and other tissues. It is reasonable to assume that for dietary carotenoids to have a physiologic effect they must, after consumption, appear in the blood and tissues to a significant extent. Therefore, evaluation of carotenoid intake may also be directed by the evaluation of the profile of carotenoids in human plasma following food consumption. Methods for the determination of carotenoid contents and profiles in human plasma and other tissues are given elsewhere in this volume.

Acute Intake

A single ingestion of a carotenoid-containing food or a β-carotene supplement is relatively slow to appear in the blood, with maximum concentrations reached at 24–48 hr after consumption. Carotenoids in foods appear to have less bioavailability than purified β-carotene capsules, for example. There is also a significant degree (3- to 4-fold) of individual variability in plasma response to a single carotenoid intake. Some individuals appear to be consistently "poor absorbers" of carotenoids, suggesting that plasma carotenoid levels may not always accurately reflect dietary intake.

Chronic Intake

Long-term ingestion of carotenoid-containing foods and β-carotene supplements show many of the same patterns as single carotenoid consumption. It may be difficult to observe a response in blood carotenoid levels in all individuals with consumption of certain carotenoid-containing foods. Carrot consumption and red palm oil ingestion, for example, appear to raise β-carotene levels, especially among individuals who usually ingest a relatively low-carotenoid diet. Other carotenoids, such as lutein and lycopene, may also be increased significantly by the consumption of certain foods.

Carotenodermia

When chronic carotenoid consumption reaches certain levels, this may be reflected in development of carotenodermia, a yellow coloration of skin. Carotenodermia may be first detected when total plasma carotenoid levels reach 4.0 mg/liter. This plasma level corresponds to dietary carotenoid consumption of 12–30 mg β-carotene per day for approximately 25 to 42 days.

Carotenoid levels as reflected in skin coloration appear to lag behind carotenoid levels in blood. Carotenodermia may also occur from hyperlycopenemia associated with excessive consumption of tomato juice. These signs may be helpful in the clinical evaluation of carotenoid intake.

Carotenodermia may be assessed by physical examination of the skin, especially that of the face (zygomatic prominence), hands, elbows, knees, and feet. Yellowing of the skin may be graded based on subjective clinical assessments that allow room for error, such as "yellowing not present," "yellowing may be present," or "yellowing present." More objective assessments may be made through the use of reflectometers.

[3] Uptake and Cleavage of β-Carotene by Cultures of Rat Small Intestinal Cells and Human Lung Fibroblasts

By GIORGIO SCITA, GREGORY W. APONTE, and GEORGE WOLF

Introduction

β-Carotene (BC) has assumed importance as an anticarcinogen, both in epidemiologic studies in human beings[1] and in animal models.[2] Whether BC itself is the anticarcinogen or is first converted to retinol or retinoic acid (RA) is still undetermined. Recent publications by Pung *et al.*[3] and Rundhaug *et al.*[4] show definitively that in the fibroblastic cell line C3H/10T1/2, BC and canthaxanthin inhibit chemically and physically induced transformation without being converted to retinol, retinal, or RA.

Another important property of BC that has been known for a long time,[5] is its lack of toxicity in contrast to retinol. When increasing concentrations of BC are given to rats, the conversion of BC to retinol levels off before the retinol produced becomes toxic. However, the control point which regulates the cleavage has not been established. Indeed, controversy even surrounds the cleavage reaction. Even though the central cleavage of BC, resulting in two molecules of retinal, had been established by Olson and Hayaishi[6] and Goodman *et al.*,[7] and recently confirmed by Lakshman *et al.*,[8] it has been disputed by Hansen and Maret.[9] Napoli and Race[10]

---

[1] C. H. Hennekens, S. L. Mayrent, and W. Willet, *Cancer* **58** (Suppl. 8), 1837 (1986).
[2] R. C. Moon, *J. Nutr.* **119**, 127 (1989).
[3] A. Pung, J. E. Rundhaug, C. N. Yoshizawa, and J. S. Bertram, *Carcinogenesis* **9**, 1533 (1988).
[4] J. E. Rundhaug, A. Pung, C. M. Read, and J. S. Bertram, *Carcinogenesis* **9**, 1541 (1988).
[5] T. Moore, "Vitamin A." Elsevier, New York, 1957.
[6] J. A. Olson and O. Hayaishi, *Proc. Natl. Acad. Sci. U.S.A.* **54**, 1364 (1965).
[7] D. S. Goodman, H. S. Huang, and T. Shiratori, *J. Biol. Chem.* **241**, 1929 (1966).
[8] M. R. Lakshman, I. Mychkovsky, and M. Attlesey, *Proc. Natl. Acad. Sci. U.S.A.* **89**, 9124 (1989).
[9] S. Hansen and W. Maret, *Biochemistry* **27**, 200 (1988).
[10] J. L. Napoli and K. R. Race, *J. Biol. Chem.* **257**, 13385 (1988).

reported that retinal is not a free intermediate in the cleavage reaction in a cell-free system of intestinal mucosa.

In a study of liver storage of vitamin A in rats and chicks fed BC, Brubacher and Weiser[11] report that for the rat "in the range of about one to a maximum of 10 times the daily vitamin A requirement, BC is completely absorbed and transformed to vitamin A, with a relationship of one molecule BC corresponding to one molecule of retinol. With a higher intake, the log of the absorption or transformation rate decreases linearly, inverse to the log of intake." They postulate two possible mechanisms of regulation: (1) the transport of BC into the mucosal cell is regulated, so that the excess is excreted in the feces; and (2) the transport of BC into the mucosal cell is not limited, and it is the cleavage reaction which is regulated. In the rat, only small amounts of BC enter the bloodstream[12]; in mechanism (2), therefore, the unconverted BC must be reexcreted into the gut. Mechanism (2) seems the more likely in view of the work of Hollander and Ruble[13] which showed that, in the rat, BC enters the intestinal mucosa by passive diffusion.

In an attempt to distinguish between these two hypotheses, we examined the uptake kinetics of BC into a line of small intestinal cells in culture, in order to find out whether uptake into these cells is regulated. At the same time, we studied cleavage kinetics to determine whether BC cleavage is limiting. No attempt was made here to investigate the mechanism of cleavage.

Although whole rats,[13] as well as cell-free systems,[10] have been used to study some of the above parameters, to our knowledge no studies have been reported for cell culture systems with cells from the small intestine, the locus of BC uptake.[14] This was made possible by the generation by one of us (G.W.A.) of an immortalized line of polarized rat small intestinal cells (hBRIE 380).[15]

Materials and Methods

β-[^{14}C]Carotene and canthaxanthin were generously provided by Hoffmann-LaRoche (Nutley, NJ). β-[^{14}C]Carotene (specific activity 53.5 μCi/mol) is purified (98%) through a reversed-phase HPLC column with methanol plus 0.5% (w/v) ammonium acetate/toluene (3:7, elution volume 17 ml). Unlabeled β-carotene was purchased from Fluka (Buchs,

[11] G. B. Brubacher and H. Weiser, *Int. J. Vitam. Nutr. Res.* **55**, 5 (1985).
[12] S. S. Shapiro, D. J. Mott, and L. J. Machlin, *J. Nutr.* **114**, 19024 (1984).
[13] D. Hollander and P. E. Ruble, *Am. J. Physiol.* **233**, E686 (1978).
[14] H. S. Huang and D. S. Goodman, *J. Biol. Chem.* **240**, 2839 (1965).
[15] G. W. Aponte, A. Keddie, G. Hallden, R. Hess, and P. Link, *Proc. Natl. Acad. Sci. U.S.A.* **88**, 5282 (1991).

Switzerland). Retinol, retinoic acid, retinyl palmitate, and lycopene were all purchased from Sigma (St. Louis, MO). Tetrahydrofuran (THF), 99.9%, and Diazald for diazomethane preparation were purchased from Aldrich (Milwaukee, WI); dimethyl sulfoxide (DMSO) was from Baxter (Muskegon, MI). All HPLC solvents were optimum grade from Fisher Scientific (Kent, WA).

Cell Culture. Human lung fibroblasts (WI-38) and human embryonic lung fibroblasts (HLF) were purchased from the American Type Culture Collection (ATCC, Rockville, MD) and maintained in Earle's salts minimum essential medium (MEM) supplemented with 10% (v/v) heat-inactivated fetal bovine serum, sodium bicarbonate (2.2 g/liter), 100 units/ml penicillin, 0.1 mg/ml streptomycin, 0.1% (w/v) lactalbumin, and 2 mM L-glutamine. Small intestinal cells (hBRIE 380) are maintained in Iscove's modified Dulbecco's medium with 10% (v/v) fetal bovine serum and sodium bicarbonate (3.04 g/liter). The hBRIE 380 cell line is a hybrid intestinal epithelial line derived from the fusion of a spontaneously transformed rat small intestinal mucosal epithelial cell line with freshly isolated mucosal epithelial cells from rat duodenum.[15]

Cells are grown in Corning dishes (35 mm diameter) in a humidified atmosphere of air/CO_2 (95:5, v/v) at 37°C.

Chromatography. HPLC is performed with a Beckman Fullerton, CA model with a reversed-phase column (0.46 × 15 cm Beckman Ultrasphere ODS, 5 μm particles). The sample is eluted at a flow rate of 1.5 ml/min with a linear gradient developed in 12 min from 100% solvent A (methanol plus 0.5% ammonium acetate) to 80% solvent A plus 20% toluene and monitored at 452 nm for the separation of carotenoids. The separation of retinoids is carried out at a flow rate of 1.5 ml/min with a mixture of water plus 2% ammonium acetate/acetonitrile (28:72)[9] (unless otherwise stated).

Addition of β-[^{14}C]Carotene to Medium. The method for delivering carotenoids in a water solution is adapted from that of Bertram *et al.*[16] The micellar-like nature of the "solution" has been demonstrated by Bertram *et al.*,[16] who suggested that when BC is "solubilized" by THF in water, it exists in a highly constrained but not solid state. Five to ten microliters of a solution of BC in THF/DMSO (1:1) (v/v) (0.5–3 mM) is rapidly injected into 1 ml of culture medium, to give a final concentration varying over a range of 1 to 28 μM, and stirred for 30 min. The BC-containing medium is then incubated with cells previously grown to confluency.

Analysis of β-Carotene Uptake and Cleavage. After incubating cells with β-[^{14}C]carotene at different concentrations for different times, the

---

[16] J. S. Bertram, A. Pung, M. Churley, T. J. Kappock IV, L. R. Wickens, and R. V. Cooney, *Carcinogenesis* **12,** 671 (1991).

medium is removed and the cell monolayer rinsed 3 times with PBS (until no longer radioactive). The cells are then detached with 0.25% w/v trypsin, pelleted, and resuspended in 0.4 ml of phosphate-buffered saline (PBS)/ ethanol (1:1) with 0.025% BHT. The cell suspension is then sonicated for a total of 30 sec in 3 bursts of 10 sec each, keeping it in ice, by an ultrasonic sonifier model Ultrasonics W 185 (Ultrasonic, Inc., Plainview, NY). Lycopene (5 μl, 0.5 mM) in THF is added to the sample before extraction as internal standard. Calculations are made on the basis of the percent recovery of the internal standard. The suspension is saturated with NaCl and extracted twice with 0.6 ml of hexane plus 0.025% BHT. When retinoid analysis is carried out, 5 μl of retinol (1 mM) and 5 μl of retinoic acid (1 mM) in chloroform are added as internal standards. The hexane phase is evaporated to dryness under N_2, reconstituted in the HPLC solvent, and injected into the HPLC column.

Each fraction is collected from the column and mixed with 10 ml of Cytoscint (ICN Biomedicals, Irvine, CA) scintillation liquid. Radioactivity is measured by a liquid scintillation counter.

Chemical Identification of Retinol and Retinoic Acid. The acetylation of retinol is done using acetyl chloride according to Fischli *et al.*[17] The reaction products are detected by HPLC with methanol plus 0.5% (w/v) ammonium acetate at a flow rate of 1 ml/min followed by measurement of the radioactivity in each HPLC fraction by liquid scintillation counting. The methylation of retinoic acid is performed by mixing diazomethane in ether with the appropriate HPLC fraction in ether as a reagent and monitoring the reaction product by HPLC with acetonitrile water plus 2% ammonium acetate (78:28) at a flow rate of 1.5 ml/min and liquid scintillation counting as described above.

Uptake of β-Carotene

The concentration and time dependence of uptake of BC into hBRIE 380 cells is shown in Fig. 1. The absence of a receptor-regulated uptake into these cells is suggested because there is no leveling off of uptake up to a concentration of 25 μM BC. Higher concentrations could not be tested because of the insolubility of BC. Unincorporated excess BC in the medium was completely removable by washing, because of the total water miscibility of the THF–DMSO solution of BC and the absence of any microscopically detectable crystalline BC. After washing 3 times with PBS, no radioactivity could be detected in the washes (data not shown).

The temperature-dependent splitting of BC to retinol and retinoic acid and the complete absence of extracellular conversion activity (no retinol or retinoic acid could be detected in the cell culture medium after 24 hr

---

[17] A. Fischli, H. Mayer, W. Simon, and H. J. Stoller, *Helv. Chim. Acta* **59**, 397 (1976).

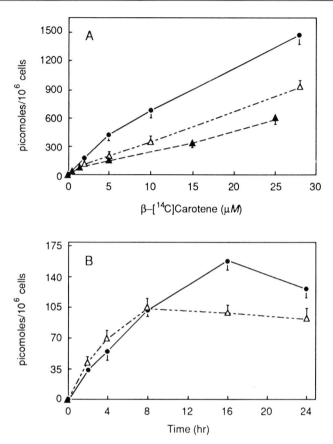

FIG. 1. Time and concentration dependence of uptake of β-[^{14}C]carotene by small intestinal cells and human lung fibroblasts. (A) Different concentrations of [^{14}C]BC in THF/DMSO (1:1) were rapidly mixed with culture medium as described in the text and incubated for 8 hr with confluent fibroblasts and 16 hr with hBRIE 380 cells at 37°. (B) [^{14}C]BC (2 nmol) was mixed with 1 ml of culture medium and incubated with confluent hBRIE 380 or WI-38 cells at 37°. At each concentration (A) and time (B) the cells were washed, collected, extracted, and analyzed for [^{14}C]BC content as described. Values are the means of 3 determinations ± S.E. Filled circles, hBRIE 380; open triangles, WI-38; filled triangles, HLF. (Reproduced with permission from Ref. 17a.)

incubation with or without cells, Fig. 2) further suggested that not only is BC incorporated, but it is also available for an enzymatic activity that accounts for its conversion.

The slight inhibition of BC uptake at 4° (20%; Table I), especially when compared to inhibition of the conversion of BC to retinol (80%; Table I), seems to exclude uptake as an enzymatic or receptor-mediated process.

[17a] G. Scita, G. W. Aponte, and G. Wolf, *J. Nutrit. Biochem.* **3**, 118 (1992).

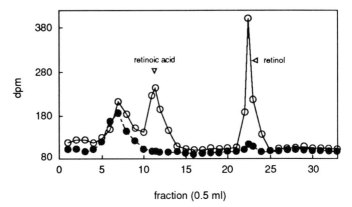

FIG. 2. Chromatographic analysis of [^{14}C]retinol and [^{14}C]retinoic acid produced from [^{14}C]BC by cells. [^{14}C]BC (1 nmol) was mixed with 1 ml of culture medium and incubated in presence (○) or in absence (●) of confluent hBRIE 380 cells at 37°. After 24 hr the cells or an aliquot of the medium were analyzed for BC cleavage products as described in the text. (Reproduced with permission from Ref. 17a.)

The standard experiment to demonstrate receptor saturation, by incubation of [^{14}C]BC with an excess of unlabeled BC, also failed to show receptor-regulated uptake (Table II). Similarly, a carotenoid closely related to BC, canthaxanthin, did not inhibit uptake (Table II). With excess unlabeled BC much more total BC entered the cells; the cleavage enzyme(s) was then presumably saturated, and hence a smaller amount of labeled cleavage products was found (Table II).

The concentration dependence of uptake of BC into two different cell lines of human lung fibroblasts (WI-38 and HLF) (Fig. 1) was similar to

TABLE I
EFFECT OF TEMPERATURE ON β-[^{14}C] CAROTENE UPTAKE AND CONVERSION TO RETINOL IN hBRIE 380 CELLS

Treatment			
Temperature (°)	Incubation time (hr)	[^{14}C]BC uptake (pmol/10^6 cells)	[^{14}C]Retinol formation (pmol/10^6 cells)
37	4	61 ± 5.2[a]	5.9 ± 1.1
4	4	51 ± 4.5	1.1 ± 0.2
37	24	119 ± 11	18.9 ± 2.1
4	24	95 ± 7.3	2.1 ± 0.3

[a] Mean ± S.E.M. of 3 incubations.

TABLE II
EFFECT OF UNLABELED CAROTENOIDS ON β-[^{14}C] CAROTENE UPTAKE AND CONVERSION
TO RETINOL AND RETINOIC ACID IN hBRIE 380 CELLS

Incubation with	[^{14}C]BC (pmol/10^6 cells)	[^{14}C]Retinol (pmol/10^6 cells)	[^{14}C]Retinoic acid (pmol/10^6 cells)
1 μM [^{14}C]BC	63 ± 6.0[a]	6.1 ± 0.7	3.4 ± 0.2
1 μM [^{14}C]BC plus 20 μM unlabeled BC	60 ± 5.4	3.2 ± 0.3	1.4 ± 0.2
1 μM [^{14}C]BC plus 20 μM canthaxanthin	65 ± 4.9	6.2 ± 1.1	2.9 ± 0.3

[a] Mean ± S.E.M. of 3 incubations.

that into hBRIE 380 cells, except that only about half the amount was taken up at the lower concentrations. The time dependence of WI-38 was similar to that found in hBRIE 380.

The mechanism of uptake of BC into hBRIE 380 cells and fibroblasts seems to be passive and confirms the observation of Hollander and Ruble[13] with rat intestinal loops *in vivo*. Uptake of BC into cells from the medium at a medium concentration of BC comparable to that used by Rundhaug *et al.*[4] was less than one-tenth as great as reported by these authors, also using fibroblast cells. Possibly the use of water-dispersible beadlets by these authors could account for the difference. We used a micellar suspension (micellar particle diameter <0.45 μm). Rundhaug *et al.*[4] detected a small loss (<4%) of medium BC by chemical oxidation in 24 hr; in our system, BC was stable over 24 hr.

Cleavage of β-Carotene

As shown in Fig. 3, small intestinal cells produce both retinol and retinoic acid from BC: the time dependence is shown in Fig. 3A and the concentration dependence (for retinol) in Fig. 3B. The production of retinol observed in 24 hr was 17.3% of the BC in the cells, of RA, 5.35%, at a BC concentration of 1 μM with 1 × 10^6 cells. The apparent K_m was 9 μM with respect to retinol formation, of the same order of magnitude as that found by Napoli and Race[10] for their cell-free system (20.6 μM). Table II shows that, as expected, excess unlabeled BC inhibited the conversion of [^{14}C]BC to labeled retinoids, whereas the closely related carotenoid canthaxanthin did not. Thus, conversion was appreciable, with 22% of the BC converted to retinol and RA in 24 hr by 1 × 10^6 hBRIE 380 cells.

Human lung fibroblasts (WI-38) or human embryonic lung fibroblasts (HLF) also converted BC to retinol and RA (time dependence, Fig. 4A;

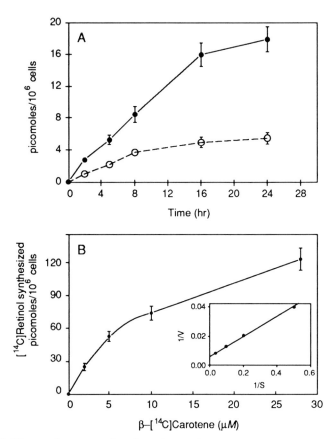

FIG. 3. Time and concentration dependence of the conversion of β-[^{14}C]carotene to retinol and retinoic acid by small intestinal hBRIE 380 cells and human lung fibroblasts. (A) Synthesis of retinol (solid line) and retinoic acid (dashed line) was analyzed at various times of incubation in the presence of a fixed amount of [^{14}C]BC (2 μM) as described in the text. (B) Increasing concentrations of β-[^{14}C]carotene were incubated with cells for 16 hr as described in the legend to Fig. 1A. The inset represents the linear transformation of the saturation curve of retinol synthesis. The HPLC analyses were performed as described in the text. Values are the means of 3 determinations ± S.E. (Reproduced with permission from Ref. 17a.)

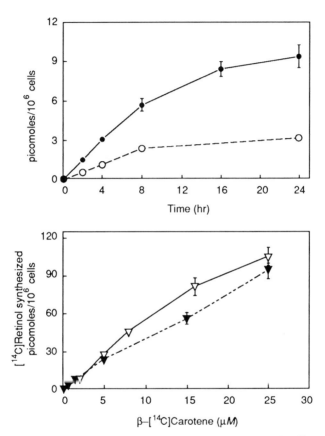

FIG. 4. Time and concentration dependence of the conversion of β-[^{14}C]carotene to retinol and retinoic acid by human lung fibroblasts. (A) Synthesis of retinol (solid line) and retinoic acid (dashed line) was analyzed at various times of incubation in the presence of a fixed amount of β-[^{14}C]carotene (2 μM) as described in the text. (B) Increasing concentrations of β-[^{14}C]carotene were incubated with cells (WI-38, open triangles; HLF, filled triangles) for 16 hr as described in the lengend to Fig. 1A. The HPLC analyses were performed as described in the text. Each point represents the mean ± S.E.M. ($n = 3$). (Reproduced with permission from Ref. 17a.)

concentration dependence for retinol formation, Fig. 4B), but at a rate about one-half that of hBRIE 380 (7.8% retinol and 2.5% RA at a BC concentration of 1 μM by 1 × 10^6 cells). It was not possible to determine a K_m for retinol formation with these cells, since the maximum achievable concentration of BC in the medium was 25 μM, at which the conversion to retinol had not yet leveled off.

Identification of retinol and RA (Figs. 5 and 6) was by retention times on HPLC compared to authentic samples in two solvent systems (water–ammonium acetate–acetonitrile; methanol–toluene) and by conversion of retinol to its acetate and RA to its methyl ester, with shifts in retention times corresponding to those observed for light absorption at 326 and 350 nm, respectively. A more polar labeled peak than either retinol or RA (Figs. 5 and 6) was detected by incubation of [^{14}C]BC which was not hydrolyzed by acid or alkali to RA or retinol, and therefore was not

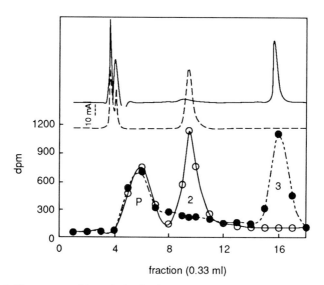

FIG. 5. Chromatographic analysis of retinyl ester produced after acetylation of the retinol formed by β-[^{14}C]carotene cleavage in small intestinal cells. [^{14}C]BC was incubated with hBRIE 380 cells for 16 hr. After collecting the cells, 5 μl of authentic retinol (2.5 nmol) in THF was added, and the cell suspension was sonicated and extracted. The hexane phase was evaporated to dryness under N$_2$, and the samples were reconstituted in chloroform. The acetylation was performed as described in the published procedure.[17] The HPLC solvent system used was methanol plus 0.5% (w/v) ammonium acetate at a flow rate of 1 ml/min. Measurements of optical density were by flow detector at 330 nm. The HPLC profile of acetylated (filled circles) and nonacetylated (open circles) [^{14}C]retinol synthesized from [^{14}C]BC in hBRIE 380 cells is shown in the lower plot. Peak P, Unidentified polar compounds; peak 2, retinol; peak 3, retinyl acetate. (Reproduced with permission from Ref. 17a.)

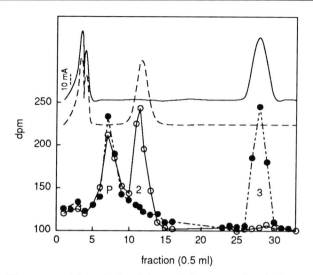

FIG. 6. Chromatographic analysis of the methyl ester produced after methylation by diazomethane of retinoic acid formed by conversion of β-carotene in small intestinal cells. [^{14}C]BC (2 μM) was incubated with hBRIE 380 cells for 6 hr as described in the text. After collecting the cells, 5 μl of authentic retinoic acid (2.5 nmol) in THF was added, and the cell suspension was sonicated and extracted with hexane. An ether solution of diazomethane was added dropwise to the hexane phase, and the mixture was stirred in ice for 15 min before adding acetic acid. After evaporating to dryness and resuspending in the HPLC solvent system, the samples were analyzed by HPLC as described in the text. Measurements of optical density were by flow detection at 350 nm. The HPLC profile of methylated (filled circles) and nonmethylated (open circles) labeled retinoic acid synthesized from [^{14}C]BC by hBRIE 380 cells is shown. Peak P, Unidentified polar compounds; peak 2, retinoic acid; peak 3, methyl retinoate. (Reproduced with permission from Ref. 17a.)

retinoylglucuronide.[18] It was not acetylated by acetyl chloride nor methylated by diazomethane. This peak also appeared in the control incubations without cells and is therefore not a metabolic product.

The temperature-dependent formation of retinol (Table I), together with the saturation curve (Fig. 3B), clearly indicates an enzymatic process. No conversion occurred on incubation without cells. Human lung fibroblasts converted about 10% BC to retinol and retinoic acid in 24 hr. This is in contrast to the findings of Rundhaug et al.,[4] who could not detect cleavage of BC in their fibroblastic cell system (C3H/10T1/2), finding neither retinol nor retinal. This discrepancy may be due to the use of water-soluble BC beads or a different cell line by these authors.

[18] A. B. Barua and J. A. Olson, *Am. J. Clin. Nutr.* **43**, 481 (1986).

TABLE III
RETINOL UPTAKE AND ESTERIFICATION IN hBRIE 380 CELLS

Incubation time (hr)	Retinol (pmol/10^6 cells)	Ester (pmol/10^6 cells)
2	139 ± 7.3	4.9 ± 0.5
4	145 ± 10.2	7.6 ± 1.0
6	120 ± 6.0	8.5 ± 0.6

The absence of detectable retinyl esters was surprising, in view of the results *in vivo* of Olson.[19] This could be explained by the low rate of conversion of the retinol produced in the cleavage reaction to esters in our cell system. As shown in Table III, when these cells were exposed to labeled retinol, only about 3.5 to 6.7% of the entering retinol was esterified in 2 hr, so that the esterification of the retinol produced by BC cleavage would be undetectable in our system. Furthermore, as shown by El-Gorab *et al.*,[20] esterification is strictly dependent on the presence of bile salts, which were not used in the present work.

We could detect no retinal or retinyl esters in hBRIE 380 cells in spite of the fact that the limit of detection of our experimental system was less than 1 pmol. The failure to detect any retinal is in agreement both with the fact that the intracellular concentration of this retinoid was found to be close to zero in different cell types (with the exception of the visual cells) and with the finding by Napoli and Race[10] of the absence of free retinal in a cell-free system of rat small intestine which converts BC to retinoic acid. Lakshman *et al.*,[8] on the other hand, showed retinal to be an intermediate in the conversion of BC to retinol in rat intestinal mucosa. Nevertheless, as the same authors suggest, it is necessary to fractionate the BC cleavage enzyme activity from other interfering activities which could carry out the rapid metabolic conversion of the retinal intermediate. Therefore, in our cell system, the occurrence of retinal might be transient.

Acknowledgments

The authors thank Dr. H. N. Bhagavan, Hoffmann-La Roche, Inc., for a gift of β-[^{14}C]carotene. Research reported in this chapter was supported by U.S. Department of Agriculture Grant 87-CRCR-1-2593 (G.W.) and National Institutes of Health Grant DK 38310 (G.W.A.).

[19] J. A. Olson, *J. Biol. Chem.* **236**, 349 (1961).
[20] M. I. El-Gorab, B. A. Underwood, and J. D. Loerch, *Biochim. Biophys. Acta* **401**, 265 (1975).

[4] Association of Carotenoids with Human Plasma Lipoproteins

By BEVERLY A. CLEVIDENCE and JOHN G. BIERI

Introduction

The majority of circulating carotenoids are associated with lipoproteins[1] which serve as plasma transport vehicles. The spectrum of carotenoids found in plasma are also present in a variety of human tissues,[2] indicating that there is effective transfer of carotenoids from plasma lipoproteins to tissues. The apolipoprotein components of lipoproteins provide the specificity that is required for transfer of lipoprotein components to specific tissues, and the array of apolipoproteins varies among the lipoprotein classes. Various apolipoproteins serve as recognition ligands for lipoprotein receptors on tissues, enabling lipoprotein uptake and catabolism at specific sites. The apolipoprotein components also serve as cofactors for certain enzymes, such as hepatic and extrahepatic lipases, which are intimately involved in lipoprotein metabolism and hence in regulating transfer of lipoprotein components to tissues. Thus, it is reasonable to infer that the final destination and ultimate metabolic fate of a carotenoid depends largely on the relative affinity of that carotenoid for specific lipoprotein classes. Additionally, carotenoids may have a role in protecting lipoproteins from oxidative modification *in vivo*.

Distribution of Carotenoids among Lipoproteins

Plasma carotenoids are associated predominantly with low-density lipoproteins (LDL), the major cholesterol-transporting lipoprotein. This view is supported by correlations of serum carotene concentrations with total cholesterol[3] and LDL–cholesterol[4] levels, but more definitively by analysis of various carotenoids from lipoprotein classes.[1,5–7] Even by gross observation the yellow pigments of carotenoids can be visualized in iso-

---

[1] N. I. Krinsky, D. G. Cornwell, and J. L. Oncley, *Arch. Biochem. Biophys.* **73**, 233 (1958).
[2] L. A. Kaplan, J. M. Lau, and E. A. Stein, *Clin. Physiol. Biochem.* **8**, 1 (1990).
[3] L. L. Adams, R. E. LaPorte, L. O. Watkins, D. D. Savage, M. Bates, J. A. D'Antonio, and L. H. Kuller, *Cancer* **56**, 2593 (1985).
[4] A. Vahlquist, K. Carlson, D. Hallberg, and S. Rossner, *Int. J. Obes.* **6**, 491 (1982).
[5] M. M. Mathews-Roth and C. L. Gulbrandsen, *Clin. Chem.* **20**, 1578 (1974).
[6] L. K. Bjornson, H. J. Kayden, E. Miller, and A. N. Moshell, *J. Lipid Res.* **17**, 343 (1976).
[7] P. P. Reddy, B. A. Clevidence, E. Berlin, P. R. Taylor, J. G. Bieri, and J. C. Smith, *FASEB J.* **3**, A955 (Abstr.) (1989).

lated, human LDL, suggesting a strong affinity of carotenoids for this lipoprotein class. LDL have been reported to carry 55 to 79% of plasma total carotenoids in healthy subjects who were not taking carotenoid supplements.[5,7]

There are few reports of the relative distribution of plasma carotenoids among lipoprotein classes.[1,5-7] Most studies of carotenoid distribution among lipoproteins have measured total carotenoids or only a few individual carotenoids using absorbance spectrophotometry. With the recent development of HPLC methods[7a] for analysis of carotenoids, the spectrum of plasma carotenoids can be analyzed concurrently. This newly developed methodology is becoming routine for analysis of carotenoids from whole plasma, but such data for individual lipoproteins are, to date, limited to a single abstract.[7]

Individual carotenoids are not uniformly distributed among lipoproteins. Table I shows the distribution of five carotenoids among lipoproteins of fasting, normolipemic men, all eating a controlled, typical American diet. In this population, LDL accounted for 55% of the total carotenoids, whereas high-density lipoproteins (HDL) had 31% and the very low-density lipoprotein (VLDL) fraction carried a relatively small proportion (10–19%) of each of the carotenoids. Three of the carotenoids, all hydrocarbons, were carried predominantly by LDL. Of the hydroxy carotenoids measured, cryptoxanthin was about equally distributed between LDL and HDL, whereas zeaxanthin and lutein, which eluted together on HPLC, were carried predominantly by the HDL fraction. Mechanisms governing the selective association of the various carotenoids with the various lipoprotein classes have not been identified.

Recently the antioxidant components of LDL have received increased attention. The types and amounts of carotenoids, as well as other endogenous antioxidants, that are carried by LDL are potential mitigating factors in the development of atherosclerosis. It is believed that atherogenesis is promoted when LDL, trapped in the extravascular space of the arterial wall, become oxidatively modified. *In vitro* studies suggest that, after oxidative modification, LDL are taken up by cells of the artery wall, resulting in formation of lipid-laden foam cells.[8] It has yet to be determined whether the carotenoid components of LDL, in concert with vitamin E, can effectively inhibit oxidative modification of LDL *in vivo* or whether carotenoids are too rapidly degraded within artery-trapped LDL to have a significant protective role. *In vitro* studies have identified α- and γ-tocopherol as well

---

[7a] F. Khachik, G. R. Beecher, M. B. Goli, and W. R. Lusby, this series, Volume 213 [18].
[8] M. T. Quinn, S. Parthasarathy, L. G. Fong, and D. Steinberg, *Proc. Natl. Acad. Sci. U.S.A.* **84,** 2995 (1987).

TABLE I
CAROTENOID AND CHOLESTEROL PROFILE OF 22 HEALTHY MEN[a,b]

Lipoprotein	Total carotenoids	Lutein/ zeaxanthin	β-Cryptoxanthin	Lycopene	α-Carotene	β-Carotene	Cholesterol
VLDL	14	16	19	10	16	11	6
LDL	55	31	42	73	58	67	65
HDL	31	53	39	17	26	22	29

[a] Values based on percentage of recovered carotenoid or cholesterol.
[b] Adapted from P. P. Reddy, B. A. Clevidence, E. Berlin, P. R. Taylor, J. G. Bieri, and J. C. Smith, *FASEB J.* **3**, A955 (Abstr.) (1989).

as three individual carotenoids (phytofluene, lycopene, and β-carotene) as important parameters for the oxidative resistance of LDL.[9] The amounts of these antioxidants varied greatly among LDL samples.

Factors Influencing Carotenoid Content of Lipoproteins

The maximum capacities of lipoproteins for binding the various carotenoids have not been determined, but binding capacities would be likely to differ among the lipoprotein classes. This question has been partially addressed by assessing the distribution of carotenes among lipoproteins from subjects supplemented with a very high dose of β-carotene. It can be reasoned that if the maximum carotenoid-binding capacity of lipoproteins differs among the lipoprotein classes, then the relative distribution of a carotenoid among VLDL, LDL, and HDL should change following supplementation with that carotenoid. β-Carotene supplementation (120–180 mg/day for 3–10 weeks) was shown to produce a marked (~10-fold) increase in plasma total carotene[5] and β-carotene[6] levels. However, it is not clear whether maximum binding of carotene within a lipoprotein class occurs, as would be suggested by a relative enrichment of one lipoprotein class at the expense of another. Mathews-Roth and Gulbrandsen[5] found an increase in the proportion of plasma carotene carried by LDL, largely at the expense of HDL, in subjects taking supplements, yet Bjornson *et al.*[6] found no change in β-carotene distribution among lipoproteins from patients with erythropoietic protoporphyria who took β-carotene.

Conditions that alter levels of lipoproteins are known to alter the distribution of tocopherols; however, there are few similar data for carotenoids. The effect of hyperlipoproteinemia on the relative distribution of carotenoids has been addressed in only one study, where the distribution of β-carotene was analyzed in lipoproteins from five subjects.[6] In two hypertriglyceridemic patients the largest proportions of β-carotene were carried by LDL even though VLDL carried an unusually high percentage of plasma cholesterol and α-tocopherol. On the other hand, in a subject with chylomicronemia and in another with dysbetalipoproteinemia, the largest proportion of β-carotene was carried in the chylomicron plus VLDL fraction and the VLDL fraction, respectively.

Postprandial Incorporation of β-Carotene into Lipoproteins

Following a test meal supplemented with β-carotene, this carotenoid reached peak levels in both chylomicron and VLDL fractions within 7 hr

[9] H. Esterbauer, M. Dieber-Rotheneder, G. Striegl, and G. Waeg, *Am. J. Clin. Nutr.* **53**, 314S (1991).

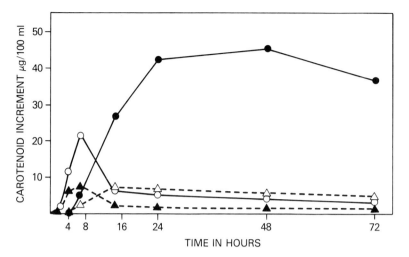

FIG. 1. Mean lipoprotein increments in β-carotene for three normal subjects: chylomicrons (▲), VLDL (○), LDL (●), HDL (△). [From D. G. Cornwell, F. A. Kruger, and H. B. Robinson, *J. Lipid Res.* **3**, 65 (1962).]

in normolipemic subjects.[10] These peaks for β-carotene appeared to lag behind the time of peak lipid response to the test meal. The β-carotene levels of LDL are reported to surpass those of VLDL within 16 hr and reach a maximum at 24 to 48 hr that is considerably higher than the β-carotene content of the other lipoprotein fractions (Fig. 1). The β-carotene content of LDL remained elevated 72 hr after the test meal. Comparable data for the other carotenoids of human plasma have not been reported.

Transfer of Carotenoids

There are no published reports on the mechanism(s) of carotenoid transfer among lipoproteins or from lipoproteins to tissues. Unlike α-tocopherol and unesterified cholesterol, β-carotene does not appear to be rapidly exchanged either among lipoproteins or between lipoproteins and red blood cells. This view is supported by the delayed peak response for β-carotene content of VLDL and LDL following administration of this carotenoid[10] and by the observation that peak levels of β-carotene in red blood cells lag substantially behind peak plasma levels following supplementation with β-carotene.[6]

[10] D. G. Cornwell, F. A. Kruger, and H. B. Robinson, *J. Lipid Res.* **3**, 65 (1962).

Lipoprotein Separation Techniques

This section is written to orient persons interested in carotenoid metabolism to the variety of methods for lipoprotein fractionation and to provide references from originators of the methods and from other comprehensive reports. The techniques used to separate lipoprotein classes typically involve ultracentrifugation, gel filtration, precipitation, affinity chromatography, or a varied combination of these approaches. The choice of a method will depend on the technical difficulty and expense of the method as well as the number of samples to be analyzed and the volume of plasma available. It should be noted that most methods separate lipoproteins without completely isolating each fraction from other plasma components. This may have relevance for lutein if, as reported, approximately 20% of this carotenoid is not associated with lipoproteins.[1]

Although individual carotenoids are fairly stable in whole plasma held at room temperature for up to 24 hr,[11] there has been no systematic evaluation of the stability of carotenoids in isolated lipoprotein fractions. Once removed from the protective environment of plasma, carotenoids are likely to be at greater risk for oxidation. Prior to the development of HPLC for carotenoid analysis, recovery of total carotenoids was reported to parallel the recovery of cholesterol from lipoproteins isolated by ultracentrifugation, thus suggesting that observed losses were due to manipulation rather than to oxidation.[1] However, this conclusion is tentative since some products of carotenoid oxidation are measured as carotenoids by nonspecific, spectrophotometric methods. Oxidation of carotenoids during lipoprotein manipulation should be inhibited by inclusion of EDTA and antioxidants that are routinely used to slow degradation of lipoproteins.[12]

To make comparisons among studies it would be useful for investigators not only to report relative and absolute carotenoid values for each lipoprotein fraction, but also to state whether the distribution is based on the percentage of total plasma values or the percentage of recovered carotenoids. The former is preferred since accurate assessment of distribution depends, to a large extent, on the ability to achieve a high degree of recovery.

It is prudent to assess the purity of lipoprotein fractions by a lipoprotein separation technique such as agarose gel electrophoresis[13] or nondenaturing polyacrylamide gel electrophoresis.[14] Simple methods for immunoas-

---

[11] N. E. Craft, E. D. Brown, and J. C. Smith, *Clin. Chem.* **34,** 44 (1988).
[12] V. N. Schumaker and D. L. Puppione, this series, Vol. 128, p. 155.
[13] R. P. Noble, *J. Lipid Res.* **9,** 693 (1968).
[14] H. K. Naito and M. Wada, *in* "Handbook of Electrophoresis" (L. Lewis and J. Opplt, eds.), Vol. 1, p. 183. CRC Press, Boca Raton, Florida, 1980.

say in gels can be used to detect specific apolipoproteins with a high degree of sensitivity.[14a]

Sequential Flotation Ultracentrifugation

Sequential ultracentrifugal flotation is the method most frequently used to isolate lipoproteins for carotenoid measurement, and it is the one that we have selected to outline. With this method, each lipoprotein class is sequentially floated to the top of the tube based on its buoyant density in a salt solution (NaCl–NaBr or NaCl–KBr). Sequential ultracentrifugation has been described in detail[15,16]; a review by Schumaker and Puppione appears in this series.[12] Useful details for collecting lipoproteins by ultracentrifugation, including use of the tube slicer for recovery of lipoprotein fractions, can be found in the Lipid Research Clinics Program *Manual of Laboratory Operations.*[17]

Materials

Three density solutions are prepared as described by Lindgren *et al.*[15] using salts that have been thoroughly dried and stored desiccated. Desired antioxidant and antibacterial agents[12] are incorporated into the solutions, including 10 mg/100 ml Na_2 EDTA. Densities should be checked by refractometric measurement or pycnometry and should be accurate within limits of ± 0.0002 g/ml.

Solution 1: 0.195 M NaCl, $d = 1.0063$ g/ml at 20°
Solution 2: 0.195 M NaCl, 2.434 M NaBr, $d = 1.1816$ g/ml at 20°
Solution 3: 0.195 M NaCl, 7.570 M NaBr, $d = 1.4744$ g/ml at 20°

The method outlined here for lipoprotein isolation by ultracentrifugation uses a Beckman 40.3 or 50.3 Ti rotor, which holds 18 tubes, each with a 6 ml capacity. Although a variety of tubes are available for use with these rotors, lipoprotein fractions are easily visualized in transparent tubes of thermoplastic polyester [Ultra-Clear (Beckman Instruments, Inc., Palo Alto, CA) or Polyclear (Seton Scientific, Sunnydale, CA)], thus facilitating

---

[14a] See this series, Vol. 73, Section IV.
[15] F. T. Lindgren, L. C. Jensen, and F. T. Hatch, *in* "Blood Lipids and Lipoproteins: Quantitation, Composition, and Metabolism" (G. J. Nelson, ed.), p. 181. Wiley (Interscience), New York, 1972.
[16] F. T. Hatch and R. S. Lees, *Adv. Lipid Res.* **6,** 1 (1968).
[17] Lipid Research Clinics Program, "Manual of Laboratory Operations." DHEW Publ. No. (NIH) 75-628, May 1974, revised 1982.

recovery of lipoproteins. These tubes can be used with the tube slicing technique. They must be capped when used in fixed-angle rotors such as the 40.3 or 50.3 Ti. Formulas for converting run conditions from one rotor type to another have been detailed by Schumaker and Puppione.[12]

Procedure

Isolation of VLDL. The VLDL fraction is floated to the top of the tube at plasma background density ($d = 1.0063$ g/ml), a process that typically employs an 18-hr centrifugation at 40,000 rpm (114,000 g) and 16-18°. When using less than 6 ml of plasma, solution 1 ($d = 1.0063$) is added to bring the volume to 6 ml. VLDL, which appear as a turbid layer at the top of the tube, are carefully removed at the meniscus with a syringe or a Pasteur pipette in a 1.0 ml volume and diluted to the desired concentration, usually the original plasma volume, with 0.195 M NaCl containing 10 mg/100 ml EDTA.

Isolation of LDL. The 4 ml infranatant fraction, containing LDL, HDL, and other plasma components, is thoroughly mixed to resuspend the sediment, transferred to a new tube, and adjusted to a final volume of 6.0 ml by the addition of 2.0 ml of solution 2 ($d = 1.1816$ g/ml NaCl-NaBr solution). The contents are mixed thoroughly by inversion. This produces a density before centrifugation of 1.065 g/ml. The LDL fraction is floated to the surface during centrifugation at 40,000 rpm and 16-18° for 20 hr. The LDL fraction, which appears as a yellow band, is removed in the top 1.0 ml. This fraction, which has a solvent density of 1.063, is reconstituted to the original plasma volume as for VLDL. A second 1-ml fraction may be taken to check background density.

Isolation of HDL. The 4 ml infranatant fraction from the previous spin is thoroughly mixed, transferred to a new tube, and adjusted to a 6.0 ml volume by adding 2.0 ml of solution 3 ($d = 1.4744$ g/ml NaCl-NaBr solution). The contents are mixed thoroughly by inversion. This produces a density before centrifugation of 1.216 g/ml. HDL are floated during a 40-hr spin at 40,000 rpm and 16-18°. The HDL fraction, a faintly pigmented band at the surface of the tube, is removed as the top 1.0 ml, which has a solvent density of 1.203 g/ml. Background density is measured from the next 0.5 ml to avoid incorporation of albumin. (If subfractionation of HDL is desired, HDL_2 can be floated at an overall density of 1.125 prior to subsequent flotation of HDL_3 at a density of 1.21 g/ml.)

Lipoprotein-Depleted Plasma. The lipoprotein-depleted plasma that remains as the infranatant fraction may be analyzed for carotenoid content along with the isolated lipoprotein fractions.

Comments

If the salt contents of the LDL and HDL fractions are to be lowered by dialysis, care should be taken to protect lipoproteins against oxidative damage by incorporating appropriate inhibitors into the dialyzing buffer. Dialysis of lipoprotein fractions is not required prior to extraction of carotenoids.

Separation of chylomicrons is generally not required when working with plasma from fasting, normolipemic subjects since chylomicrons have a short half-life in circulating plasma. However, isolation of chylomicrons may be important for the study of postprandial association of carotenoids with lipoproteins. Chylomicrons can be removed by an initial spin in which 2 ml of $d = 1.006$ g/ml solution is layered over 4 ml of plasma and chylomicrons are floated during a 30 min to 1 hr spin at 19,000 rpm (26,000 g) at room temperature.[16]

Advantages and Disadvantages

Sequential ultracentrifugation has the advantage of being widely known and accepted for human lipoprotein separation, but the method is also lengthy and costly. Quantitative recovery of lipoproteins can be difficult for the novice, but with experience recoveries of at least 95% are typical, based on cholesterol recovery. Although this is the classic method against which other methods of lipoprotein separation are judged, the method may produce some structural alterations of lipoproteins. Plasma lipids are subject to oxidation during this prolonged separation process, but oxidative modification can be inhibited by the appropriate use of antioxidants and chelating agents. A density gradient procedure,[18] which allows separation of lipoprotein fractions in a single spin, requires less time for physical separation of lipoproteins but also requires more technical skill.

Gel-Filtration Chromatography

Gel-filtration chromatography separates lipoproteins by size as they pass through a column of agarose beads that act as a sieve. The technique has been described in detail by Rudel *et al.*[19,20] Although whole plasma can be chromatographed, it is common to use total lipoproteins isolated from plasma by a single ultracentrifugation at a density of 1.225 g/ml. This

---

[18] T. G. Redgrave, D. C. K. Roberts, and C. E. West, *Anal. Biochem.* **65**, 42 (1975).
[19] L. L. Rudel, C. A. Marzetta, and F. L. Johnson, this series, Vol. 129, p. 45.
[20] L. L. Rudel, J. A. Lee, M. D. Morris, and J. M. Felts, *Biochem. J.* **139**, 89 (1974).

combination of ultracentrifugation and gel filtration has been used successfully by Behrens et al.[21] in preparation for tocopherol analysis of lipoprotein fractions and could prove useful for assessing carotenoids from individual lipoprotein classes.

Materials

Columns with internal diameters of 1.6 cm are prepared using a support medium of agarose gel beads, typically BioGel A-5m [6% (w/v) agarose], 200–400 mesh, at a bed height of 90 cm. The column elution buffer contains (w/v) 0.9% NaCl, 0.01% EDTA, and 0.01% NaN_3, pH 7.4. Columns should be stored at 4°.

Procedure

1. Total plasma lipoproteins are isolated by ultracentrifugal flotation. Plasma, containing appropriate chelating agents and antioxidants, is brought to $d = 1.225$ g/ml with solid KBr, 0.3517 g/ml. The sample is overlayed with a solution of equal density and centrifuged in a 70.1 Ti rotor at 50,000 rpm and 15° for 24 hr, in a SW-40 rotor at 40,000 rpm and 15° for 40 hr,[19] or in a 40.3 or 50.3 Ti rotor at 40,000 rpm and 18° for 40 hr.[15] The total lipoprotein fraction at the top of the tube is recovered essentially as described above.

2. A 2 to 4 ml volume of total lipoprotein, containing up to 75 mg of lipoprotein cholesterol, is applied to the top of the column (dialysis is not required) using an in-line loop or a sample application cup.

3. The lipoproteins are sequentially eluted from the column bed at a flow rate of 5–8 ml/hr. VLDL elute first followed by LDL and then HDL. The column eluate is monitored at 280 nm, and fractions, collected with a fraction collector, are pooled by lipoprotein class.

Advantages and Disadvantages

Gel-filtration chromatography is an inexpensive technique that is gentle and nondestructive for lipoproteins; however, only one sample can be separated during a column run that lasts 1 day. Lipoprotein classes are well separated with this method, and fractionated lipoproteins are reported to be chemically and physically the same as their counterparts separated by ultracentrifugal flotation.[20] Lipoprotein fractions become diluted during chromatography but can be concentrated in filter devices, if desired. However, this step is not a prerequisite for extracting carotenoids from frac-

[21] W. A. Behrens, J. N. Thompson, and R. Madere, *Am. J. Clin. Nutr.* **35**, 691 (1982).

tions. Column recovery, based on cholesterol, is reported to be 89.5% (S.D. 8.8%).[19]

Lipoprotein Precipitation

Carotenoid components of HDL can be measured from a supernatant solution following precipitation of other lipoproteins which contain apolipoprotein B. A varied combination of sulfated polysaccharides and divalent cations have been used to precipitate apolipoprotein B-containing lipoproteins, the most common being heparin–Mn^{2+}. Only limited use of precipitation techniques have been made to fractionate lipoproteins for carotenoid analysis,[1] but the speed and simplicity of these methods make them attractive choices for separating VLDL and LDL from HDL.

A combination of lipoprotein precipitation and ultracentrifugation, a modification[17] of the method of Burstein and Samaille,[22] is widely used to assess lipoprotein cholesterol and could be used to assess lipoprotein carotenoids. This method is outlined below as modified by Warnick and Albers[23] and reviewed by Bachorik and Albers,[24] particularly with regard to optimum Mn^{2+} concentrations for precipitating apolipoprotein B-containing lipoproteins without excessive precipitation of HDL.

Materials

A heparin solution is made from powdered heparin (from porcine intestinal mucosa) dissolved in 0.15 M NaCl at a concentration of 35 mg/ml. A 2.0 M solution of $MnCl_2 \cdot 4H_2O$ is prepared in distilled water. Both solutions are stable for at least 1 month when stored at 4°.[24]

Procedure

1. Eighty microliters of heparin solution is added to 2.0 ml of plasma and mixed thoroughly, avoiding foaming. Then 100 μl of 2.0 M $MnCl_2$ is added, and the contents are again thoroughly mixed. *Note:* This concentration of $MnCl_2$ (92 mM final concentration) is not appropriate for use with serum as discussed below.

2. Samples are allowed to stand at room temperature for 10 min, then centrifuged for 30 min at 1500 g to sediment the precipitate. The supernatant fraction, which contains HDL, is retained for carotenoid analysis. Supernatant fractions should be clear; cloudiness indicates incomplete

---

[22] M. Burstein and J. Samaille, *Clin. Chim. Acta* **5**, 609 (1960).
[23] G. R. Warnick and J. J. Albers, *J. Lipid Res.* **19**, 65 (1978).
[24] P. S. Bachorik and J. J. Albers, this series, Vol. 129, p. 78.

sedimentation of apolipoprotein B-containing lipoproteins. Carotenoid values should be corrected for dilution by the precipitating reagents.

3. From a separate plasma sample, LDL plus HDL are separated from VLDL by ultracentrifugation at $d = 1.006$ g/ml. The two fractions are brought to plasma volume for carotenoid analysis.

4. LDL carotenoid is calculated as the difference between LDL plus HDL carotenoid (from the $d = 1.006$ g/ml infranatant fraction) and HDL carotenoid (from the supernatant fraction following precipitation).

Advantages and Disadvantages

The precipitation method assumes that plasma contains 1 mg/ml EDTA. Since EDTA chelates with polyvalent anions and lowers the effective Mn^{2+} concentration, higher concentrations may lead to incomplete precipitation of apolipoprotein B-containing lipoproteins. On the other hand, lower EDTA concentrations can cause significant precipitation of HDL. When using serum (without added EDTA) rather than plasma, the concentration of $MnCl_2$ should be lowered by one-half (46 mM final concentration) to prevent partial precipitation of HDL.[24] Samples with high triglyceride contents can give falsely high HDL values due to incomplete precipitation of VLDL. Solutions to this problem have been discussed in detail.[23,24] They include removal of VLDL by ultracentrifugation, filtration to remove unsedimented lipoprotein complexes, dilution prior to heparin–Mn^{2+} precipitation, and high-speed centrifugation to increase sedimentation.

A dual precipitation method has been described by Gidez *et al.*[25] for measuring components of HDL subclasses. Total HDL are separated from other lipoproteins by a heparin–Mn^{2+} precipitation procedure. HDL_2 is then precipitated from the total HDL fraction with dextran sulfate, leaving HDL_3 in the second supernatant fraction. This procedure has recently been outlined in detail.[24]

Heparin–Sepharose Affinity Chromatography

The heparin–Sepharose affinity column technique uses agarose-bound heparin as a ligand for retaining apolipoprotein B- and E-containing lipoproteins.[26,27] This results in retention of VLDL and LDL while HDL pass through the column with the unretained fraction. The retained fraction is

[25] L. I. Gidez, G. J. Miller, M. Burstein, S. Slagle, and H. A. Eder, *J. Lipid Res.* **23**, 1206 (1982).
[26] K. H. Weisgraber and R. W. Mahley, *J. Lipid Res.* **21**, 316 (1980).
[27] C. L. Bentzen, K. J. Acuff, B. Marechal, M. A. Rosenthal, and M. E. Volk, *Clin. Chem.* **28**, 1451 (1982).

subsequently dissociated and eluted with a higher ionic strength NaCl solution. If the carotenoid content of VLDL is to be assessed, this fraction can be separated prior to affinity chromatography using ultracentrifugation at plasma density. Although commercially available heparin affinity columns are marketed as a preparatory step for lipoprotein cholesterol determination, the columns have been successfully used in preparation for assessing the tocopherol content of lipoproteins.[28] The potential extension of this separation method for assessing the carotenoid content of lipoproteins is clear.

Materials

The method outlined here uses the commercially available heparin affinity column (LDL-Direct) from Isolab, Inc. (Akron, OH). Each column contains approximately 1 g of agarose beads covalently bound to heparin. Stabilization and elution buffers are supplied with the columns.

Procedure

1. The column is equilibrated with 1.0 ml of low ionic strength buffer (buffer A), the eluant is discarded, and a collection tube is placed under the column.
2. Two hundred microliters of plasma (or serum) is applied to the column. After the sample has migrated into the matrix, 1 ml of buffer A is applied to the column. The 1.2 ml eluate collected is the unbound fraction and contains the bulk of plasma HDL as well as nonlipoprotein plasma components.
3. A second tube is placed under the column and the bound fraction is eluted by applying 1.2 ml of buffer B, a high ionic strength buffer. This fraction contains lipoproteins that have apolipoprotein B or apolipoprotein E components, namely, VLDL, LDL, and also a minor subfraction of HDL.

Advantages and Disadvantages

Heparin–Sepharose chromatography is inexpensive and allows rapid lipoprotein separation of multiple samples. The method has the added advantage of yielding high recoveries: 97.5% (S.D. 6.1%) based on recovery of total cholesterol in lipoprotein fractions.[27] Cholesterol levels of lipoproteins separated on heparin affinity columns were reported to be in reasonable agreement with cholesterol levels of lipoproteins separated by ultracentrifugation or precipitation procedures.[27] The heparin affinity method

[28] B. A. Clevidence and J. Lehmann, *Lipids* **24**, 137 (1989).

is appropriate for human plasma because HDL without apolipoprotein E accounts for approximately 94% of total HDL,[26] but the method is problematic for plasma from species that have apolipoprotein E-rich HDL. A major disadvantage of the technique is that VLDL and LDL fractions are not separated.

Extraction of Carotenoids for Chemical Analysis

Mixtures of polar and nonpolar organic solvents that have traditionally been used to extract fat-soluble compounds from biological samples have also been used to extract carotenoids from plasma. Regardless of which extraction method is chosen, precautions should be taken against oxidation of carotenoids. It is especially important that solvents to be used for extraction and HPLC are routinely monitored to ensure that they are free of peroxides. Individual carotenoids are reported to be stable for at least 18 hr in an HPLC solvent of acetonitrile/dichloromethane/methanol (70:20:10, v/v) making automated analysis a feasible option.[11] HPLC techniques used to separate and quantitate individual carotenoids from isolated lipoprotein fractions are the same as those described for whole plasma.[7a]

The technique outlined here uses a combination of ethanol and hexane for carotenoid extraction.[29] Plasma (or a lipoprotein fraction) is mixed with an equal volume of absolute ethanol to precipitate proteins. If an internal standard (e.g., echinenone, 0.4–1.0 μg/ml) is used, it is conveniently incorporated into the ethanol. Two volumes of hexane are then added, and the contents are mixed vigorously for 45 sec, then centrifuged at low speed (1040 g, 5 min) to separate the layers. Half of the hexane layer is removed, evaporated under N_2, and mixed thoroughly (30 sec) with HPLC solvent. When using whole plasma of normal carotenoid concentration, 100 μl is appropriate for both the starting plasma volume and the HPLC solvent volume. When using lipoprotein fractions that have been reconstituted to plasma volume, extract approximately 200 μl of LDL and HDL, and 500 μl of VLDL. Sample and solvent volumes must be scaled proportionately to compensate for the added dilution of lipoprotein fractions isolated by gel-filtration or affinity chromatography.

---

[29] J. G. Bieri, E. D. Brown, and J. C. Smith, *J. Liq. Chromatog.* **8**, 473 (1985).

[5] Translocation of Carotenoid Droplets in Goldfish Xanthophores

By T. T. TCHEN and JOHN D. TAYLOR

Introduction

Organelle motility is an active area of research in cell biology and relies on both biochemical and microscopic techniques. The diverse experimental systems may be divided into different groups by different criteria: regulated versus spontaneous; microtubule- versus microfilament (F-actin)-dependent; and using whole cells versus reconstituted systems. One large group of regulated (hormonally or neurally) organelle translocation is the translocation of pigment organelles, including carotenoid droplets, in many fishes, amphibians, and invertebrates. Two radically different systems of carotenoid droplet translocation are known. In the squirrel fish erythrophores, the translocation of carotenoid droplets is extremely rapid (a few seconds for complete dispersion or aggregation), microtubule-dependent, and regulated by calcium concentration (see Ref. 1). Thus far, this rapid translocation appears to be unique for the squirrel fish erythrophore, and little biochemical information is available. In the goldfish xanthophores, the translocation of carotenoid droplets is much slower, requiring over 30 min for complete dispersion or aggregation. Such kinetic behavior appears to be more widespread, present in other fishes as well as in amphibians. The successful isolation of goldfish xanthophores[2] allowed parallel microscopic and biochemical studies on the process of carotenoid droplet dispersion.[3-10] In this chapter, we describe the isolation and main-

---

[1] M. A. McNiven and J. B. Ward, *J. Cell Biol.* **106,** 111 (1988).
[2] S. J. Lo, S. M. Grabowski, T. J. Lynch, D. G. Kern, J. D. Taylor, and T. T. Tchen, *In Vitro* **18,** 356 (1982).
[3] S. J. Lo, J. D. Taylor, and T. T. Tchen, *Biochem. Biophys. Res. Commun.* **86,** 748 (1979).
[4] S. J. Lo, J. D. Taylor, and T. T. Tchen, *Cell Tissue Res.* **210,** 371 (1980).
[5] T. J. Lynch, S. J. Lo, J. D. Taylor, and T. T. Tchen, *Biochem. Biophys. Res. Commun.* **102,** 127 (1981).
[6] G. R. Walker, J. Matsumoto, J. D. Taylor, and T. T. Tchen, *Biochem. Biophys. Res. Commun.* **133,** 873 (1985).
[7] T. J. Lynch, J. D. Taylor, and T. T. Tchen, *J. Biol. Chem.* **261,** 4204 (1986).
[8] T. J. Lynch, J. D. Taylor, and T. T. Tchen, *J. Biol. Chem.* **261,** 4212 (1986).
[9] T. T. Tchen, R. D. Allen, J. Hayden, S. J. Lo, T. J. Lynch, R. E. Palazzo, G. R. Walker, and J. D. Taylor, *Ann. N.Y. Acad. Sci.* **466,** 887 (1986).
[10] T. T. Tchen, S. J. Lo, T. J. Lynch, R. E. Palazzo, G. Peng, G. R. Walker, B.-Y. Wu, F.-X. Yu, and J. D. Taylor, *Cell Motil. Cytoskeleton* **10,** 143 (1988).

tenance of xanthophores, the preparation of a permeabilized system capable of carotenoid droplet dispersion, and the identification of carotenoid droplet protein p57/pp57 and a cytosolic protein anterogin as important determinants for carotenoid droplet dispersion.

Isolation of Xanthophores

Common xanthic goldfish of 4–6 inches in length are decapitated, and the scales are removed and washed 4 or 5 times with 20 volumes of phosphate-buffered saline (PBS: $0.137\ M$ NaCl, $1.1\ mM$ KH_2PO_4, $8.1\ mM$ Na_2HPO_4, $0.1\ mM$ $CaCl_2$, and $0.5\ mM$ $MgCl_2$, pH 7.3–7.6). The scales are then used to isolate xanthophores as follows.

Step 1: Removal of Epidermis. Incubate scales with 20 volumes of EDTA medium (3% EDTA, 0.4% $NaHCO_3$, pH 7.3–7.6) for 20 min with one change of medium and gentle stirring (all stirrings are with a magnetic bar). Discard the loosened epidermis, and wash scales (bright orange and iridescent) 4 or 5 times with 20 volumes of PBS.

Step 2: Removal of Iridophores and Other Dermal Nonpigment Cells. Incubate scales from Step 1 with 10 volumes of collagenase medium (1 mg/ml collagenase, Worthington type III, 10 mg/ml soybean trypsin inhibitor, 0.05 mg/ml DNase I, and 10 mg/ml bovine serum albumin in medium 199). The duration of incubation varies from 15–30 min with occasional stirring. (The duration of incubation depends on the strength of the collagenase. The collagenase medium can be sterilized by filtration and reused for 4 or 5 experiments, requiring progressively longer periods of incubation.) The end point is judged by eye as follows. As the iridophores come off the scales, the scales become less iridescent while shiny specks become abundant in suspension. Xanthophores come off the scales later

---

[11] R. E. Palazzo, S. J. Lo, T. J. Lynch, J. D. Taylor, and T. T. Tchen, *Cell Motil. Cytoskeleton* **13**, 9 (1989).

[12] R. E. Palazzo, S. J. Lo, T. J. Lynch, J. D. Taylor, and T. T. Tchen, *Cell Motil. Cytoskeleton* **13**, 21 (1989).

[13] G. R. Walker, J. D. Taylor, and T. T. Tchen, *Cell Motil. Cytoskeleton* **14**, 458 (1989).

[14] F.-X. Yu, B.-Y. Wu, J. D. Taylor, and T. T. Tchen, *Biochem. Biophys. Res. Commun.* **161**, 626 (1989).

[15] C.-F. Yang, Z.-C. Zeng, S.-C. Chou, F.-X. Yu, J. D. Taylor, and T. T. Tchen, *Pigm. Cell Res.* **2**, 408 (1989).

[16] Z.-C. Zeng, J. D. Taylor, and T. T. Tchen, *Cell Motil. Cytoskeleton* **14**, 485 (1989).

[17] F.-X. Yu, J. D. Taylor, and T. T. Tchen, *Cell Motil. Cytoskeleton* **15**, 139 (1990).

[18] B.-Y. Wu, F.-X. Yu, T. J. Lynch, J. D. Taylor, and T. T. Tchen, *Cell Motil. Cytoskeleton* **15**, 147 (1990).

than iridophores and the incubation is terminated when the scales lost most of the iridescence while retaining the bright orange color.

Step 3: Dissociation of Xanthophores from Scales. Wash the scales from Step 2 with 20 volumes of PBS (5-7 times) and incubate in 10 volumes of collagenase medium for 1.5 hr followed by 30 min of gentle stirring. The scales should become almost colorless. Approximately 30% of the cells in suspension are xanthophores.

Step 4: Further Purification of Xanthophores by Density Centrifugation. Collect cells from Step 3 by centrifugation (500 g, 10 min); resuspend in 38% Percoll (in 20 mM Tris buffer, pH 7-7.2, and 0.154 M NaCl) at a concentration of 1 × 10^6 cells/ml. Place the cell suspension in centrifuge tubes, overlay with 2-3 ml of PBS, and centrifuge at 1000 g for 10 min. Collect xanthophores (orange band at Percoll-PBS interphase) and wash once with PBS.

Comments

The xanthophore preparations have 70-90% xanthophores and 30-10% nonpigment cells. Collection of the entire orange band gives better cell yield (1-3 × 10^6 xanthophores per fish) but lower purity. Collecting only the top of the orange band gives higher purity but lower yield.

The xanthophores as isolated have aggregated pigment and will respond to adrenocorticotropin (ACTH) [or melanocyte-stimulating hormone (MSH) or cyclic AMP (cAMP)] by pigment dispersion. They can be used directly for biochemical experiments. For studying pigment dispersion/aggregation, however, it is necessary to plate these cells and give them time to flatten and to become dendritic as described below.

Maintenance of Xanthophores

Xanthophores can be plated on a variety of surfaces: T250 culture flasks, 24- and 96-well test plates, glass, or Formvar-coated grids (for electron microscopy). The medium is medium 199 with 10% fetal calf serum and 100 units (U) of penicillin, 10 units of Fungizone, and 0.1 mg each of gentamycin, streptomycin, kanamycin, and neomycin per milliliter. Cell attachment is facilitated by the presence of ACTH (or MSH), with attachment within a few hours and gradual flattening of the cells for 1-3 days. Xanthophores can be maintained (they grow very slowly, if at all) for over 3 weeks. However, as the nonpigment cells continue to grow, the percentage of xanthophores in the cultures diminishes with time. For experiments where contaminating nonpigment cells may be a problem (such as studies on protein phosphorylation), xanthophores should be used no later than 2 days after isolation. For other experiments where the purity of the culture is less important but where more dendritic cells are desired (such as monitoring movement of pigment), 3- to 5-day-old cultures are more suitable.

Comments

For longer durations of culture, microbial contamination can be a serious problem. As precaution, we routinely plate the cells in aliquots in 24-well test plates so that if a few wells are contaminated, they can be discarded. (If the entire preparation were plated in one flask, the whole preparation would be lost.) To further reduce the risk of contamination, we introduced the various steps of repetitive washing during cell preparation and the addition of a combination of antibiotics in the culture medium as described above. Even so, there are about 1-2 months during late summer when more than half of the cell preparations are lost completely due to bacterial and/or fungal contamination. In contrast, completely uncontaminated cultures are often obtained during winter and early spring.

We also noticed that xanthophores become progressively paler with prolonged culture in medium containing fetal calf serum so that, after 10-14 days, they become faint yellow in color, instead of bright orange. We presume that this is due to the lack of carotenoids in the fetal calf serum and therefore use carp serum instead of fetal calf serum for such long-term cultures. However, most experiments require shorter durations of culture, and the use of carp serum is therefore not necessary. (We should also note that, on rare occasions, all xanthophores of a preparation fail to undergo pigment redistribution for unknown reason.)

Preparation of Xanthophores with Aggregated Pigment

Xanthophores maintained in the presence of ACTH have dispersed pigment and are highly dendritic. To obtain cultured xanthophores with aggregated pigment, two methods have been used.

In the first method, xanthophores are maintained in the absence of ACTH for 12-24 hr. The vast majority of cultures have xanthophores with tightly aggregated pigment. However, occasionally, pigment remained in the dispersed state for reasons unknown. These latter cells can be induced to undergo pigment aggregation by the second method below, which can also be applied directly to xanthophores with fully dispersed pigment.

The second method entails washing cells once with PBS and then incubating in PBS. This leads to rapid aggregation of pigment (complete within 1 hr).

Comments

Although the second method is more convenient and in fact more reliable, we have usually used the first method as we do not know what metabolic perturbation(s) may be induced by the drastic change of medium in the second method.

Permeabilized Xanthophores for Studies on Pigment Dispersion

Treat cultured xanthophores with aggregated pigment for 3 min with permeabilization buffer [1.3% digitonin, 1.3% polyethylene glycol (M_r 6000), in PEM buffer consisting of 0.1 M PIPES, 2 mM EGTA, and 2 mM MgSO$_4$, pH 6.9], and wash twice with PEM buffer.

Induction of Pigment Dispersion in Permeabilized Xanthophores. The permeabilized xanthophores retain tightly aggregated pigment for over 2 hr and still respond to dispersion medium [PEM buffer with 5 mM MgSO$_4$, 1 mM GTP, 1 μM cAMP, 1 mM ATP, and different concentrations (final dilutions 1 : 10 to 1 : 50) of cytosol (150,000 g supernatant of homogenates in PEM buffer of xanthophores, liver, or yeast)]. Full pigment dispersion requires 30 min or longer (see below).

Quantitation of Pigment Dispersion/Aggregation

Plated xanthophores at a density of approximately 500 per well in a 96-well test plate and incubate for 1–4 days. Induce pigment aggregation as described above. Select a field of over 50 monodispersed xanthophores; mark and photograph using color slides. Add ACTH medium to induce pigment dispersion and photograph the same field at different time periods (typically from 5 to 60 min). Project slides and score the xanthophores for the distribution of pigment, with tightly aggregated pigment given a score of 1 and fully dispersed pigment a score of 5. Cells with intermediate degrees of pigment dispersion are scored as 2–4. The average score of the cells in the field is the xanthophore index *(XI)*.

Comments

Typically, a culture of xanthophores with aggregated pigment will have a *XI* of 1.1–1.2, whereas a culture of xanthophores with dispersed pigment will have a *XI* above 4.5. Although the scoring is somewhat arbitrary, it has been our experience that the *XI* values of different cultures determined by two different persons are very consistent.

In the case of permeabilized xanthophores, the end point of pigment dispersion is *XI* slightly under 4, compared to *XI* above 4.5 with intact xanthophores. This lesser degree of pigment dispersion is probably an artifact resulting from the failure to detect the faintly pigmented cell periphery in the permeabilized cells. (Intact cells respond to ACTH or cAMP also by thickening of dendrites.[4] This response is absent in permeabilized xanthophores. Thus, even if pigment has migrated into the flat dendrites, the color of the dendrites may be too faint to be scored.) The dispersion of pigment in permeabilized xanthophores, represented as XI values at different time periods after the initiation of pigment dispersion, is dependent on the concentration of cytosol in the dispersion medium. With optimal concentrations of cytosol (e.g., xanthophore cytosol at a final dilution of 1:20 or less), the time required to achieve complete pigment dispersion in the permeabilized system is 30–60 min, similar to that seen with intact cells.[17]

The permeabilized system provides a bioassay for the purification of the active factor (anterogin) in the cytosol, as was indeed applied to the purification of this factor from calf liver.[16] It also allows an examination of the cytoskeletal component that is responsible for pigment dispersion. Agents acting on microtubules or intermediate filaments do not effect pigment dispersion, whereas several agents acting on the actin system (antiactin, DNase I, and phalloidin) inhibit the dispersion of carotenoid droplets.[17] The identity of the motor for pigment dispersion is, however, still uncertain.[18]

We should add at this time that we have been unsuccessful in developing a reconstituted system for the translocation of carotenoid droplets on microtubules or F-actin, nor have we been able to develop a permeabilized system that can carry out pigment aggregation.

Phosphorylation of Carotenoid Droplet Protein p57

Biochemical studies indicate that the primary effect of cAMP or ACTH is the phosphorylation of two polypeptides collectively referred to as p57. p57 can be labeled in two ways.

Labeling of p57 in Intact Xanthophores

Preload xanthophores with aggregated pigment (either freshly isolated or preferably cultured in 24-well test plates, $3-5 \times 10^5$ cells per well) by incubating for 15 min with ortho[^{32}P]phosphate (0.5 mCi/ml, 1 mCi/μmol in medium 199 with a reduced phosphate concentration of 0.5 mM). Incubate for another 30 min (or different intervals of time if desired) without (control wells) or with ACTH, 5 IU/ml (experimental wells). Remove the medium, wash once with PBS (0.14 M NaCl, 2.7 mM KCl, 1.1 mM KH$_2$PO$_4$, 8.1 mM Na$_2$HPO$_4$, 0.9 mM CaCl$_2$, 0.5 mM MgCl$_2$, pH 7.3–7.5), and dissolve proteins in 25–40 μl of sodium dodecyl sulfate (SDS) sample buffer (containing, in a total volume of 200 ml, 25 ml glycerol, 10 ml mercaptoethanol, 4.6 g SDS, and 1.52 g Tris buffer, pH 6.8) for SDS-PAGE.[19] Quantitate the labeled pp57 by scanning the fluorograms of SDS gels.

Comments. When xanthophore proteins are analyzed by two-dimensional electrophoresis and fluorography[7,20] (10^6 cells give good signals), it is seen that ACTH stimulates the phosphorylation of a protein of M_r 57,000. In the absence of ACTH, pp57 is already one of the major ^{32}P-labeled proteins in the cell. In the presence of ACTH, pp57 becomes the dominant labeled protein in these cells. Indeed, when the gels are underexposed, one sees essentially only labeled pp57, in the form of two streaks of slightly different M_r with isoelectric points of 4.9–4.0. (The two streaks correspond to two sets of discrete charge variants of phosphoproteins when [^{3}H]leucine-labeled cells are used.[7]) The maximal rate of p57 phosphorylation is reached at 0.1 mM cAMP or 6 IU/ml of ACTH, corresponding to the concentrations required to give optimal pigment dispersion. Kinetically, phosphorylation of p57 precedes pigment dispersion, whereas pp57 dephosphorylation precedes pigment aggregation.

Labeling of p57 by Cell-Free Systems

Subcellular fractionation studies showed that p57 is associated with carotenoid droplets and labeled pp57 can be prepared by incubating isolated carotenoid droplets with the catalytic subunit of cAMP-dependent protein kinase or with xanthophore cytosol and cAMP.[16]

Procedure for Isolation of Carotenoid Droplets. Suspend xanthophores with aggregated pigment in 20 volumes of ice-cold 0.25 M sucrose, 10 mM Tris-HCl, pH 7.5, and homogenize in a Teflon–glass Duall homogenizer. Centrifuge for 10 min at 1000 g; collect the supernatant and bring to 1 M sucrose with the addition of 2.6 M sucrose. Layer 2-ml aliquots over 6 ml of 2 M sucrose and overlay with 2 ml each of 0.7 and 0.25 M sucrose. Centrifuge for 2 hr at 100,000 g in an SW 41 rotor and collect the band of carotenoid droplets (bright orange) at the interface of the 0.25/0.7 M sucrose solutions.

Comment on Light and Heavy Carotenoid Droplets. For reasons not entirely clear, the density of carotenoid droplets depends on the state of pigment distribution in the cell. If one overlays xanthophore homogenates with 0.25, 0.4, and 0.7 M sucrose solutions, one collects two bands of carotenoid droplets, namely, at the interfaces of 0.25/0.4 and 0.4/0.7 M

---

[19] U. K. Laemmli, *Nature (London)* **227**, 680 (1970).
[20] P. H. O'Farrell, *J. Biol. Chem.* **250**, 4007 (1975).

sucrose. We have termed these light and heavy carotenoid droplets, respectively. If one begins with xanthophores with aggregated pigment, the light carotenoid droplet band is the major band. In contrast, if one begins with cells with dispersed pigment, the heavy carotenoid drops form the major band. Biochemically, the heavy carotenoid droplets have relatively higher ATPase activity and more actin.[18]

Labeling of p57. Incubate carotenoid droplets (5–20 ng of protein in a total reaction volume of 200–500 µl) with 0.25 mM [γ-^{32}P]ATP (100 mCi/mmol), 12 mM phosphate buffer, pH 7.0, 10 mM MgCl$_2$, and either (1) 200 ng of the catalytic subunit of cAMP-dependent protein kinase (bovine heart) or (2) xanthophore cytosol (final dilution 1:20) and 5 mM cAMP with out or with 40 mM NaF. Maximum phosphorylation is achieved after approximately 30 min. The extent of p57 phosphorylation is determined by SDS-PAGE and fluorography.

Comments. When isolated carotenoid droplets are phosphorylated by the catalytic subunit of cAMP-dependent protein kinase, pp57 is the only labeled protein. On storage before SDS-PAGE, some labeled peptides of lower M_r are formed, indicating the presence of some protease activity, probably contaminants in the carotenoid droplet preparations. The total extent of labeling can still be determined by summation of label in all these labeled peptides. This procedure, although cumbersome, is more reliable than counting the total amount of label in trichloroacetic acid (TCA) precipitates (with or without additional carrier protein such as bovine serum albumin) which in our experience gives poor and unreliable recovery. When more complex labeling systems are used, it is essential to perform SDS-PAGE separation of labeled polypeptides as soon as possible.

Labeling by intact cells or by a mixture of carotenoid droplets and cytosol without fluoride do no fully phosphorylate p57. Instead, the labeled pp57 consists of similar amounts of pp57s with 1–5 (or 6) phosphates per polypeptide. If labeling was by the catalytic subunit of cAMP-dependent protein kinase, or by cytosol in the presence of fluoride, the extent of maximal labeling is increased about 3-fold. The resulting pp57 has little or no mono- or diphosphorylated species and consists primarily of pp57 with 5 or 6 phosphates per polypeptide.[16]

Unanswered Questions

We have described briefly the methods used for studying the dispersion of carotenoid droplets in goldfish xanthophores. Procedures for electron microscopy and immunofluorescence have been described elsewhere[11,13] and are omitted here. The biochemical and microscopic studies have provided some answers, but they leave a large number of unresolved questions. What is the motor for carotenoid dispersion?[18] What are the exact roles of anterogin and pp57? Why are there two closely related (L.-Y. Gu, unpublished, 1992) polypeptides in p57/pp57? How do the various components interact for carotenoid dispersion? What is the mode of carotenoid aggregation? How does p57 interact with microtubules?[14] What are the relationships between the redistribution of carotenoid droplets and pterinosomes and cytoskeletal rearrangement?[13,15] Does anterogin play a role in secretion in other cells?[10,17] Finally, it is our impression that the cultured xanthophores can carry out carotenoid uptake from carp serum as they do not fade as rapidly in the presence of carp serum. In this respect, these cells may also be useful for studying the cellular uptake of carotenoids.

[6] Solubilization, Cellular Uptake, and Activity of β-Carotene and Other Carotenoids as Inhibitors of Neoplastic Transformation in Cultured Cells

By ROBERT V. COONEY, T. JOSEPH KAPPOCK, IV, AO PUNG, and JOHN S. BERTRAM

Introduction

The role of carotenoids in electron transport in photosynthesis and in providing protection from oxidative damage in plants has been well documented.[1] While the nutritional value of provitamin A carotenoids such as β-carotene (β-C) has long been recognized,[2] the more general function and possible importance of other dietary carotenoids in humans are less clear. Although many epidemiologic studies have identified inverse correlations between carotenoid (particularly β-C) consumption and cancer incidence rates (reviewed in Ref. 3), the mechanism by which carotenoids lower cancer risk has not been established. The common view that carotenoids are active via conversion to retinoids, which have known cancer-preventive properties,[3] restricts activity to a minority of dietary carotenoids.

---

[1] T. W. Goodwin, "The Biochemistry of the Carotenoids," 2nd Ed., Vol. 1. Chapman & Hall, London, 1980.
[2] R. C. Moon and L. M. Itri, *In* "The Retinoids" (M. B. Sporn, A. B. Roberts, D. S. Goodman, eds.), Vol. 2. pp. 327–371. Academic Press, New York, 1984.
[3] J. S. Bertram, L. N. Kolonel, and F. L. Meyskens, Jr., *Cancer Res.* **47**, 3012 (1987).

However, activity could be independent of such conversion; this would have mechanistic importance and would call for a reevaluation of much epidemiologic data.

The resolution of this question has been hampered by the lack of an appropriate experimental model system. Research in animals has the drawback that the mouse and the rat are poor models for absorption and tissue distribution of carotenoids in humans.[1] Moreover, research using cell culture systems has previously been hindered by difficulties in formulating these highly lipophilic compounds into a bioavailable form. In the cases of β-C and canthaxanthin (CTX),[4] this difficulty was overcome by incorporating the carotenoid into a beadlet (a proprietary method developed by Hoffmann-LaRoche in which the carotenoid is dispersed in an oil–protein matrix which is miscible with water). This method is limited in two ways: carotenoids other than β-C and CTX are not available in beadlet form, and the beadlet preparations contain many other chemical constituents that may have effects independent of the carotenoid. Other methods have utilized various types of liposomes[5,6]; however, they are rather limited with regard to the concentration of carotenoid that can be achieved, and we have observed toxic effects of various liposome constituents on cells in culture. We have investigated other solvents: hexane partitioned, evaporated, and resulted in poor bioavailability of carotenoids; chloroform was cytotoxic; dimethyl sulfoxide (DMSO) and acetone/DMSO mixtures were effective at low concentrations of β-C but did not allow sufficiently high carotenoid concentrations without precipitation of the β-C. We finally evaluated tetrahydrofuran (THF), which has been reported to be one of the best solvents for β-C.[7] We report here that this solvent was effective in delivering carotenoids of interest to cells without apparent toxicity and with biological activity as evidenced by the ability of structurally diverse carotenoids to inhibit methylcholanthrene (MCA)-induced neoplastic transformation.

Preparation of Carotenoid Solutions

Reagents. Tetrahydrofuran, 99.9% [with and without 0.025% (w/v) butylated hydroxytoluene (BHT)], was purchased from Aldrich (Milwaukee, WI). Lycopene and lutein were from Sigma (St. Louis, MO). β-Caro-

---

[4] J. E. Rundhaug, A. Pung, C. M. Read, and J. S. Bertram, *Carcinogenesis* **9**, 1541 (1988).
[5] S. M. Anderson and N. I. Krinsky, *Photochem. Photobiol.* **18**, 403 (1973).
[6] M. Murakoshi, J. Takayasu, O. Kimura, E. Kohmura, H. Nishino, A. Iwashima, J. Okazumi, T. Sakai, T. Sugimoto, I. Imanishi, and R. Iwasaki, *J. Natl. Cancer Inst.* **81**, 1649 (1989).
[7] F. Khachik, G. R. Beecher, J. T. Vanderslice, and G. Furrow, *Anal. Chem.* **60**, 807 (1988).

tene, α-carotene, and canthaxanthin were gifts from Hoffmann-LaRoche (Basel, Switzerland). The purity of carotenoids is determined by HPLC analysis after storage and prior to use. All HPLC solvents were Optima grade from Fisher Scientific (Kent, WA).

Routinely, solutions of 2 mM carotenoid are prepared in THF (99.9% containing 0.025% BHT as preservative). Although higher aqueous concentrations of β-C are possible with a larger percentage of THF, concentrations described above were found to be generally most useful under the conditions used for cell culture. The THF is opened in a glove box under a nitrogen atmosphere and dispensed to smaller vials sealed with Neoprene septa. The THF and THF/carotenoid solutions are stored in the dark at $-35°$ under N_2. Under these conditions no decay of carotenoid is detected for up to 2 weeks. Samples are removed by hypodermic needle taking care not to introduce room air into the sealed vials. Similar precautions are taken with the THF used as solvent and control. When stored in this manner, THF does not influence MCA-induced transformation frequencies or have toxic effects on cells (up to 0.5% THF in culture medium).

Twenty-five microliters of a THF solution of carotenoid (at 4° to minimize evaporation) is rapidly injected into 5 ml of either water or culture medium to give a final carotenoid concentration of 10^{-5} M (0.5% THF). In instances where larger volumes are needed, the process is scaled up, and the medium is vigorously stirred with a magnetic stirrer while the carotenoid/THF solution is quickly injected into the aqueous phase. Slow addition to the aqueous phase, particularly with more concentrated solutions (>2 mM β-C in THF), tends to produce slight cloudiness, indicating that some precipitation has occurred. This can be overcome by increasing the percentage of THF in the final solution; however, THF levels above 1% can affect cell growth. It also was observed that lower temperatures reduce the formation of turbidity. At 4° or room temperature generally no turbidity occurs, whereas at 50° slight precipitation of β-C is observed.

Cells and Culture Conditions

Assays for neoplastic transformation are performed with C3H/10T1/2 cells essentially as described by Pung et al.,[8] except for the use of tetrahydrofuran (THF) as solvent for the carotenoids. Cells are cultured in basal Eagle's medium (BME) supplemented with 5% calf serum (HyClone Laboratories, Inc., Logan, UT) and 25 μg/ml gentamicin as described.[8] In the transformation assays, cultures are treated for 24 hr with 1 or 3 μg/ml MCA (Sigma) and then, 7 days after removal of MCA, with the appropriate

[8] A. Pung, J. Rundhaug, C. N. Yoshizawa, and J. S. Bertram, *Carcinogenesis* **9**, 1533 (1988).

carotenoid. At this time, and for the duration of the experiment, the serum concentration is reduced to 3%. Carotenoids are readded on a weekly basis for 4 weeks after refeeding cultures. Cultures are then fixed, stained, and quantitated for transformation.

Carotenoid Analysis

Carotenoids are assayed by HPLC for initial purity and to measure stability and uptake by cells. Cells and media are extracted as described previously.[4] Dried samples are dissolved in 0.3 ml mobile phase (acetonitrile/dichloromethane/methanol/hexane, 60:20:10:5, v/v), containing 0.025% BHT immediately prior to use. A volume of 20 μl is then injected into a Beckman System Gold HPLC equipped with a Phenomenex Spherex 5 μm particle size reversed-phase (C_{18}) column (4.6 × 250 mm) coupled to a diode array detector module. The sample is eluted using premixed mobile phase at a flow rate of 1.5 ml/min.[9] The effluent is monitored at the appropriate wavelength (see below), and the absorption spectrum for each peak (200–500 nm) is compared to freshly prepared standards and published spectra for identification. Peak areas are compared to those of authentic standards for quantitation. Purities are calculated from published extinction coefficients. Calculated purities for the other compounds as they were received and the wavelengths at which they were monitored are as follows: β-C, 90%, 454 nm; α-carotene, 86%, 447 nm; CTX, 96%, 474 nm; lycopene, 96%, 473 nm; lutein, 85%, 446 nm.

Characterization of Tetrahydrofuran as a Drug-Delivery Vehicle

Tetrahydrofuran was found to be well tolerated by C3H/10T1/2 cells at the 0.5% final concentration normally used in the studies described below. The addition of 10^{-5} M β-C appeared to slightly decrease the growth rate and saturation density (Fig. 1). While inhibitor-free THF and THF containing BHT were well tolerated, THF lacking an inhibitor was found on occasion to be toxic to cells after extended storage, presumably as a result of peroxide formation.[10] As a consequence, THF containing 0.025% BHT was routinely used for our studies. The final BHT concentration in the media was 0.000125% or about 5.7×10^{-6} M. At this concentration, BHT did not affect the transformation studies.

[9] J. S. Bertram, A. Pung, M. Churley, T. J. Kappock IV, L. R. Wilkens, and R. V. Cooney, *Carcinogenesis* **12**, 671 (1991).
[10] D. F. Shriver, "The Manipulation of Air-Sensitive Compounds." McGraw-Hill, New York, 1969.

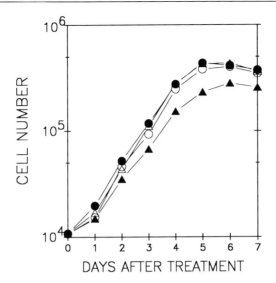

FIG. 1. Growth curves of C3H/10T1/2 cells treated with THF alone or with THF and β-carotene. C3H/10T1/2 cells were seeded at 10^4 cells/dish in 5 ml BME plus 3% calf serum into replicate 60-mm culture dishes. Cultures were treated 24 hr postseeding with 25 μl THF alone or 25 μl of a β-C solution in THF. Controls were untreated. Cells were counted with a Coulter counter at the indicated times. Results represent the means of duplicate counts on two individual cultures. ●, No treatment; ○, 0.5% THF; △, 10^{-6} M β-C; ▲, 10^{-5} M β-C. (Reprinted from Ref. 9, by permission of the Oxford University Press.)

Physical State

When β-C in THF was added to water, the resulting solution was observed to be colored yet showed no turbidity as seen with liposome preparations. Solutions appeared stable and homogeneous. In an effort to better characterize the nature of this aqueous "solution" of β-C, the following experiments were conducted. It was found that aqueous β-C solutions could be filtered through Millipore (Bedford, MA) Millex GV filters (Table I) with good efficiency, indicating β-C was not present as a precipitate. This property is beneficial in that it allows efficient bacterial sterilization. Physicochemical measurements made of β-C when solubilized in a water/THF solution indicated a highly ordered state for the β-C molecule. For example, on addition of a THF/β-C solution to water (0.5% THF, 10^{-5} M β-C), the absorption spectrum shifted markedly toward the UV (Fig. 2). This shift was reversible in the presence of increasing concentrations of THF (Fig. 2) or by extraction of β-C into hexane (Fig. 3), indicating that no chemical transformation occurs. Although the absorption spec-

TABLE I
FILTRATION OF AQUEOUS β-CAROTENE
SOLUTIONS[a]

Treatment	A_{400}	% Control
None	0.779 ± 0.016	100
0.8 μm Filtration	0.794 ± 0.018	102
0.2 μm Filtration	0.603 ± 0.081	77

[a] An aqueous solution of β-C (1.2×10^{-5} M, 1.0% THF) was filtered through Millipore microfilters of the indicated size and the absorbance spectrum of the filtered solution measured with a Beckman DU 60 spectrophotometer.

trum for 10^{-5} M β-C in aqueous solution (0.5% THF) differs markedly from that observed for 10^{-5} M β-C formulated in beadlets, the spectrum for the extracted β-C from each of these solutions is identical (Fig. 3). The observed spectra and HPLC retention times of extracted β-C show that no trans-to-cis isomerization had occurred. The NMR spectrum of β-C (10^{-4} M) dissolved in deuterated THF showed clear resolution of the

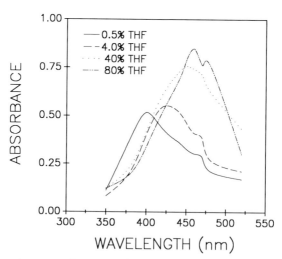

FIG. 2. Absorption spectra for aqueous β-C solutions containing varying amounts of THF. Solutions containing 10^{-5} M β-C were prepared by adding 5 μl of freshly prepared 2 mM β-C in THF to aqueous solutions containing varying concentrations of THF (total volume 5 ml). Absorption spectra for each solution were then obtained with a Beckman DU 60 spectrophotometer.

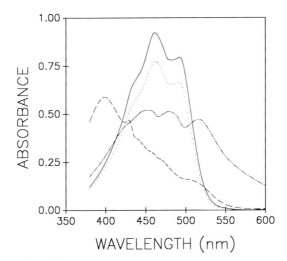

FIG. 3. Comparison of aqueous and extracted organic phase absorption spectra for β-C delivered when solubilized by beadlet (Hoffmann-LaRoche formulation) or by THF as described here. Spectra for aqueous 10^{-5} M β-C (0.5% THF) (- - -) and beadlet β-C (- $\cdot$ -) are shown. Each of the aqueous solutions was then extracted into chloroform and the corresponding spectra measured (———, 0.5% THF; $\cdot\cdot\cdot$, beadlet).

expected protons. However, when a β-C solution in THF was added to D_2O (deuterated THF 5%, β-C 10^{-4} M), there was a total loss of the expected signals.[9]

A possible explanation for these observations is that THF forms a cage around one or more β-C molecules that are maintained in a constrained state within the cage, yet the cage itself is fluid within the aqueous phase. Presumably the oxygen atoms of THF molecules are directed outward, toward the aqueous environment, while the nonpolar hydrocarbon ends solubilize the carotenoid. A similar explanation has been proposed for microemulsions (reviewed by Kahlweit[11]) involving the creation of miscible oil–water mixtures with small concentrations of an amphiphile. In those studies, lower temperature was found to enhance formation of the microemulsion, similar to what we found for the β-C/THF aqueous solutions. In our model, increasing the amount of THF serves to increase the size of the cage and consequently the fluidity of β-C within the cage, changing its spectral characteristics as was observed (Fig. 2).

[11] M. Kahlweit, *Science* **240**, 167 (1988).

Bioavailability of Solubilized Carotenoids

When presented to cells, THF-solubilized β-C was in a bioavailable state. Cellular uptake was rapid and was comparable to that previously obtained with beadlet-administered β-C.[4] Peak cellular levels of β-C were generally obtained within 24 hr after treatment.[9] After 7 days cellular levels of β-C decreased more rapidly in cells treated at 10^{-5} M than in cells treated at 10^{-6} M β-C (Table II). Analysis of the culture medium showed concentrations of β-C after 7 days to be similar in cultures which had received 10^{-6} or 10^{-5} M β-C. We attribute the more rapid decay of concentrated solutions of β-C, and the relative stability of 10^{-6} M solutions, to the possible saturation at high β-C concentrations of protein binding sites that protect β-C against degradation. Pure aqueous solutions of β-C were found to be less stable than those containing medium. In addition, solutions kept at 37° decayed 5 times faster than those kept at 4°.

Differences in the observed potency of the tested carotenoids may be due to intrinsic differences in biological activity or, alternatively, could result from pharmacokinetic differences in, for example, chemical stability, cellular uptake, or, as we found previously with all-*trans*-retinoic acid,

TABLE II
UPTAKE AND HALF-LIFE OF SELECTED CAROTENOIDS DELIVERED
TO CELLS IN CULTURE[a]

Carotenoid	Concentration (M)	Cellular uptake (nmol/10^6 cells)		Half-life in culture medium (hr)
		1 day	7 days	
CTX	10^{-5}	0.88	0.68	101
CTX	10^{-6}	0.06	0.09	Stable
β-C	10^{-5}	0.43	0.19	31
β-C	10^{-6}	0.12	0.04	Stable
α-C	10^{-5}	0.10	0.10	7.7
α-C	10^{-6}	0.01	0.003	16
Lycopene	10^{-5}	0.25	0.03	354
Lycopene	10^{-6}	0.06	0.02	Stable
Lutein	10^{-5}	0.52	2.20	Stable
Lutein	10^{-6}	0.39	1.36	Stable

[a] At time 0, carotenoid in THF (0.5% final concentration) was added to confluent cultures of C3H/10T1/2 cells in 60-mm dishes at the final concentration indicated. Cellular concentrations of carotenoid after 1 day and 7 days of incubation were assayed by HPLC. Data from Ref. 9.

cellular metabolism.[12] To examine these other variables, we measured by HPLC the stability of selected carotenoids in cell culture medium and their uptake into target C3H/10T1/2 cells. As shown in Table II, the stability of carotenoids varied markedly. Lutein was the most stable of the five carotenoids tested, showing no measurable reduction in concentration over 7 days. The order of stability of the other carotenoids tested at 10^{-5} M was lycopene, CTX, β-C, and α-carotene, with half-lives of 354, 101, 31, and 7.7 hr, respectively. It is of interest that carotenoids were much more stable at 10^{-6} M than when tested at 10^{-5} M; at the lower concentration only α-carotene showed appreciable decay.

Bioactivity of Carotenoids

When delivered in THF, β-C was biologically active as indicated by its ability to inhibit MCA-induced transformation of C3H/10T1/2 cells. In Fig. 4, we compare the activity of β-C delivered in THF with β-C formulated in beadlets.[8,9] Because of experiment-to-experiment differences in absolute yield of transformants, results are normalized to respective controls exposed only to MCA. As can be seen, the two methods of delivery result in essentially identical biological activity.

Cellular uptake of carotenoids did not always correlate with their respective stabilities, nor with their biological activity in transformation assays. Lutein, for example, having the highest concentration in medium and cells after 7 days (Table II), was only weakly active in the transformation assay (Fig. 5). α-Carotene when added at 10^{-5} M achieved lower initial cellular levels than β-C (Table II), presumably because of its more rapid degradation. However, comparable cellular levels were observed after 7 days; this we attribute to stabilization by cellular components. At this concentration these compounds were equiactive in the transformation assay (Fig. 5). However, when α-carotene was added at 10^{-6} M, at which concentration it was inactive, cellular levels were 5- to 10-fold lower than those of β-C. At this concentration β-C retained activity in the transformation assay. Thus, differences in activity of diverse carotenoids in C3H/10T1/2 cells may in part be due to reasons of instability or differential uptake. These variations in pharmacokinetics complicate the interpretation of structure–activity correlations.

As shown in Fig. 6,[13] THF had no effect on the transformation assay in the absence of carotenoid, and a 0.5% solution of THF did not reduce the

[12] J. Rundhaug, M. L. Gubler, M. I. Sherman, W. S. Blaner, and J. S. Bertram, *Cancer Res.* **47**, 5637 (1987).
[13] R. L. Merriman and J. S. Bertram, *Cancer Res.* **39**, 1661 (1979).

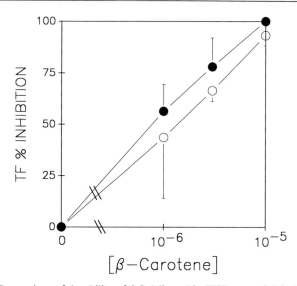

FIG. 4. Comparison of the ability of β-C delivered in THF versus β-C in beadlet form to inhibit MCA-induced transformation. C3H/10T1/2 cultures were exposed for 24 hr to 3 μg/ml MCA, then after 7 days to the indicated concentrations of β-C delivered as a THF solution (●). Treatments were continued at weekly intervals for 4 weeks until cultures were fixed, stained, and scored for transformed foci. Results are expressed as the percent inhibition of the transformation frequency *(TF)* observed in carotenoid-treated cultures compared to the *TF* observed in cultures exposed to MCA and then to 0.5% THF alone. The *TF* in carcinogen-treated controls, that is, the number of foci/number of surviving cells at risk × 100, was 1.01%. Also plotted are historical data[8] obtained using β-C formulated in beadlets (○) using virtually identical protocols. The slopes are not significantly different. (Reprinted from Ref. 9, by permission of the Oxford University Press.)

proliferation rate of cells (Fig. 1). This is of importance since the detection of morphological transformation of C3H/10T1/2 cells depends on the ability of transformants to proliferate within a confluent monolayer of postconfluent growth-arrested nontransformed cells. While the highest concentrations of carotenoids tested (10^{-5} M) caused marginal reductions of the proliferation rate and final saturation density, this effect was shared by carotenoids that were both active and inactive in the transformation assays.[9] Representative dishes showing the effects of diverse carotenoids on transformation are shown in Fig. 7.

Discussion

Tetrahydrofuran is a nontoxic solvent for the highly lipophilic carotenoids that allows them to be placed into an aqueous medium in a readily

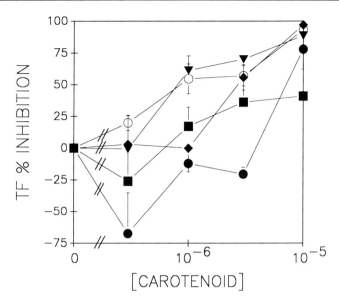

FIG. 5. Effects of diverse carotenoids on MCA-induced neoplastic transformation. The carotenoids were added in THF as described to yield the final stated concentration. Cultures were retreated for a total of 4 weeks beginning 7 days after removal of the MCA. Cultures were then fixed, stained, and the number of type II and III foci determined. Results represent the mean percent inhibition of transformation frequency ± S.E. for the following number of experiments (number of dishes): ▼, β-carotene, 3 (36); ○, canthaxanthin, 4 (47); ◆, α-carotene, 3 (36); ■, lycopene, 3 (35); ●, lutein, 2 (24). For each experiment, control cultures of 24 dishes were exposed to MCA and then to THF vehicle as in the experimental dishes. The mean percent transformation frequency (i.e., number of foci/number of surviving cells × 100) in the four carcinogen controls was 1.02 ± 0.3. Typically, 60% of carcinogen control cultures contained foci and the mean number of foci/dish was 0.9 for all dishes combined. For each experiment 24 dishes did not receive carcinogen or carotenoid but were treated with appropriate solvents. One type II focus would typically be observed in these 24 dishes. This background frequency has not been subtracted from the experimental results, and the experiment was rejected if more than three foci were observed (Data extracted from Ref. 9.)

bioavailable form. The development of this method provides a much needed way in which to study the effects of various carotenoids on cells in culture. Advantages include the minimal number of components found in the delivery system, the lack of effect of the solvent on transformation assays, the low toxicity of the solvent, the generally high solubility of all carotenoids tested in THF, the volatility of the solvent, and the ability to achieve relatively high carotenoid concentrations in aqueous solution. We have used this method to deliver numerous carotenoids, in addition to the carotenoids described above, to cells growing in culture; these include

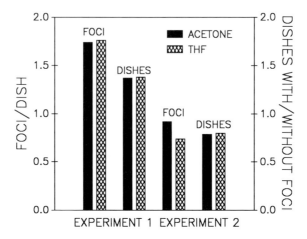

FIG. 6. Lack of effect of THF on MCA-induced transformation. Results show the means of two separate experiments in which acetone (0.5%) or THF (0.5%) were added to dishes treated 8 days previously with 3 μg/ml MCA. THF or acetone was readded at each medium change for the remaining 4 weeks in culture. Dishes were then fixed, stained, and scored for transformed foci. Results are expressed as the mean number of foci/dish or the ratio of dishes with/without foci. Experiment 1 is derived from 24 and 36 dishes treated with acetone or THF, respectively. Experiment 2 is from 36 and 72 dishes treated with acetone or THF, respectively. In control cultures not exposed to MCA, one focus was seen in 24 dishes. Acetone alone does not influence transformation rates.[13]

trans-methylbixin, renierapurpurin, β-cryptoxanthin, phytoene, and a number of synthetic carotenoids.

We have demonstrated that diverse carotenoids possess the ability to inhibit methylcholanthrene-induced neoplastic transformation in C3H/10T1/2 cells. Compounds exerted activity when added 7 days after carcinogen exposure; they must therefore be inhibitory to events occurring postinitiation. We have previously demonstrated, in the case of X-ray-induced transformation of C3H/10T1/2 cells, that treatment with β-C or CTX was only effective when applied during this period. When, in contrast, treatment was applied prior to, during, or immediately after irradiation, or after transformation had occurred, treatment was not effective in inhibiting transformation.[8] Consistent with an action on postinitiation events, inhibition of transformation by β-C and CTX was reversible on drug removal.[8] We assume, but have not demonstrated, that the other active carotenoids examined here also exhibit activity during the promotional phase of carcinogenesis. Retinoids also exert activity in this phase of carcinogenesis.[13]

The similarity of action between retinoids and carotenoids in C3H/10T1/2 cells suggests that carotenoids function by conversion to retinoids. However, in previous studies, we had been unable to detect the products of

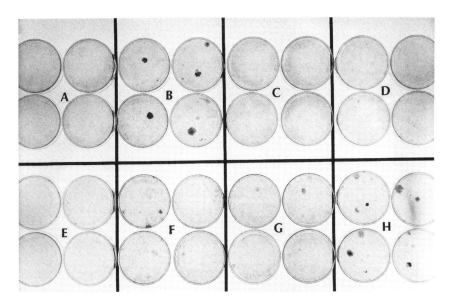

FIG. 7. Representative culture dishes from the transformation assay. All dishes were fixed and stained 5 weeks after seeding and after receiving drug treatments as described in the text. A, THF alone; B, MCA (3 μg/ml) then THF; C, MCA then β-carotene (10^{-5} M); D, MCA then α-carotene (10^{-5} M); E, MCA then canthaxanthin (10^{-5} M); F, MCA then lycopene (10^{-5} M); G, MCA then lutein (10^{-5} M); H, MCA then trans-methylbixin (10^{-5} M). trans-methylbixin was previously shown to have no effect on transformation.[9]

conversion of β [^{14}C]carotene to expected retinoids[4]; furthermore, there is no evidence that CTX can act as a provitamin A source in mammals (though it can in fish).[14] Nevertheless, the carotenoids which demonstrated the greatest ability to inhibit transformation (i.e., β-C, CTX, and α-carotene) all have the potential on cleavage to yield retinoids with chemopreventive activity (retinal or retinoic acid in the case of β- and α-carotene,[15] and 4-ketoretinoic acid in the case of CTX). 4-Ketoretinoic acid was demonstrated to be highly active in the hamster trachea as an inhibitor of squamous metaplasia.[16] The discovery of a family of nuclear retinoic acid receptors (RAR) with high affinity for the ligand (10^{-9} M), and low abundance of receptor molecules,[17] raises the possibility that conversion of only 1 in every 10^4 molecules of β-C to retinoids could activate the RAR. This

[14] J. C. Bauernfeind, Agric. Food Chem. 20, 456 (1972).
[15] J. S. Bertram, Cancer Res. 40, 3141 (1980).
[16] D. Newton, W. R. Henderson, and M. B. Sporn, Cancer Res. 40, 3413 (1980).
[17] A. Zelent, A. Krust, M. Petkovich, P. Kastner, and P. Chambon, Nature (London) 339, 714 (1989).

level of conversion would not have been detectable in our previous studies because of the low specific activity of available labeled β-C.[4] In the studies reported here, we did not detect by HPLC products of conversion of β-C (or other carotenoids) to retinoids, It is of interest that conversion need not necessarily occur via enzymatic attack; oxidative damage followed by cleavage could lead to biologically active fragments.[18] Recently we have shown that canthaxanthin, a potent inhibitor of transformation, does not induce the retinoic acid receptor β gene.[19]

Carotenoids have been reported to have antiproliferative effects[20] that could directly inhibit transformation. Recent studies do not support such an action in C3H/10T1/2 cells.[9] When added to cultures in a protocol identical to that used in the transformation experiments, all carotenoids, both active and inactive as inhibitors of transformation, caused a dose-dependent reduction in saturation density.[9] When added to logarithmically growing cultures at a concentration of 10^{-5} M, β-C marginally reduced the proliferation rate of cells (Fig. 1). However, in the transformation assays, carotenoids were not added until 7 days after removal of carcinogen, when initiation and most of the cellular proliferation are complete. In other studies we have shown that β-C and CTX do not suppress the ability of established transformed cells to grow on a contact-inhibited monolayer of C3H/10T1/2 cells,[8] and that carotenoids reversibly inhibit transformation. We thus conclude that any antiproliferative effects of carotenoids per se cannot explain their ability to inhibit transformation in C3H/10T1/2 cells. The same conclusion was previously reached in studies of retinoid action in C3H/10T1/2 cells.

In any study of the comparative biological effects of carotenoids, a uniform method of delivery is essential for elucidating structure–activity relationships. The use of THF as vehicle for carotenoid delivery appears to meet this requirement, as it has been effective for all types of carotenoids we have examined to date. Because of the low reported systemic toxicity of THF, further research into the use of this method for the delivery of carotenoids or other lipophilic drugs intravenously or orally may be warranted.

Acknowledgments

Research reported in this chapter was supported by Grant BC 686 from the American Cancer Society and Grant ES04302 from the National Institute of Environmental Health Sciences.

[18] J. A. Olson, *J. Nutr.* **119**, 105 (1989).
[19] L. X. Zhang, R. V. Cooney, and J. S. Bertram, *Cancer Res.* **52**, in press (1992).
[20] E. Seifter, G. Rettura, J. Padawer, F. Stratford, P. Goodwin, and S. M. Levenson, *J. Natl. Cancer Inst.* **71**, 409 (1983).

[7] Differentiation between Central and Excentric Cleavage of β-Carotene

By GUANGWEN TANG *and* NORMAN I. KRINSKY

Introduction

In 1931, when Karrer et al.[1] first elucidated the structure of retinol (vitamin A), it was apparent that it represented a hydrolytic cleavage product of the central 15,15'-double bond of β-carotene. Then, in the late 1940s, investigators discovered that the reaction was oxidative in nature, forming retinal which was subsequently reduced to retinol. The intestine appeared to be the major site for β-carotene metabolism. As no other products were observed, a consensus developed that the original hypothesis of central cleavage was correct. However, Glover and associates, starting in 1954,[2-4] presented a series of observations that culminated in the hypothesis that β-carotene could undergo both central and excentric cleavage.[5] This hypothesis was partially supported by observations that the apocarotenals (8'-, 10'-, and 12'-) were effectively converted to retinol in rats, as well as being oxidized to their corresponding carotenoic acids.[2,3,6-9] In addition, these apocarotenals were identified in the intestine of chickens after the administration of β-carotene,[10] and could serve as a substrate for a partially purified enzyme preparation from both rabbit and guinea pig intestine.[11-13] Ganguly and Sastry claimed that the "weight of the evidence so far is overwhelmingly in favor of random cleavage and is against any specific attack at the central double bond of the carotene molecule."[14] However, even recent reviews of both pathways are ambiguous in their

---

[1] P. Karrer, R. Morf, and K. Schöpp, *Helv. Chim. Acta* **14**, 1431 (1931).
[2] J. Glover and E. R. Redfearn, *Biochem. J.* **58**, xv (1954).
[3] S. Fazakerley and J. Glover, *Biochem. J.* **65**, 38p (1957).
[4] M. J. Fishwick and J. Glover, *Biochem. J.* **66**, 36p (1957).
[5] J. Glover, *Vitam. Horm. (Leipzig)* **18**, 371 (1960).
[6] P. Karrer and U. Solmssen, *Helv. Chim. Acta* **20**, 682 (1937).
[7] G. Brubacher, U. Gloor, and O. Wiss, *Chimia* **14**, 19 (1960).
[8] S. M. Al-Hasni and D. B. Parrish, *J. Nutr.* **94**, 402 (1968).
[9] R. V. Sharma, S. N. Mathur, and J. Ganguly, *Biochem. J.* **158**, 377 (1976).
[10] R. V. Sharma, S. N. Mathur, A. A. Dimitrovskii, R. Das, and J. Ganguly, *Biochim. Biophys. Acta* **486**, 183 (1977).
[11] M. R. Lakshmanan, H. Chansang, and J. A. Olson, *J. Lipid Res.* **13**, 477 (1972).
[12] M. R. Lakshmanan, J. L. Pope, and J. A. Olson, *Biochem. Biophys. Res. Commun.* **33**, 347 (1968).
[13] H. Singh and H. R. Cama, *Biochim. Biophys. Acta* **370**, 49 (1974).
[14] J. Ganguly and P. S. Sastry, *World Rev. Nutr. Diet.* **45**, 198 (1985).

conclusions, one claiming "it appears that evidence of excentric cleavage of β-carotene in mammalian systems is undeniable"[15] and the other stating "whether excentric cleavage occurs in mammals is still debated."[16] Olson has stated that "to remain viable as an alternative pathway of vitamin A formation from carotenoids in mammals, the excentric cleavage hypothesis clearly requires unambiguous direct supporting evidence."[17] Three recent publications now offer this direct supporting evidence,[18-20] and the methods that they used are described below.

Assay Method: General

Purified β-carotene is incubated with homogenates of small intestine and other tissues to form a series of homologous cleavage products, including β-apo-15- (retinal), β-apo-14'-, 12'-, 10'-, 8'-carotenals, β-apo-13-carotenone, and retinoic acid. The incubation is carried out under red light. The production of β-apocarotenoids and retinoids is optimized in the presence of β-NAD and dithiothreitol (DTT) and is inhibited by disulfiram. Control vials are run lacking either β-carotene or the tissue homogenates. The separation and quantitation of products is achieved by HPLC analysis. The identifications of metabolites are made by comparison of retention times with authentic standards, by absorbance spectra, and by chemical derivatization, with final confirmation by mass spectral analysis.[18-20]

Purification of β-Carotene. β-Carotene is purified through a 5% (v/w) water-weakened alumina column using hexane as a solvent[21,22] immediately before use. The purity of β-carotene in the eluant is checked by HPLC. The purified β-carotene is evaporated under nitrogen, dissolved in propylene glycol (or 10% ethanol in propylene glycol, v/v), and added to the incubation mixtures.

Tissue Homogenates. Small intestines from fasted subjects are flushed out with ice-cold HEPES buffer (pH 7.35, 20 mM), and the mucosa is scraped with a glass slide. The intestinal scrapings are homogenized in a

---

[15] L. E. Gerber and K. L. Simpson, this series, Vol. 189, p. 433.
[16] J. A. Olson and M. R. Lakshman, this series, Vol 189, p. 425.
[17] J. A. Olson, *J. Nutr.* **119**, 105 (1989).
[18] X.-D. Wang, G.-W. Tang, J. G. Fox, N. I. Krinsky, and R. M. Russell, *Arch. Biochem. Biophys.* **285**, 8 (1991).
[19] G. W. Tang, X. D. Wang, R. M. Russell, and N. I. Krinsky, *Biochemistry* **30**, 9829 (1991).
[20] X.-D. Wang, N. I. Krinsky, G.-W. Tang, and R. M. Russell, *Arch. Biochem. Biophys.* **293**, 298 (1992).
[21] N. I. Krinsky, D. G. Cornwell, and J. L. Oncley, *Arch. Biochem. Biophys.* **73**, 233 (1958).
[22] D. S. Goodman and J. A. Olson, this series, Vol. 15, p. 462.

Brinkmann Polytron homogenizer with HEPES buffer (weight:volume, 1:5). The homogenate is centrifuged in a Sorval RT6000 refrigerated centrifuge at 800 g for 30 min at 4°. The supernatant is collected and stored at −70° until used.[18] Liver, kidney, lung, and fat tissues are minced and homogenized in a similar fashion. There is no loss of enzymatic activity over a period of 7 days.

Incubation of β-Carotene with Homogenates. The tissue homogenates (100 µl, 0.5–1.5 mg protein) are incubated with 10 µl of purified β-carotene in propylene glycol at 37° in a shaking water bath in air under red light for 1 hr in the presence of 20 mM HEPES buffer (pH 7.35), 150 mM KCl, 2 mM β-NAD, and 2 mM DTT, in a total volume of 1.0 ml.[18] Control vials are run lacking either β-carotene or the tissue homogenate. The incubation is stopped by the addition of 6.0 ml of $CHCl_3/CH_3OH$ (2:1, v/v).

Chromatography and Characterization of Retinoids and β-Apocarotenoids

Chromatography System. For quantitation of metabolites, retinyl acetate and γ-carotene in ethanol are added to the incubation mixture as internal standards just before extraction.[18,19] The incubation mixture containing the internal standards is centrifuged at 4° for 10 min at 320 g, and the chloroform layer is removed. After the chloroform is evaporated to dryness under nitrogen, the residue is redissolved in 100 µl ethanol. A 50-µl aliquot of the extract is injected onto the HPLC system. Metabolites are detected after HPLC separation using a Pecosphere-3 C_{18} cartridge column (3 µm, 0.46 × 8.3 cm, Perkin-Elmer, Norwalk, CT) with mobile phases of CH_3CN/tetrahydrofuran (THF)/water. (50:20:30, v/v, 1% ammonium acetate in water as solvent A; 50:44:6, v/v, as solvent B). The gradient procedure at a flow rate of 1 ml/min is as follows: 100% solvent A is used for 3 min followed by a 7-min linear gradient to 100% solvent B, a 17-min hold at 100% solvent B, then a 2-min gradient back to 100% solvent A. A multiwavelength spectrophotometer detector is set at 340 nm for retinoids and at 450 nm for carotenoids. The amounts of metabolites are quantitated by determining peak areas calibrated against known amounts of standards. In addition, a diode array liquid chromatograph is used to record the absorption spectra of the metabolites.

Identification of β-Apocarotenoids. The retention times of various metabolites of β-carotene from the HPLC chromatogram have been shown to be identical to those of authentic samples of retinoic acid, retinal, β-apo-12′-, 10′-, 8′-carotenals, and β-apo-5,6-epoxide (Fig. 1). Furthermore, these compounds have been confirmed by absorption spectra recorded

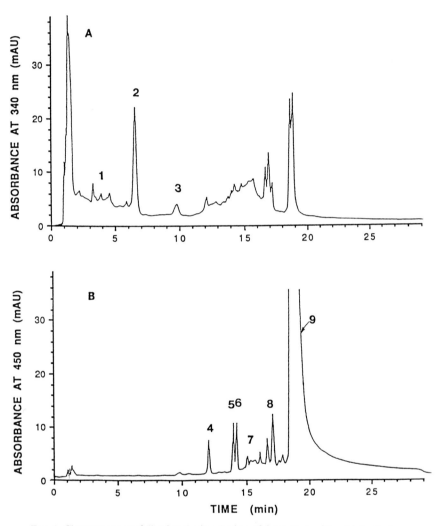

FIG. 1. Chromatograms following the incubation of β-carotene with rat intestinal mucosa homogenate. The detector was set at 340 nm (A) and 450 nm (B) to monitor cleavage products. Peaks are as follows: (1) retinoic acid, (2) β-apo-13-carotenone, (3) retinal, (4) β-apo-14'-carotenal, (5) β-apo-12'-carotenal, (6) β-apo-10'-carotenal, (7) β-apo-8'-carotenal, (8) β-carotene-5,6-epoxide, and (9) β-carotene. Column: Pecosphere-3 C_{18} 0.46 $\times$ 8.3 cm cartridge column. Solvents: CH_3CN/THF/water (50:20:30, v/v, 1% ammonium acetate in water, solvent A) and CH_3CN/THF/water (50:44:6, v/v, solvent B). Flow rate: 1 ml/min. [Reprinted with permission from Tang et al., Biochemistry **30**, 9829 (1991). Copyright 1991 by American Chemical Society.]

with the diode array liquid chromatographic detector. Along with these compounds, two major unidentified metabolites appeared on the chromatogram,[18] and they have been characterized by spectral properties, chemical derivatization, and GC/MS analysis. Reduction by KBH_4 and formation of the O-ethyloxime[23] provide evidence that these metabolites are conjugated carbonyl compounds. Judging from their absorption spectra, the metabolites are carbonyls with five and seven conjugated double bonds, namely, the lower and higher homologs of retinal, β-apo-13-carotenone and β-apo-14'-carotenal. These compounds are formed by the cleavage of the 13,14- or 13',14'-double bond of β-carotene, and their identifications have also been confirmed by GC/MS analysis.[19]

Enzyme Properties

The formation of β-apocarotenoids and retinoids from the incubation of β-carotene with crude homogenates is increased significantly by the addition of NAD and DTT, which demonstrates concentration-dependent production of the β-apocarotenoids. However, the addition of 2 mM NADH and 2 mM DTT to the incubation mixture did not stimulate retinoic acid and retinal formation from β-carotene.[18] Only small amounts of retinol were detected in the incubation with rat intestinal homogenates. The pH optimum for β-carotene metabolism is between 6.5 and 7.0, and the addition of the sulfhydryl inhibitor disulfiram inhibits cleavage activity.

The enzymatic activity for the conversion of β-carotene to the cleavage products in other organs (lung, kidney, liver, and fat) is higher than in the intestine. The homogenates from primates (human and monkey) produce more β-apocarotenals and retinoids than those from rats and ferrets.[18] However, this correlation is not observed in the production of β-apo-13-carotenone.[19]

Biological Significance

Apart from the intrinsic interest in the fact that β-carotene can be enzymatically cleaved at several locations along the polyene chain, there is now a strong possibility that excentric cleavage may play an important role in the formation of retinoic acid. It has been shown that retinoic acid is an important metabolite of β-carotene in mammalian tissues, including the

---

[23] G. J. Handelman, E. A. Dratz, C. C. Reay, and F. J. G. M. van Kuijk, *Invest. Ophthalmol. Visual Sci.* **29**, 850 (1988).

small intestine,[24] as well as liver, kidney, lung, and testes.[25] The latter study is particularly significant, for retinal was not identified as an intermediate in the conversion of β-carotene to retinoic acid. Could the excentric cleavage mechanism play a role in retinoic acid formation? A recent report has clearly demonstrated that human intestinal mucosal homogenates can carry out the conversion of β-carotene to retinoic acid under conditions where the oxidation of retinal to retinoic acid has been completely blocked by the addition of citral.[20] These observations, along with the earlier demonstration of retinoic acid formation from β-carotene in a series of mammalian tissues,[25] may begin to offer some understanding of the significance of an excentric pathway for β-carotene metabolism.

[24] F. D. Crain, F. J. Lotspeich, and R. F. Krause, *J. Lipid Res.* **8**, 249 (1967).
[25] J. L. Napoli and K. R. Race, *J. Biol. Chem.* **263**, 17372 (1988).

[8] Carotenoid–Protein Complexes

By M. R. Lakshman and Chitua Okoh

Introduction

β-Carotene and other carotenoids have been reported to exert biological actions in animals distinct from their role as precursors of vitamin A,[1] without the toxicity associated with high vitamin A doses.[2,3] β-Carotene is enzymatically converted to vitamin A in the liver and intestines.[4-6] Although very little or no intact β-carotene is normally absorbed by rodents, pigs, and chickens,[7-9] investigators have demonstrated that rats, chicks, ferrets, mice, and guinea pigs, like humans, are capable of absorbing intact β-carotene and storing quantifiable amounts in the liver and other organs.[10-13] In these studies, animals fed β-carotene had higher concentra-

[1] A. Bendich and J. A. Olson, *FASEB J.* **3**, 1927 (1989).
[2] G. W. Burton, *J. Nutr.* **119**, 109 (1989).
[3] G. W. Burton and K. U. Ingold, *Science* **224**, 569 (1984).
[4] D. S. Goodman and H. S. Huang, *Science* **149**, 879 (1965).
[5] J. A. Olson and O. Hayaishi, *Proc Natl. Acad. Sci. U.S.A.* **54**, 1364 (1965).
[6] M. R. Lakshman, I. Mychkovsky, and M. Attlesey, *Proc. Natl. Acad. Sci. U.S.A.* **86**, 9124 (1989).
[7] J. Ganguly, J. W. Mehl, and H. J. Deuel, Jr., *J. Nutr.* **50**, 59 (1953).
[8] L. Woytkiw and N. C. Esselbaugh, *J. Nutr.* **43**, 451 (1951).
[9] T. W. Goodwin and R. A. Gregory, *Biochem. J.* **43**, 505 (1948).

tions of β-carotene in all the tissues, and in serum, relative to control animals. The transportation of absorbed β-carotene by low-density lipoproteins has been reported.[14,15] However, the mechanism of tissue uptake, storage, and transport of intact β-carotene in a mammalian system is not fully understood. In cyanobacteria, carotenoids are generally classified into carotenoproteins that are either water- or detergent-soluble. Water-soluble carotenoproteins have been isolated from three genera of cyanobacteria,[16] and the detergent-soluble carotenoproteins were isolated from the cytoplasmic membranes of *Synechocystis* sp. PCC 6714[17] and *Synechococcus* strain PCC 7942[18] and later identified as carotenoid-binding proteins. The gene *(cpbA)* coding for a carotenoid-binding protein of the *Synechococcus* sp. (*Anacystis nidulans* R2) has been cloned and sequenced.[19] Carotenoproteins containing the carotenoids in stoichiometric proportions as prosthetic groups have been isolated from certain crustaceans mainly in the form of glycoproteins.[20,21] The characteristics of many of these isolated carotenoproteins have been summarized.[22] A carotenoid–protein complex was isolated from the pulp of mangos *(Mangifera indica).*[23]

Until recently, no carotenoprotein has been reported for vertebrates in general or mammals specifically. The difficulty involved in characterizing such a protein complex from the tissues may be due to a lack of understanding regarding the mechanism of tissue processing of intact β-carotene in higher organisms. If quantifiable amounts of intact β-carotene can be stored in the livers of β-carotene-fed rats, as has already been reported,[10-13] it is likely that some protein may be involved in this liver storage. Based on this rationale, and with the application and modification of known techniques, we have been able to characterize partially a carotenoprotein from the livers of rats fed β-carotene. This chapter presents the detailed method

[10] M. M. Mathews-Roth, *Nutr. Rep. Int.* **117**, 419 (1977).
[11] S. S. Shapiro, D. J. Mott, and L. J. Machlin, *J. Nutr.* **114**, 1924 (1984).
[12] J. D. Ribaya-Mercado, S. C. Holmgren, J. G. Fox, and R. M. Russel, *J. Nutr.* **119**, 665 (1989).
[13] S. T. Mayne and R. S. Parker, *Lipids* **21**, 164 (1986).
[14] N. I. Krinsky, D. G. Cornwell, and J. L. Oncley, *Arch. Biochem. Biophys.* **73**, 233 (1958).
[15] S. Ando and M. Hatano, *J. Lipid Res.* **29**, 1264 (1988).
[16] T. K. Holt and D. W. Krogmann, *Biochim. Biophys. Acta* **637**, 408 (1981).
[17] G. S. Bullerjahn and L. A. Sherman, *J. Bacteriol.* **167**, 396 (1986).
[18] K. Masamoto, H. C. Reithman, and L. A. Sherman, *Plant Physiol.* **84**, 633 (1987).
[19] K. J. Reddy, K. Masamoto, M. M. Sherman, and L. A. Sherman, *J. Bacteriol.* **171**, 3486 (1989).
[20] W. L. Lee and P. F. Zagalsky, *Biochem. J.* **101**, 9c (1966).
[21] D. F. Cheesman, W. L. Lee, and P. F. Zagalsky, *Biol. Rev.* **42**, 132 (1967).
[22] P. F. Zagalsky, E. E. Eliopoulos, and J. B. C. Findlay, *Comp. Biochem. Physiol.* **97B**, 1 (1990).
[23] C. Subbarayan and H. R. Cama, *Indian J. Biochem.* **3**, 225 (1966).

of isolation used for the rat liver carotenoprotein and discusses briefly the methods of previous researchers who have worked with other carotenoproteins.

Assay Method

Principle. To achieve quantifiable liver storage of β-carotene, male Wistar rats weighing approximately 200 g are placed on a diet containing 0.2% (w/w) β-carotene for a reasonable length of time. The liver samples have to be fresh; all manipulations are done in yellow light at refrigeration temperatures, and all the buffers should contain appropriate detergents (for solubilization of membrane proteins), an antioxidant, and protease inhibitors.

β-Carotene Feeding. The 0.2% β-carotene diet is made by mixing 20 g of β-carotene beadlets (gift from Hoffmann-LaRoche, Basel, Switzerland) with 980 g of ground rat chow (Linguard F. Klein Co., Inc., Baltimore, MD). The diet is fed to the experimental animals for 12 weeks while control rats are fed only regular rat chow for the same length of time.

Isolation of Fresh Livers. At the end of the feeding period, all the rats are sacrificed by aortic exsanguination under pentobarbital anesthesia (50 mg/kg, i.p.), and the livers are isolated after perfusion with ice-cold 0.85% saline. For best results, livers are processed immediately after isolation or, when this is not possible, are stored at $-80°$ only for 3-4 days before processing.

Controlling for Ultraviolet Light and High Temperatures. All manipulations are carried out on ice, in foil-wrapped containers, and under F40 Gold fluorescent light to prevent both photoreduction of the carotenoids and denaturation of the proteins. All unprocessed liver tissues are also stored at $-80°$ in foil-wrapped containers.

Protease Inhibitors. Phenylmethylsulfonyl fluoride (PMSF), leupeptin, and aprotinin are added to the processing buffer to prevent possible proteolytic breakdown of proteins.

Antioxidants. Butylated hydroxytoluene (BHT) is used in the processing buffer to prevent oxidative breakdown of the carotenoid–protein complex.

Detergents. Triton X-100, a peroxide-free nonionic detergent, n-octyl-β-D-glucopyranoside (OBDG), a nonionic detergent, and 3-[(3-cholamidopropyl)dimethylammonio]-1-propanesulfonate (CHAPS), a zwitterionic detergent, are used in the processing buffer as membrane protein solubilizers.

Preparation of Aqueous Extracts. A homogenization buffer is made up of the following constituents: 50 mM morpholinoethanesulfonic acid

(MES), 1 mM EDTA, 20% (w/v) glycerol, 0.2% (w/v) OBDG, 5 mM CHAPS, 0.5% Triton X-100, BHT at a concentration of 50 µg/ml, 2 mM PMSF, and aprotinin and leupeptin, each at a concentration of 1 µg/ml. One gram of fresh liver from either the experimental or control group is first macerated with a pair of surgical scissors and then homogenized with a sonicator (Heat Systems–Ultrasonics, Inc., Plainview, NY) for 30 sec at knob setting 7 in 20 ml of the homogenization buffer. The homogenate is spun at 105,000 g for 1 hr at 4°, and the supernatant is saved. The pellet is homogenized again in a 10-ml volume of the buffer and respun. The two supernatant fractions are pooled and processed as described below.

Isolation of Carotenoid–Protein Complex. To the pooled supernatant fraction is added solid ammonium sulfate to 50% saturation with constant stirring for 30 min. The mixture is spun at 25,000 g for 30 min. The yellow pellet fraction is redissolved in a known volume of the homogenization buffer and dialyzed extensively at 4° against the elution buffer, which differs from the former only by the fact that the glycerol concentration is reduced to 5% while CHAPS and OBDG are omitted. The dialyzate is stored at 4° for spectrophotometric, chromatographic, and electrophoretic studies and also for quantitative analyses.

The ammonium sulfate fraction is further fractionated on a DEAE-Sephacel column (1.5 × 25 cm bed; Sigma, St. Louis, MO), already equilibrated in the elution buffer. The column is initially washed with 2 bed volumes of the buffer containing 10 mM NaCl, then with 2 bed volumes of the buffer containing 100 mM NaCl. Finally the same buffer containing 350 mM NaCl is used as the elution buffer at a flow rate of 0.4 ml/min. The carotenoid–protein complex is eluted from the column as a yellow band. The fractions corresponding to the yellow band are pooled, concentrated by the use of a Speed-Vac concentrator (Forma Scientific, Inc., Marieta, OH), and stored at 4° for further studies.

Protein and Carotenoid Assay. The Pierce bicinchoninic acid method of protein assay (Pierce, Rockford, IL) is used for measuring total proteins in the crude ammonium sulfate extract and the DEAE-Sephacel concentrated fraction. The total carotenoid concentration of the liver tissue is quantified by homogenizing 1.0 g of liver with 20 volumes of chloroform/methanol (2:1, v/v) and filtering under reduced pressure over anhydrous sodium sulfate on a sintered glass filter. For the crude ammonium sulfate extract and the partially purified protein, β-carotene extraction is achieved by shaking a 1-ml aliquot with an equal volume of ethanol or acetone before extracting with 8 ml of hexane. The organic extracts are evaporated under nitrogen and redissolved in 1 ml of hexane for carotenoid analysis.

Subcellular Distribution of Hepatic Carotenoid–Protein Complex. The subcellular distribution of the carotenoid–protein complex is investigated by liver fractionation in detergent-free homogenization buffer, pH 6.5, according to the method described by Mayne and Parker.[13] On isolation, appropriate amounts of the detergents are added to each subcellular fraction, which is then sonicated and processed as described above for the extraction of the carotenoid–protein complex. The distribution of the complex is assessed by high-performance liquid chromatography (HPLC) analysis of the lipid extract of a 1-ml aliquot of each 0–50% ammonium sulfate fraction, dissolved in 1 ml of hexane.

Chromatographic Analysis. The β-carotene concentration is assayed by a reversed-phase HPLC method, using Gilson HPLC systems (Gilson Medical Electronics, Inc., Middleton, WI). Separation is achieved with a mobile phase of methanol/water (97:3, v/v) containing 0.5% ammonium acetate for the first 8 min and 100% methanol thereafter, for the elution of β-carotene. The mobile phase is monitored at a flow rate of 1 ml/min, with pressure at 2000 psi. The detection of β-carotene is monitored with a Kratos Model 783 detector at 460 nm. β-Carotene has a retention time of 11.5 min under these conditions.

Gel Electrophoresis. The yellow fraction which is eluted from the DEAE-Sephacel column is electrophoresed on an 8–12% linear gradient polyacrylamide vertical gel (Bio-Rad, Richmond, CA) in the presence of sodium dodecyl sulfate (SDS-PAGE), essentially as described by Laemmli and Favre,[23a] and stained with Coomassie Brilliant Blue.

Major Findings

A carotenoid–protein complex has been partially characterized from aqueous liver extracts of male Wistar rats fed 0.2% (w/w) β-carotene but not from the livers of control rats which did not receive any β-carotene. The carotenoid–protein complex was precipitated at 4° from the aqueous 105,000 g supernatant fraction between 0 and 50% saturation with ammonium sulfate (AS fraction). Ion-exchange chromatography of the AS fraction on a DEAE-Sephacel column resulted in the elution of a major yellow fraction which had a characteristic carotenoid absorption spectrum, with a primary peak at 460 nm and two shoulders at 432 and 489 nm (Fig. 1A). This spectrum was not abolished by treatment with $NaBH_4$ (Fig. 1B), indicating that it does not arise from a flavoprotein. The carotenoid component of the complex was quantitatively extracted with hexane only after denaturing the complex with an equal volume of ethanol or acetone. The organic extract exhibited the characteristic absorption spectrum of β-caro-

---

[23a] U. K. Laemmli and M. Favre, *J. Mol. Biol.* **80**, 575 (1973).

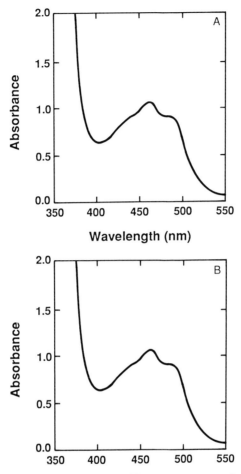

FIG. 1. Absorption spectra in elution buffer of the carotenoid protein complex before (A) and after (B) treatment with sodium borohydride.

tene (Fig. 2). SDS-PAGE of the DEAE-Sephacel fraction gave four major bands with molecular weights ranging between 40,000 and 60,000.

The subcellular distribution of the carotenoid-protein complex (Table I) showed that 84.5% of the recovered β-carotene was localized in the mitochondrial and lysosomal fractions, while the nuclear, microsomal, and cytosolic fractions had negligible amounts of β-carotene. These results imply that a significant portion of the carotenoid-protein complex exists in the membrane fraction of the liver cell.

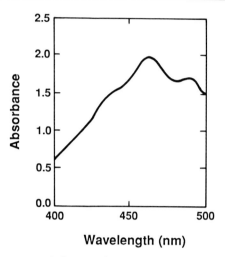

FIG. 2. Absorption spectrum in hexane of the lipid extract of the yellow fraction from the DEAE-Sephacel column.

Precautions

The freshness of the liver tissue is the most essential factor for successful extraction of the carotenoid–protein complex. Thus, it is necessary to isolate the liver soon after the rat is sacrificed. The livers should be thoroughly perfused with ice-cold 0.85% saline solution to remove blood before isolation, weighed immediately, and then suspended in the same solution until needed. The homogenization buffer must contain all the listed ingredients in the right proportions in order to ensure good recovery of the complex.

The carotenoid-binding protein reported by earlier investigators is a membrane protein. Therefore, sonication of the liver is absolutely necessary for the release of the protein. Conventional homogenization procedures for liver do not work well for the release of the protein complex as evidenced by the absence of a carotenoid spectrum.

The partially purified carotenoid–protein complex is sensitive to (1) temperatures higher than 4° (2) prolonged liver storage at temperatures between -20 and $-4°$ for periods longer than 1 month, (3) the presence of proteases, (4) freezing and thawing, and (5) bright light. Each of these conditions causes an initial disappearance of the 432 nm peak and a gradual hypsochromic shift of the 460 nm peak.

The complex is thus fairly stable in the presence of protease inhibitors, antioxidants, high ionic concentrations, high salt concentrations, refrigera-

TABLE I
SUBCELLULAR DISTRIBUTION OF CAROTENOID-PROTEIN
COMPLEX[a]

0-50% Ammonium sulfate fraction of liver component	Concentration (nmol/g)	Distribution (%)
Total extractable	10.07 ± 0.68	100
Nucleus	Not detectable	—
Mitochondria	5.77 ± 0.69	57.3
Lysosomes	2.74 ± 1.02	27.2
Microsomes	0.35 ± 0.08	3.5
Cytosol	0.13 ± 0.01	1.3
All subcellular fractions	8.99 ± 1.80	89.3

[a] Four rats were fed a diet containing 0.2% (w/w) β-carotene for 12 weeks. For total extractable, the 0-50% ammonium sulfate pellet of the supernatant fraction of each liver homegenate in detergent-containing buffer was analyzed. All subcellular fractions were isolated from each fresh liver homogenate in non-detergent-containing buffer. Each fraction was redissolved in homogenization buffer and subjected to 0-50% ammonium sulfate (AS) fractionation. Each AS pellet was dispersed in the detergent-containing buffer and analyzed for β-carotene absorption spectrum and for concentration by HPLC as described in the text. Each value is the mean ± S.E.

tion temperatures, and in the absence of bright light. It is advisable to leave the complex as a 0-50% ammonium sulfate pellet at refrigeration temperature, until needed for further studies. Dialyzing briefly against the elution buffer shortly before use is necessary, especially for chromatographic and electrophoretic purposes. All manipulations should be carried out in a yellow or fairly dark room, and containers need to be prewrapped in aluminum foil.

Previous Studies

The properties of isolated carotenoproteins, mostly from marine invertebrates, have been well discussed.[21,22]

Isolation and Purification of Invertebrate Carotenoproteins

Throughout the invertebrate phyla and in some lower organisms, there is a common group of carotenoid-protein complexes whose apparent function is that of protective coloration.[21,22] The varied colors of marine animals, especially in the Crustacea, is mostly due to the presence of carotenoproteins that are found mainly in the exoskeleton, ectoderm, eggs, and ovaries. The commonly occurring carotenoids astaxanthin (3,3′-dihydroxy-4,4′-diketo-β-carotene) and canthaxanthin (4,4′-diketo-β-carotene) form complexes with proteins, resulting in various colors. This is due to the fact that the apoproteins are able to effect batho- or hypsochromic spectral shifts as the protein interacts with the carotenoid chromophore. Some kind of color is always associated with biological materials containing carotenoproteins. This facilitates monitoring of these proteins during the isolation process. The ratio of absorption at about 280 nm (maximum for proteins in the absence of interfering substances) and a suitable wavelength in the visible range helps to indicate the progress of purification. Furthermore, the isolation of these proteins by standard techniques of protein chemistry is facilitated by their characteristic spectral properties.

Eggs and Soft Tissues. The carotenoproteins are isolated and purified generally by the following steps: (1) homogenization with water or dilute salt solutions, (2) fractional precipitation with ammonium sulfate, and (3) purification by chromatography on ion-exchange cellulose or by selective adsorption on calcium phosphate or aluminum hydroxide, or a combination of these processes.[24,25]

Carapace Carotenoproteins. Preliminary decalcification of the finely ground material is necessary to bring the protein into solution. This is accomplished by the use of citrate buffer, pH 5.0,[26] EDTA, pH 7.5,[27] and ammonium sulfate.[28] Further purification of the protein is thereafter effected by a combination of the procedures outlined above.[27,29] All extraction procedures are carried out in the cold room with reduced illumination in order to avoid fading. Apparently, no carotenoprotein is completely insensitive to light. It is also observed that carotenoproteins are best preserved in solutions of high ionic strength or as precipitates in strong ammonium sulfate solutions.[30,31]

---

[24] D. F. Cheesman, *Proc. R. Soc. London Ser. B* **149**, 571 (1958).
[25] D. A. Norden, Ph.D. Thesis, University of London, 1962.
[26] G. Wald, N. Nathanson, W. P. Jencks, and E. Tarr, *Biol. Bull. (Woods Hole, Mass.)* **95**, 249 (1948).
[27] W. P. Jencks and B. Buten, *Arch. Biochem. Biophys.* **107**, 511 (1964).
[28] H. J. Ceccaldi and B. H. Allemand, *Recl. Trav. Stn. Mar. Endoume* **35**, 3 (1964).
[29] D. F. Cheesman, P. F. Zagalsky, and H. J. Ceccaldi, *Proc. R. Soc. London, Ser. B* **164**, 130 (1966).
[30] K. G. Stern and K. Salomon, *J. Biol. Chem.* **122**, 461 (1938).
[31] E. G. Ball, *J. Biol. Chem.* **152**, 627 (1944).

Carotenoid-Binding Proteins of Cyanobacteria

Carotenoids are virtually ubiquitous among microorganisms, plants, and animals, affording protection from photosensitizers and oxygen radicals.[32-35] Water-soluble carotenoproteins have been isolated from three genera of cyanobacteria,[16] and detergent-soluble forms were isolated from the cytoplasmic membranes of *Synechocystis* strain PCC 6714[17] and *Synechococcus* strain PCC 7942.[18]

Isolation Method. The carotenoid-binding protein of *Synechocystis* sp. is isolated by first pelleting the membrane prepared from 5 liters of autotrophically grown culture at 115,000 g and then suspending the pellet in 1.5 ml of 50 mM MES buffer (pH 6.5) containing 0.2% (w/v) Triton X-100 and 0.2% (w/v) dodecyl-β-D-maltoside.[17] The membrane suspension is passed through an Affi-Gel 501 organomercury column equilibrated with the buffer. The eluate is next applied to a DEAE-Sephacel column and washed with the same buffer containing increasing concentrations of NaCl. The carotenoid-binding protein elutes from the column as an orange fraction.

Characterization. The polypeptide demonstrated a characteristic mobility shift when electrophoresed in lithium dodecyl sulfate–polyacrylamide gels. The protein migrated with an apparent molecular mass of 35 kDa when solubilized at 0°; after solubilization at 70°, however, the protein migrated as a 45-kDa species. The carotenoid-binding protein accumulated only in autotrophically grown cells. Cytoplasmic membranes prepared from photoheterotrophically grown cells lacked this component.

The absorption spectrum of the native carotenoid-binding protein at 25° revealed a characteristic carotenoid spectrum having three peaks at 450, 476, and 516 nm. The chemical structure of the associated carotenoid pigment has not yet been determined.

Genetic Studies. The gene *(cbpA)* coding for a carotenoid-binding protein of the cyanobacterium *Synechococcus* sp. strain PCC 7942 (*Anacystis nidulans* R2) has been cloned and sequenced.[19] A polyclonal antibody against the protein was used to identify immunoreactive clones from λgt11 expression library of *Synechococcus* strain PCC 7942.

---

[32] M. Griffiths, M. R. Sistrom, and G. Cohen-Bazire, *Nature (London)* **176**, 1211 (1955).
[33] N. I. Krinsky, *Pure Appl. Chem.* **51**, 649 (1979).
[34] M. M. Mathews and W. R. Sistrom, *Nature (London)* **184**, 1892 (1959).
[35] D. Siefermann-Harms, *Physiol. Plant.* **69**, 561 (1987).

Carotenoid-Protein Complexes of Plants

A colored protein containing β-carotene as a component was isolated from green parsley leaves.[36] A carotenoid-protein complex has also been isolated from the pulp of ripe *Mangifera indica* (mango fruit) and partially characterized.[23] Partial purification of the complex was achieved by fractionation with ammonium sulfate (60-70% saturation). The complex was shown to be a lipoprotein, and the carotenoid in the complex was identified as β-carotene. The ratio of carotene to protein in the partially purified complex was 1:500.

The absorption spectrum of the complex, having peaks corresponding to the proteins in the ultraviolet region (280 nm) and to carotenoids in the visible region (445 and 475 nm), has been reported. The complex has been partially characterized by disk electrophoresis. The presence of minute quantities of phospholipids and the absence of sterols in the complex indicated that β-carotene was bound directly to the protein moiety of the lipoprotein complex.[23]

A carotenoprotein from carrot chromoplasts has been recently isolated and partially purified.[37] The native complex has been reported to have a molecular weight of 2 million with a subunit molecular weight of 54,000. HPLC analysis showed that 1 mol of α-carotene and 2 mol of β-carotene were bound to each 54 KDa subunit. Amino acid analysis showed the protein subunits to have a high percentage of α-helical conformation with reverse turns.

Carotenoid-Protein Reconstitution

Only a few carotenoproteins have been obtained in a sufficiently pure and homogeneous state to merit detailed studies. All the carotenoproteins examined so far are readily split into carotenoid and apoprotein components by treatments of an aqueous solution with acetone or ethanol. This suggests that covalent bonding may not be involved in the carotenoid-protein linkage. In all of the cases, the procedure precipitated a colorless apoprotein which could be completely freed from the carotenoid component with several repetitions of the process. The process, however, would involve some degree of denaturation. The prosthetic group of ovorubin, ovoverdin, and crustacyanin has been identified as astaxanthin.[29] With the exception of ovoverdin, the apoproteins have been prepared, and the carotenoproteins reconstituted in the native form, by mixing an acetone

---

[36] M. Nishimura and K. Takamatsu, *Nature (London)* **180,** 699 (1957).

[37] J. D. Bryant, J. D. McCord and J. W. Erdman, Conference on Molecular and Comparative Nutrition, Bethesda, Maryland, July 22-24, Abstr. No. 5 (1991).

solution of carotenoid with an aqueous solution of protein, diluting with water, and dialyzing away the acetone.[24,29] However, with ovoverdin, as with other carotenoproteins, gentle heating results in a partial dissociation of carotenoid from protein, which may be reversed on cooling.[30,31]

The specificity of the carotenoid-protein linkage in crustacyanin was studied by reconstituting apocrustacyanin with various carotenoids and comparing the resultant chromoproteins electrophoretically and by gel filtration with the original apoprotein and β-crustacyanin.[20,29] This carotenoprotein has been successfully crystallized.[29] Resonance Raman spectroscopy studies on the reconstitution of astaxanthin-binding proteins with natural and synthetic carotenoids have been discussed.[22]

Storage Conditions

Most carotenoproteins are exceedingly labile, even in the dark, and this creates a problem in terms of reproducibility or in carrying out prolonged experiments on the same preparation.[31] However, crustacyanin and ovorubin are relatively stable for long periods of storage in the refrigerator (either in solution or in freeze-dried form) and also to light.[24,27] In the freeze-dried form, they undergo slow changes on exposure to daylight. Irradiation for a period of 1 week caused partial denaturation of crustacyanin, whereas ovorubin showed changes in its absorption spectrum which suggested an alteration in the carotenoid-protein linkage.[24] Solutions of ovoverdin faded rapidly in daylight.[30] Better stability was achieved when the carotenoproteins were stored in strong salt solutions.[30,31]

Summary

This chapter provides an updated report of carotenoprotein-related research. Very few carotenoproteins have been purified; however, their presence in aqueous extracts may be indicated by spectroscopic evidence. Carotenoproteins have been isolated, purified, and characterized from the ectoderm, exoskeleton, eggs, and ovaries of marine invertebrates, especially crustaceans. Water-soluble and detergent-soluble carotenoid-protein complexes have also been isolated from the cytoplasmic membrane of some cyanobacteria, *Mangifera indica*, and carrots. Recently, we have been able to partially purify a β-carotene-protein complex from fresh livers of rats fed β-carotene. Studies are currently in progress to purify and characterize the protein. This is the first successful isolation of a vertebrate carotenoprotein. The isolation of carotenoproteins is generally by standard techniques of protein chemistry. Purification, crystallization of the complex, and reconstitution of apoprotein and carotenoid components have been achieved for some crustacean carotenoproteins. The complex is very sensitive to bright light and to temperatures above that of refrigeration. However, it is best preserved in solutions of high ionic strength or as a precipitate in strong ammonium sulfate solutions.

[9] Analysis of Carotenoids in Human Plasma and Tissues

By ROBERT S. PARKER

Introduction

Interest in the determination of concentrations of specific carotenoid pigments in human plasma and tissues has intensified in recent years because of findings regarding the potential biological effects of these compounds or their metabolites. In spite of the recent attention given to these phytopigments, however, relatively little is known of concentrations of specific pigments in human solid tissues, particularly pigments other than β-carotene, and the factors which influence or regulate these levels. Because many of the postulated effects of carotenoids in tissues, such as antioxidant activity,[1] retinoid precursors,[2] and gene regulation,[3] are likely to be both structure specific and concentration specific, methods which distinguish among the many carotenoids usually present are most useful. Earlier reports of concentrations of total carotenes or carotenoids in human liver[4] and adipose tissue[5,6] using nonspecific spectrophotometric analysis of tissue lipid extracts are thus of limited value in this regard. More recently, Parker,[7,8] Tanumihardjo *et al.*,[9] Kaplan *et al.*,[10] and Schmitz *et*

[1] G. W. Burton, *J. Nutr.* **119**, 109 (1989).
[2] J. L. Napoli and K. R. Race, *J. Biol. Chem.* **263**, 17372 (1988).
[3] M. Z. Hossain, L. R. Wilkins, P. P. Mehta, W. Loewenstein, and J. S. Bertram, *Carcinogenesis* **10**, 1743 (1989).
[4] H. Klaui and J. C. Bauernfeind, *in* "Carotenoids as Colorants and Vitamin A Precursors" (J. C. Bauernfeind, ed.), p. 48. Academic Press, New York, 1980.
[5] A. W. Pierce, *Med. J. Aust.* **1**, 589 (1954).
[6] J. M. Dagadu, *Br. J. Nutr.* **21**, 453 (1967).
[7] R. S. Parker, *Am. J. Clin. Nutr.* **47**, 33 (1988).
[8] R. S. Parker, *J. Nutr.* **119**, 101 (1989).
[9] S. A. Tanumihardjo, H. C. Furr, O. Amedee-Manesme, and J. A. Olson, *Int. J. Vitam. Nutr. Res.* **60**, 307 (1990).
[10] L. A. Kaplan, J. M. Lau, and E. A. Stein, *Clin. Physiol. Biochem.* **8**, 1 (1990).

al.[11] have employed high-performance liquid chromatographic (HPLC) techniques to quantify individual carotenoid pigments in various human solid tissues. The method of Schmitz et al.[11] is described in detail elsewhere in this volume.

Comprehensive analysis of polar and nonpolar carotenoids in solid tissues is complicated by three conspiring factors. First, comprehensive resolution of both polar and nonpolar pigments requires gradient HPLC elution in which the initial solvent is relatively weak. Second, most tissue extracts will contain appreciable quantities of lipids which are insoluble in the initial eluting solvent, such as triglycerides and cholesterol esters. Third, some highly polar carotenoids containing multiple functional groups may have poor stability during base saponification, or poor recovery from the postsaponification solution. Although saponification can be avoided by using HPLC elution solvents strong enough to solubilize the total lipid extract (as most published procedures for tissue carotenoids have involved), the price of this convenience is loss of resolution of many polar carotenoids. Therefore we have developed a procedure in which polar carotenoids are first partitioned from nonpolar carotenoids and lipids, thus permitting their analysis without need of saponification. Nonpolar carotenoids are analyzed following a normal saponification procedure, and the comprehensive carotenoid composition of the original extract is obtained as the sum of the results of the two HPLC analyses.

Because fatty tissues of adults normally contain substantial amounts of carotenoids, such that adipose tissue represents a significant deposition site for these pigments, we have recently developed methodology suitable for measurement of a large number of individual carotenoids in this tissue. Also, since fatty tissues can often be removed during several types of routine surgery, high-quality specimens can be obtained from living individuals and compared with the kinds and concentrations of pigments circulating in the bloodstream of the same individuals. This is not possible with autopsy specimens, with which there is the added unknown factor of postmortem pigment stability, which is likely to differ with both tissue type and carotenoid type. In addition, clarification of the blood–tissue relationship (if any exists) may allow inferences of tissue concentration in instances where only plasma (or serum) samples are available. Discussed below are suitable methods for quantitative extraction and HPLC analysis of carotenoid pigments from human plasma and fatty tissues.

---

[11] H. H. Schmitz, C. L. Poor, E. T. Gugger, and J. W. Erdman, Jr., this volume [11].

Materials and Methods

Extraction of Carotenoids from Fatty Tissues

The method involves the preparation of a total lipid extract, followed by solvent partition of carotenoids into polar and nonpolar classes. The nonpolar preparation will contain any triglycerides present in the total lipid extract and will usually require saponification prior to HPLC analysis. The polar preparation can be chromatographed directly.

A 0.5-g tissue sample (e.g., adipose tissue) is ground in a chilled mortar with 5 ml hexane–acetone (1:1); the extract is passed through a 0.22 μm filter and evaporated to dryness under a stream of nitrogen. A suitable internal standard can be added to the extract in the mortar prior to filtration. The residue is redissolved in 9 ml cyclohexane, then washed with 1 ml of an aqueous solution containing 1% ethylenediaminetetraacetic acid (EDTA) and 4% ascorbic acid. The cyclohexane phase is evaporated to dryness under nitrogen, 8 ml *n*-hexane and 2 ml dimethylformamide (DMF) are added, and the tube contents are shaken after the headspace is flushed with nitrogen. The tube is then centrifuged and the hexane (upper) phase transferred to a fresh tube and reextracted with 2 ml DMF. The combined DMF extracts are reextracted with 4 ml *n*-hexane, and the hexane extracts are combined and reserved for saponification. The DMF is removed under a nitrogen stream or under reduced pressure, using azeotropes as needed, and redissolved in an appropriate volume of methanol for gradient HPLC analysis as described below.

The combined hexane extracts, containing nonpolar lipids (including nearly all triglycerides present in the original tissue extract), are evaporated to dryness under a nitrogen stream. The residue is saponified with 2 ml of 5% methanolic sodium hydroxide, under nitrogen atmosphere at 65° for 15 min in the dark. The tube is cooled and extracted a first time with 5 ml *n*-hexane. One milliliter of water is then added to the methanol phase and the hexane extraction repeated. The combined hexane extracts are reduced in volume under a nitrogen stream, and any sodium soaps which precipitate are removed by centrifugation. The hexane phase is then diluted to 5–6 ml, washed with water, and evaporated to dryness under nitrogen. The residue is dissolved in 2 ml HPLC solvent A (see below), and any remaining soaps which precipitate at this point can be removed by centrifugation. This step ensures that no soaps will precipitate on the HPLC column, since their removal from the column is normally a laborious procedure. The clear extract is finally dried under nitrogen and redissolved in an appropriate volume of methanol for gradient HPLC analysis as described below.

Extraction of Carotenoid and Bilirubin Pigments from Human Plasma

Several excellent procedures have been developed for the extraction of carotenoids from plasma or serum, and many can be found in other contributions to this volume. Essentially all involve some form of solvent partition, between water (in the sample) containing a denaturing alcohol (usually ethanol) and an organic solvent (usually hexane). Although aqueous ethanol-hexane partitions have been shown to extract major plasma carotenoids (including lutein and β-carotene) with relatively high efficiency, they do not extract bilirubin. While this is an advantage in most cases, applications necessitating recovery of both bilirubin and carotenoids require a different procedure. In addition, highly polar minor plasma carotenoids or carotenoid metabolites with appreciable solubility in aqueous ethanol but low solubility in hexane may exist. The partition procedure of Barua and Furr[12] utilizes ethyl acetate in place of hexane as the extracting solvent, and this procedure is shown to extract bilirubin. We have developed a procedure which does not involve any partition and thus will recover all lipids, regardless of polarity. This method is described below.

A 0.5-ml plasma or serum sample is added to 2 ml of ice-cold methanol-acetone-dichloromethane (1:4:1, v/v, containing 0.01% butylated hydroxytoluene) while vortexing. The tube is flushed with nitrogen and stored at 4° in the dark for 1 h to precipitate proteins. The tube is then centrifuged at 4° for 3 min at 1500 g, and the supernatant is transferred to a 9-ml tube and dried under a nitrogen stream. Because the extract will contain all water present in the original plasma sample, the time to complete dryness is somewhat lengthened, but the presence of alcohol and acetone minimizes the additional time needed. Four milliliters of cyclohexane-diethyl ether (6:1, v/v) is added, then the tube is flushed with nitrogen and sonicated for 1 min in a bath sonicator to dislodge adherent material (hydrophobic protein). Two milliliters of water is added; the tube is shaken, then centrifuged 4 min at 1500 g to separate phases. The organic (top) layer is transferred to a fresh tube, dried under nitrogen, redissolved in cyclohexane-diethyl ether (6:1, v/v), and transferred to a 1.5-ml brown glass vial.

Prior to HPLC analysis the sample is again dried under nitrogen and redissolved in 30 μl dichloroethane plus 90 μl methanol. Probably due to the presence of bilirubin, we have noted that such extracts are less stable (in terms of carotenoid and bilirubin concentration) than those obtained with the ethanol-hexane partition procedure, and thus should be analyzed

---

[12] A. B. Barua and H. C. Furr, this series, Vol. 213 [24].

within 1 day of their preparation. Recovery of zeaxanthin, apo-β-8′-carotenal (internal standard), β-cryptoxanthin, and β-carotene exceed 90% with this technique.

Chromatographic HPLC Analysis of tissue and Plasma Carotenoid Extracts

The HPLC system for the chromatograms shown here consists of a Rheodyne 7125 injector valve with a 50-μl loop, Perkin-Elmer (Wilton, CT) 250 binary pump, Perkin-Elmer LC-95 variable wavelength UV–VIS detector, Perkin-Elmer 1020X dual channel data acquisition, analysis, and storage system, and a Hewlett-Packard (Avondale, PA) Laser Jet IIP printer. Solvents A (acetonitrile–methanol–dimethylformamide–ammonium acetate, 81:15:4:0.06, v/v) and B (2-propanol–methanol, 85:15, v/v) are maintained under N_2 atmosphere in a pressurized solvent manifold. The column arrangement consists of a Perkin-Elmer C_{18} guard cartridge and a 25 cm Ultremex 3 μm C_{18} analytical column (Phenomenex, Inc., Torrance, CA). Titanium frits are used exclusively on all columns. The solvent gradient profile is as follows: 100% A for 14 min at 0.7 ml/min; 7-min linear gradient to 80% A/20% B at 0.8 ml/min; 80% A/20% B for 24 min at 0.9 ml/min, after which all carotenoids have eluted; step change to 100% A and reequilibration for 12 min at 1.0 ml/min. Under these conditions, bilirubin, lutein, zeaxanthin, apo-8′-carotenal, α-cryptoxanthin, β-cryptoxanthin, lycopene (four isomers), α-carotene, and all-*trans*-β-carotene elute at approximately 6.9, 10.2, 10.7, 14.2, 22.9, 24.5, 29.8–31.0, 38.7, and 40.3 min respectively.

Comments

HPLC chromatograms of carotenoid pigments extracted from thigh and breast adipose tissue biopsies obtained from a healthy 44-year-old female on the same day are shown in Figs. 1 and 2, respectively. In each case the solid trace represents the DMF-soluble material and the dotted trace the hexane-soluble material. It can be seen that essentially no pigments eluting prior to lycopene are recovered in the hexane phase, while nearly all of these more polar compounds are found in the DMF fraction. Conversely, nearly all the β-carotene is recovered in the hexane fraction, with lycopene partitioning into both fractions. If the two fractions are combined prior to HPLC analysis, a substantial loss in resolution occurs in the first half of the chromatogram, with several peaks coeluting. This is presumably due to interference by free fatty acids coextracted into the hexane fraction following saponification, which then have the effect of increasing the strength of solvent A or competing for binding sites on the stationary phase. Therefore, it is necessary to analyze each fraction separately. While this increases the total analysis time, the additional information on composition of the polar carotenoid fraction obtained with the procedure is substantial. Comparison of Figs. 1 and 2 indicates a high degree of similarity in both composition and concentration of carotenoids between the two tissue sites examined, a unique finding to date. Although we have applied these extraction and chromatographic procedures primarily to adipose tissue, they should also prove useful for analysis of other solid tissues.

A typical chromatogram of plasma carotenoids obtained with the extraction procedure described above and the same HPLC conditions as

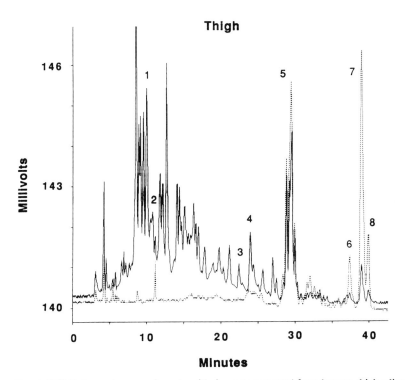

FIG. 1. HPLC chromatogram of carotenoid pigments extracted from human thigh adipose tissue. Detection is at 460 nm. Peak identities: 1, zeaxanthin; 2, lutein; 3, α-cryptoxanthin; 4, β-cryptoxanthin; 5, lycopene; 6, α-carotene; 7, all-*trans*-β-carotene; 8, *cis*-β-carotene. Solid trace, DMF-soluble fraction; dotted trace, hexane-soluble fraction. No internal standard was used in this sample.

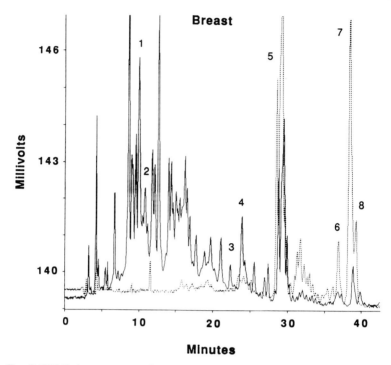

FIG. 2. HPLC chromatogram of carotenoid pigments extracted from human breast tissue (same donor as for sample shown in Fig. 1). Detection is at 460 nm. Peak identities are as described in Fig. 1. Solid trace, DMF-soluble fraction; dotted trace, hexane-soluble fraction. No internal standard was used in this sample.

described for tissue extracts is shown in Fig. 3. Bilirubin, several polar xanthophylls, four lycopene isomers, and two β-carotene isomers are resolved with this technique. In contrast to the tissue extracts, plasma extracts typically contain relatively small amounts of polar carotenoids which elute prior to zeaxanthin (compare Fig. 3 with Figs. 1 and 2). This has been the case with all samples of human plasma and adipose tissue we have analyzed to date, including blood and tissue samples obtained from the same individual donors. The reason(s) for the relative enrichment of adipose tissue in these highly polar compounds, reported here for the first time, is unknown at present, but it may represent either preferential uptake or retention of these pigments. Because the plasma extraction procedure used here is complete in that no aqueous–organic partition is involved, this observation is not the result of poor recovery of these highly polar pigments from the plasma.

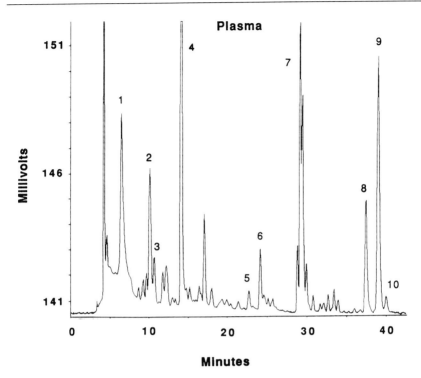

FIG. 3. HPLC chromatogram of carotenoid pigments extracted from human plasma. Detection is at 460 nm. Peak identities: 1, bilirubin; 2, zeaxanthin; 3, lutein; 4, apo-β-8′-carotenal (internal standard); 5, α-cryptoxanthin; 6, β-cryptoxanthin; 7, lycopene (four isomers); 8, α-carotene; 9, all-*trans*-β-carotene; 10, *cis*-β-carotene.

Summary

Procedures are described for the extraction and analysis of carotenoid pigments from human tissues and plasma. The method allows resolution of many polar pigments, including several such compounds more polar than zeaxanthin. While the method requires two separate HPLC runs when applied to tissues containing significant quantities of triglyceride, the additional information made available on the polar carotenoid composition of such samples is substantial.

Acknowledgments

The author wishes to thank Dr. James Marshall, Southern Tier Plastic Surgery Associates, Elmira, New York, for providing the adipose tissue samples for the analyses shown in Figs. 1 and 2.

[10] Assessing Variability in Quantitation of Carotenoids in Human Plasma: Variance Component Model

By ANNA R. GIULIANO, MONIKA B. MATZNER, and LOUISE M. CANFIELD

Introduction

Several epidemiological studies have shown an inverse association between the concentration of β-carotene in plasma and the relative risk of cancer.[1-3] These data have led to the initiation of a growing number of studies investigating the role of plasma β-carotene in the etiology and prevention of cancer at various sites.[4-6] The interpretation of such studies will rely on the accuracy and reliability of β-carotene measurements in populations of individuals. Several reports have shown that plasma β-carotene varies significantly between individuals.[7-10] Given the chemical properties of β-carotene (i.e., light sensitivity, reactivity with active oxygen species, and insolubility in aqueous systems), the variability in plasma values both within and between individuals will be influenced by sample handling prior to and during HPLC quantitation.

We describe here a variance component model which can be used to identify the sources of technical variability and to assign the relative contribution of each of these sources to the overall variability in the method for quantitating β-carotene in human plasma. Although β-carotene was analyzed in this system, the method presented here for determining technical variability in a multistep preparation assay can be applied to the analysis of other compounds found in plasma that require extensive sam-

[1] H. B. Stahelin, E. Buess, and G. Brubacher, *J. Natl. Cancer Inst.* **73**, 1463 (1984).
[2] A. M. Nomura, G. N. Stemmerman, L. K. Heilbrun, R. M. Salkeld, and J. P. Viulleumier, *Cancer Res.* **45**, 2369 (1985).
[3] M. S. Menkes, G. W. Cornstock, J. P. Vuilleumier, K. J. Helsing, and R. Brookmeyer, *N. Engl. J. Med.* **315**, 1250 (1986).
[4] H. F. Stich, M. P. Rosin, and M. O. Villijara, *Lancet* **2**, 1204 (1984).
[5] H. F. Stich, A. P. Hornby, and B. P. Dunn, *Int. J. Cancer* **36**, 321 (1985).
[6] E. R. Greenberg, J. A. Baron, T. A. Stukel, M. M. Stevens, J. S. Mandel, S. K. Spencer, et. al., *N. Engl. J. Med.* **323**, 789 (1990).
[7] W. C. Willett, M. J. Stampfer, B. A. Underwood, J. O. Taylor, and C. H. Hennekens, *Am. J. Clin. Nutr.* **38**, 559 (1983).
[8] J. C. Meyer, H. P. Grundman, B. Seeger, and U. W. Schynder, *Dermatologica* **171**, 76 (1985).
[9] N. V. Dimitrov, C. W. Boone, M. B. Hay, P. Whetter, M. Pins, G. J. Kelloff, and W. Malone, *J. Nutr. Growth Cancer* **3**, 227 (1986).
[10] C. C. Tangney, R. B. Shekelle, W. Raynor, M. Gale, and E. P. Betz, *Am. J. Clin. Nutr.* **45**, 764 (1987).

ple processing. In addition, problems unique to the assay of carotenoids in plasma, including standardization, are discussed.

Methods

Sample Protocol. Blood is collected by venipuncture into 10-ml Vacutainer tubes containing heparin as an anticoagulant. One pool of plasma is obtained by combining the plasma derived from four separate Vacutainer tubes (Becton Dickenson, Lincoln Park, NJ) following separation by centrifugation. Aliquots of 0.60 ml are stored in separate microcentrifuge tubes at $-80°$ until analysis. On each of 5 consecutive days, four aliquots of plasma were extracted in duplicate, and duplicate injections were performed from each extraction for a total of 16 sample injections per day (Fig. 1).

Materials. β-Carotene and β-apo-8'-carotenal obtained from Fluka (Buchs, Switzerland) are dissolved in tetrahydrofuran (THF) and stored at $-80°$ in darkened vials. Organic solvents for HPLC are filtered through an 0.45-μm Fluoropore filter (Millipore, Bedford, MA) prior to use. Cyclohexane (Baker Analyzed) is purchased from J. T. Baker, Inc. (Phillipsburg, NJ), and acetonitrile, dichloromethane, and hexane are purchased from Fisher Scientific (Fair Lawn, NJ). All solvents used for chromatography are HPLC grade or better.

Extraction Procedure. Aliquots of plasma are thawed at 37° for 1 min and mixed with gentle shaking. Two 0.25-ml aliquots are removed to separately labeled microfuge tubes. Spectrophotometric grade ethanol (0.25 ml) containing 0.25 g/liter butyrated hydroxytoluene (BHT) and

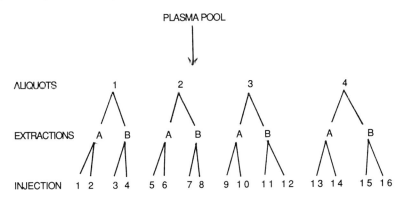

FIG. 1. Schematic representation of protocol followed on each of 5 consecutive days. From one plasma pool four aliquots of plasma were thawed. Two extractions from each stored aliquot and two injections per extraction were analyzed. The protocol was repeated on each of 5 consecutive days for a total of 80 injections.

0.50 ml hexane is added to each tube. Samples are mixed by vortexing for 1 min followed by centrifugation at 16,000 g for 2 min (Beckman Model E Microcentrifuge, Beckman Instruments, Inc., Palo Alto, CA). The resulting hexane layer is retained, and a second hexane extraction of the aqueous phase is performed. The combined hexane layers are evaporated to dryness with nitrogen and redissolved in 150 μl HPLC running solvent (see below), and 50 μl is analyzed by HPLC.

Chromatographic Analysis. HPLC analysis is performed using a Waters Model 510 pump, a Milton Roy Programmable Detector Model SM 4000, a 50-μl loop, and a Waters Maxima 820 version 3.02 system controller (Waters Associates, Milford, MA). A Waters Resolve reversed-phase C_{18} column [10 μm, 4.6 mm inner diameter (1D) $\times$ 100 mm] in a 8 mm $\times$ 10 cm Radial Pak cartridge holder is used for all analyses. The running solvent is acetonitrile–dichloromethane (80:20, v/v) containing 0.1% cyclohexane and 0.25 g/liter BHT at a flow rate of 2.0 ml/min with a total run time of 18 min. β-Carotene is detected at 452 nm.

β-Carotene Standards. For the quantitation of β-carotene concentrations in plasma, both external neat standards and authentic standards added to plasma (spiked plasma standards) are utilized. Neat external standards of authentic β-carotene are prepared by serial dilution into HPLC running solvent from a stock β-carotene solution of approximately 1 mg/ml THF. Three neat standards containing 0.6, 1.1, and 2.2 μg β-carotene/ml are assayed by HPLC daily. Spiked plasma standards are prepared daily by the addition of 12.5 μl of a 1.1, 2.2, and 4.5 μg/ml β-carotene standard to each of three 250-μl aliquots of thawed plasma. β-Carotene-spiked plasma samples undergo identical extraction and HPLC analysis as do the pooled plasma samples.

Statistical Analysis. Variance analyses are conducted using peak area expressed as millivolts-seconds. Plasma β-carotene is analyzed statistically by a hierarchical, random effects analysis of variance using SAS PC Proc Varcomp.[11] This procedure separates the total variance into components associated with differences between aliquots, extractions, days, and injections. The log transformation of β-carotene is used for computation of variance estimates and confidence intervals to better approximate normality. Approximate confidence intervals for the variances of vials, extractions, and days are computed using the method of Searle.[11] This same method gives an exact confidence interval for the error variance.

---

[11] S. R. Searle, "Linear Models," p. 413. Wiley, New York, 1971.

TABLE I
RANDOM EFFECTS ANALYSIS OF VARIANCE[a]

Parameter	Variance	
	Estimate	Standard error
Day	0.0098	0.0074
Aliquot	0.0016	0.0018
Extraction	0.0004	0.0009
Random error (including injection)	0.0112	0.0019

[a] Variance is given as ln transformation.

Results and Discussion

β-Carotene analysis in plasma requires sample processing to extract the lipid layer containing β-carotene prior to injection onto the HPLC. Several extraction procedures have been described for extracting β-carotene and other carotenoids from plasma.[12-14] We utilized a two-step hexane extraction procedure since this procedure had previously been shown to minimize carotenoid degradation while extracting approximately 98% of the endogenous β-carotene in plasma.[14] Given that multiple steps are required in processing a plasma sample prior to its analysis by HPLC, we were interested in quantitating the variability accrued in each of the various steps leading up to obtaining the integral from the HPLC, while utilizing a standard extraction procedure. To eliminate between-technician variability, a single technician with extensive HPLC experience carried out all aspects of the experiment. As automatic injectors for HPLC are not universally available, samples were injected manually. In addition, peak area (millivolts-seconds) was used in the analysis to eliminate any variability due to analytical manipulation. β-Carotene eluted with a k' of 8.0. In agreement with others, the limit of detection of our method was 1.1 μg/dl.[12,14] We define a limit of detection as a signal-to-baseline ratio of at least 3.

The variance due to each of the various components of the analysis is listed in Table I: the effect of assaying on different days, number of aliquots, number of separate extracts per aliquot, and random error which

[12] Y.-M. Peng, J. Beaudry, D. S. Alberts, and T. P. Davis, *J. Chromatogr.* **273**, 410 (1983).
[13] C. S. Yang and M.-J. Lee, *J. Nutr. Growth Cancer* **4**, 19 (1987).
[14] L. R. Cantilena and D. W. Nierenberg, *J. Micronutr. Anal.* **6**, 127 (1989).

includes injections onto the HPLC. Using this method we identified random error (including variability in injection) as the largest source of variability. Day-to-day variability was the next largest source of variability, while the lowest source of variability was from the extraction procedure. These results are somewhat unexpected as the extraction procedure consists of several steps. It should be noted however, that the extractions were performed on the same plasma. For extractions performed on plasma from separate individuals with varying lipid composition, we might expect higher extraction variability.

The variance associated with random error in this analysis is the sum of the error of each of the steps in the processing, instrumental analysis of plasma β-carotene including injection, and peak integration. The sources of variability in the random error term are likely due to the variability in organic phase volume that is collected from the two-step extraction procedure and the variability in evaporation of the redissolved sample prior to injection into the column. Although these effects are expected to be minor, to correct for these volume changes an internal standard can be added to the plasma samples prior to the extraction step. Several different internal standards have been used with varying degrees of success, including as α-tocopherol,[13] torulene and echinenone,[14] and β-apo-8-carotenal. Problems with internal standards have ranged from poor commercial availability and impurities to coelution with peaks of interest. β-Apo-8-carotenal has been used in our laboratory as an internal standard. In our experience, the percent coefficient of variation (%$C.V.$) of β-apo-8-carotenal over 10 separate extractions was 7.9, which demonstrates the variability in volume lost during the extraction and prior to the injection steps.

The precision of injection of a sample onto the column is itself determined by two factors: the amount of sample loaded into an injection loop and the injection loop capacity (Rheodyne Manufacturer's Manual, Cotati, CA). In this study, approximately 52 μl of sample was loaded into a 50-μl injection loop. Using this method of equal volume of sample to loop capacity, sample is both lost from the loop as well as diluted by residual mobile phase along the loop wall. Together these problems compound variability in the assay. Ideally the volume loaded into an injection loop should be greater than 3 times the loop capacity so that only pure sample is injected from the loop onto the column for analysis. When the volume loaded onto the loop is between one-half and twice the loop capacity, the peak area of the substance measured will be underestimated. To test this, an external β-carotene standard (3.0 μg/ml) was injected 6 times using each of two methods: either (1) exactly 50 μl of sample or (2) 200 μl of sample was loaded onto a 50-μl loop and the resulting integrals compared. The mean peak area for β-carotene appeared higher (403 ± 6.45 versus

TABLE II
ERROR RATES FOR MEAN β-CAROTENE[a] ACCORDING TO NUMBER OF
DAYS, ALIQUOTS, EXTRACTIONS, AND INJECTIONS

Number of days	Number of aliquots	Extractions per aliquot	Injections per extract	Percent error
				<30%
1	1	1	2-10	26.35-22.70
1	1	2	1-10	26.20-22.04
1	2	1	1-10	25.74-21.48
1	2	2	1-10	23.30-21.04
				<20%
1	11	10	8-10	20.00-20.00
1	12	8	8-10	20.00-20.00
1	12	9	5-7	20.00-19.99
2	1	1	2-10	19.66-17.24
2	1	2	1-10	19.46-16.68
2	1	3	1-10	18.41-16.49
2	2	1	1-10	18.83-15.94
2	2	2	1-10	17.07-15.52
2	2	3	1-10	16.43-15.38
				<10%
5	4	4	3-10	10.00-9.93
5	4	5	2-10	9.98-9.89
5	4	10	1-10	9.90-9.80

[a] β-Carotene was measured as millivolts-seconds.

374 ± 17.83) in the procedure where 4 times the loop volume was loaded. In addition, the %$C.V.$ decreased from 4.76 to 1.60 when 200 μl was loaded onto the loop instead of 50 μl. To reduce the variance associated with the manual injection step, we recommend that a volume 3 or more times the injection loop capacity should be loaded. Alternatively, the use of an autoinjector, if available, will greatly reduce variability.

The second greatest source of variability encountered in the procedure was day. Day-to-day variability is a consequence of the additive effects of differences in mobile phase preparation, temperature, variability in equipment performance, humidity, and column variations.[15]

To determine the optimal combination of days, aliquots, extractions, and injections required to minimize the error in determining plasma β-carotene, Table II lists these combinations by percent error associated with

[15] L. M. Canfield, J. M. Hopkinson, A. F. Lima, G. S. Martin, K. Sugimoto, J. Burr, C. Clark, and D. L. McGee, *Lipids* **25**, 406 (1990).

any one combination, using a method similar to that of McShane et al.[16] The percent error in our protocol measuring a plasma pool on each of 5 days with four aliquots per day, two separate extractions per aliquot, and two injections per extract gave a value of 10.37. The mean plasma β-carotene of the pool was 30.2 ± 2.3 μg/dl. This stringent protocol is not possible when analyzing plasma β-carotene from large numbers of subjects. Given the maximum level of variability which could be associated with each of the various components of β-carotene analysis, it was determined that an error greater than 30% could be associated with a β-carotene measurement if the sample was run only on 1 day from one aliquot with one extraction and one injection. To reduce this percent error to below 30%, at least duplicate injections from one extraction would need to be made. Interestingly, as can be seen from Table II, the common practice of increasing the number of injections per extract to decrease variability does not dramatically reduce the percent error even when as many as 10 injections are made.

Given the constraints in handling a large number of samples for β-carotene analysis, a determination of the mean β-carotene concentration with 95% confidence to within 20% of the true mean will suffice for most clinical studies. To achieve this level of variability, a minimum of 11 aliquots for one individual sample would need to be processed if the sample was run on only 1 day. However, if the sample is run on 2 days instead of 1, the error may be reduced to less than 20% using only one aliquot with one extraction and two injections. Increasing the number of aliquots, extractions, and injections beyond this combination decreases the associated error. However, the time and labor input required do not justify the small precision gained in this system. A combination of 2 days, one aliquot, one extraction, and one or two injections is the optimal combination for the example presented here. To reduce the error to below 10%, samples would need to be run on 5 days, with four aliquots, a minimum of four separate extractions per aliquot, and a minimum of three injections per extract. Greater precision may be obtained by measuring β-carotene in serum as compared with plasma, since serum is a less complex matrix (N. Craft, personal communication, 1991). However, plasma more nearly approximates the *in vivo* system and is preferred for some analyses.

For the quantitation of micronutrients in plasma where the organic matrix may interfere with analysis of micronutrients, authentic standards are typically added to plasma to produce a standard curve. This method corrects for the effect of the organic matrix on both the percent recovery of the nutrient during extraction and the instrumentation.

---

[16] L. M. McShane, L. C. Clark, G. F. Combs, and B. W. Turnbull, *Am. Stat. Assoc.*, in press (1992).

To determine the appropriate method of calibration, we prepared fresh standard curves on each of 5 days using both standards added to plasma and external neat standards. We found a 35% C.V. from day to day with standards added to plasma compared with a 7.5% C.V. for external neat standards. This was true whether authentic standards were added on the day of an assay to pooled plasma samples or when added to plasma samples immediately after they were drawn and then stored $-80°$. Thus exogenous β-carotene does not equilibrate with endogenous β-carotene in plasma. This suggests that β-carotene may be binding to specific plasma lipoproteins[17]; however, the cause of this anomaly is not known. Therefore, the use of an external calibration procedure using neat standards is recommended for quantitating β-carotene in plasma. We recognize that this method does not control for extraction efficiency. In previous reports, however, the recovery of β-carotene from plasma using this extraction procedure was shown to be essentially complete.[14]

In summary, using a variance components model, we have estimated the variance for each major source of variability in the assay of β-carotene from human plasma, including effects of day and numbers of aliquots, extractions, and injections. This method provides a mechanism to identify sources of variability in the assay of micronutrients so that procedures may be modified accordingly prior to undertaking the analysis of large numbers of biological samples. From such an analysis, a table of variables and their respective contributions to the overall error of measuring β-carotene in plasma (Table II) was generated. Using Table II, we determined that for the analysis of β-carotene in plasma using our procedure, a single sample, extracted on 2 separate days with duplicate injections, provided acceptable precision (<20% error) with a reasonable investment of time and labor. Where manual injections are made, a 3- to 4-fold ratio of injection volume to loop capacity is recommended to further reduce variability in the method. β-Apo-8-carotenal is useful in controlling for volume losses. However, as exogenous β-carotene could not be equilibrated with endogenous β-carotene, concentrations could not be quantitated by reference to authentic calibration standards added to plasma.

Acknowledgments

This work was supported in part by National Institutes of Health Contract NO1-6-2922. We wish to thank Martha Stephenson for expert technical assistance and Neal Craft (National Institute of Standards and Technology) for valuable suggestions in the analysis of plasma β-carotene, and many helpful comments on the manuscript.

---

[17] J. V. O'Fallen and B. P. Chew, *Proc. Soc. Exp. Biol. Med.* **177**, 406 (1984).

[11] Analysis of Carotenoids in Human and Animal Tissues

By HAROLD H. SCHMITZ, CHRISTOPHER L. POOR, ERIC T. GUGGER, and JOHN W. ERDMAN, JR.

Introduction

Interest in the carotenoid content and the biological importance of carotenoids in human tissues is increasing.[1,2] The analysis of carotenoids in nonhuman tissues is becoming more important as various species are being investigated as models of carotenoid absorption and metabolism by humans.[3-6] Accordingly, it is important to determine accurately the amounts and types of carotenoids present in various human and animal tissues.

Several laboratories have successfully separated and quantitated carotenoids in biological tissues.[1,7-10] In the past two decades, the majority of this work has employed high-performance liquid chromatography (HPLC) to affect the separation and quantitation of carotenoids. Indeed, reversed-phase HPLC is currently regarded as the method of choice for carotenoid analysis because of the speed and efficiency afforded when compared to either thin-layer or open-column chromatography.[11] The amount of information regarding the separation and identification of carotenoids in human tissues other than serum or plasma is, however, relatively small.[1]

---

[1] R. S. Parker, *J. Nutr.* **119**, 101 (1989).
[2] A. Bendich and J. A. Olson, *FASEB J.* **3**, 1927 (1989).
[3] S. S. Shapiro, D. J. Mott, and L. J. Machlin, *J. Nutr.* **114**, 1924 (1984).
[4] N. I. Krinsky, M. M. Mathews-Roth, S. Welankiwar, P. K. Sehgal, N. C. G. Lausen, and M. Russett, *J. Nutr.* **120**, 81 (1990).
[5] J. D. Ribaya-Mercado, S. C. Holmgren, J. G. Fox, and R. M. Russell, *J. Nutr.* **119**, 655 (1989).
[6] C. L. Poor, T. L. Bierer, N. R. Merchen, G. C. Fahey, Jr., M. R. Murphy, and J. W. Erdman, Jr., *J. Nutr.* **122**, 115 (1992).
[7] F. Khachik, G. R. Beecher, and W. R. Lusby, *J. Agric. Food Chem.* **37**, 1465 (1989).
[8] T. Matsuno, *in* "Carotenoids: Chemistry and Biology" (N. I. Krinsky, M. M. Mathews-Roth, and R. F. Taylor, eds.), p. 59. Plenum, New York, 1989.
[9] H. H. Schmitz, C. L. Poor, R. B. Wellman, and J. W. Erdman, Jr., *J. Nutr.* **121**, 1613 (1991).
[10] C. R. Broich, L. E. Gerber, and J. E. Erdman, Jr., *Lipids* **18**, 253 (1983).
[11] K. L. Simpson, S. C. S. Tsou, and C. O. Chichester, *in* "Methods of Vitamin Assay" (J. Augustin, B. P. Klein, D. A. Becker, and P. B. Venugopal, eds.), p. 202. Wiley, New York, 1985.

This chapter describes the extraction of carotenoids from human and animal tissues and subsequent analysis using HPLC. Carotenoids will be indicated by their commonly used trivial names throughout this chapter, accompanied at first mention by the semisystematic name recommended by the IUPAC-IUB Commission on Biochemical Nomenclature.[12]

General Considerations for Analysis of Carotenoids in Animal Tissues

There are several aspects of the extraction process which deserve special attention when analyzing human and animal tissues for carotenoids.

Physicochemical Characteristics. Carotenoids are thermally labile compounds.[13] Therefore, samples should not be exposed to excess heat during extraction and storage processes. For example, tissue samples should be kept on ice during the homogenization step because the shear forces involved in sample homogenization generate excess heat. Saponification procedures, when necessary, should be as mild as possible. The procedure used here calls for a saponification temperature of 60° for 30 min. Recovery studies conducted with β-carotene (β,β-carotene) under these conditions showed no significant loss at this temperature. Saponification at room temperature ($\sim 22-25°$) may also be used.[11] However, this method requires a significantly longer time increment (at least several hours) and may not yield complete hydrolysis of esterified carotenoids,[11] resulting in inaccurate quantitation if carotenoid esters are present in the tissue.

Saponification procedures should be used with caution during analysis of the more polar carotenoids. The loss of minor carotenoid peaks eluting prior to lutein has been observed following the use of certain saponification procedures.[14] A possible alternative to saponification is the use of lipase,[14] which eliminates the risk of carotenoid degradation and/or isomerization as a result of excessive heat treatment or exposure to the extremely basic environment required for saponification. In any case, standards should always be treated and analyzed under the same conditions as the sample to determine if any compounds are being lost as a result of sample treatment.

In addition to thermal degradation, carotenoids are also susceptible to degradation by light and oxygen.[11] Preferably, carotenoid extraction and analysis should be conducted under yellow or red lighting to minimize

---

[12] O. Isler, *in* "Carotenoids" (O. Isler ed.), p. 851. Birkhauser, Basel, 1971.
[13] R. F. Taylor and B. H. Davies, *J. Chromatogr.* **103**, 327 (1975).
[14] N. I. Krinsky, M. D. Russett, G. J. Handelman, and D. M. Snodderly, *J. Nutr.* **120**, 1654 (1990).

photodegradation and isomerization. If yellow lighting is not available, tissue samples should be protected from white light by using amber glass or by covering all containers with an opaque material such as aluminum foil, and work should be conducted under indirect lighting with sunlight excluded from the laboratory.

Degradation by oxygen can be minimized by limiting the time carotenoids are exposed to air and by utilizing inert gases such as nitrogen or argon during evaporation and storage of samples. Argon is preferable because it is both an inert gas and heavier than air, resulting in better sample protection at the gas–liquid interface. In addition, tissue samples should not be allowed to sit uncovered at room temperature for time periods which are longer than necessary. The addition of antioxidants, such as butylated hydroxytoluene (BHT), during extraction procedures is another important precaution against oxidation.

The choice of solvents for sample extraction, storage, and analysis can affect carotenoid degradation and/or isomerization of carotenoids. Tetrahydrofuran is readily oxidized in the presence of oxygen to yield highly reactive peroxides, and its use as both an organic extraction matrix and an HPLC solvent has been shown to promote significant loss of β-carotene during serum and plasma analysis.[15] Prolonged exposure of carotenoids to chlorinated solvents, such as dichloromethane and chloroform, should also be avoided. Trace amounts of hydrochloric acid are often present in these solvents,[16] which can degrade and/or isomerize carotenoids.[11,16,17]

Analytical Variables. The amounts and types of solvent used may also affect the efficiency of carotenoid extraction from human and animal tissues. The solvent used for extraction of tissue carotenoids in this study was hexane. An extraction using dichloromethane and ethyl acetate in sequence was also performed to determine if better yields could be obtained. Hexane alone resulted in a greater yield of carotenoids from human liver tissue (see Table I). Alternatively, Barua *et al.*[18] demonstrated that extraction of serum by ethyl acetate followed by hexane provides a greater yield than hexane alone for nonpolar carotenoids, suggesting that different solvent systems may be required to obtain optimum carotenoid yields depending on the tissue type and class of carotenoid.

The effects of saponification on carotenoid yield were also examined. Saponified human liver samples yielded a greater quantity of carotenoids

---

[15] D. W. Nierenberg, *J. Chromatogr.* **339**, 273 (1985).
[16] F. Khachik, G. R. Beecher, J. T. Vanderslice, and G. Furrow, *Anal. Chem.* **60**, 807 (1988).
[17] S. Liaaen-Jensen, *in* "Carotenoids: Chemistry and Biology" (N. I. Krinsky, M. M. Mathews-Roth, and R. F. Taylor, eds.), p. 149. Plenum, New York, 1989.
[18] A. B. Barua, R. O. Batres, H. C. Furr, and J. A. Olson, *J. Micronutr. Anal.* **5**, 291 (1989).

TABLE I
EFFECT OF EXTRACTION METHOD AND SAPONIFICATION ON
CAROTENOID RECOVERY FROM HUMAN LIVER

Subject	Carotenoids recovered (nmol extracted/g of liver tissue)		
	Hexane[a] following saponification	Hexane[b] without saponification	Dichloromethane[b,c] and ethyl acetate without saponification
1	7.9	10.2	5.2
2	36.7	26.1	13.1
3	2.5	2.0	NA[d]
4	16.1	11.9	5.0
5	8.6	8.6	NA[d]
6	24.0	14.1	6.6
7	13.4	8.9	9.0

[a] Tissue saponified with 10% NaOH and extracted 3 times with hexane.
[b] Tissue ground with Na_2SO_4 prior to extraction.
[c] Tissue extracted with dichloromethane followed by ethyl acetate. Extraction procedure repeated twice.
[d] Tissues not analyzed.

on extraction with hexane than did unsaponified tissue samples (see Table I). Saponification removes excess noncarotenoid lipids from the sample, which can confound chromatographic analysis, and deesterifies lipid-bound carotenoids. It is not known whether esterified carotenoids are present in mammalian tissues, although their presence has been well documented in several plant tissues.[19] It should be clear that the degree of esterification of carotenoids, if any, cannot be determined after saponification.

Tissue Variability. Carotenoid concentrations may differ within a single tissue. Variation in vitamin A concentrations with respect to location in liver tissue in both humans and animals has been reported.[20] Effort should therefore be made to obtain samples from similar locations within the tissue whenever possible.

Ease of homogenization varies according to tissue type. Liver and spleen tissues are readily homogenized, whereas lung and muscle tissues are more difficult due to the relatively high concentrations of insoluble

[19] F. Khachik and G. R. Beecher, *J. Agric. Food Chem.* **36**, 938 (1988).
[20] B. A. Underwood, *in* "Retinoids" (M. B. Sporn, A. B. Roberts, and D. S. Goodman, eds.), Vol. 1, p. 327. Academic Press, Orlando, Florida, 1984.

structural proteins present. Adipose tissue also presents difficulties due to its hydrophobic nature which prevents rapid dispersion into the hydrophilic phase used during homogenization. Therefore, procedures should be optimized to yield complete homogenization of the tissue, while minimizing exposure of the sample to excess heat and oxygen as a result of the homogenization procedure.

Another consideration in tissue saponification is the amount of esterified lipid present in the tissue sample and the corresponding amount of base which should be added to hydrolyze all of the ester bonds present. Adipose tissue, for example, will require more base than liver or lung tissue.

Chromatographic Analysis. Identification and quantitation of carotenoids are routinely accomplished for many types of plant tissues using HPLC techniques, and careful examination of these methods is helpful in developing approaches to analyze human and animal tissues for carotenoids. Knowledge concerning the amount and type (e.g., polar or nonpolar) of carotenoids likely to be present in the tissue of interest, along with an understanding of general HPLC protocol and the separation capabilities of different columns, is critical for the rapid development of satisfactory HPLC procedures.

The occurrence of peak splitting and false peaks may result from inappropriate injection procedures used during HPLC analysis. The use of dichloromethane and chloroform as injection solvents has been shown to distort peaks, promote chromatographic artifacts,[16] and decrease peak resolution.[14] Similar effects were observed in this study when samples reconstituted in dichloromethane were injected in volumes exceeding 35–40 μl. Injection volumes not exceeding 20–25 μl did not result in peak splitting or the presence of false peaks. Dichloromethane and chloroform are useful for the reconstitution of carotenoid samples because these solvents ensure complete solubilization of the compounds. One way to gain the benefits of the high solubility of carotenoids in these compounds while avoiding the potential problems which they may cause is to first solubilize the sample in a small quantity of dichloromethane or chloroform, then bring the sample to volume in the HPLC mobile phase.

Identification of carotenoids present in a given sample should also be considered here. It is generally accepted that a minimum of corresponding retention times with authentic standards and spectral analysis in agreement with corresponding standard spectra are required for the identification of a carotenoid using HPLC.[21] This is emphasized by Barua *et al.*[18] in their

[21] R. F. Taylor, P. W. Farrow, L. M. Yelle, J. C. Harris, and I. G. Marenchic, *in* "Carotenoids: Chemistry and Biology" (N. I. Krinsky, M. M. Mathews-Roth, and R. F. Taylor, eds.), p. 105. Plenum, New York, 1989.

reversed-phase HPLC analysis of human serum, which yielded a peak in the more polar range of the chromatogram, possibly indicating the presence of a xanthophyll. Spectral analysis revealed that the peak was bilirubin. Resolution of the more polar carotenoid region of a typical reversed-phase chromatogram from human blood and solid tissues has proven to be quite difficult. This region has been further resolved in human serum samples using a nitrile-bonded HPLC column, indicating that this fraction may contain 12 or more carotenoids and their isomers.[22] The probability that these carotenoids are present should therefore be considered when presenting values for lutein ([3R,3'R]-β,β-carotene-3,3' diol) in tissues. In addition, the presence of phytoene (15-cis-7,8,11,12,7',8',11', 12'-octahydro-ψ,ψ-carotene) and phytofluene (15-cis-7,8,11,12,7',8'-hexahydro-ψ,ψ-carotene), less polar carotenoids which do not exhibit strong absorbance in the visible spectrum, have also been shown to be present in human serum.[23] These carotenoids may not be detected unless the column eluent is monitored at an appropriate wavelength in the ultraviolet spectrum.

Experimental Design

Human Tissues. Human liver, kidney, and lung tissues obtained at autopsy are analyzed for five different carotenoids, namely, α-carotene ([6'R]-β,ϵ carotene), β-carotene, lycopene (ψ,ψ-carotene), cryptoxanthin ([3R]-β,β-carotene-3-ol), and lutein.[9] Autopsy is performed within 24 hr of death, and tissue samples are stored at $-20°$ prior to extraction.

Calf Tissues. Tissues are collected from 2- to 4-week-old preruminant Holstein bull calves 24 hr after receiving a single oral dose (20 mg) of β-carotene dissolved in milk replacer.[6] The calves are sacrificed by stunning with a captive bolt followed by exsanguination. Liver, kidney, adrenal, lung, spleen, heart, adipose (perirenal white fat), and muscle (left biceps femoris) are collected and frozen ($-20°$) immediately.

Ferret Tissues. Tissues are collected from young (12- to 15-week-old) male ferrets (Marshall Farms, North Rose, NY) and analyzed for carotenoid content.[24] The ferrets are fed a cat chow diet which was shown by HPLC to contain significant amounts of β-carotene and two unidentified polar carotenoids. In addition, the ferrets are dosed with β-carotene (10 mg/kg body weight) by dissolving water-soluble β-carotene beadlets

[22] R. S. Parker, *Am. J. Clin. Nutr.* **47**, 33 (1988).
[23] F. Khachik, G. R. Beecher, M. B. Goli, W. R. Lusby, and J. C. Smith, *Anal. Chem.* **64**, 2111 (1992).
[24] E. T. Gugger, T. L. Bierer, T. M. Henze, W. S. White, and J. W. Erdman, Jr., *J. Nutr.* **122**, 115 (1992).

(Hoffmann-LaRoche, Nutley, NJ) in 0.5 ml of distilled water, then mixing the solution with 6 g of moistened diet. This meal is fed to the animals, followed by the normal morning meal. Animals are sacrificed 16 hr after the dose by anesthesia with i.m. ketamine followed by heart puncture. Liver, kidney, lung, spleen, adipose (perirenal), and muscle (left biceps femoris) tissue samples are collected and promptly frozen at $-20°$ prior to analysis.

Experimental Procedures

Tissue Extraction. Duplicate tissue samples are weighed (0.5–1.0 g of liver and adrenal tissue and 1.0–5.0 g of other tissue types) and transferred to a 50-ml centrifuge tube on ice. Tissues are suspended in 5–10 ml of either an ethanol–water mixture (50:50, v/v) or 95% (v/v) ethanol, each with 0.1% (w/v) BHT. Tissues are then held in a beaker of ice and homogenized for 45–90 sec using a Polytron (Brinkmann Instruments, Westbury, NY) homogenizer.

Subsequently, 5–10 ml of either a 10% (w/v) NaOH in ethanol solution or 1–2 ml of a saturated aqueous KOH solution is added to the samples, and the samples are vortexed and saponified in a 60° water bath for 30 min. Samples are immediately placed on ice, and a volume of distilled water equal to the volume of base is added. All samples are then extracted 3 times with 10 ml of hexane. The extracts are evaporated to dryness under argon, and the sealed sample vial is stored at $-20°$. HPLC analysis of each sample is completed within 48 hr of extraction. Inaccurate quantitation is observed in analyses when plastic tubes are used repeatedly during extraction procedures, necessitating the use of glassware.

Chromatography Equipment. The HPLC system for human tissue analysis consists of a Tracor 950 chromatographic pump and 970A variable wavelength detector (Austin, TX), a Rheodyne 7125 sample injector (Cotati, CA), and a Shimadzu (Kyoto, Japan) C-R3A Chromatopac integrator. In addition, a photodiode array detector (PDAD) system consisting of an HP 1040A High Speed Spectrophotometric Detector controlled by an HP 85 Personal Computer operating in conjunction with an HP 9121 disk drive (all from Hewlett-Packard, Corvallis, OR) is used to obtain peak spectra. A Milton Roy Constametric III (Riviera Beach, FL) pump and Bio-Rad Model 1790 UV/VIS monitor (Richmond, CA), in combination with the columns and solvents below, are used for calf and ferret tissue analysis.

Two columns, a monomeric Supelco LC-18 (Bellefonte, PA) HPLC column [25 cm × 4.6 mm inner diameter (ID)] packed with 5 μm ODS and a polymeric Vydac TP201 (Hesparia, CA) HPLC column (15 cm ×

4.6 mm ID) packed with 5 μm C_{18} bonded to silica, are used separately in conjunction with the above systems. The two columns are protected by a precolumn (Upchurch Scientific, Oak Harbor, WA) packed with 30–38 μm ODS C_{18}. Injections are performed using a 50-μl Hamilton gastight syringe (Hamilton Co., Reno, NV), which is rinsed with chloroform between injections.

Standards. Crystalline α-carotene, β-carotene, lutein, and lycopene are purchased from Sigma Chemical Company (St. Louis, MO). Crystalline cryptoxanthin was a gift from Hoffmann-LaRoche, Inc.

Chromatography Procedures. Carotenoids are chromatographed on the Supelco column using a solvent system consisting of methanol–acetonitrile–chloroform (47:47:6, v/v) at a flow rate of 2 ml/min.[10] A solvent system consisting of methanol–acetonitrile–water (88:9:3, v/v) at a flow rate of either 1 or 2 ml/min is developed to separate carotenoids using the Vydac column.[6,9,24] Both columns are flushed periodically with a more nonpolar mobile phase consisting of dichloromethane–acetonitrile (75:25, v/v), especially when there is an apparent loss of resolution (presumably due to the binding of extremely nonpolar compounds to the stationary phase). All solvents are filtered (0.2 μm) and degassed by sonication under reduced pressure for 15 sec, and are of spectrograde quality.

Carotenoid standards and tissue extracts are dissolved in dichloromethane immediately prior to injection. Primary standards are stored in stock solutions of hexane. Identification of carotenoids is accomplished by comparison with appropriate standard retention times and comparison of spectra obtained by the PDAD with spectra of the appropriate carotenoid standards.

Carotenoids are quantified by comparing appropriate peak area counts to corresponding standard curves for each carotenoid of interest. The purity of standards is confirmed daily by HPLC analysis, and carotenoid standards are quantified daily using a spectrophotometer in conjunction with appropriate extinction coefficients.[11] Standard curves are adjusted, if necessary, following daily calibration of the HPLC system. HPLC analysis is done in duplicate for each extracted sample.

Results and Discussion

Qualitative Results. Chromatograms obtained from various human and animal tissues using the Vydac column are given in Figs. 1–3. For human liver, kidney, and lung the major peaks identified by retention time and peak spectra in order of elution were lutein, cryptoxanthin, α-carotene, β-carotene, and lycopene. In addition to these major carotenoids, several smaller peaks representing both polar and nonpolar carotenoids

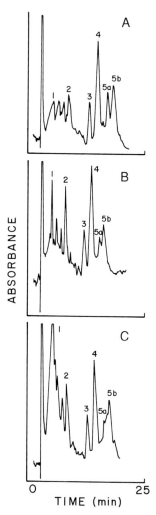

FIG. 1. Chromatographic separation of carotenoids present in human liver (A), kidney (B), and lung tissues (C) using the Vydac column. Peak 1 is lutein, peak 2 is cryptoxanthin, peak 3 is α-carotene, peak 4 is β-carotene, and peaks 5a and 5b are lycopene and its cis isomer.

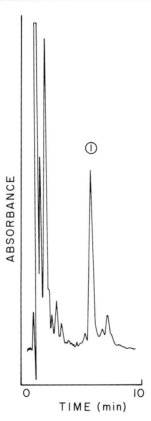

FIG. 2. Separation of carotenoids present in ferret liver using a Vydac column. Peak 1 is β-carotene.

were present in many samples. The multiple designation for lycopene represents all-*trans*-lycopene and at least one cis isomer of lycopene, which was confirmed by spectral analysis yielding increased absorption in the 330–340 nm range. This elution pattern is representative of reversed-phase carotenoid analysis where more polar carotenoids elute before less polar carotenoids.

In general, polymeric columns have been shown to provide improved resolution of certain structurally similar carotenoids when compared to monomeric columns.[25] The chromatographic results of the current work support this observation, shown by the superior resolution of α- and

[25] C. S. Epler, N. Craft, L. C. Sander, S. A. Wise, and W. E. May, "Report of Analysis #552-90-058," U.S. Dept. of Commerce, National Institute of Standards, Gaithersburg, Maryland, 1990.

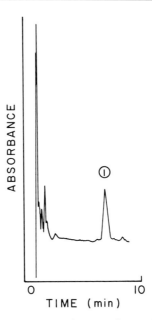

FIG. 3. Separation of carotenoids present in preruminant calf liver using a Vydac column. Peak 1 is β-carotene.

β-carotene and the separation of lycopene isomers by the polymeric column (Figs. 1 and 4). The polymeric column retained some of the more polar carotenoids, such as lutein, longer than the monomeric column, resulting in greater separation of these carotenoids from the solvent front and allowing for more accurate quantitation.

The carotenoid profile obtained from a given tissue is dependent on a number of factors, including the extraction and HPLC procedures, the ability of different species to absorb different carotenoids, and the amount and type of carotenoids which are present in a given diet. The carotenoid profiles shown in Figs. 1 and 4 are typical for human tissues, including serum and plasma,[1,15,26] although some tissues from other subjects contained carotenoid profiles that were somewhat different from the normal profiles given here. Carotenoid profiles did not differ significantly between tissues from the same subject.

The Vydac TP201 column and isocratic solvent system described here adequately resolved the major carotenoids of interest present in the human

[26] L. A. Kaplan, J. M. Lau, and E. A. Stein, *Clin. Physiol. Biochem.* **8**, 1 (1990).

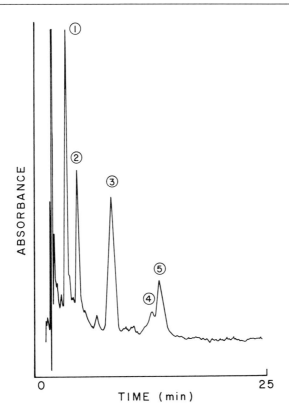

FIG. 4. Separation of carotenoids present in human liver tissue using the Supelco column. Peak 1 is lutein, peak 2 is cryptoxanthin, peak 3 is lycopene, peak 4 is α-carotene, and peak 5 is β-carotene.

and animal tissues analyzed. This, coupled with the simplicity and shorter overall analysis time of isocratic HPLC systems when compared to gradient systems, demonstrates the utility of this method for analysis of carotenoids in these tissues.

Quantitative Results. Quantitatively, the major carotenoids found in human tissues were α- and β-carotene, cryptoxanthin, lutein, and lycopene, with β-carotene and lycopene generally predominating. Ferret tissues contained the dosed β-carotene and at least two more polar carotenoids, probably as a result of corn-based feeds. Preruminant calf tissues contained β-carotene exclusively as a result of their controlled, low carotene diet.

Of the human tissues studied, liver always contained the highest carotenoid concentration, ranging from 10- to 30-fold higher than either lung or kidney tissue from the same patient. Previous research has indicated that other human tissues, including adipose, adrenal, and reproductive tissues,[1,24,22] often contain carotenoid concentrations which are equivalent to or exceed that of liver tissue. The mean carotenoid concentrations for human liver, kidney, and lung tissues are given in Table II. Total carotenoid concentration (representing the sum of the five carotenoids quantified in this study) for adult liver tissue varied approximately 30-fold between individuals (2.5–77.1 nmol/g tissue), 60-fold in kidney tissue (0.2–12.7 nmol/g tissue), and 85-fold in lung tissue (0.1–8.4 nmol/g tissue).

Adrenal tissue contained the highest concentration of carotenoids in preruminant calf tissue, followed closely by liver (Table III). Lung, kidney, and spleen also displayed elevations in β-carotene concentration following dosing. Calves displayed a wide range of variation in β-carotene concentrations among individuals. Tissues requiring more extensive saponification, such as adipose tissue, displayed a standard error of the mean which was large relative to the β-carotene value, indicating the difficulties experienced in extracting carotenoids from fatty tissues.

Ferret liver contained the highest concentration of β-carotene (Table IV). Kidney, spleen, and lung contained 10- to 30-fold lower levels of β-carotene when compared to liver, whereas muscle and adipose β-carotene concentrations were 300-fold lower.

TABLE II
INDIVIDUAL CAROTENOID AND VITAMIN A CONCENTRATIONS OF HUMAN TISSUES

Tissue	Concentration (nmol/g tissue)[a,b]						
	BC	AC	CX	LYC	LUT	TC	VA
Liver	4.4 ± 5.0	1.8 ± 2.6	2.0 ± 4.6	5.7 ± 6.4	3.2 ± 3.6	17.1 ± 19.1	294.8 ± 284.3
Range	0–19.4	0–10.8	0–20.0	0–17.2	0–12.2	2.5–77.1	8.7–1102.2
($n = 18$)							
Kidney	0.6 ± 0.7	0.3 ± 0.4	0.4 ± 1.1	0.6 ± 0.6	1.2 ± 2.8	3.1 ± 4.2	52.4 ± 90.0
Range	0.1–2.8	0–1.5	0–3.9	0.1–2.4	0–10.4	0.2–12.7	3.5–343.9
($n = 13$)							
Lung	0.4 ± 0.5	0.2 ± 0.3	0.4 ± 0.8	0.6 ± 1.1	0.5 ± 0.7	2.0 ± 2.8	51.4 ± 106.8
Range	0–1.6	0–1.0	0–2.5	0–4.2	0–2.3	0.1–8.4	0.7–404.6
($n = 13$)							

[a] Mean ± standard deviation.
[b] BC, β-Carotene; AC, α-carotene; CX, cryptoxanthin; LYC, lycopene; LUT, lutein; TC, total carotenoids; VA, preformed vitamin A.

TABLE III
β-CAROTENE CONCENTRATIONS IN TISSUES OF
PRERUMINANT CALVES[a]

Tissue	β-Carotene concentration (nmol/g tissue)[b]
Adrenal	0.48 ± 0.11
Liver	0.45 ± 0.12
Spleen	0.24 ± 0.03
Kidney	0.11 ± 0.02
Lung	0.11 ± 0.01
Heart	0.09 ± 0.05
Adipose	0.07 ± 0.05
Muscle	0.03 ± 0.01

[a] Twenty-four hours after a single oral dose of β-carotene (20 mg).
[b] Means ± S.E.M. ($n = 3$).

Conclusions

Accurate qualitative and quantitative analysis of carotenoids present in human and animal tissues is dependent on careful selection of extraction and analytical procedures. Strict adherence to laboratory protocol which minimizes degradation and/or isomerization of carotenoids during both

TABLE IV
FERRET TISSUE β-CAROTENE CONCENTRATIONS[a]

Tissue	β-Carotene concentration (nmol/g tissue)[b]
Liver	0.95 ± 0.35
Kidney	0.09 ± 0.01
Spleen	0.08 ± 0.01
Lung	0.04 ± 0.01
Adipose	0.02 ± 0.02
Muscle	0.01 ± 0.01

[a] Sixteen hours after receiving a single oral dose of β-carotene (10 mg/kg body weight).
[b] Means ± S.E.M. ($n = 3$).

extraction and analysis procedures should be observed to ensure accurate results. The procedures and modifications presented in this chapter provide a method by which the major carotenoids present in human and animal tissues may be accurately identified and quantified.

[12] Plasma Carotenoid Levels in Anorexia Nervosa and in Obese Patients

By CHERYL L. ROCK and MARIAN E. SWENDSEID

Introduction

Approximately 30–60% of patients with anorexia nervosa are hypercarotenemic, as described first by Dally in 1959[1] and later by others,[2] but the cause is unknown. Patients usually deny excessive intake of carotenoid-rich foods, and hypercarotenemia is not associated with other medical conditions characterized by cachexia.[3]

Several theories have been proposed to explain increased plasma carotenoids in these patients: an acquired error in metabolism, enhanced absorption, increased dietary intake, increased release from fat stores, or reduced tissue storage capacity. Reduced clearance and degradation of low-density lipoprotein (LDL), a major carrier of carotenoids in the plasma, has also been hypothesized to play a role. Elevated blood cholesterol levels, associated with increased LDL, are occasionally found in patients with anorexia nervosa.[4]

The usual clinical laboratory methods used to diagnose hypercarotenemia in anorexia patients have relied on a simple spectrophotometric assay that does not permit determination of individual carotenoid levels.[1,2] Also, the role of dietary carotenoid intake in these patients has been disputed because of inconsistent evidence.[5,6]

We measured plasma levels of individual carotenoids and assessed carotenoid intake in patients with anorexia nervosa. These data were

[1] P. J. Dally, *Br. Med. J.* **1**, 1333 (1959).
[2] M. A. Pops and A. D. Schwabe, *J. Am. Med. Assoc.* **205**, 533 (1968).
[3] M. S. Robboy, A. S. Sato, and A. D. Schwabe, *Am. J. Clin. Nutr.* **27**, 362 (1974).
[4] R. C. Casper, *Annu. Rev. Nutr.* **6**, 299 (1986).
[5] E. Kemman, S. A. Pasquale, and R. Skaf, *J. Am. Med. Assoc.* **249**, 926 (1983).
[6] J. Curran-Celentano, J. W. Erdman, R. A. Nelson, and S. J. Grater, *Am. J. Clin. Nutr.* **42**, 1183 (1985).

compared with data from normal subjects and from obese patients who had recently lost weight because of reduced-calorie diets of natural foods and exercise. The plasma response to a single oral dose of purified β-carotene was measured to ascertain if a difference in absorption of carotenes might exist.

Procedures

Subjects. Three types of subjects between 18 and 60 years of age were recruited from a university medical center community: patients with anorexia nervosa, obese patients who had recently lost weight, and normal subjects. Patients with anorexia nervosa had body weights less than 85% of desirable weight and met other DSM-III-R criteria for this diagnosis.[7] These patients were clinically stable and had been recently admitted to inpatient or outpatient units to begin refeeding protocols.

Obese patients had attained a body weight greater than 120% of desirable weight but had lost at least 15% of original body weight within the preceding 6 months by consuming a low-fat, reduced-calorie diet and exercising regularly. Normal subjects were within 10% of ideal weight and had maintained a stable body weight during the preceding year. Exclusion criteria included use of carotene supplements or medications known to affect nutrient metabolism or gastrointestinal function, presence of comorbid disease, and active bulimic or purging behavior.

Methods. Subjects complete a food-frequency questionnaire designed to estimate food consumption during the preceding year[8] and are instructed to continue their usual eating and exercise patterns throughout the study period. None of the subjects were smokers at the time of the study. All reported that they consumed alcohol rarely or occasionally and that they had sun exposure rarely or occasionally. Height, weight, midarm circumference, and triceps skinfold measurements are obtained.

A fasting blood sample is collected between 0800 and 1000 hours, and subjects then ingest 50 mg purified β-carotene in two capsules (Hoffmann-La Roche, Nutley, NJ) under observation. The capsules contain dry gelatin beadlets compounded with butylated hydroxytoluene (BHT), butylated hydroxyanisole (BHA), and sodium benzoate preservatives. A second blood sample is obtained 30 hr later.

Approximately 35 ml of blood is collected by venipuncture (using heparinized tubes), and plasma and erythrocytes are immediately sepa-

---

[7] American Psychiatric Association, "Diagnostic and Statistical Manual of Mental Disorders." American Psychiatric Association, Washington, D.C., 1987.
[8] W. C. Willett, R. D. Reynolds, S. C. Hoehner, L. Sampson, and M. L. Browne, *J. Am. Diet. Assoc.* **87**, 43 (1987).

rated by centrifugation. Exposure to light is minimized, and samples are stored at −70°.

Plasma Carotenoid Measurements. Plasma carotenoids are determined by the method of Bieri et al.[9] This method permits separation and quantification of four distinct carotenoids plus an additional peak containing both zeaxanthin and lutein. Plasma samples with β-carotene of known concentration from the U.S. Department of Commerce National Institute of Standards and Technology (formerly the National Bureau of Standards) are analyzed with each assay, and day-to-day coefficients of variation are generally less than 10%.

Crystalline α-carotene (Sigma, St. Louis, MO), β-carotene (Sigma), lycopene (Sigma), zeaxanthin (Hoffmann-La Roche), cryptoxanthin (Hoffmann-La Roche), and echinenone (Hoffmann-La Roche) are used as standards. Crystalline standards are weighed and dissolved in hexane (β-carotene, α-carotene, echinenone) or chloroform (lycopene, zeaxanthin, cryptoxanthin) for stock solutions. Working standards are prepared by diluting 0.500 ml of stock standard in 25 ml of 100% ethanol, and concentrations are determined with molar extinction coefficients and measurements of absorption at λ_{max}. HPLC-grade methanol, ethanol, acetonitrile, chloroform, and hexane (Fisher, Pittsburgh, PA) are used. Optima-grade dichloromethane (Fisher) is used.

One milliliter echinenone in ethanol is added as an internal standard to each 1 ml of plasma sample, and the tube is vortexed for 10 sec at medium setting. One milliliter hexane is then added and the sample vortexed for 45 sec, followed by centrifugation for 10 min at full speed on an IEC (Needham Heights, MA) tabletop centrifuge. The clear hexane layer is removed, transferred to a vial, and evaporated under a stream of nitrogen gas. Another 1 ml of hexane is added to the plasma sample and the process repeated to ensure maximal lipid extraction.

Samples are reconstituted with methanol–dichloromethane (3:1, v/v). The HPLC system used is a Varian LC5000 (Sunnyvale, CA) with a DS600 Data Station and a Varichrom UV/VIS Variable Wavelength Detector (set at 464 nm with a bandpass filter of 8 nm). The isocratic mobile phase consists of acetonitrile–methanol–dichloromethane (70:10:20, v/v), with a flow rate of 1.7 ml/min. The column used is a Supelco (Bellefonte, PA) Supelcosil LC-18 (25 cm × 4.6 mm, 5 µm).

With this method, the peak designated zeaxanthin is assumed to also contain the isomeric dihydroxycarotenoid lutein. Values for this peak are calculated as zeaxanthin and are comparable to zeaxanthin/lutein data

[9] J. G. Bieri, E. D. Brown, and J. C. Smith, *J. Liq. Chromatogr.* **8**, 473 (1985).

TABLE I
DESCRIPTIVE CHARACTERISTICS OF SUBJECTS[a]

Characteristic	Normal subjects ($n = 18$)	Anorexia nervosa patients ($n = 9$)	Obese patients ($n = 12$)
Gender (female:male)	15:3	8:1	7:5
Age (year)[b]	31 ± 7	26 ± 5	46 ± 13
Height (cm)	163 ± 8	163 ± 5	171 ± 9
Weight (kg)[c]	60.8 ± 11.3	42.4 ± 4.5	91.4 ± 23.1
Midarm circumference (cm)[c]	27.2 ± 2.4	19.5 ± 1.9	34.6 ± 5.3
Triceps skinfold (mm)[c]	15.8 ± 5.6	5.7 ± 3.2	20.5 ± 11.5
Body mass index (kg/m^2)[c]	22.8 ± 2.4	15.9 ± 1.4	31.2 ± 5.6
Total plasma cholesterol (mM)	4.42 ± 0.65	3.88 ± 0.96	5.15 ± 0.85

[a] Mean ± S.D.
[b] Significantly higher in obese patients compared to either anorexia nervosa patients or normal subjects ($p < 0.001$).
[c] Significantly lower in anorexia nervosa patients compared to either obese patients or normal subjects ($p < 0.001$).

reported by other investigators applying this method.[9-11] Response factors for analytes are generated based on peak-area ratios versus peak-concentration ratios, with echinenone used as internal standard.

Plasma Cholesterol Determinations. Total plasma cholesterol in the fasting blood collection is analyzed in duplicate by an enzymatic method (Sigma) and with an automated Beckman Analyzer (Palo Alto, CA).

Statistical Analysis. The data are evaluated by analysis of variance, and Student's t test is performed when data from two groups are compared. Linear regression and correlation computations are also performed. Differences associated with p values below 0.05 are regarded as statistically significant.

Results

Characteristics of subjects in the various subgroups are presented in Table I. Gender distribution was not significantly different, but the obese

[10] E. D. Brown, M. S. Micozzi, N. E. Craft, J. G. Bieri, G. Beecher, E. K. Edwards, A. Rose, P. R. Taylor, and J. C. Smith, *Am. J. Clin. Nutr.* **49**, 1258 (1989).
[11] N. E. Craft, E. D. Brown, and J. C. Smith, *Clin. Chem.* **34**, 44 (1988).

group was significantly older. Body mass index *(BMI)* is calculated by the following formula:

$$BMI = \text{weight (kg)/height (m)}^2$$

Weight, midarm circumference, triceps skinfold, and *BMI* were significantly lower in the anorexia nervosa group.

Analysis of dietary intake data from the food-frequency questionnaire revealed no statistically significant differences. Normal subjects averaged $17,840 \pm 21,090$ (mean $\pm$ S.D.), anorexia nervosa patients averaged $17,048 \pm 18,367$, and obese patients averaged $10,365 \pm 4977$ IU/day provitamin A carotene within the preceding year. With this questionnaire, provitamin A carotene is the term used to describe the calculated figure because it is based on early analytical methods and the presumed vitamin A activity of hydrocarbon carotenoids. The values are described as representing most of the β-carotene, half of the α-carotene, and a small fraction of the lycopene in the diet.[8] A trend toward fewer calories in the anorexia nervosa group was evident, with that group consuming 1426 ± 534 kcal/day (mean $\pm$ SD), compared with 1708 ± 511 among normal subjects and 1826 ± 540 among obese patients.

Mean plasma β-carotene was significantly higher in the anorexia nervosa group compared to either the normal or obese patient groups (Table II). A trend toward higher levels of α-carotene and zeaxanthin/lutein in the anorexia nervosa group was observed, but the differences were not statistically significant. Comparison of data for females exclusive of males yields similar results (data not shown).

TABLE II
MEAN PLASMA CAROTENOID LEVELS[a]

Plasma carotenoid	Carotenoid levels (μM)		
	Normal subjects ($n = 18$)	Anorexia nervosa patients ($n = 9$)	Obese patients ($n = 12$)
β-Carotene	0.471 ± 0.319	0.889 ± 0.401[b]	0.382 ± 0.257
α-Carotene	0.149 ± 0.166	0.278 ± 0.244	0.099 ± 0.065
Zeaxanthin/lutein	0.429 ± 0.214	0.916 ± 0.606	0.529 ± 0.376
Cryptoxanthin	0.304 ± 0.230	0.440 ± 0.331	0.327 ± 0.210
Lycopene	1.032 ± 0.358	0.812 ± 0.397	1.017 ± 0.462

[a] Mean $\pm$ S.D.
[b] Significantly different than in normal subjects and obese patients ($p < 0.01$).

TABLE III
MEAN PLASMA β-CAROTENE LEVELS BEFORE AND AFTER INGESTING TEST DOSE OF β-CAROTENE[a]

Measurement	β-Carotene level (mM)		
	Normal subjects ($n = 18$)	Anorexia nervosa patients ($n = 9$)	Obese patients ($n = 12$)
β-Carotene			
0 hr	0.471 ± 0.319	0.889 ± 0.401[b]	0.382 ± 0.257
30 hr	0.650 ± 0.369	1.224 ± 0.445[b]	0.602 ± 0.486
Difference in β-carotene following test dose (mean % increase ± SD)[c]	0.183 ± 0.212 (58 ± 68)	0.335 ± 0.340 (50 ± 67)	0.220 ± 0.363 (52 ± 57)

[a] Values represent mean ± S.D. plasma β-carotene levels, at 0 hr (initial) and 30 hr after ingesting a single dose of purified β-carotene.
[b] Significantly different than in normal subjects and obese patients ($p < 0.01$).
[c] Numbers in parentheses indicate percent increase, based on the initial level.

Pre- and postdosage levels of plasma β-carotene, and the percent increase in response to a single oral dose of purified β-carotene supplement, are presented in Table III. The mean change in plasma concentration of β-carotene was greater in the anorexia nervosa patients than in the normal subjects and obese patients. However, because considerable variation in plasma response was observed in all groups, the difference was not statistically significant. Mean percent increase in plasma levels at 30 hr after the oral dose of purified β-carotene was similar among patients with anorexia nervosa, normal subjects, and obese patients.

Plasma β-carotene levels did not significantly correlate with age or triceps skinfold. However, current BMI in the total group, over the wide range represented by these subjects, was negatively correlated with plasma β-carotene ($r = -0.59$, $p < 0.0001$). The plasma β-carotene level was also inversely related to the total plasma cholesterol level ($r = -0.56$, $p < 0.05$). Estimated dietary intake of provitamin A carotene and other dietary components of interest (i.e., zinc) did not correlate with levels of plasma carotenoids or with response to the test dose of purified β-carotene.

Comments

Data from this study indicate that patients with anorexia nervosa have elevated levels of plasma β-carotene, which confirms the results reported by others.[1-4] Although mean plasma levels of α-carotene and zeaxanthin/lutein were higher among the anorexia nervosa patients, the differences were not significantly different from those of the normal subjects and obese patients.

We found that dietary intake of provitamin A carotene was not different in anorexia nervosa patients compared with normal subjects and obese patients. The means and ranges of intake in this study were similar to data reported from other studies in which this questionnaire was used to estimate intake.[8] These data indicate that differences in dietary intake of carotenoid-rich foods do not explain the elevated plasma levels which are characteristically observed among anorexia nervosa patients.

To compare absorption among these groups, we used an approach previously reported by Brown et al.[10]: the plasma response to an oral dose of β-carotene. In normal subjects, maximum plasma β-carotene concentrations have been found to occur 24–48 hr following a single oral dose.[10] Data from the present study suggest that the immediate plasma response of anorexia nervosa patients is not significantly different from that observed in normal subjects and obese patients, at 30 hr following a single test dose. There was a wide variation in plasma response within each group, however.

It has been hypothesized that weight loss, especially loss of adipose tissue, might promote elevated plasma carotenoid levels in anorexia nervosa. We found that elevated plasma levels occurred in the anorexia nervosa patients but not in the obese patients who had been losing weight yet still had increased body mass. Previously, patients with disease-related cachexia who had been hospitalized for severe illness (i.e., cancer, renal failure) were observed to have lower levels of serum carotene than normal subjects and patients with anorexia nervosa, but the diets and patterns of weight loss are likely to be quite different among those groups.[3] Patients with anorexia nervosa typically lose weight slowly over a period of months, in a similar pattern and induced by the same behaviors which had produced weight loss in the obese subjects in the present study.[4,12] Calorie-dense, high-fat foods are usually the primary items eliminated from the diet, and exercise is pursued to promote loss of adipose tissue.

Elevated plasma cholesterol levels were not observed in the anorexia nervosa patients. In an early study by Banji et al.[13] elevated plasma carotene levels which were observed did not appear to be associated with hypercholesterolemia in anorexia nervosa patients.

[12] C. L. Rock and J. Yager, *Int. J. Eating Dis.* **6**, 267 (1987).
[13] S. Banji and D. Mattingly, *Br. J. Psychiatry* **139**, 238 (1981).

In an evaluation of the nutritional status of 30 patients with anorexia nervosa, Casper et al.[14] observed moderately reduced serum zinc levels and hypogeusia, but these findings did not correlate with observations of hypercarotenemia. Curran-Celentano et al.[6] investigated the relationship between hypercarotenemia, vitamin A, and altered metabolic status in 27 patients with anorexia nervosa. Fifteen were found to have elevated carotenoid levels, with the most significant increase in the β-carotene fraction, as was observed in the present study. Relative weight and loss of original body weight were not significantly different among hypercarotenemic versus normocarotenemic anorexia nervosa patient groups. However, the hypercarotenemic group had a significant elevation in plasma retinyl esters and a significant decrease in levels of the active thyroid hormone triiodothyronine.

Results from the present study suggest that when similar types of diets and amounts of carotene-rich foods are consumed over a period of time, individuals with reduced body mass exhibit higher plasma β-carotene levels than those with greater body mass. This finding is corroborated by results from epidemiologic studies which suggest that lower relative weight is associated with increased plasma β-carotene levels in the normal population.[15]

We examined the relationship between plasma carotenoid levels and *BMI* over a wider range than has been previously reported in studies of hypercarotenemia. *BMI* describes the degree of adiposity and also the amount of other body compartments into which β-carotene can distribute. Increased body mass involves not only an increased amount of adipose tissue, but also increased blood volume and increased extracellular fluid volume. The present evidence supports the theory that body mass, and possibly tissue storage capacity, may play a role in the hypercarotenemia of anorexia nervosa.

Acknowledgments

The authors thank Dr. Hemmige Bhagavan, Hoffmann-La Roche, Inc., Nutley, New Jersey, for providing the β-carotene supplements and samples of pure carotenoid standards.

---

[14] R. C. Casper, B. Kirschner, H. H. Sandstead, R. A. Jacob, and J. M. Davis, *Am. J. Clin. Nutr.* **33**, 1801 (1980).

[15] W. S. Stryker, L. A. Kaplan, E. A. Stein, M. J. Stampfer, A. Sober, and W. C. Willett, *Am. J. Epidemiol.* **127**, 283 (1988).

[13] Psoralen Photosensitization and Plasma and Cutaneous β-Carotene Concentrations in Hairless Mice

By WENDY S. WHITE, HAROLD B. FAULKNER, and DAPHNE A. ROE

Introduction

The photoprotective actions of carotenoids include direct quenching of the excited triplet states of photosensitizers, scavenging of radical intermediates, and quenching of excited oxygen species formed by reaction of photodynamic triplet sensitizers with molecular oxygen, as previously reviewed.[1] Psoralens are a class of naturally occurring, tricyclic furocoumarins, some of which are powerful photosensitizers when excited by longwave ultraviolet (UVA) light. Reported protective effects of β-carotene against the harmful effects of psoralen photosensitization include protection against 8-methoxypsoralen (8-MOP) photomutagenicity in bacteria,[2] and 8-MOP phototoxicity[3] and photocarcinogenicity[4] in rodents. The 8-MOP-photosensitized induction of erythema or DNA lesions, as assessed by autoradiographical measurement of unscheduled DNA synthesis, was not reduced in human subjects after administration of combined oral doses of β-carotene and canthaxanthin.[5]

Protection against the 8-MOP-induced skin photosensitization reaction in rodents by β-carotene[3] and other singlet oxygen quenchers[6] was thought to suggest involvement of reactive oxygen intermediates or direct quenching of the psoralen excited triplet state and prevention of its cross-linking with DNA. Quenching of the triplet states of psoralens by β-carotene was demonstrated spectroscopically,[7] and β-carotene significantly reduced the photobinding of 8-MOP to DNA *in vitro*.[8] Although their photobiological effects are generally attributed to formation of covalent photoadducts with DNA via anoxic type I reactions,[9,10] psoralens also participate in oxygen-dependent type II photosensitization reactions.[11,12] The photobiological

[1] N. I. Krinsky, *Prev. Med.* **18**, 592 (1989).
[2] L. Santamaria, L. Bianchi, A. Bianchi, R. Pizzala, G. Santagati, and P. Bermond, *Med. Biol. Environ.* **13**, 123 (1985).
[3] A. Giles, W. Wamer, and A. Kornhauser, *Photochem. Photobiol.* **41**, 661 (1985).
[4] L. Santamaria, A. Bianchi, L. Andreoni, G. Santagati, A. Arnaboldi, and P. Bermond, *Med. Biol. Environ.* **12**, 531 (1984).
[5] C. Wolf, A. Steiner, and H. Honigsmann, *J. Invest. Dermatol.* **90**, 55 (1988).
[6] M. A. Pathak and P. C. Joshi, *Biochim. Biophys. Acta* **798**, 115 (1984).
[7] R. V. Bensasson, E. J. Land, and C. Salet, *Photochem. Photobiol.* **27**, 273 (1978).
[8] F. Bordin, M. T. Conconi, and A. Capozzi, *Photochem. Photobiol.* **46**, 301 (1987).
[9] R. S. Cole, *Biochim. Biophys. Acta* **254**, 30 (1971).

importance of the latter is controversial.[13] It was suggested that formation of active oxygen species (singlet oxygen and superoxide anion radicals) may mediate some psoralen photosensitization effects, such as the erythema of the skin phototoxic response[14] and the photomutagenicity of psoralens in bacteria.[15]

This chapter section describes methods used to investigate effects of 8-MOP photosensitization and UV irradiation on plasma and cutaneous β-carotene concentrations in albino hairless mice (Skh/HR-1) fed a β-carotene-fortified diet. Acute UV light-induced effects upon nonenzymatic and enzymatic antioxidant systems, including reductions of concentrations of tocopherol and ubiquinones and activities of glutathione reductase and catalase, were recently reported in hairless mouse skin irradiated *in vivo* or after excision.[16,17] We observed significant decreases of the plasma total carotenoid concentrations of human subjects after daily exposure to a UV light source in two crossover trials.[18] It was hypothesized that irradiation-induced reductions of plasma carotenoid concentrations might result from interactions with reactive species such as those produced during photosensitization reactions.

Animals

Female albino hairless mice of the Skh/HR-1 outbred stock were obtained from the Temple University School of Medicine (Philadelphia, PA) and cared for according to U.S. Department of Health and Human Ser-

---

[10] M. A. Pathak, D. M. Kramer, and T. B. Fitzpatrick, *in* "Sunlight and Man: Normal and Abnormal Photobiologic Responses" (M. A. Pathak, L. C. Harber, M. Seiji, A. Kukita, and T. B. Fitzpatrick, eds.), p. 335. Univ. of Tokyo Press, Tokyo, 1974.
[11] W. Poppe and L. I. Grossweiner, *Photochem. Photobiol.* 22, 217 (1975).
[12] N. J. de Mol and G. M. J. Beijersbergen van Henegouwen, *Photochem. Photobiol.* 30, 331 (1979).
[13] E. Sage, T. Le Doan, V. Boyer, D. E. Helland, L. Kittler, C. Helene, and E. Moustacchi, *J. Mol. Biol.* 209, 297 (1989).
[14] N. J. de Mol and G. M. J. Beijersbergen van Henegouwen, *Photochem. Photobiol.* 33, 815 (1981).
[15] N. J. de Mol, G. M. J. Beijersbergen van Henegouwen, G. R. Mohn, B. W. Glickman, and P. M. van Kleef, *Mutat. Res.* 82, 23 (1981).
[16] J. Fuchs, M. E. Huflejt, L. M. Rothfuss, D. S. Wilson, G. Carcamo, and L. Packer, *J. Invest. Dermatol.* 93, 769 (1989).
[17] L. Packer, J. Fuchs, J. Leipala, E. Witt, K. Marenus, D. Maes, and W. P. Smith, *Clin. Res.* 38, 672A (1990).
[18] W. S. White, C. Kim, H. J. Kalkwarf, P. Bustos, and D. A. Roe, *Am. J. Clin. Nutr.* 47, 879 (1988).

vices guidelines.[19] On arrival, mice were 8 weeks of age and weighed 21–30 g. Their health status was monitored by daily observation and weekly measurement of body weights. Diet and water were consumed *ad libitum*. Room lighting was furnished by a single 300-W incandescent bulb on an automated diurnal cycle (light and dark periods were each 12 hr).

Diet

To produce an accumulation of β-carotene in blood and skin, the mice were fed a pelleted 1% (w/w) β-carotene-fortified purified diet for an initial 8-week period. Hairless mice, rats, and guinea pigs have previously been shown to absorb β-carotene intact when fed sufficient quantities.[20,21] The β-carotene source was beadlets formulated by Hoffmann-LaRoche (Nutley, NJ) to contain 10% β-carotene (w/w), gelatin, sucrose, food starch, and peanut oil, with ascorbyl palmitate and all-*rac*-α-tocopherol added as antioxidants. The β-carotene beadlets were substituted for an equal weight of sucrose in the AIN-76 purified diet for rats and mice.[22] The mean carotene content of three diet samples sent to Nutrition International, Inc. (East Brunswick, NJ) for initial determination of carotene content by the open-column chromatography procedure of the Association of Official Analytical Chemists (AOAC)[23] was $1.07 \pm 0.02\%$ (S.E.M.) (w/w). The mice tolerated the diet well with no morbidity or apparent growth failure.

Experimental Design

Factorial experimentation allowed concurrent study of the effects of irradiation alone and irradiation with 8-MOP photosensitization on plasma and cutaneous β-carotene concentrations. After the initial 8-week feeding period, 84 mice were randomly assigned to one of four groups of 21. Each group received one of the following treatments according to a 2×2 factorial design: 8-MOP with irradiation, 8-MOP without irradiation, corn oil vehicle with irradiation, and corn oil without irradiation.

---

[19] U.S. Department of Health and Human Services, "Guide for the Care and Use of Laboratory Animals." National Institutes of Health, Bethesda, Maryland, 1985.
[20] M. M. Mathews-Roth, D. Hummel, and C. Crean, *Nutr. Rep. Int.* **16**, 419 (1977).
[21] S. S. Shapiro, D. J. Mott, and L. J. Machlin, *J. Nutr.* **114**, 1924 (1984).
[22] American Institute of Nutrition, *J. Nutr.* **107**, 1340 (1977).
[23] Association of Official Analytical Chemists, "Official Methods of Analysis." AOAC, Washington, D.C., 1984.

Psoralen Administration

Because of its poor water solubility (0.23 mM)[24] and increased bioavailability in the dissolved state,[25] 8-MOP (Sigma Chemical, St. Louis, MO) was solubilized in corn oil and administered by gavage. The 8-MOP was estimated to be 100% pure by high-performance liquid chromatographic (HPLC) analysis.[26] Psoralen-treated mice received a single oral dose of 6 mg of 8-MOP/kg body weight in 0.25 ml of tocopherol-stripped corn oil. Between 4.5 and 7.0 mg of 8-MOP, as determined by the body weight of the animal, was dissolved in 2 ml of acetone and diluted in 8 ml of tocopherol-stripped corn oil; the acetone was subsequently removed by rotary evaporation at 40°. Fresh batches were prepared daily. Acetone-treated tocopherol-stripped corn oil formulations were prepared similarly and used to dose the vehicle control animals. Mice were fasted 2 hr before dosing to reduce variability in 8-MOP bioavailability due to food ingestion such as that reported in humans.[27]

Irradiations

The light source was a Blue Light 2002 PUVA phototherapy unit (Dr. Honle, Munich, Germany) with two 400-W high-pressure metal halide lamps. Emitted light was filtered (filter H2) to remove wavelengths of less than 295 nm. The most intense ultraviolet emission band occurred within the action spectrum for 8-MOP skin photosensitization which extends from 320 to 380 nm in the UVA region.[28]

The UVB, UVA, and blue visible light intensities at the skin level of the mice were measured daily before light treatments. A calibrated International Light (Newburyport, MA) IL 443 radiometer with SEE1240 probe and UVB filter was used to determine the UVB irradiance. The UVA and blue visible light intensities were measured with an International Light IL 1700 research radiometer with SED038 broadband silicon detector, wide

---

[24] D. R. Bickers and M. A. Pathak, *Natl. Cancer Inst. Monogr.* **66**, 77 (1984).
[25] U. Busch, J. Schmid, F. W. Koss, H. Zipp, and A. Zimmer, *Arch. Dermatol. Res.* **262**, 255 (1978).
[26] M. Van Boven, P. Adriaens, R. Roelandts, and P. Daenens, *Photodermatology* **1**, 241 (1984).
[27] M. J. Herfst and F. A. de Wolff, *Eur. J. Clin. Pharmacol.* **23**, 75 (1982).
[28] Y. Nakayama, F. Morikawa, M. Fukuda, M. Hamano, K. Toda, and M. A. Pathak, *in* "Sunlight and Man: Normal and Abnormal Photobiologic Responses" (M. A. Pathak, L. C. Harber, M. Seiji, A. Kukita, and T. B. Fitzpatrick, eds.), p. 591. Univ. of Tokyo Press, Tokyo, 1974.

TABLE I
IRRADIATION-INDUCED EDEMA RESPONSES FOR PHOTOSENSITIZED AND CONTROL MICE[a]

	Time after UV exposure[b]							
	Ear thickness (mm)				Dorsal skinfold thickness (mm)[c]			
Mice	0 min	30 min	12 hr	24 hr	0 min	30 min	12 hr	24 hr
Vehicle controls								
1	0.5	0.5	0.5	0.8	0.9	0.9	0.9	1.3
2	0.5	0.5	0.5	0.7	0.9	0.9	0.9	1.2
8-MOP-photosensitized								
3	0.5	0.5	0.8	0.9	0.9	0.9	1.3	1.5
4	0.5	0.5	0.8	0.9	0.9	0.9	1.3	1.5

[a] Skin responses were observed during preliminary phototesting in Skh/HR-1 mice fed an unrefined diet without β-carotene fortification.
[b] Shown are the time periods elapsed after exposure to 4.00 J/cm^2 UVA and 0.23 J/cm^2 UVB.
[c] The thickness was measured at the longitudinal midpoint when the midline skin was pinched upward at the neck and base of the tail.

eye quartz diffuser, and UVA phototherapy or TBLU blue-color filter. Separate calibrations were performed for each detector–filter unit. The quartz diffuser was used to provide an 180° field of view and cosine-weighted response for the detector–filter combination. Measured UVB, UVA, and blue visible light irradiances at skin level remained constant at 0.29, 5.01, and 5.28 mW/cm^2, respectively.

Mice were restrained using a small polypropylene container (8.5 × 4.5 × 4.0 cm) with a wire-mesh top and individually irradiated on a dull black surface 61 cm below the lower glass filter of the phototherapy unit. Light treatments were administered 90 min after dosing with 8-MOP or corn oil vehicle. Tritium radioactivity in the blood of female Skh/HR-1 mice was reported to peak 90 min after administration of 6 mg of [^{3}H]8-MOP/kg body weight in corn oil.[29] Irradiated mice received a 4.00 J/cm^2 UVA dose during a 13-min 18-sec exposure. The corresponding UVB and blue visible light doses received were 0.23 and 4.21 J/cm^2, respectively. Control mice not receiving light treatments (dark controls) were placed beneath the unlit phototherapy unit for an identical period.

The phototoxic responses of hairless mice to the UV treatment were evaluated over a 24-hr period in a preliminary study (Table I). Edema was

[29] I. A. Muni, F. H. Schneider, T. A. Olsson, and M. King, *Natl. Cancer Inst. Monogr.* **66**, 85 (1984).

TABLE II
β-CAROTENE CONCENTRATIONS IN EPIDERMAL/DERMAL
HOMOGENATES FOR TWO SKIN SEPARATION METHODS

	β-Carotene concentration (nmol/g wet tissue weight)	
Skin sample	Heat separation[a]	Mechanical separation[b]
1	0.7	0.6
2	0.6	0.4
3	0.3	0.4
	$\bar{x} = 0.5 \pm 0.1$ (S.E.M.)	$\bar{x} = 0.5 \pm 0.1$ (S.E.M.)

[a] 50°; 30 sec.
[b] Sharp dissection.

detected by a change in the right ear or dorsal skinfold thickness measured with a caliper. Edema was apparent in psoralen-photosensitized mice but not irradiated vehicle controls at 12 hr after UV exposure, and in all mice at 24 hr. A pale pink erythema was observed in all mice at 12 and 24 hr after irradiation. The phototoxic response was not expressed in human erythema units because of a species difference in erythema development.[30] Erythema or edema were not detected in the mice at 30 min after UV light exposure.

Blood and Tissue Collection

Blood and skin samples were collected 30 min after light or dark control treatments. During the 30-min period before sample collection, the mice continued to fast but consumed water *ad libitum*. A maximum of 600 μl of blood is collected periorbitally from each mouse under ether anesthesia using heparinized Pasteur pipettes. Plasma was obtained after a 10-min centrifugation at 2000 g at 4°, and then transferred to 1.0-ml glass microvials with Teflon-lined screw caps (Wheaton, Inc., Millville, NJ). After the animals were sacrificed by exsanguination, dorsal skin samples (~2 g wet tissue weight) were excised, rinsed with cold isotonic saline, and blotted. Plasma and skin samples were stored under nitrogen at −70° in the dark.

Before analysis, thawed skin samples were immersed in 50° distilled water for 30 sec, and the epidermal/dermal skin sheets were separated from

[30] P. D. Forbes and F. Urbach, *in* "The Biologic Effects of Ultraviolet Radiation" (F. Urbach, ed.), p. 279. Pergamon, Oxford, 1969.

the underlying subcutis by blunt dissection. Histologic examination revealed intact epidermis and dermis with no contamination by subcutaneous tissue. The skin sheet was rinsed with isotonic saline, blotted, minced finely with a pair of surgical scissors, and weighed.

The effect of heat separation on cutaneous β-carotene concentrations was investigated using dorsal skin samples from three mice fed the β-carotene-fortified diet. Each skin sheet was cut longitudinally into two halves of approximately identical size, and the subcutis was separated from one sample by heat treatment, and from the other by scraping with a scalpel. The β-carotene concentrations in the tissue samples were then determined in duplicate by the method described below. No loss of β-carotene was apparent when heat-treated and shaved epidermal/dermal skin sheets were compared (Table II).

Analytical Methods

Instrumentation

The HPLC system used to determine β-carotene concentrations in plasma and tissue extracts included a Beckman (Berkeley, CA) 110B single-piston reciprocating pump, a Rheodyne (Cotati, CA) 7125 sample injection valve with a 100-μl loop, a Kratos (Applied Biosystems, Ramsey, NJ) Spectroflow 783 variable wavelength UV–visible absorbance detector set at 440 nm and 0.007 AUFS, and a Hewlett-Packard (Avondale, PA) 3390A reporting integrator. Separations were performed on a 25 cm × 4.6 mm inner diameter (i.d.) 5-μm particle size Ultrasphere ODS C_{18} column (Beckman) protected by a Brownlee (Santa Clara, CA) 7-μm RP-18 NewGuard cartridge. The solvent system was acetonitrile–methylene chloride–methanol (70:20:10, v/v/v)[31] run at 1.7 ml/min. Solvents were HPLC grade (J.T. Baker, Phillipsburg, NJ). Prior to use, solvents were dried over molecular sieve (3 Å, Fisher, Fair Lawn, NJ),[32] filtered through a 0.2-μm Nylon 66 membrane (Schleicher and Schuell, Keene, NH), and degassed under reduced pressure from a water aspirator.

---

[31] J. G. Bieri, E. D. Brown, and J. C. Smith, *J. Liq. Chromatogr.* **8**, 473 (1985).

[32] Removal of trace amounts of water from solvents was necessary to prevent the appearance of extraneous chromatogram peaks after injection of β-carotene standards. The additional peaks appeared to be degradation products of β-carotene as their appearance was accompanied by significant reduction of the β-carotene peak area, but may have been chromatographic artifacts [(F. Khachik, G. R. Beecher, J. T. Vanderslice, and G. Furrow, *Anal. Chem.* **60**, 807 (1988)]. A need for dry solvents appears to be unique to use of this particular HPLC column and solvent system.

Standardization

Synthetic β-carotene (Fluka Chemical, Ronkonkoma, NY) and lycopene extracted from tomatoes (Sigma) were obtained for use as standards. Vials of lycopene from several lots were discarded due to the presence of insoluble contaminants. Standards used were estimated to be at least 98% pure by HPLC analysis. Solvents used to prepare standards were dried over molecular sieve and, with the exception of the absolute ethanol, were HPLC grade.

Stock solutions of β-carotene or lycopene were prepared in hexane or a combination of hexane and toluene and diluted in absolute ethanol to prepare working standards. Stock solutions of β-carotene were prepared by dissolving 1 mg in 10 ml of hexane containing 0.01% (w/v) butylated hydroxytoluene (BHT) (Sigma). These were diluted with absolute ethanol (0.01% BHT) to give working standards with a concentration of 1.0 μg/ml. Lycopene stock solutions were prepared by dissolving 1 mg in 10 ml of toluene (0.01% BHT) followed by 10 ml of hexane (0.01% BHT). These were diluted with absolute ethanol (0.01% BHT) to give working standards with a concentration of 0.4 μg/ml.

The concentrations of working standards were determined initially and checked biweekly by spectrophotometry using published absorptivity ($E^{1\%}_{1cm}$) values[33]: β-carotene, 2620 at 453 nm; and lycopene, 3450 at 472 nm. Lycopene standards were found to be unstable at concentrations above 0.4 μg/ml due to precipitation from ethanol at −20°. Stacewicz-Sapuntzakis et al.[34] had previously observed recrystallization of lycopene from ethanol at concentrations above approximately 0.2 μg/ml at −20°. New lycopene standards were prepared every 10 to 14 days. The β-carotene working standards were stable for at least 1 month at −20° when protected from light.

Calibration curves for β-carotene were prepared for each day of HPLC analysis using lycopene as an internal standard. A series of β-carotene concentrations with a constant concentration of lycopene internal standard were injected in a fixed volume (50 μl), and the ratios of the peak areas were plotted against the β-carotene concentration. The amount of β-carotene injected ranged from 5 to 25 ng; the amount of lycopene injected was 10 ng. All procedures are performed under dim red light.

---

[33] E. De Ritter and A. E. Purcell, in "Carotenoids as Colorants and Vitamin A Precursors: Technological and Nutritional Applications" (J. C. Bauernfeind, ed.), p. 883. Academic Press, New York, 1981.

[34] M. Stacewicz-Sapuntzakis, P. E. Bowen, J. W. Kikendall, and M. Burgess, *J. Micronutr. Anal.* **3**, 27 (1987).

Plasma Extraction

Using an Eppendorf digital pipette (Brinkmann Instruments, Westbury, NY), duplicate 100-μl aliquots of thawed plasma were transferred to 2-ml glass microcentrifuge tubes fitted with glass stoppers. Plasma was deproteinized by adding 50 μl of internal standard solution (0.4 μg lycopene/ml) followed by 50 μl of absolute ethanol containing 0.01% BHT (w/v). The mixture was extracted twice with 700 μl of hexane containing 0.01% BHT (w/v), vortexed for 3 min and centrifuged for 5 min at 3000 rpm in a clinical centrifuge (Model HN-S, International Equipment Co., Needham Heights, MA). The combined hexane layers were transferred to a glass microvial and evaporated to dryness under nitrogen at room temperature using a Wheaton 412800 microrotary evaporator. The residue was reconstituted with 100 μl of mobile phase, mixed for 30 sec, and a 50-μl aliquot was injected into the HPLC system. Injection of 50% of the volume of the reconstituted extract was necessary to ensure injection of a quantity of β-carotene that exceeded the limit of detection.

Recovery studies were performed by adding β-carotene and lycopene standards to rat plasma. Like mice, rats do not accumulate carotenoids in blood and tissues unless fed high concentrations in the diet.[21] Analyses of extracts of rat plasma that did not have added β-carotene or lycopene revealed no interfering peaks on the HPLC chromatograms. Recoveries of β-carotene were determined by comparing the peak area ratios obtained from rat plasma spiked with 20 ng of β-carotene standard and 20 ng of lycopene standard to the peak area ratios of the internal standard calibration curve. The amount of lycopene added to rat plasma (20 ng) was twice the amount of lycopene standard injected for the calibration curve (10 ng) to account for the injection of only 50% of the reconstituted plasma extract into the HPLC system. The mean recovery of β-carotene from five replicates of spiked plasma was 91.6% with a coefficient of variation (CV) of 6.9% when lycopene was used as an internal standard.

Absolute recoveries of β-carotene and lycopene from spiked rat plasma were calculated as described by Kaplan et al.[35] using direct proportion of the chromatogram peak areas of β-carotene and lycopene in plasma extracts with those of their respective standards. The mean absolute recoveries from five replicates were 89.0% with a CV of 15.7% for lycopene and 87.1% with a CV of 8.3% for β-carotene. The comparable absolute recoveries from spiked rat plasma indicate that lycopene is an appropriate internal standard for β-carotene. The recovery of a carotenoid internal standard added to plasma samples may not be representative of that of

[35] L. A. Kaplan, J. A. Miller, and E. A. Stein, *J. Clin. Lab. Anal.* **1**, 147 (1987).

endogenous β-carotene which must be extracted from lipoproteins.[31] However, a potential underestimation of plasma β-carotene concentrations due to incomplete extraction accompanies the use of either internal or external standards.

Epidermal Tissue Extraction

The first ten epidermal/dermal samples collected in each treatment group are saponified, extracted and analyzed for β-carotene content by HPLC. Weighed epidermal/dermal samples (0.30–0.68 g wet weight) were homogenized in 13 parts 1.15% KCl for 6 min over ice using a Polytron homogenizer (Brinkmann Instruments). Homogenates were aliquoted, weighed, and stored overnight under nitrogen at −20° in the dark. Saponifications and extractions were performed in duplicate by the method of Taylor *et al.*[36] as modified by Shapiro *et al.*[21] In a test tube with a Teflon-lined screw cap, 0.9 ml of homogenate was mixed with 0.3 ml of 25% sodium ascorbate, 0.6 ml of absolute ethanol, and 67 μl of internal standard solution (0.4 μg lycopene/ml). The mixture was preincubated under nitrogen at 70° for 5 min. Following the addition of 0.6 ml of 10 N KOH, samples were saponified under nitrogen for 30 min at 70°.

After cooling over ice, the saponified mixtures were extracted by adding 4.0 ml of hexane containing 0.01% (w/v) BHT followed by 3 min of vortexing. The phases were separated by centrifugation for 10 min at 2500 rpm in the clinical centrifuge. A 3-ml aliquot of the hexane layer was transferred to a 5-ml glass microvial, concentrated to a volume of approximately 700 μl by rotary evaporation under nitrogen at room temperature, and quantitatively transferred to a 1-ml microvial. Vials were flushed with nitrogen, capped, and frozen overnight at −20° in the dark. The hexane was evaporated to dryness the next day under nitrogen. The residue was reconstituted in 100 μl of mobile phase, mixed for 30 sec, and a 50-μl aliquot was injected into the HPLC system.

Recovery studies demonstrated that lycopene was relatively stable in the described saponification and extraction procedures. The mean absolute recovery of lycopene from mouse epidermal/dermal homogenates spiked with 27 ng was 100.9% with a CV of 5.3% in five replicates. Recoveries that exceeded 100% probably resulted from either concentration of the hexane extract by evaporative losses prior to removal of the 3-ml aliquot or inclusion of an area corresponding to a shoulder on the lycopene peak during integration of chromatograms of skin extracts. The shoulder was most likely an artifact of saponification and extraction such as an unre-

---

[36] S. L. Taylor, M. P. Lamden, and A. L. Tappel, *Lipids* **11**, 530 (1976).

solved isomer. Both isomerization and evaporative losses would be expected to affect the analyte and internal standard to a similar extent, so that the accuracy of β-carotene determinations by the internal standard method would not be significantly altered.

Statistical Analysis

Plasma and epidermal/dermal β-carotene concentrations were compared between groups by analysis of variance (ANOVA) using the general linear model of the Statistical Analysis System (SAS) (SAS Institute Inc., Cary, NC). The body weights of the mice at the time of plasma and skin sample collection were investigated as potential covariates with plasma or epidermal/dermal β-carotene concentrations. Statistical tests were performed at the $p = 0.05$ level of significance.

The statistical significance of the relationship between plasma and epidermal/dermal β-carotene concentrations was tested by analyzing the partial (Type III) sums of squares (SS) contributed by inclusion of plasma β-carotene concentration in the general linear model for epidermal/dermal β-carotene concentration. Analysis of the partial sums of squares excluded the potential confounding effects of treatment-induced correlation between plasma and epidermal/dermal β-carotene concentrations.

Effects of 8-MOP Photosensitization and Irradiation

There were no statistically significant differences in plasma β-carotene concentrations between treatment groups. Group means for plasma β-carotene levels are shown in Table III. Data are presented for 74 of the 84 mice that began the study. Plasma samples from the remaining mice were either too small to allow accurate analysis of β-carotene concentrations in

TABLE III
PLASMA β-CAROTENE CONCENTRATIONS

Mice	Concentration (μmol/liter)[a]	
	8-MOP-photosensitized	Vehicle controls
Irradiated	0.471 ± 0.035	0.443 ± 0.035
	(n = 18)	(n = 17)
Dark controls	0.425 ± 0.034	0.391 ± 0.034
	(n = 19)	(n = 20)

[a] Group means ± S.E.M. calculated from the pooled S.D. for the model.

duplicate or were lost during analysis. The profiles of the HPLC chromatograms obtained from analysis of plasma extracts with detection at 440 nm appeared identical for irradiated and nonirradiated mice (Fig. 1).

There were no significant differences between treatment groups in the β-carotene concentrations of the epidermal/dermal skin homogenates. The mean epidermal/dermal β-carotene concentration was 1.1 ± 0.1 (S.E.M.) nmol/g wet tissue weight for the nonirradiated 8-MOP-photosensitized mice; the mean epidermal/dermal β-carotene concentrations were 1.0 ± 0.1 (S.E.M.) nmol/g for each of the other three treatment groups. The profiles of the HPLC chromatograms obtained from analysis of epidermal/dermal extracts with detection at 440 nm were identical for irradiated and nonirradiated mice.

There was a significant relationship between plasma and epidermal/dermal β-carotene concentrations ($p = 0.02$). There was no significant relationship between the body weights of the mice and either plasma or epidermal/dermal β-carotene concentrations.

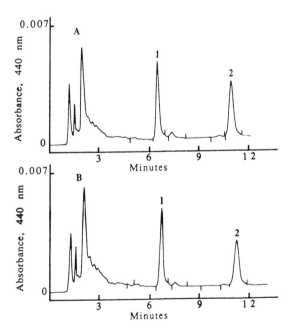

FIG. 1. Representative HPLC chromatograms of extracts of plasma from (A) irradiated 8-MOP-photosensitized Skh/HR-1 mice and (B) nonirradiated 8-MOP-photosensitized Skh/HR-1 mice. Lycopene internal standard (1) and β-carotene (2) peaks were tentatively identified by coelution with standards and the absence of coeluting peaks in plasma extracts from mice fed an unrefined diet without β-carotene fortification.

Conclusions

Irradiation with or without 8-MOP photosensitization had no apparent acute effects on β-carotene concentrations in the plasma or epidermal/dermal skin layers of the albino hairless mice. The plasma and cutaneous β-carotene concentrations of the mice were not significantly related to their body weights. There was a significant relationship between plasma and epidermal/dermal β-carotene concentrations which may be attributable to blood present in the highly vascular dermal tissue rather than equilibration between plasma and tissue pools.

Irradiation-induced reductions of plasma and cutaneous β-carotene concentrations may have been prevented by ingestion of α-tocopherol and ascorbyl palmitate as antioxidants in the β-carotene beadlets. Dietary supplementation with placebo beadlets containing an antioxidant formulation of butylated hydroxytoluene (BHT), butylated hydroxyanisole (BHA), and ascorbyl palmitate was reported to reduce psoralen phototoxicity in rats, although supplementation with β-carotene beadlets provided greater protection.[3]

Ultraviolet light-induced reductions of plasma carotenoid concentrations may be a delayed response or may not occur in this animal model because of species differences in carotenoid metabolism[37] and erythema development.[30] Irradiation-induced reductions of the plasma total carotenoid concentrations of human subjects were observed after repeated UV exposures, and were most apparent in subjects who had higher carotenoid concentrations.[18] The plasma β-carotene concentrations of the hairless mice were within the range of those reported for humans,[34,38] but were much lower than the total plasma carotenoid concentrations of healthy individuals consuming a varied diet.[39] Blood and skin concentrations of other carotenoids may be more responsive to UV light exposure. Ultraviolet light-induced reductions of epidermal lycopene concentrations were recently reported in human subjects 24 hr after a single irradiation.[40]

It is important to note that the onset and peak of the cellular damage and erythema produced by 8-MOP photosensitization and UVA irradiation are significantly delayed as compared with the sunburn reaction to

---

[37] H. B. Jensen and T. K. With, *Biochem. J.* **33**, 1771 (1939).
[38] D. W. Nierenberg, T. A. Stukel, J. A. Baron, B. J. Dain, and E. R. Greenberg, *Am. J. Epidemiol.* **130**, 511 (1989).
[39] G. Arroyave, C. O. Chichester, H. Flores, J. Glover, L. A. Mejia, J. A. Olson, K. L. Simpson, and B. A. Underwood, *in* "Biochemical Methodology for the Assessment of Vitamin A Status." International Vitamin A Consultative Group, Washington, D.C., 1982.
[40] J. D. Ribaya-Mercado, M. Garmyn, J. Bhawan, B. A. Gilchrest, and R. M. Russell, *FASEB J.* **6**, A1646 (1992).

UVB exposure.[41] In the current study, blood and skin samples were collected 30 min after hairless mice were exposed to UV light. In a previous report, peak reductions of the epidermal and dermal glutathione concentrations of hairless mice were not observed until 24 to 48 hr after UVA irradiation with 8-MOP photosensitization.[42]

Acknowledgments

This research was supported by a National Science Foundation Graduate Fellowship awarded to W. S. White and U.S. Department of Agriculture Grant 87-CRCR-1-2338.

[41] J. A. Parrish, R. S. Stern, M. A. Pathak, and T. B. Fitzpatrick, in "The Science of Photomedicine" (J. D. Regan and J. A. Parrish, eds.), p. 595. Plenum, New York, 1982.
[42] L. A. Wheeler, A. Aswad, M. J. Connor, and N. Lowe, J. Invest. Dermatol. **87**, 658 (1986).

[14] Human Metabolism of Carotenoid Analogs and Apocarotenoids

By SIHUI ZENG, HAROLD C. FURR, and JAMES A. OLSON

Introduction

Studies of the absorption and metabolism of carotenoids have been stimulated by claims of their anticarcinogenic properties independent of their provitamin A activity. Many epidemiological studies have indicated a possible association between increased dietary consumption of foods rich in carotenoids and a lower cancer risk.[1] Carotenoids have been shown to have potent antioxidant activities,[2] to stimulate the immune response,[3] and to protect cells and animals from cancerous events.[4] However, little is known about the metabolism of carotenoids in humans, other than the cleavage of β-carotene to retinal in the gut, liver, and some other tissues.[5]

Studies on the metabolism of carotenoids in humans are complicated by the presence of dietary carotenoids and the slow clearance of endogenous carotenoids from human serum. Therefore, we have used carotenoid

[1] R. G. Ziegler, J. Nutr. **119**, 116 (1989).
[2] G. W. Burton, J. Nutr. **119**, 109 (1989).
[3] A. Bendich, J. Nutr. **119**, 112 (1989).
[4] N. I. Krinsky, J. Nutr. **119**, 123 (1989).
[5] J. A. Olson, J. Nutr. **119**, 105 (1989).

analogs, not normally found in blood, as an effective way of studying the metabolism of carotenoids in humans. This chapter describes the procedures used in our study for the preparation of carotenoid analogs, for the administration of doses, for blood sampling, for HPLC analysis of serum samples, and for the characterization of metabolites.

Selection of Carotenoid Analogs

Three carotenoid analogs, β-apo-8'-carotenal, ethyl β-apo-8'-carotenoate, and 4,4'-dimethoxy-β-carotene, have been selected (Fig. 1). β-Apo-8'-carotenal has been suggested as a possible intermediate in the conversion of β-carotene to retinal, but ethyl β-apo-8'-carotenoate, a derivative of β-apo-8'-carotenal, has no provitamin A activity. 4,4'-Dimethoxy-β-carotene, a synthetic analog, is not known to be cleaved to 4-methoxyretinal by β-carotenoid 15,15'-dioxygenase or by other enzymes.

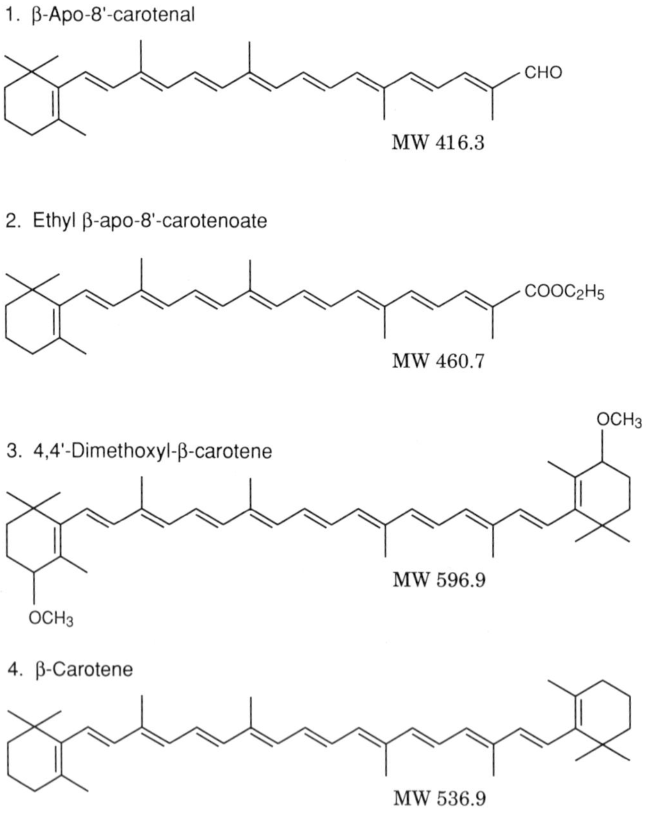

1. β-Apo-8'-carotenal

MW 416.3

2. Ethyl β-apo-8'-carotenoate

MW 460.7

3. 4,4'-Dimethoxyl-β-carotene

MW 596.9

4. β-Carotene

MW 536.9

FIG. 1. Structures of three carotenoid analogs and β-carotene.

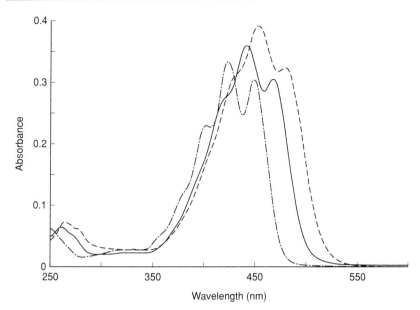

FIG. 2. UV/visible absorption spectra of β-apo-8'-carotenal (---), β-apo-8'-carotenol (— · —), and ethyl β-apo-8'-carotenoate (——) (in hexanes).

These three analogs can serve a very useful role in analyzing the absorption and kinetic behavior of carotenoids in humans.

Preparation of Three Carotenoid Analogs

β-Apo-8'-carotenal and ethyl β-apo-8'-carotenoate are commercially available in high purity (Fluka, Ronkonkoma, NY). To confirm their purity, the UV/visible absorption spectra and reversed-phase HPLC and TLC behavior were determined before use. The absorption maximum of β-apo-8'-carotenal is at 457 nm ($E^{1\%}_{1cm}$ 2640, ε 109,900) in petroleum ether,[6] and that of ethyl β-apo-8'-carotenoate is at 445 and 470 nm ($E^{1\%}_{1cm}$ 2500 and 2110, ε 115,000 and 97,000) in petroleum ether.[7] The UV/visible absorption spectra of β-apo-8'-carotenal and ethyl β-apo-8'-carotenoate are shown in Fig. 2.

4,4'-Dimethoxy-β-carotene is prepared by reduction of a commercially available starting material, canthaxanthin, with sodium borohydride, fol-

[6] R. Ruegg, M. Montavon, G. Ryser, G. Saucy, U. Schwieter, and O. Isler, *Helv. Chim. Acta* **42**, 854 (1959).
[7] U. Schwieter, H. Gutmann, H. Lindlar, R. Marbet, N. Rigassi, R. Ruegg, S. F. Schaeren, and O. Isler, *Helv. Chim. Acta* **49**, 369 (1966).

lowed by methylation with methyl iodide. Canthaxanthin (500 mg) (Fluka) is dissolved in a 1:1 (v/v) mixture (~250 ml) of methanol and tetrahydrofuran (HPLC grade). The UV/visible spectrum and the TLC behavior of canthaxanthin are checked before starting the reduction. The UV/visible absorption spectrum of canthaxanthin is shown in Fig. 3. The R_f value of canthaxanthin on silica gel TLC with 15% acetone in hexane as the mobile phase is 0.5. (All solvent mixtures are expressed as v/v.)

Sodium borohydride (~3 g) is then gradually added to the canthaxanthin solution (under argon) and stirred magnetically at room temperature for 4 hr. The reaction is then checked by absorption spectroscopy and by TLC. The UV/visible absorption spectrum of isozeaxanthin, the major expected reduction product, is also shown in Fig. 3. The R_f value of isozeaxanthin under the same TLC conditions is 0.3. An equal volume of water is then added to the mixture. After extraction with ethyl acetate (3 times, 300 ml each), the pooled extract is washed with water (2 times, 300 ml each), dried over anhydrous sodium sulfate, and then evaporated to dryness in a rotary evaporator. The residue is further dried by blowing argon into the flask for 1 hr to assure absolute dryness and to prevent oxidation by oxygen.

The next reaction is methylation. All steps involved in the reaction are

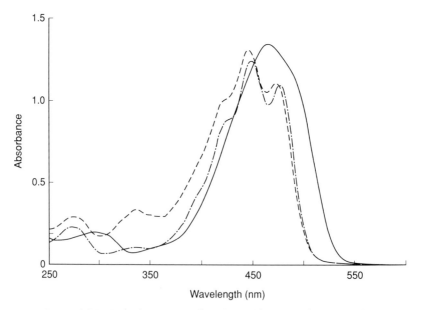

FIG. 3. UV/visible absorption spectra of canthaxanthin (——), isozeaxanthin (---), and 4,4'-dimethoxy-β-carotene (—·—) (in hexanes).

done in a moisture-free argon atmosphere. All glassware is gently flame-dried. Sodium hydride (1 g, 80% in mineral oil) is washed with dry hexane (2 times, 10 ml each) to remove the mineral oil, then dissolved in 200 ml tetrahydrofuran (HPLC grade, treated with sodium hydride to remove moisture). The isozeaxanthin is also dissolved in tetrahydrofuran (120 ml, HPLC grade, dried over sodium hydride), transferred to a dropping funnel, and slowly added to the sodium hydride solution (with magnetic stirring). After all of the isozeaxanthin is added, the mixture is stirred for 30 min longer. Methyl iodide (30 ml) is then added to the reaction mixture. The flask is wrapped with aluminum foil and left for 18 hr at room temperature with stirring. Thereafter, the formation of 4,4'-dimethoxy-β-carotene is checked by UV/visible spectroscopy (Fig. 3) and by TLC. The R_f value of 4,4'-dimethoxy-β-carotene on silica gel TLC is 0.8. Then methanol (~20 ml) is added to destroy the unreacted sodium hydride. After the solvent mixture is largely evaporated in a rotary evaporator, an equal volume of water is added to the residue. After extraction with ethyl acetate (3 times, 200 ml each), the pooled extract is washed with water (2 times, 200 ml each), dried over anhydrous sodium sulfate, and then evaporated to dryness in a rotary evaporator.

The crude 4,4'-dimethoxy-β-carotene preparation is dissolved in a minimum volume of hexane and then subjected to alumina column chromatography. Alumina (neutral, Brockman Activity I, 75 g) is thoroughly mixed with water (6 ml) to reduce its activity and then left for several hours before packing a column. The column is slurry packed with hexane; thereafter, about 15 ml of the crude 4,4'-dimethoxy-β-carotene preparation is applied to the column. Development of the column with hexane containing 2% ethyl ether results in the movement of two colored bands. The eluates are collected in fractions of about 6 ml. After elution of the first minor band, hexane containing 4% ethyl ether is used as mobile phase to elute the second 4,4'-dimethoxy-β-carotene band. Fractions are collected continuously until the second band is eluted. All fractions that show the UV/visible absorption spectrum of 4,4'-dimethoxy-β-carotene (as shown in Fig. 3), a single peak on reversed-phase HPLC, and a single spot (R_f ~0.8) on silica TLC are pooled and evaporated to dryness in a rotary evaporator. The resultant 4,4'-dimethoxy-β-carotene is further dried under a stream of argon to remove any trace amount of hexane. 4,4'-Dimethoxy-β-carotene has extinction coefficients ($E_{1\,cm}^{1\%}$) of 2240 and 1960 (ε 133,700 and 117,000) at absorption maxima of 450 and 478 nm, respectively, in petroleum ether.[8]

[8] P. Zeller, F. Bader, H. Lindler, M. Montavon, P. Muller, R. Ruegg, G. Ryser, G. Saucy, S. F. Schaeren, U. Schwieter, K. Stricker, R. Tamm, P. Zurcher, and O. Isler, *Helv. Chim. Acta* **42**, 841 (1959).

β-Apo-8'-carotenol and its palmitate ester can be prepared for comparison of their chromatographic retention times and absorption spectra with those of carotenoid metabolites. β-Apo-8'-carotenal (1 mg) is dissolved in methanol and then reduced to the alcohol by addition of solid $NaBH_4$. Completion of the reaction is assessed by absorption spectroscopy (see Fig. 2 for the UV/visible absorption spectrum of β-apo-8'-carotenol) and by TLC (β-apo-8'-carotenal, R_f 0.35; β-apo-8'-carotenol, R_f 0.3 on silica gel with benzene as the mobile phase). The apocarotenol is extracted into hexane after adding water to the methanolic solution, and the hexane solution is dried over anhydrous sulfate.

For esterification, a portion of the solution is evaporated under a gentle stream of argon, and the β-apo-8'-carotenol residue is immediately dissolved in 1 ml redistilled triethylamine. Ten drops of palmitoyl chloride are added, and the reaction mixture is allowed to stand overnight at room temperature. The β-apo-8'-carotenyl palmitate reaction product is extracted with hexane after the addition of water, washed extensively with water, and dried over anhydrous sodium sulfate (R_f 0.9 on silica TLC, benzene mobile phase). Maximum absorbance peaks for β-apo-8'-carotenol and its esters are at 426 and 453 nm (hexane; $E_{1cm}^{1\%}$ 2690 and 2440, ε 112,500 and 102,100).[6]

Subjects, Dosing, and Blood Sampling

Healthy male human subjects who are within 10% of average body weights for age are required to take a general physical examination and to provide information on their diet. The average estimated serum volume is 4% of body weight.

Each subject is given 100 μmol of the selected carotenoid in oil as part of a light breakfast. Thereafter, blood samples (7 ml) are taken from the antecubital vein by a registered medical technician at appropriate intervals. Because serum concentrations of the analogs rise and then decline rapidly in the first 1 or 2 days after dosing, frequent blood samples are needed in the early part of the experiment to trace accurately the time course of plasma concentrations. Blood samples continue to be taken at much longer time intervals until the dosed analogs disappear from the blood.

Extraction and Chromatographic Analysis of Serum Samples

All extractions and analyses are conducted under yellow fluorescent lighting to avoid isomerization. Blood samples are allowed to clot for 2 hr at room temperature and then are centrifuged at 2000 rpm for 15 min to obtain the serum. Two 1-ml portions of serum are kept frozen at $-20°$

until analysis. For extraction, 1-ml samples are thawed and an internal standard (retinyl acetate) added in order to monitor extraction efficiency. Samples are then treated with methanol (1 ml) to denature serum proteins and are extracted with hexane (2 times, 1 ml). After acidification with dilute acetic acid (10%, 100 µl), samples are further extracted with ethyl acetate (1 ml) and then with hexane (1 ml) once again for the most complete extraction of carotenoids.[9] Pooled organic extracts of each sample are evaporated to dryness under argon and redissolved in a mixture of 2-propanol and dichloroethane (2:1, 60 µl) for HPLC analysis.

A reversed-phase gradient HPLC apparatus is used for all serum samples, with slightly different solvent systems. The HPLC apparatus consists of a Waters (Milford, MA) automated gradient controller, two Waters Model 510 pumps, two ISCO (Lincoln, NE) Model V[4] absorbance detectors set at 325 and 450 nm, respectively, a Waters Model 712 WISP automatic sample injection system, a Shimadzu (Kyoto, Japan) Model CR4-A integrator, and a Waters Resolve 5-µm spherical C_{18} column (3.9 mm × 15 cm), preceded by an Upchurch (Oak Harbor, WA) C-130 guard column packed with C_{18} pellicular material.

Three different solvent systems are necessary to optimize the separation of the three analogs and their major metabolites from endogenous carotenoids in human serum. For ethyl β-apo-8′-carotenoate, a linear gradient of mobile phase is programmed from acetonitrile–water (85:15, v/v, containing 0.1% ammonium acetate) (solvent A) to acetonitrile–dichloroethane (80:20, v/v, containing 0.1% cyclohexene) (solvent B) over 15 min at a flow rate of 1.5 ml/min. Solvent B is continuously run for 17 more min after the 15-min gradient is finished. Ammonium acetate is added to solvent A to sharpen free acid peaks, which are usually broad on HPLC chromatograms. Addition of cyclohexene may suppress deleterious effects arising from free radical formation from dichloroethane, a chlorinated hydrocarbon solvent. Under these conditions, ethyl β-apo-8′-carotenoate is well separated from all other major carotenoids such as lycopene, β-carotene, the cryptoxanthins, and lutein. An HPLC chromatogram of human serum after a dose of ethyl β-apo-8′-carotenoate is shown in Fig. 4A.

For β-apo-8′-carotenal, a slightly different solvent system is used due to the appearance of many metabolites after the dose. This system is a 20-min linear gradient from acetonitrile–methanol–water (75:10:15, v/v, containing 0.1% ammonium acetate) (solvent A) to acetonitrile–dichloroethane (75:25, v/v, containing 0.1% cyclohexene) (solvent B) at a flow rate of 1.5 ml/min. At the end of the gradient program, solvent B is

[9] A. B. Barua, R. O. Batres, H. C. Furr, and J. A. Olson, *J. Micronutr. Anal.* **5**, 291 (1989).

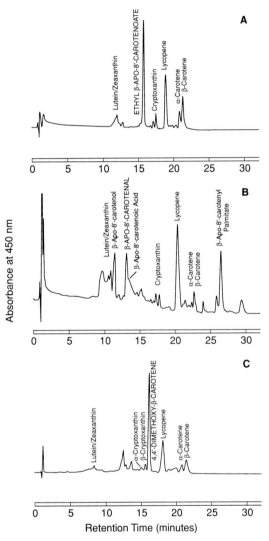

FIG. 4. Chromatograms obtained with gradient reversed-phase HPLC and absorption detection at 450 nm, using a Waters Resolve 5-μm C_{18} column. (A) HPLC chromatogram of a serum sample taken at 144 hr after the ethyl β-apo-8'-carotenoate dose (AUFS 0.256). HPLC conditions: 15-min linear gradient from acetonitrile–water (80:20, v/v, with 0.1% ammonium acetate) to acetonitrile–dichloroethane (80:20, v/v with 0.1% cyclohexene) at 1.5 ml/min. (B) HPLC chromatogram of a serum sample taken at 5 hr after the β-apo-8'-carotenal dose (AUFS 0.064). HPLC conditions: 20-min linear gradient from acetonitrile–methanol–water (75:10:15, v/v, with 0.1% ammonium acetate) to acetonitrile–dichloroethane (75:25, v/v, with 0.1% cyclohexene) at 1.5 ml/min. (C) HPLC chromatogram of a serum sample taken at 48 hr after the 4,4'-dimethoxy-β-carotene dose (AUFS 0.064). HPLC conditions: 15-min linear gradient from acetonitrile–methanol–water (75:15:10, v/v) to acetonitrile–dichloroethane (85:15, v/v, with 0.1% cyclohexene) at 1.5 ml/min.

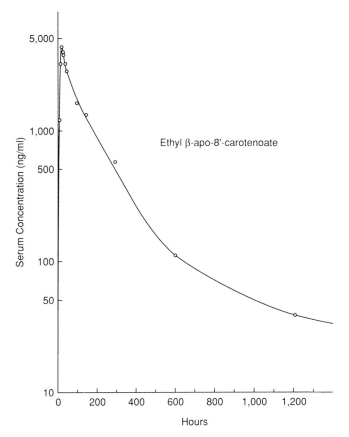

FIG. 5. Concentration versus time curve for ethyl β-apo-8′ carotenoate in a human subject.

run for an additional 15 min. Introduction of methanol to solvent A helps resolve β-apo-8′-carotenol, a metabolite of β-apo-8′-carotenal, from the lutein/zeaxanthin peak. The less polar solvent B is needed to elute β-apo-8′-carotenyl esters, also metabolites of β-apo-8′-carotenal, within a reasonable time period. An HPLC chromatogram of human serum for this experiment is shown in Fig. 4B.

4,4′-Dimethoxy-β-carotene and its metabolities can be resolved from other endogeneous carotenoids in human serum by use of a 15-min linear gradient from acetonitrile–methanol–water (75 : 15 : 10, v/v), (solvent A) to acetonitrile–dichloroethane (85 : 15, v/v, containing 0.1% cyclohexene) (solvent B) at a flow rate of 1.5 ml/min, followed by a 15-min period of isocratic elution with solvent B at the same flow rate. The methanol present in solvent A plays an important role in fostering the separation of 4,4′-dimethoxy-β-carotene from β-cryptoxanthin and lycopene. All major carotenoids in human serum, following the 4,4′-dimethoxy-β-carotene dose, are separated by the use of this solvent system (see Fig. 4C).

Several metabolites appear in human serum after doses either of β-apo-8′-carotenol or of 4,4′-dimethoxy-β-carotene, as shown in Fig. 4. For the purpose of characterization of metabolites, HPLC retention time and the absorption spectrum of a metabolite found in blood are compared with those of a standard. Chemical modifications such as saponification, acetylation, methylation, and reduction/oxidation usually facilitate the process of identification. β-Apo-8′-carotenyl palmitate, for example, is identified by comparison of its HPLC retention time with that of a synthetic standard, by comparison of its absorption spectrum, and by its susceptibility to saponification to yield a compound with the retention time of β-apo-8′-carotenol.

For quantitation, a standard curve for each of the three analogs and their major metabolites is constructed. In each case, the amount of the standard injected is plotted against its peak area under the same HPLC conditions as used for serum sample analysis. All data are corrected for the recovery of the internal standard. A concentration versus time curve for each subject is thus obtained. An example of the serum concentration versus time curve for ethyl β-apo-8′-carotenoate in a human subject is given in Fig. 5.

Kinetic Analysis

To analyze the time courses obtained from the human experiments, individual data for the concentration of the analogs or their metabolites versus time after dosing are initially fitted by weighted, nonlinear least-squares regression[10] to a five-component exponential equation of the form

$$y(t) = \sum_{i=1}^{n} I_i \exp(-g_i t)$$

where $y(t)$ is concentration of dose in serum at time t, I_i, and g_i are constants equal to the intercept and slope, respectively, of the semilog plot of each component i of the equation, and n is the number of exponential terms. One or two of the components may drop out during nonlinear regression. A fractional standard deviation (FSD, standard deviation/mean) of 0.1 is assumed as the weighting factor for all data. The process of curve fitting is facilitated by the conversational version of the Simulation,

---

[10] M. Berman and M. F. Weiss, *SAAM Manual*, DHEW Publ. No. 78-180 (NIH). U.S. Govt. Printing Office, Washington, D.C., 1978.

Analysis, and Modeling (SAAM) computer program (CONSAAM). The mechanics of this process are described in the SAAM tutorials.[11]

With an exponential equation, the area under the concentration–time curve (AUC) is readily calculated as

$$AUC = \int_0^\infty y(t)\, dt$$

$$= \int_0^\infty \sum_{i=1}^n [I_i \exp(-g_i t)]\, dt$$

$$= \sum_{i=1}^n I_i/g_i$$

AUC has units of (concentration × time). The area under the moment curve ($AUMC$) is calculated as

$$AUMC = \int_0^\infty ty(t)\, dt$$

$$= \int_0^\infty t\left\{\sum_{i=1}^n [I_i \exp(-g_i t)]\right\} dt$$

$$= \sum_{i=1}^n I_i/g_i^2$$

Units on $AUMC$ are (concentration × time^2).

Mean sojourn time (MST), the total time an average molecule spends in the body, after introduction into plasma but before irreversible loss,[12,13] is calculated from AUC and $AUMC$:

$$MST = AUMC/AUC$$

$$= \left[\int_0^\infty ty(t)\, dt\right] \Big/ \left[\int_0^\infty y(t)\, dt\right]$$

$$= \left(\sum_{i=1}^n I_i/g_i^2\right) \Big/ \left(\sum_{i=1}^n I_i/g_i\right)$$

Acknowledgments

This work was supported by Grant No. CA 46406 from the National Institutes of Health.

[11] D. M. Foster, R. C. Boston, J. A. Jacquez, and L. A. Zech, "The SAAM Tutorials: An Introduction to Using Conversational *SAAM* Version 29." Resource Facility for Kinetic Analysis (RFKA), Seattle, Washington, 1988.
[12] M. H. Green, J. B. Green, and K. C. Lewis, *J. Nutr.* **117**, 694 (1987).
[13] R. A. Shipley and R. E. Clark, "Tracer Methods for *in Vivo* Kinetics: Theory and Applications." Academic Press, New York, 1972.

[15] Metabolism of Carotenoids and *in Vivo* Racemization of (3S,3'S)-Astaxanthin in the Crustacean *Penaeus*

By KATHARINA SCHIEDT, STEFAN BISCHOF, and ERNST GLINZ

Introduction

In the shrimp *Penaeus* of the order Decapoda, as in most crustaceans, astaxanthin (3,3'-dihydroxy-β,β-carotene-4,4'-dione)[1,2] is the major carotenoid accumulated in the carapace; in the living animal, astaxanthin is bound noncovalently to a protein in a stoichiometric ratio. The carotenoprotein is water-soluble and may vary in color from blue to green to brown. Only after boiling or solvent extraction, astaxanthin is liberated by turning into its typically red-orange color.

Animals cannot synthesize carotenoids *de novo;* however, in the early 1970s it was also found that the shrimp *Penaeus* is able to convert dietary carotenoids to astaxanthin. Like most crustaceans they belong to the class of aquatic animals that can oxidize carotenoids at the 3,3' as well as the 4,4' position, regardless of whether the β-end groups are primarily oxidized in one of the two positions or not at all. This implies that the shrimp can convert β,β-carotene, zeaxanthin (β,β-carotene-3,3'-diol), and canthaxanthin (β,β-carotene-4,4'-dione) to astaxanthin, but they can absorb astaxanthin as well. Feeding experiments with *Penaeus monodon*,[3] *P. japonicus*,[4] and *P. vannamei,* three species of great interest in aquaculture, have confirmed the above theory.

In the 1980s, based on improved analytical methods, an almost fully "racemic" mixture of all three stereoisomers of astaxanthin, namely (3R,3'R), (3R,3'S; *meso*), and (3S,3'S) in an approximate ratio of 1:2:1 was observed in various species of *Penaeus*[5,6] and also in other orders of higher crustaceans.[5] Besides astaxanthin, minor amounts of yellow carot-

---

[1] Y. Tanaka, *Kogakubu Kenkyu Hokoku (Yamaguchi Daigaku) (Mem. Fac. Fish. Kagoshima Univ.)* **27**, 355 (1978).
[2] T. W. Goodwin, "The Biochemistry of the Carotenoids, Volume II: Animals," 2nd Ed. Chapman & Hall, London and New York, 1984.
[3] P. Menasveta, W. Worawattanamateekul, T. Latscha, and J. S. Clark, *Rovithai Int. Shrimp Semin.*, Bangkok, July 1990.
[4] S. Yamada, Y. Tanaka, M. Sameshina, and Y. Ito, *Aquaculture* **87**, 323 (1990).
[5] T. Matsuno, T. Maoka, M. Katsuyama, M. Ookubo, K. Katagiri, and H. Jimura, *Nippon Suisan Gakkaishi (Bull. Jpn. Soc. Sci. Fish.)* **50**, 1589 (1984).
[6] K. Schiedt, in "Carotenoids, Chemistry and Biology" (N. I. Krinsky, M. M. Mathews-Roth, and R. F. Taylor, eds.), Plenum, New York, 1990.

enoids, assumed to be reduction products of astaxanthin, have also been found in *P. japonicus*[6,7] and in *P. vannamei*.[6] One of these compounds was 3,4,3′,4′-tetrahydroxypirardixanthin(5,6,5′,6′-tetrahydro-β,β-carotene-3,4,3′,4′-tetrol). Moreover, the yellow isoastaxanthin (4,4′-dihydroxy-ε,ε-carotene-3,3′-dione) with the novel 3-oxo-4-hydroxy ε-end group has been identified in our laboratories.[6,8] The relative and absolute configurations of the two compounds were assigned using ^{1}H NMR and circular dichroism (CD). Two diastereomers of tetrahydroxypirar-dixanthin were characterized; the most relevant one was identical with the optically active (3R,4S,5R,6R,3′R,4′S,5′R,6′R) isomer synthesized by Hengartner.[9] In our recent paper,[8] we reported the *in vivo* racemization of optically active (3S,3′S)-[15,15′-^{3}H$_2$]astaxanthin in *Penaeus japonicus*.

Objectives and Strategy

The aim of this study was to confirm the *in vivo* racemization of optically active (3S,3′S)-[15,15′-^{3}H$_2$]astaxanthin in *Penaeus japonicus* with regard to the astaxanthin esters representing the bulk of carotenoids in the carapace. Previously,[8] we had analyzed the "free astaxanthin fraction" only. The mono- and diester fractions containing a 6-fold amount of carotenoids and a 3-fold dose of radioactivity had to be further investigated. These fractions also contained the yellow metabolites isoastaxanthin and the tetrols, all in esterified form. Besides astaxanthin, the so-called monoacetylenic and diacetylenic asterinic acids (7,8-didehydroastaxanthin and 7,8,7′,8′-tetradehydroastaxanthin) also had to be analyzed as to their chirality in analogy to that of the free "astaxanthin fraction." Moreover, which of the other carotenoids occurring in the carapace are metabolites of astaxanthin should be investigated by radiolabeling.

The relevant end groups and structures of carotenoids identified in *P. vannamei*[6,8] are depicted in Fig. 1. Some reference samples and numerous spectroscopic and chromatographic data for those compounds were available for comparison with those now isolated from *P. japonicus*.

A detailed methodology is presented that made it possible to unravel the very complex mixture of labeled and unlabeled carotenyl esters, their saponification, and identification. The following difficulties had to be addressed. In part, the carotenoid moities of some esters, for instance, those with one or two astaxanthin end groups, are oxidized to the astacene

---
[7] K. Katagiri, Y. Koshino, T. Maoka, and T. Matsuno, *Comp. Biochem. Physiol. B: Comp. Biochem.* **87B**, 161 (1987).
[8] K. Schiedt, S. Bischof, and E. Glinz, *Pure Appl. Chem.* **63**, 89 (1991).
[9] U. Hengartner, Abstract 5-4, *9th Int. Symp. Carotenoids, Kyoto, May 20–25, 1990*.

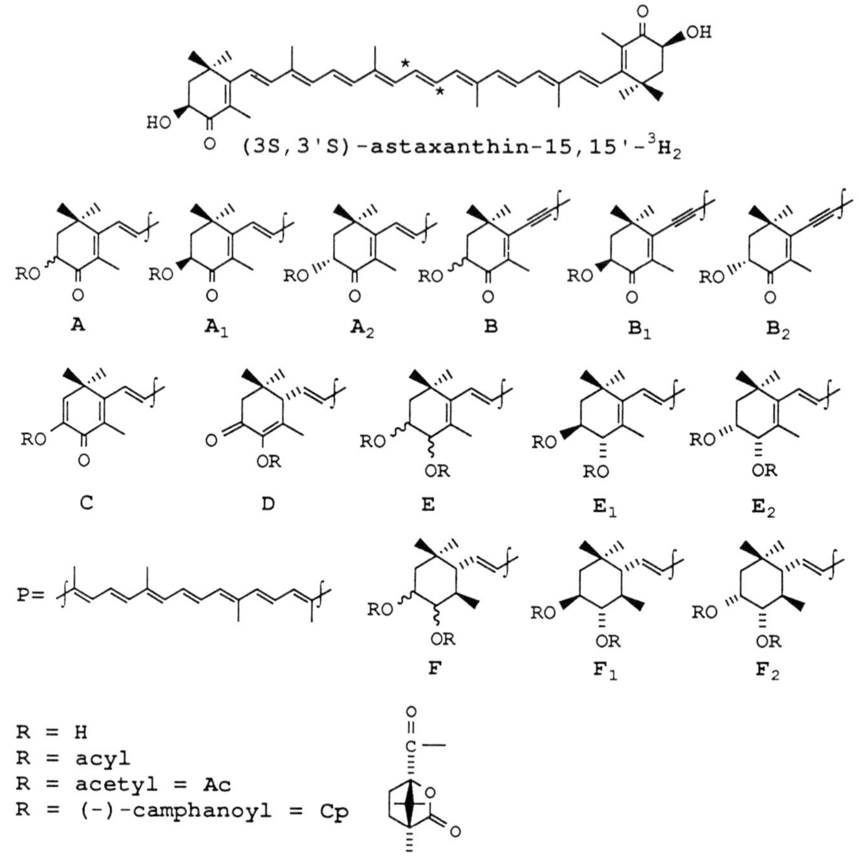

(3S,3'S)-astaxanthin-15,15'-³H₂

A, A₁, A₂, B, B₁, B₂

C, D, E, E₁, E₂

P=

F, F₁, F₂

R = H
R = acyl
R = acetyl = Ac
R = (−)-camphanoyl = Cp

FIG. 1. Structures and derivatives mentioned in the text.

A–P–A,	R = H, Astaxanthin; R = Cp, astaxanthin (−)-dicamphanate
C–P–C,	R = H, Astacene; R = Ac, astacene diacetate
C–P–A,	R = H, Semiastacene
B–P–A,	R = H, "Monoacetylenic asterinic acid," i.e., 7,8-didehydroastaxanthin; R = Cp, 7,8-didehydroastaxanthin dicamphanate
B–P–B,	R = H, "Diacetylenic asterinic acid," i.e., 7,8,7',8'-tetradehydroastaxanthin; R = Cp, 7,8,7',8'-tetradehydroastaxanthin dicamphanate
D–P–D,	R = H, Isoastaxanthin; R = Ac, isoastaxanthin diacetate
E–P–E,	R = H, Crustaxanthin; R = Ac, crustaxanthin tetraacetate
F–P–E,	R = H, 5,6-Dihydrocrustaxanthin; R = Ac, 5,6-dihydrocrustaxanthin tetraacetate
F–P–F,	R = H, Tetrahydroxypirardixanthin
F₁–P–F₁,	R = Ac, Tetraacetoxypirardixanthin, 3,4-*trans*/3',4'-*trans*
F₂–P–F₂,	R = Ac, (3S,4S,5R,6R,3'R,4'S,5'R,6'R)-Tetraacetoxypirardixanthin, 3,4-*cis*/3',4'-*cis*
F₁–P–F₂,	R = Ac, Tetraacetoxypiradixanthin, 3,4-*trans*/3',4'-*cis*
A₁/A₂,	(3S) and (3R) end groups
B₁/B₂,	(3S) and (3R) end groups
E₁/E₂,	(3S,4S) and (3R,4S) end groups

end group (3-hydroxy-2,3-didehydro-4-oxo-β-carotene end group), thus losing their chirality. Other carotenoids are lost during hydrolysis owing to alkali instability, and others are not stable in the free form. To recognize unwanted reactions, it is essential to determine the carotenoid content and radioactivity before each reaction and each step of the isolation procedure and recovery thereafter.

Strictly anaerobic alkaline saponification was preferred to the enzymatic one because of the better solubility of the lipid extract and because of the need to avoid a possible preferential enzymatic hydrolysis of one particular stereoisomer.

Various systems of adsorption and reversed-phase (RP) chromatography on columns (CC), thin layer chromatography (TLC), and high-performance liquid chromatography (HPLC) were used for the separation of the different carotenoids. The yellow carotenoids such as isoastaxanthin and the tetrols were acetylated immediately after saponification to improve their stability. Moreover, favorable conditions for HPLC were obtained by acetylation. Not only were the retention times (R_t) of these polar polyhydroxy compounds reduced, but tailing of isoastaxanthin, caused by its enolic hydroxy groups, was also abolished. The stereochemical composition of astaxanthin, 7,8-didehydroastaxanthin, and 7,8,7',8'-tetradehydroastaxanthin was analyzed via the dicamphanate derivatives, and resolution of the diastereoisomers was achieved by HPLC.

Methods and Instruments

Precautions should be taken to protect light- and oxygen-sensitive carotenoids. All operations are carried out in subdued light and the solvents removed on a Rotavapor at water-bath temperatures not exceeding 40°. Reagent and/or HPLC grade solvents are used. The dried residues of the samples are flushed with argon and stored deep-frozen at −20° (or for longer periods of several weeks at −80°).

Quantification of Carotenoids

The apparatus used is a UVIKON 810 spectrophotometer (Kontron, Zurich, Switzerland) combined with an UVIKON 21 Recorder. The carotenoids of extracts and of purified fractions are quantified by UV/VIS spectrophotometry. Because of the high polarity of the relevant oxycarotenoids, the residues are dissolved in traces of ethanol and/or dichloromethane. The final solution contains 2–3% ethanol in n-hexane in order to prevent adhesion of the carotenoids to the glass wall. Not more than 2% dichloromethane is allowed. The estimation of the carotenoid content of crystalline preparations is based on the extinction coefficient of the all-trans isomer, whereas in biological fractions a "cis-share" of about 20% is taken into account. The extinction coefficients ($E_{1\,cm}^{1\%}$) at λ_{max} in n-hexane tabulated below are used:

Compound	All-trans		Cis/trans
	$E_{1\,cm}^{1\%}$	λ_{max} (nm)	$E_{1\,cm}^{1\%}$
Astaxanthin	2120	472	1910
7,8-Didehydroastaxanthin	2280	474	—
7,8,7',8'-Tetradehydroastaxanthin	2440	474	—
Yellow oxycarotenoids	—	438–448	2300

Carotenoid esters are expressed in microgram carotenol equivalents. When using HPLC, the carotenoids and their cis and trans isomers are quantified by peak area and expressed as the percentage of the whole fraction.

Counting of Tritium Radioactivity

The samples are dissolved in a mixture consisting of 3 ml methanol plus 9 ml LIPOTRON (Kontron), a commercial lipophilic scintillator cocktail, counted in a Kontron BETAmatic II scintillation counter, and the results corrected for quenching and half-life decay. The counting efficiency for tritium is approximately 38%.

Experimental

Animals

Shrimp. Penaeus japonicus, average body weight 6–7 g, are purchased from a hatchery at the Mediterranean Sea of France.

Rearing Conditions. At the start, 100 animals are kept in glass tanks of 200 liters, whose bottom is covered with sand.

Seawater. Artificial seawater is used in a recycling system, with oxygen saturation and a temperature of 22°.

Feeding. In our study, the commercial feed Gold Coin was supplied for the first 4 weeks but was not readily accepted. Therefore, fresh mussels *Mytilus edulis* were fed for 3 months. Growth of the shrimps improved, but mortality was still high, until the number of shrimps was reduced drastically. The remaining 6 animals developed quite well (weight range 8–14 g) and were again trained to commercial feed.

Astaxanthin Additive

(3S,3'S)-[15,15'-^{3}H$_2$]Astaxanthin is synthesized with a specific activity of 198 μCi/mg, corresponding to 440,000 disintegrations/min (dpm)/μg. The radiopurity is 98%.

Feed Supplementation. ^{3}H-Labeled astaxanthin (12 mg) is dissolved in 25 ml dichloromethane, then pooled with 2.7 g cod liver oil containing 12% lecithin and freed carefully from dichloromethane in a Rotavapor at 38°. The mixture is kept warm in order to avoid crystallization of astaxanthin.

For dietary supplementation, 2.25 g (10 mg astaxanthin) of the oily solution is added to 47.8 g Gold Coin pellets and mixed immediately, resulting in a level of 200 ppm astaxanthin. The feed intake is estimated at 2% of body weight per day. Daily portions are weighed into brown bottles, stored under argon at 6°, and added to the water. The dosage per 6 prawns per day is 2 g feed, equivalent to 400 μg astaxanthin (83.6 μCi). The duration of the feeding period is 21 days. The size of the shrimp at termination ranged from 7 to 14 g (live weight), length 10.5–12.5 cm. The animals are sacrificed by freezing and the carapaces removed immediately before extraction.

Extraction

The pooled integuments are ground and extracted exhaustively with acetone; the lipids are transferred into ether/hexane (1:1, v/v) after addition of ethanol and water. The organic phase is washed and the solvent evaporated, yielding 295 mg lipids (0.9%), containing 1.6 mg astaxanthin equivalents. The radioactivity is 28.16 × 10^6 dpm, equivalent to 17,300 dpm/μg carotenoids.

Anaerobic Saponification

An apparatus has been developed permitting saponification of microgram amounts of astaxanthin esters in complete absence of oxygen, yielding 80% astaxanthin along with minor amounts of semiastacene (3,3'-dihydroxy-2,3-didehydro-β,β-carotene-4,4'-dione) and traces of astacene (3,3'-dihydroxy-2,3,2',3'-tetradehydro-β,β-carotene-4,4'-dione). The modified Schlenk tube with appendix is suitable for lipid residues of at least 10 to 100 mg containing 5 μg to milligram amounts of carotenoids. The alkali equivalents are adjusted to the lipid content (assumed molecular weight of fatty acids, 300).

Operations. The reduced pressure in the empty system has to be checked (0.001 mm Hg). To remove oxygen from all connections, the system is then flushed with argon.

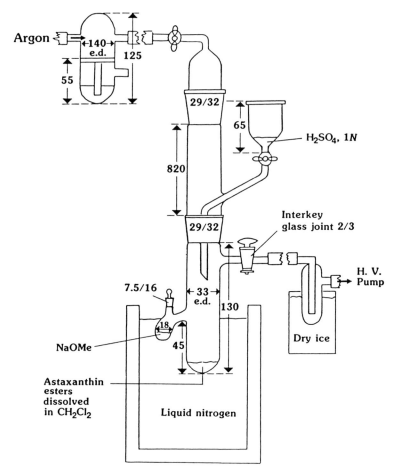

FIG. 2. Apparatus and equipment for anaerobic saponification, measurements given in millimeters.

1. Lipids (≤100 mg, ≤0.3 mmol) are dissolved in 1 ml dichloromethane and transferred onto the bottom of the tube containing a small magnetic rod.
2. One milliliter sodium methylate (0.6 mmol Na; the solution contains 1.5 g sodium dissolved in 100 ml methanol) is added into the appendix.
3. Flush the tube with argon and close the system.
4. The solutions in tube and appendix are frozen in liquid nitrogen and evacuated (0.001 mm Hg). Close the connection to the high vacuum pump.

5. Remove the system from the liquid nitrogen and thaw carefully at room temperature in order to degas the solutions. Repeat Steps 4 and 5 twice.
6. The sodium methylate is then pooled with the ester solution under vacuum and stirred for 10 min at room temperature.
7. The reaction is stopped by acidification under vacuum with 1 ml of 0.5 M H_2SO_4 (0.5 mmol) from the funnel and stirred vigorously. The system can then be opened by flushing with argon.
8. The reaction mixture is transferred with ethanol and water to a separatory funnel, the carotenoids extracted with ether/n-hexane (1:1, v/v) washed to neutral, and the solvents evaporated.

Conventional Saponification

We have observed that, apparently under the conditions used for the anaerobic saponification, the esters of astaxanthin and asterinic acids are hydrolyzed, whereas some unpolar yellow fractions, at first considered to be carotenes, resist saponification. By various TLC systems (see Table II) these fractions proved to be esters of the hydroxycarotenoids. These esters are saponified as follows. Lipid residues (10–30 mg) are dissolved in 5 ml ether, and 10 ml of a 5% KOH solution in 90% ethanol/water is added. After 2 hr at room temperature, the carotenoids are extracted with ether/n-hexane after acidification with 10% acetic acid and the organic phase washed to neutral.

Under these conditions, the reference substances tetraacetoxypirardixanthin and isoastaxanthin diacetates yield 95% recovery of the carotenols. It should be noted, however, that the acetylenic compounds 7,8-didehydroastaxanthin and 7,8,7′,8′-tetradehydroastaxanthin tend to form considerable amounts of 9-*cis* isomers by both types of saponification. An example of the HPLC resolution of cis and trans isomers of astacene and the corresponding 2,3,2′,3′-tetradehydro artifacts of mono- and diacetylenic asterinic acids via their diacetates is shown in the HPLC section [for R_t values, see system (4)].

Preparation of (−)-Dicamphanates of Astaxanthin, 7,8-Didehydroastaxanthin, and 7,8,7′,8′-Tetradehydroastaxanthin

All-trans isomers of astaxanthin, 7,8-didehydroastaxanthin, and 7,8,7′8′-tetradehydroastaxanthin obtained after anaerobic saponification of the esters are isolated by HPLC and reacted with (−)-camphanoyl chloride as described by Vecchi and Müller[10] and later modified for bio-

---

[10] M. Vecchi and R. K. Müller, *J. High Resolut. Chromatogr. Chromatogr. Commun.* **2**, 195 (1979).

logical samples containing minute amounts of carotenoids in the presence of excess of lipids.[11]

In the experiment with crustaceans, the conditions are as follows. A mixture containing 10–40 µg carotenoid (all-*trans*-astaxanthin, 7,8-didehydroastaxanthin, or 7,8,7',8'-tetradehydroastaxanthin isolated by HPLC), 0.15 ml dry pyridine, and 15 mg (−)-camphanoyl chloride is reacted under argon at room temperature for 10 min. The reaction mixtures are transferred with ether to separatory funnels containing ice water and about 5 ml each of 10% acetic acid, ethanol, and *n*-hexane. The organic phases containing the camphanates are washed with water and evaporated. Further purification and isolation of dicamphanates are achieved by preparative TLC (2) (see below; for R_f values, see Table II).

Acetylation of Yellow Hydroxycarotenoids

After saponification of the esters, the carotenols are dried in a desiccator; 0.2 ml dry pyridine, 0.1 ml acetic anhydride, and some crystals of the catalyst 4-dimethylaminopyridine (DMAP) are added.[12] The catalyst enhances the reaction rate so that tetraacetates are yielded, whereas without catalyst a mixture of di-, tri-, and tetraacetates is obtained. The optimum reaction time has not been tested, but the mixture is left overnight at 4° under argon. The samples are transferred to 0.1% acetic acid with ether/*n*-hexane. After addition of some ethanol and extraction of the carotenoids, the organic phase is washed and evaporated [for R_f values of the acetylated compounds on TLC (3b), see Table II].

Chromatographic Systems

Column Chromatography

Three different CC systems are used, two employing silica and one magnesium oxide.

Silica CC (1) and (1a). Silica (Kieselgel 60, Merck 9385, 0.040–0.063 mm) is used for flash chromatography [CC (1)] (0.2 bar, nitrogen or argon). Larger particle size silica (Kieselgel 60, Merck 7734, 0.063–0.2 mm) is used for ordinary column chromatography [CC (1a)], should the flash system not be available.

The columns are prepared with *n*-hexane; the substance is dissolved in a few drops of dichloromethane, *n*-hexane added, and the sample applied onto the column. The eluents for CC (1) and (1a) are *n*-hexane–ether–

[11] K. Schiedt, F. J. Leuenberger, and M. Vecchi, *Helv. Chim. Acta* **64**, 449 (1981).
[12] G. Höfle, W. Steglich, and H. Vorbrüggen, *Angew. Chem.* **90**, 602 (1978).

methanol mixtures of increasing polarity. The following fractions are eluted.

20–30% Ether/hexane: "astaxanthin diesters"
40% Ether/hexane: "astaxanthin monoesters"
3% Methanol/ether: "free astaxanthin"

These main fractions are submitted to preparative TLC (1a) as indicated below. Subsequently, the corresponding zones are pooled. R_f values and VIS spectra are presented in Table I.

The shape of the spectra of fractions 2 and 3 (Table I) indicate a relatively large share of yellow carotenoids. These fractions contain the yellow metabolites reported recently.[8] Fraction 4 does not exhibit the typical, symmetrical astaxanthin spectrum either, but it is characterized by an inflection at 502 nm due to 7,8-didehydroastaxanthin and 7,8,7',8'-tetradehydroastaxanthin, which are inseparable from astaxanthin by CC and TLC, but can be resolved by HPLC as shown below (see Figs. 3 and 4).

Magnesium Oxide CC (2). Chromatography on MgO of the tetraacetoxycarotenoids prior to HPLC favors the resolution and isolation of the single pure compounds and stereoisomers with regard to the 3,4- and 3',4'-*cis* and *trans* configuration.

Two grams of MgO (British Drug House, Poole, U.K., BDH Prod. 15038) and 1 g of Dicalite Speedex are mixed, and the column is prepared with *n*-hexane. The following fractions are eluted.

Fraction 1.
30% Ethyl acetate/*n*-hexane: tetraacetoxypirardixanthin
Fraction 2.
50% Ethyl acetate/*n*-hexane: 5,6-dihydrocrustaxanthin tetraacetate
Fraction 3. 100% Ethyl acetate: crustaxanthin tetraacetate

These three fractions are submitted to HPLC for resolution (see Fig. 5).

TABLE I
MAJOR CAROTENOID FRACTIONS OF LIPID EXTRACT[a]

No.	Fraction	R_f	VIS λ_{max} (nm)		
1	Yellow esters	0.87	(sh)422	445	471
2	"Astaxanthin diesters"	0.80	444	467	(sh)504
3	"Astaxanthin monoesters"	0.62	446	471	(sh)502
4	"Astaxanthin"	0.35		471	(sh)502

[a] From integuments obtained by CC on silica, followed by preparative TLC (1a). Shoulder or inflection (sh).

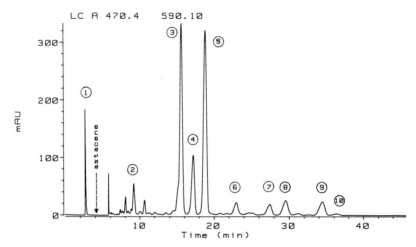

FIG. 3. HPLC system (1). Peaks and R_t: 1, not identified; 2, semiastacene, 9.23 min; 3, all-*trans*-diacetylenic asterinic acid, 15.68 min; 4, all-*trans*-monoacetylenic asterinic acid, 17.26 min; 5, all-*trans*-astaxanthin, 18.89 min; 6, diacetylenic asterinic acid, unidentified cis isomer, 23.02 min, 7, diacetylenic asterinic acid, unidentified cis isomer, 27.48 min; 8, 9-*cis*-astaxanthin, 29.59 min; 9, 13-*cis*-astaxanthin, 34.50 min; 10, 15-*cis*-astaxanthin, 36.10 min.

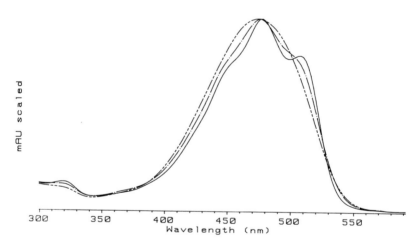

FIG. 4. VIS spectra of all-trans peaks recorded by HPLC system (1). (Peak 3, Fig. 3) Diacetylenic asterinic acid (——), λ_{max} (sh)452,476,(sh)506 nm; (peak 4, Fig. 3) monoacetylenic asterinic acid (—·—), λ_{max} 476(sh)508 nm; and (peak 5, Fig. 3) astaxanthin (-----), λ_{max} 474 nm.

Thin-Layer Chromatography and Preparative Thin-Layer Chromatography

TLC (1). Silica plates are Kieselgel 60 F_{254} (Merck 5715, 0.25 mm, 20/20 cm).

TLC (2). Silica plates HPTLC (high-potency thin-layer chromatography, Kieselgel 60 F_{254}, Merck 5629, 10/10 cm) are used for qualitative and preparative use on a microscale.

TLC (3). Silica plates coated with citric acid (CA) are prepared by submerging silica plates indicated under TLC (1) in a solution of 2.5% citric acid in methanol; the plates are air-dried and then activated in a vacuum drier at 120° for 1 hr. Coating with citric acid prevents tailing of astacene, semiastacene, and isoastaxanthin with their enolic hydroxy groups exhibiting acidic characteristics. Thus, sharp separation and precise quantification of the single zones are achieved.

TLC (4). Reversed-phase plates are KC_{18} F (Whatman, Clifton, NJ, No. 4803 800, 20/20 cm, 0.2 mm), and the eluent is 5% water in acetone for the separation of carotenyl esters and carotenes.

Eluents for TLC (1)–(3). Three eluents are prepared: (a) ethyl acetate/n-hexane (1:1, v/v), (b) ethyl acetate/n-hexane (2:1, v/v), and (c) 3% ethyl acetate in n-hexane.

Desorption of Carotenoid Zones Resolved by Preparative TLC. Ethyl acetate/ether/ethanol (50:50:3, v/v) is used for desorption of carotenoids. When CA-coated plates [TLC (3)] are used, the citric acid has to be washed out with water before evaporation of the solvent.

The R_f values of some carotenoids and their derivatives in various systems are compiled in Table II.

High-Performance Liquid Chromatography

The instrument is an HP 9000/300 Pascal work station that runs HP 79995 software, Revision 4.2. HP 1040M and 1090M diode array detectors are used. The columns are stainless steel, inner diameter 4 mm, length 25 cm.

System (1). System (1) separates astaxanthin, semiastacene, astacene, and mono- and diacetylenic asterinic acid (see Figs. 3 and 4). The essential point of this modified method is the use of an acid-coated column as described by Vecchi *et al.*[13] The stationary phase is Nucleosil 100-3, coated with 1% H_3PO_4; the mobile phase is 10% dichloromethane–1% 2-propanol in n-hexane, flow rate 1 ml/min. The detector is set at λ 470 nm.

---

[13] M. Vecchi, E. Glinz, V. Meduna, and K. Schiedt, *J. High Resolut. Chromatogr. Chromatogr. Commun.* **10**, 348 (1987).

TABLE II
R_f Values of Carotenoids and Their Derivatives on Thin-Layer Chromatography

Compound	(1a)	(2a)	(2c)	(3a)	(3b)	(4)
Astaxanthin	0.3	0.37		0.4		
Astaxanthin dicamphanate		0.45				
7,8-Didehydroastaxanthin	0.3					
7,8,7′,8′-Tetradehydroastaxanthin	0.3					
Astaxanthin monopalmitate	0.6					
Astaxanthin dipalmitate	0.8					
Zeaxanthin dipalmitate	0.85	0.85	0.33			0
β,β-Carotene	0.85	0.9	0.8	0.8		0.3
Semiastacene				0.55	0.42	
Astacene				0.70	0.65	
Astacene diacetate	0.45	0.45				
Isoastaxanthin				0.71		
Isoastaxanthin diacetate	0.55	0.55				
Crustaxanthin	0				0.07	
Crustaxanthin tetraacetate	0.5					
Tetraacetoxypirardixanthin	0.45–0.6				0.07	
Tetrahydroxycarotenoids	0					
Mytiloxanthin	0.13					
Mytiloxanthin triacetate	0.58					
2,3-Didehydro-4-oxomytiloxanthin	—				0.44	
2,3-Didehydro-4-oxomytiloxanthin triacetate	0.42					

System (2). Dicamphanates of astaxanthin and asterinic acid are separated by HPLC system (2). The conditions described by Vecchi and Müller[10] have been modified slightly. The stationary phase is Spherisorb S3-CN, and the mobile phase is 17% isopropyl acetate–7% acetone in n-hexane. The flow rate is 0.9 ml/min, and the detector is set at λ 470 nm. Retention times (R_t) are tabulated for the following dicamphanates.

	R_t (min)		
All-trans isomer	(3R,3′R)	(3R,3′S)	(3S,3′S)
7,8,7′,8′-Tetradehydroastaxanthin	13.2	15.9	19.3
7,8-didehydroastaxanthin	15.5	18.8	22.8
Astaxanthin	17.3	21.1	26.0

System (3). Carotenyl tetraacetates of different configurations are separated with HPLC system (3). The stationary phase is Spherisorb S3-CN. The mobile phase is 2% acetone, 10% dichloromethane, and 0.1% N-ethyldiisopropylamine in n-hexane, at a flow rate of 1 ml/min. Detection is at λ 445 nm.

Resolution of carotenyl tetraacetates is shown in Fig. 5. In Fig. 6, VIS spectra of the various tetraacetates are compared with that of isoastaxanthin diacetate, which exhibits a similar fine structure.

System (4). Separation of isoastaxanthin diacetate, astacene diacetate, 7,8-didehydroastacene diacetate, and 7,8,7′,8′-tetradehydroastacene diacetate is achieved with HPLC system (4). The astacene derivatives with their 2,3-didehydro β-end groups are artifacts due to saponification under aerobic conditions. They are side products during the isolation of isoastaxanthin. The formation of large amounts of 9-*cis* and 9,9′-di-*cis* isomers of the acetylenic compounds is expressed in relative peak area.

The stationary phase is Spherisorb S3-W; the mobile phase is 20% ethyl acetate, 0.1% N-ethyldiisopropylamine in n-hexane, at a flow rate of 1 ml/min. Detection is at λ 470 nm.

Retention times (R_t) and relative peak area of the above mentioned diacetates are tabulated below.

Diacetate	Isomer	R_t (min)	Relative area (%)
Diacetylenic astacene	9,9′-di-*cis*	11.3	40
Diacetylenic astacene	9-*cis*	12.9	44
Isoastaxanthin	all-*trans*	14.1	—
Diacetylenic astacene	all-*trans*	14.4	16
Monoacetylenic astacene	9-*cis*	16.1	46
Monoacetylenic astacene	all-*trans*	18.3	54
Astacene	all-*trans*	22.2	—

Results and Discussion

We have described various methods suitable not only for the separation and quantification of some carotenoids occurring in the crustacean *Penaeus,* but also for the isolation of pure compounds and their identification by physicochemical methods (VIS, MS, ^{1}H NMR, CD). It is beyond the scope of this chapter to present all the spectroscopic data. The assignment of the absolute configuration of isoastaxanthin and tetraacetoxypirardixanthin isolated previously from *P. vannamei* is intended for publication.

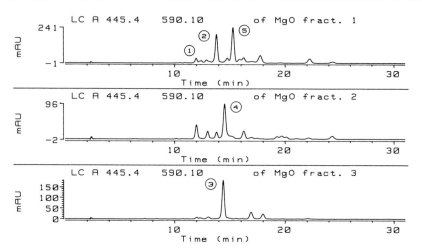

FIG. 5. Resolution of MgO fractions 1-3 by HPLC system (3). Assignment of relative configurations was by comparison to reference samples. (Peaks 1) Tetraacetoxypirardixanthin, 3,4-*trans*/3',4'-*trans*; (peak 2) tetraacetoxypirardixanthin, 3,4-*trans*/3',4'-*cis*; (peak 3) crustaxanthin; (peak 4) 5,6-dihydrocrustaxanthin; (peak 5) tetraacetoxypirardixanthin, 3,4-*cis*/3',4'-*cis*.

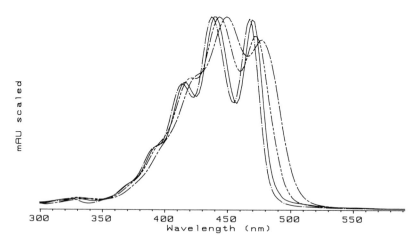

FIG. 6. VIS spectra recorded in eluents of HPLC systems (3) and (4). all-*trans*-Isoastaxanthin diacetate (——), tetraacetoxypirardixanthin (— · —), 5,6-dihydrocrustaxanthin tetraacetate (— — — —), and crustaxanthin (— — —).

In this experiment with the small number of shrimp (*P. japonicus*) fed (3S,3'S)-[15,15'-^{3}H$_2$]astaxanthin and the minute amount of carotenoids extracted from the carapace, we had to rely, at least partly, on comparison with the reference substances isolated and identified previously from *P. vannamei*.[8] This applies in particular to tetrahydroxypirardixanthin and isoastaxanthin isolated from *P. japonicus* in microgram amounts only. However, it was possible to quantify these compounds by VIS spectrophotometry and to measure their specific radioactivity for comparison with that of astaxanthin. The results regarding absorption, distribution, and metabolism are compiled in Tables III, IV, and V.

Absorption

The ^{3}H-labeled astaxanthin fed to the shrimp (specific activity 420,000 dpm/μg) had been absorbed satisfactorily (Table III). From the specific radioactivity of the isolated astaxanthin (100,000 dpm/μg) an endogenous dilution of 1:4 could be calculated. The bulk of the carotenoids were esterified. The ratio of astaxanthin, 7,8-didehydroastaxanthin, and 7,8,7',8'-tetradehydroastaxanthin, the major carotenoids in the carapace of these shrimps, was similar in all three fractions (i.e., the free carotenols, the monoesters, and the diesters). The relatively low specific activity in the mono- and diacetylenic asterinic acids was considered to be a possible contamination by some cis isomers of astaxanthin that, unexpectedly, were not removed by further purification and derivatization.

In Vivo Racemization of Astaxanthin

The endogenous racemization (Table IV) of optically active ^{3}H-labeled astaxanthin had been reported for the free astaxanthin fraction only.[8] The ester fractions were saponified anaerobically and the stereochemical composition of astaxanthin, 7,8-didehydroastaxanthin, and 7,8,7',8'-tetradehydroastaxanthin analyzed via the dicamphanate derivatives. All three stereoisomers, namely, (3R,3'R), (3R,3'S; *meso*), and (3S,3'S), were found again, both in astaxanthin and in the corresponding mono- and diacetylenic compounds. The almost "racemic" ratio of 1:2:1 observed in the free astaxanthin fraction was slightly shifted in favor of the (3S,3'S) isomers in the esters. The almost equal specific radioactivity in all three configurational isomers of astaxanthin has unequivocally proved an endogenous racemization of the optically active astaxanthin in *Penaeus*.

Carotenoid Composition and Metabolites

We have suggested that isoastaxanthin with the novel 3-oxo-4-hydroxy ε-end group might be a key compound for an *in vivo* racemization[6,8];

TABLE III
ISOLATION PROCEDURE, RATIO, AND SPECIFIC RADIOACTIVITY OF ASTAXANTHIN AND MONO- AND DIACETYLENIC ASTERINIC ACID[a]

Column and thin-layer chromatographies

Fraction	µg	dpm/µg
Xanthophyll esters	173	6900
"Astaxanthin" diesters	980	15,700
"Astaxanthin monoesters	190	26,000
"Astaxanthin," free	170	45,600

Compound	Relative area (%)	µg	dpm/µg
HPLC: Isolation of all-trans isomers, from free astaxanthin fraction			
7,8,7′,8′-Tetradehydroastaxanthin	28	20	3000
7,8-Didehydroastaxanthin	19	15	7000
Astaxanthin	53	43	94,400
Anaerobic saponifiction of ester fractions			
HPLC: Isolation of all-trans isomers, from monoesters			
7,8,7′,8′-Tetradehydroastaxanthin	42	11	2650
7,8-Didehydroastaxanthin	14	4	10,500
Astaxanthin	44	15	108,200
HPLC: Isolation of all-trans isomers, from diesters			
7,8,7′,8′-Tetradehydroastaxanthin	38	37	5070
7,8-Didehydroastaxanthin	15	9	14,900
Astaxanthin	47	37	104,400

[a] In the main fractions of the acetone extract of the carapace of *Penaeus japonicus* fed (3S, 3′S)-[15,15′-^{3}H$_2$] astaxanthin (specific activity 420,000 dpm/µg) for 3 weeks. The total carotenoids in the acetone extract (astaxanthin equivalent) were 1600 µg/28 × 10^6 dpm.

however, our hypothesis could not be confirmed in this experiment. The low specific radioactivity (Table V) found in isoastaxanthin shows that this compound was not involved directly in the racemization of astaxanthin. Also, tetrahydroxypirardixanthin, crustaxanthin, and 5,6-dihydrocrustaxanthin, with their low specific radioactivities, cannot be considered as the reduction products of the ^{3}H-labeled astaxanthin fed to the shrimp.

TABLE IV
RATIO AND SPECIFIC RADIOACTIVITY OF STEREOISOMERS OF ASTAXANTHIN, 7,8-DIDEHYDROASTAXANTHIN, AND 7,8,7',8'-TETRADEHYDROASTAXANTHIN VIA DIASTEREOMERIC (−)-DICAMPHANATE DERIVATIVES[a]

Compound, stereoisomer (dicamphanate)	All-trans isomers					
	Relative area (%)			Astaxanthin equivalents (dpm/µg)		
	Free	Monoesters	Diesters	Free	Monoesters	Diesters
7,8,7',8'-Tetradehydroastaxanthin						
(3R,3'R)	25	30	39	500	800	1110
(3R,3'S; meso)	46	49	49	50	530	1630
(3S,3'S)	29	21	12	50	980	4820
7,8-Didehydroastaxanthin						
(3R,3'R)	15	28	36	1470	5540	7140
(3R,3'S; meso)	37	49	50	990	6830	11,400
(3S,3'S)	48	23	14	1180	12,900	25,900
Astaxanthin						
(3R,3'R)	13	24	28	75,000	91,200	83,400
(3R,3'S; meso)	43	49	50	88,000	105,000	98,600
(3S,3'S)	44	27	22	113,000	139,000	147,000

[a] Specific radioactivity of (3S,3'S) [15,15'-³H₂] astaxanthin in feed was 420,000 dpm/µg.

TABLE V
RELATIVE ABUNDANCE OF CAROTENOIDS IN CARAPACE OF
Penaeus japonicus[a]

Compound	%	μg	dpm/μg	dpm × 10⁶
Astaxanthin	17.5	273	100,000	27
7,8,7′,8′-Tetradehydroastaxanthin	20.1	314	3000	0.9
7,8-Didehydroastaxanthin	9.7	151	10,000	1.5
Isoastaxanthin	2.7	42	6900	0.3
Tetrahydroxypirardixanthin	0.6	9	3000	0.03
5,6-Dihydrocrustaxanthin	0.2	3	—	
Crustaxanthin	1.6	47	50	
4-Oxomytiloxanthin	3.6	56	—	
Carotenoids, not identified	44	690		

[a] Fed *Mytilus edulis* and commercial feed supplemented with 200 ppm (3S,3′S)-[15,15′-³H₂]astaxanthin. Total carotenoids in the acetone extract were 1560 μg/28 × 10⁶ dpm.

The mono- and diacetylenic compounds 7,8-didehydro- and 7,8,7′,8′-tetradehydroastaxanthin have never been considered as metabolites of astaxanthin, but rather as the 4,4′-dioxo derivatives of diatoxanthin [(3R,3′R)-7,8-didehydro-β,β-carotene-3,3′-diol] and of alloxanthin [(3R,3′R)-7,8,7′,8′-tetradehydro-β,β-carotene-3,3′-diol] originating from algae via the food chain. Carotenoid analysis of the mussels *Mytilus edulis*,[14] fed to the shrimp preexperimentally, proved diatoxanthin and alloxanthin to be carotenoid constituents of these molluscs (5 and 34% of total carotenoids, respectively). The same compounds have previously been characterized with absolute configurations in the same organism by Hertzberg *et al.*[15] It may be concluded that, in the shrimp, the optically active (3R)-3-hydroxy β-end group found in *Mytilus edulis* was not only oxidized at the 4 position but also racemized at C-3, thus apparently contributing to the large share of 7,8-didehydroastaxanthin and 7,8,7′,8′-tetradehydroastaxanthin in our experimental animals.

It remains an intriguing phenomenon why 7,8-didehydroastaxanthin isolated from the esters exhibits a relatively high specific radioactivity, in contrast to that originating from the free "astaxanthin" fraction analyzed immediately after termination of the experiment. During the 3 months of

[14] K. Schiedt, unpublished results.
[15] S. Hertzberg, V. Partali, and S. Liaaen-Jensen, *Acta Chem. Scand.* **42**, 495 (1988).

storage of the solid samples at $-80°$ under argon, a free radical-mediated process involving a mixed aggregate of molecules that might result in ^3H transfer from astaxanthin to asterinic acid can hardly be imagined. It would be a challenge to establish experimental conditions for the elucidation of the observed phenomenon. This could also be useful for the interpretation of further metabolic studies using tritium.

Mytiloxanthin, the major carotenoid in *Mytilus edulis* (50% of total) may analogously be the precursor of 4-oxomytiloxanthin identified as triacetate (VIS, MS) in *P. japonicus*. The chirality of the astaxanthin end group was not analyzed but may also be racemic, in contrast to that of 4-oxomytiloxantin isolated from starfish *Asterias rubens* and structurally assigned by ^1H NMR (with relative configuration only).[16]

In conclusion, the *in vivo* racemization of optically active $(3S,3'S)$-astaxanthin in *Penaeus japonicus* could be demonstrated. However, the site of this mechanism and its biological significance remain unknown. The fact that crustaxanthin and tetrahydroxypirardixanthin, the proposed reduction products of astaxanthin, were not labeled does not necessarily prove that these compounds are not involved in the metabolism of astaxanthin in *Penaeus*. We should keep in mind that only one-fourth of the astaxanthin in the body was labeled and absorbed during the experimental period. The bulk of astaxanthin must have been biosynthesized or absorbed during the preexperimental period. Moreover, we have to take into account that several molting cycles occurred during the postlarval phase; obviously only a small part of the carotenoids are lost with the exuviae.

Although numerous scientists are investigating the biology of crustaceans and a rich literature is available on development, endocrine, and pigmentary relations, there exists only limited knowledge regarding the absorption, site of esterification, and formation of carotenoproteins. It is unknown whether metabolic conversions of carotenoids occur at the level of carotenoproteins or in free or esterified form, and we do not know whether the yellow tetrahydroxypirardixanthin, crustaxanthin, and isoas taxanthin are precursors or reduction products of astaxanthin. It is possible that our experiment was just not performed at the appropriate life phase of these crustaceans comprising 12 larval and 22 postlarval stages. Further studies with ^{14}C-labeled compounds are planned in our laboratories. Specialists in aquaculture, endocrinology, and protein chemistry may be encouraged to further interdisciplinary research on the biology of these interesting animals.

[16] K. Bernhard, G. Englert, W. Meister, M. Vecchi, B. Renstrøm, and S. Liaaen-Jensen, *Helv. Chim. Acta* **65**, 2224 (1982).

Acknowledgments

We thank the following scientists of F. Hoffman-La Roche, Basel, for their valuable contributions: Dr. G. Englert (NMR), Dr. K. Noack (CD), Mr. W. Meister (MS), Dr. U. Hengartner and Dr. H. Mayer (synthesis of reference substances), Mr. N. Moser and Dr. P. Preiswerk (synthesis of the tritiated astaxanthin), and Dr. J. Gabaudan and Mr. P. Horne (conducting the animal experiment). Discussions with and stimulating support by Dr. G. Britton, University of Liverpool, Dr. B. H. Davies, University College of Wales, Aberystwyth, U.K., and Prof. S. Liaaen-Jensen, University of Trondheim, are also gratefully acknowledged.

[16] Assay for Carotenoid 15,15′-Dioxygenase in Homogenates of Rat Intestinal Mucosal Scrapings and Application to Normal and Vitamin A-Deficient Rats

By LAURENCE VILLARD-MACKINTOSH and CHRISTOPHER J. BATES

Introduction

β-Carotenoid 15,15′-dioxygenase (EC 1.13.11.21), an oxygenase which cleaves β-carotene and certain related carotenoids, is the only known enzyme to be responsible for the formation of vitamin A (retinol) in vertebrates. It is of central importance in vitamin A nutrition for humans and for those animals that depend wholly or largely on plant foods. Its presence in the intestinal mucosa is consistent with its role in converting dietary carotenoids to vitamin A *in vivo*. The enzyme has been demonstrated in the intestinal mucosa, in the liver, and in the corpus luteum, and it has been found in the rat, hog, chick, guinea pig, cow, rabbit, tortoise, monkey, and in fish.[1-6] It has not, however, been detected in the intestine of the cat,[6] a species which appears to be entirely dependent on the presence of preformed vitamin A in its diet.[6] A wide range of carotenoids can be cleaved, at least by the mucosal enzyme,[7] although at considerably different rates, and there is substrate specificity both for the environment of the β-ionone ring, and for the polyene chain.[7]

[1] J. A. Olson, *J. Nutr.* **119**, 105 (1989).
[2] N. H. Fidge, F. R. Smith, and D. S. Goodman, *Biochem. J.* **114**, 689 (1969).
[3] D. Sklan, *Br. J. Nutr.* **50**, 417 (1983).
[4] A. M. Gawienowski, M. Stacewicz-Sapuncakis, and R. Longley, *J. Lipid Res.* **15**, 375 (1974).
[5] X. D. Wang, G.-W. Tang, J. G. Fox, N. I. Krinsky, and R. M. Russell, *Arch. Biochem. Biophys.* **285**, 8 (1991).
[6] M. R. Lakshmanan, H. Chansang, and J. A. Olson, *J. Lipid Res.* **13**, 477 (1972).
[7] H. Singh and H. R. Cama, *Biochim. Biophys. Acta* **370**, 49 (1974).

In addition to the two essential substrates (a carotenoid and molecular oxygen), the enzyme also requires ferrous iron and a thiol such as glutathione,[1,7] and it is stimulated by bile salts or other detergents.[2] Nicotinamide, magnesium, and lecithin are also stimulatory.[8,9] The enzyme obtained from chick mucosa appears to contain copper and zinc.[3] All preparations that have been tested are inhibited by iron-chelating or sulfhydryl-binding agents.[2,7] Endogenous inhibitors may exist, at least in the intestinal preparation, since various purification procedures generally result in apparent recoveries above 100%.[7]

A number of serious controversies have arisen over this enzyme. In 1988, Hansen and Maret[10] claimed that the accepted, standard assay[8] for carotenoid dioxygenase was flawed, and that it could not be verified when the products were subjected to rigorous analysis by high-performance liquid chromatography (HPLC). Their claim was later reexamined by Lakshman et al.,[11] who used mass spectrometry combined with HPLC of a derivative (O-ethyloxime) of the retinal product. These workers were able to verify that the original standard procedure yielded the expected product. It was suggested that Hansen and Maret may have used either an inactive enzyme preparation or inappropriate reaction conditions. Another ongoing controversy continues to evolve around "central" versus "eccentric" cleavage of the carotene molecule; the evidence regarding this has been summarized by Olson.[1]

Little is yet known about the factors which influence activity of the enzyme in vivo, or about their effect on rates of carotenoid conversion to vitamin A in vivo. Some studies[12–14] have indicated that variations in dietary protein levels can affect the activity of the intestinal enzyme in rats, resulting in a reduced rate of conversion of β-carotene to vitamin A in protein-deficient animals.[13,14] Dietary carotenoids may possibly increase the activity of the enzyme when dietary protein is limiting, and studies in our laboratory[13,16] have indicated that the enzyme can be induced by a dietary deficiency of vitamin A in rats. Two recent studies bear indirectly on this observation. Mittal[9] found that rats fed carotene before a period of

---

[8] D. S. Goodman and J. A. Olson, this series, Vol. 15, p. 462.
[9] P. C. Mittal, Nutr. Rep. Int. 28, 181 (1983).
[10] S. Hansen and W. Maret, Biochemistry 27, 200 (1988).
[11] M. R. Lakshman, I. Mychkovsky, and M. Attlesey, Proc. Natl. Acad. Sci. U.S.A. 86, 9124 (1989).
[12] A. Gronowska-Senger and G. Wolf, J. Nutr. 100, 300 (1970).
[13] D. S. Deshmukh and J. Ganguly, Indian J. Biochem. 1, 204 (1964).
[14] S. K. Kamath and L. Arnrich, J. Nutr. 103, 202 (1973).
[15] L. Villard and C. J. Bates, Br. J. Nutr. 56, 115 (1986).
[16] J. Napoli and K. R. Race, J. Biol. Chem. 263, 17372 (1988).

vitamin A depletion subsequently utilized carotene more efficiently than those fed vitamin A before an equivalent period of vitamin A depletion. Krinsky et al.[17] found that rats fed elevated amounts of β-carotene subsequently exhibited a reduced storage of [^{14}C]retinol, derived from β-[^{14}C]carotene, in their livers, compared to rats previously fed a diet without added β-carotene. One possible explanation is a decreased conversion of β-carotene to retinol in the former group. These two studies appear at first to be contradictory, but they may not be. Takruri and Thurnham[18] reported briefly that zinc deficiency may reduce the conversion of β-carotene to retinol in rats. Napoli and Race[16] have reported that the cytosol from various rat tissues can convert β-carotene to retinoic acid, possibly by a different pathway from that involving retinal. Clearly, further studies are needed in this area.

The following methodology describes a procedure based on that of Goodman and Olson[8] that has been used in our studies, and comments are made about its application to our studies of vitamin A-deficient rat intestine.

Enzyme Assay

Preparation of Enzyme Extract

The proximal 60 cm of the small intestine from mature rats, fasted overnight and then sacrificed by ether anesthesia and exsanguination, is removed rapidly, slit lengthwise, and spread on an ice-cooled glass plate. Adherent mucus is removed by rinsing with saline. The mucosa are scraped free with a microscope slide, yielding 2.5–3.0 g wet weight of scrapings per animal, and homogenized in 6 ml of buffer, pH 7.7, containing 100 mM potassium phosphate, 30 mM nicotinamide, and 4 mM magnesium chloride. (The nicotinamide and magnesium perform essential but unknown functions.[2,8,19])

The homogenate is centrifuged at 104,000 g for 15 min at 4° to give a supernatant fraction which contained sufficient amounts of a stimulatory particulate component[19] to avoid the need for subsequent recombination of two fractions. The protein content of the supernatant is 13.0–30.0 mg/ml, as measured by the biuret reaction.

---

[17] N. I. Krinsky, M. M. Mathews-Roth, S. Welankiwar, P. K. Sehgal, N. C. G. Lausen, and M. Russett, *J. Nutr.* **120**, 81 (1990).
[18] H. R. H. Takruri and D. I. Thurnham, *Proc. Nutr. Soc.* **41**, 53A (1981).
[19] D. S. Goodman, H. Huang, M. Kanai, and T. Shiratori, *J. Biol. Chem.* **242**, 3543 (1967).

The homogenate can be stored frozen at $-20°$ for several weeks with retention of more than 95% of its original activity. Goodman and Olson[8] likewise observed that the purified dioxygenase could be stored without activity loss.

Assay

Enzyme. The maximum yield of the product (labeled retinol) is obtained with 15-20 mg crude enzyme protein and a 75-min incubation at 37°. To compare activities of different mucosal extracts, a smaller amount of extract protein (e.g., 5 mg) is preferable, and the optimum incubation time is 20-35 min. Use of even smaller amounts of enzyme, or shorter incubations, yield insufficient labeled retinol for precise measurement of the product.

Substrate. β-[15,15'-^{3}H$_2$]Carotene, a gift from Hoffmann-LaRoche (Basel, Switzerland), is checked for purity by column chromatography, using 10% water-deactivated alumina (Woelm, Eschwege, Germany) which is eluted with *n*-hexane. The main peak containing radioactivity and absorbing light at 450 nm is collected and evaporated under nitrogen in the dark. It is then redissolved in acetone to yield a β-carotene concentration of 16 μg/ml, based on the extinction coefficient, $E_{1cm}^{1\%}$ of 2592 at 453 nm in *n*-hexane. Fifty microliters (0.8 μg) containing 15×10^4 disintegrations/min (dpm) is used for each assay.

Other Assay Components. The following components are included in the assay mixture (2 ml final volume):

Potassium phosphate, 200 μmol, pH 7.7 (the pH optimum is narrow)
Detergent: Sodium dodecyl sulfate (12 μmol/2 ml) is used here, although others claim that it may be inhibitory;[7] detergents such as bile salts can be used, as shown by Goodman and Olson[8]
L-α-Phosphatidylcholine (Sigma, St. Louis, MO, lecithin grade III E), 0.4 mg, is added in 100 μl hexane (see Ref. 19); it is best not to premix the lecithin with the detergent, before adding them to the reaction vessel
Magnesium chloride, 4 μmol
Nicotinamide, 30 μmol
Glutathione, 10 μmol (to maintain essential SH groups in the reduced state[2])
α-Tocopherol, 0.25 mg dissolved in 100 μl acetone (used as an antioxidant to prevent peroxidative side reactions)

Incubation. A shaking water bath in the dark at 37°, with shallow reaction vessels (glass vials) and air as the gaseous phase, are used.

Extraction

After reaction, 60 µg unlabeled β-carotene and 100 µg retinal are added as carriers in 3.0 ml of a chilled 2:1 (v/v) chloroform-methanol mixture. After thorough vortex-mixing, 5 ml of 0.01 N H_2SO_4 is added; the samples are mixed again, and the two phases are separated by centrifugation at 4° in glass centrifuge tubes for 10 min at 1500 g. The chloroform layer is collected and evaporated under nitrogen in the dark; 60 µl toluene and 500 µl n-hexane are added to dissolve the lipid products.

Chromatographic Separation

Pasteur pipettes with 0.7 g of 10% water-deactived alumina (Woelm), column height 2 cm, are used. The alumina is packed in toluene, a 50-µl aliquot of the redissolved sample is added, and unchanged β-carotene is eluted with 5 ml hexane. A further 4 ml hexane is passed to ensure complete separation, then the retinal is eluted with 7 ml 1:1 (v/v) hexane-toluene. A final 5 ml toluene is used to check possible conversion to retinol, but in practice essentially all of the product is in the retinal fraction, as has been previously reported.[12] This procedure represents a 4-10 times scaling down of earlier reported methods,[8,9] and thus substantially increases the throughput of samples.

Scintillation Counting

Determination of radioactivity is performed by standard procedures, with a scintillant containing 2,5-diphenyloxazole (PPO) (5 g) and 1,4-di-2-(5-phenyloxazolyl)benzene (POPOP) (0.3g) per liter of toluene, and an external-standard quench-correction.

Recovery

The recovery of added substrate (radioactivity) varies from 85% in the absence of any conversion to retinol to 70% when maximum enzyme-catalyzed conversion has occurred. The remaining radioactive material cannot be removed from the column by elution with the chloroform, and may be present in degradation products, or some of the ^3H may have exchanged with water.

Blanks

If the reaction is stopped without any incubation at 37° (zero time blank) or if the enzyme is omitted (no enzyme blank), then 4-7% of the applied radioactivity appears in the "retinol" faction. A no-enzyme blank is therefore included with each run.

Unit of Activity

Activity is expressed as micromoles of retinal formed per milligram of protein added per minute of incubation. Complete conversion of the β-carotene added would yield a theoretical maximum of 2.6×10^{-3} μmol retinal, under the conditions described.

Comparison between Control and Vitamin A-Deficient Rat Intestine

Induction of Vitamin A Deficiency and Its Interactions with Pregnancy in Rats

We chose to study rats during pregnancy partly because pregnancy increases the demands for most nutrients, including vitamin A, and is associated with vitamin A deficiency symptoms in marginally supplied human subjects. We wished to test the hypothesis that adaptive mechanisms, particularly during pregnancy, might result in increased efficiency of use of dietary precursors of a critical micronutrient such as vitamin A.[10]

The rats employed in our studies[15,20] are female Norwegian hooded or Sprague-Dawley animals. There is some evidence for subtle differences between the different rat strains, which requires further investigation. The animals are around 50 g body weight (weanlings) at the outset, and they are maintained for 80–100 days on the following purified diets (before extraction of the enzyme), in g/kg: sucrose, 706; casein "low in vitamins" (British Drug House, Poole, Dorset, U.K.), 210; salt mixture, 50; arachis oil, 30; choline chloride, 2; cystine, 1.5; calcium pantothenate, 0.02; thiamin hydrochloride, 0.004; riboflavin, 0.015; pyridoxine hydrochloride, 0.009; nicotinamide, 0.025; biotin, 0.001; folic acid, 0.001; cyanocobalamin, 5×10^{-5}; α-tocopherol, 0.16; menadione, 0.009; vitamin D_2, 7.5×10^{-6}. The retinol content of the casein, as measured by HPLC is 0.38 μg/g, and a typical daily food intake of 16 g is therefore equivalent to 1.3 μg retinol. No dietary carotene is detectable. Control animals receive 1.55 mg retinol as retinyl acetate per kilogram diet (i.e., ~30 μg retinol/day) and are pair-fed as a group with the deficients until they are mated. In one of the experiments, growth faltering is evident after 75–91 days; here the intake of retinol is increased to 1 μg/day to reverse the growth faltering. The deficient animals exhibit moderate irritability, hair roughness, and a slight ataxia, but there is no evidence of any severe deficiency (e.g., eye lesions).

When mated, all the animals are able to support pregnancy for at least 7 days, but those receiving only the basal diet are unable to continue to support it to day 20 (i.e., near term). However, the provision of 1 μg retinol

---

[20] L. Villard and C. J. Bates, *Proc. Nutr. Soc.* **44**, 15A (1985).

(as the acetate) per day enables nearly all the animals to support pregnancy until they are sacrificed on day 20, and the weights of the products of conception are then identical between groups.

Biochemical Evidence of Vitamin A Status

High-performance liquid chromatography is used for the analysis of vitamin A in plasma after extraction with 2 volumes of ethanol, 1 volume of water, and 4 volumes of *n*-hexane, then evaporation of the hexane extract under N_2. Retinyl acetate is used as internal standard, and separation is achieved on a 25 × 0.3 cm column of Waters (Milford, MA) microBondapak C_{18} eluted with methanol–water (95:5, v/v) flowing at 3.5 ml/min. The vitamin A peaks are detected and quantitated (peak height) by their absorption at 328 nm. Liver samples are saponified in a 1:1 homogenate in water by adding 2.5 volumes of ethanol containing potassium hydroxide, 50 g/liter, and incubating for 30 min at 60° under N_2. Retinol is then extracted into 5 ml cyclohexane and quantitated by fluorimetry at 480 nm emission (40 nm slit) and 330 nm excitation (4 nm slit) with a Perkin-Elmer MPF3 spectrofluorimeter.

Typical values for plasma retinol concentrations in our experiments are 100–260 μg/liter (deficients) and 420–640 μg/liter (controls); levels in liver are 3–4 μg/g wet weight (deficients) and 78–96 μg/g (controls).

The mean specific activity of carotenoid dioxygenase in the mucosal extracts of the 21 deficient animals is 17.3 ± 0.9 (S.E.M.) % of substrate converted per 20-min incubation, and that of the 25 control animals 22.3 ± 1.3%/20 min. The difference between these is significant (*t*-test) at $p < 0.001$.[20] The conversion rates expressed as conventional enzyme activities range between 2.4 and 6.7×10^{-7} μmol retinal formed/mg protein/min. Thus the vitamin A-deficient animals appear to have adapted by an increase in enzyme activity, and may thus have increased their efficiency of utilization of dietary β-carotene. It is not known whether the adaptation is achieved by an increase in the amount of active enzyme per unit of extracted protein, by a reduction in amount of an inhibitor, or by some other unknown mechanism. It also remains to be determined whether the increased activity *in vitro* really is accompanied by increased carotenoid conversion *in vivo,* and whether it also occurs in species other than rats, and in organs other than intestine.

[17] Control of Carotenoid Synthesis by Light

By KLAUS HUMBECK and KARIN KRUPINSKA

Introduction

Biosynthesis of carotenoids is reported to be affected by light in bacteria, fungi, algae, and higher plants. However, the extent to which synthesis is regulated by light differs from organism to organism. In some fungi like *Phycomyces*, certain amounts of carotenoids are produced in the dark, and light increases the level of pigments.[1] In other fungi, such as *Fusarium aquaeductuum*, carotenogenesis is strictly light dependent.[2] Normally, higher plants and algae are able to synthesize carotenoids during growth in darkness. But, in addition to this, light induces new synthesis of carotenoids and changes in the carotenoid composition in many plants.[3,4]

The photoregulation of carotenoid biosynthesis poses several questions. For example, what is the photoreceptor(s) absorbing the external signal light, and which step(s) of the biosynthetic pathway is under photocontrol? To identify the molecule which serves as a photoreceptor for carotenogenesis in a particular organism, action spectra have to be determined. The procedure for obtaining such action spectra is described below. Furthermore, procedures for illumination of samples and direct measurement of light-induced pigment formation via spectroscopic analysis are discussed.

Several levels of photoregulation of carotenogenesis have been reported in literature. Evidence for a photoinduced synthesis of a poly(A)-containing mRNA in *Fusarium aquaeductuum* was found by Schrott and Rau.[5] However, as pointed out by these investigators, further studies will be required to determine whether the induced mRNA synthesis is part of the photoinduction of carotenogenesis or whether it is part of an unknown photoregulatory process in *Fusarium aquaeductuum*. Evidence that the photoinduction of carotenogenesis may be regulated at the translational level has also been obtained. In *Verticillium agaricinum*, parallel to the light-dependent carotenoid synthesis, light induces protein synthesis, as shown as an increased activity of 80 S ribosomes.[6] Finally, direct activation

---

[1] M. Jayaram, D. Presti, and M. Delbrück, *Phycomyces Exp. Mycol.* **3**, 42 (1979).
[2] W. Rau, *Pure Appl. Chem.* **47**, 237 (1976).
[3] K. H. Grumbach, *Physiol. Plant.* **60**, 389 (1984).
[4] W. Rau and E. L. Schrott, in "Blue Light Responses: Phenomena and Occurrence in Plants and Microorganisms" (H. Senger, ed.), Vol. 1, p. 43. CRC Press, Boca Raton, Florida, 1987.
[5] E. L. Schrott and W. Rau, *Planta* **136**, 45 (1977).
[6] L. R. Valadon, R. L. Travis, and J. L. Key, *Physiol. Plant.* **34**, 196 (1975).

of a step in the carotenogenic pathway by phototransformation of a compound which leads to the onset of activity of preexisting enzymes has been demonstrated in algal mutants of *Chlorella*[7] and *Scenedesmus*.[8] In the investigation of such phenomena methods for separation and quantitation of carotenoids and for enzymatic and molecular biology analysis of carotenogenesis also play an important role. Since these methods are described elsewhere in this volume, this chapter does not deal with these procedures.

Light Sources and Measurement of Light Energy

Light from the ultraviolet and the visible part of the electromagnetic spectrum has been reported to be effective in carotenogenesis.[4] The wavelengths applied in these experiments range from UVB (280–320 nm) and UVA (320–390 nm) to the far-red portion at 800 nm.

The light used in each experiment has to be well defined as to both quantity and quality. Several types of artificial light sources normally used in the laboratory differ drastically in their spectral energy distribution. Incandescent lamps show a continuous spectrum with a low blue and a high far-red part. Their spectral energy distribution depends on the operating temperature which itself depends on the voltage applied. It is therefore better to reduce the quantity of light by using neutral density filters which do not alter the spectral characteristics than to reduce the voltage. Luminescent and fluorescent light sources show distinct emission maxima. The exact analysis of the spectral output of these lamps can be determined with a spectroradiometer.

Quite often light of a very narrow bandwidth has to be used in the investigation of photoregulated carotenoid synthesis. For this purpose filters must be put in the path of rays. Plexiglas filters yield a high output but often show transmission over a wide range of the visible spectrum. Here, care must be taken to ensure that no undesired wavelengths are transmitted. Better suited are selective absorption filters and the highly precise interference filters. For precise transmission, such filters must be exactly perpendicular to a colliminated light beam.

Light quantity can be measured in three different ways: photometrically, radiometrically, and in quantum terms.[9] The sensor of a photometric cell converts radiant energy to an electric current. However, this photocell is designed so that the spectral sensitivity of the instrument matches that of

[7] H. Claes, *Photochem. Photobiol.* **5**, 515 (1966).
[8] K. Humbeck, *Planta* **182**, 204 (1990).
[9] J. W. Hart, "Light and Plant Growth." Unwin Hyman, London, 1988.

the human eye, with maximum sensitivity in the green region of the spectrum at 550 nm. In this case the basic unit is the lumen. The term lux which is also used equals 1 lumen per square meter. Because these units are based on the sensitivity of an average human eye and are not sensitive in the wavelength range of plant photoreceptors, such photometric methods are inappropriate for the investigation of light-induced carotenogenesis.

Radiometric methods measure total radiant energy in joules. The instruments are also sensitive to infrared radiation. For measurements of only the visible part, appropriate filters are available. For measuring the total amount of energy received by a sample, the term fluence (J m^{-2}) is used. The flow rate, which reflects the quantity per time (J sec^{-1} = W), and the fluence rate, which gives the flow rate per area (W m^{-2}), are also frequently used terms.

In contrast to the radiometric methods quantum measurements give the number of actual photons (integrated over time and area). The unit is the mole. One mole of light equals 6.023×10^{23} photons. Depending on the purpose, the terms fluence (μmol m^{-2}), rate (μmol sec^{-1}), and fluence rate (μmol m^{-2} sec^{-1}) are used. The quantum sensors are normally corrected for a quantum response between 400 and 700 nm.

If the wavelengths involved are known, the conversion between energy and quantum measurements can be obtained by using Eq. (1).

$$W\ m^{-2} = moles\ m^{-2}\ sec^{-1}\ (Nhc/\lambda) \qquad (1)$$

In Eq. (1) N is Avogadro's number, h is Planck's constant, c is the velocity of light, λ is the wavelength in meters, and Nhc is equal to 0.1196.

Illumination Protocols

The following factors have to be considered in each experiment.

Length of Illumination Period. Two possibilities are found in nature. induction of carotenogenesis by a short light pulse which does not need prolonged illumination and carotenoid synthesis which needs permanent illumination.[4] For the investigation of light-inducible carotenogenesis the exact timing of illumination is of great importance. The duration of short pulses and the timing of possible second and third pulses have to be planned in each experiment.

Use of Inhibitors during Illumination. To determine whether newly synthesized proteins affect the photoregulatory process inhibitors of cytoplasmic (e.g., cycloheximide) or plastid (e.g., chloramphenicol) protein biosynthesis may be added prior to or after onset of illumination. The application of inhibitors of specific steps in carotenoid synthesis (like nicotine or SAN 9789) during illumination allows further statements concerning the mechanism of photoregulation.

Temperature during Illumination. When organisms like *Fusarium, Neurospora,* and others[10] are illuminated at a low temperature, that is, near 0°, no carotenoids are formed. When transferred to room temperature they begin to synthesize the pigments even in the dark. Thus, by this method it is possible to separate the first photoprocess taking place at low temperatures from the following reactions which need higher temperatures.

Use of Mutants

Valuable results in the investigation of photoregulated carotenoid synthesis have been obtained with mutants, especially from fungi and algae. In fungi such as *Phycomyces* and *Neurospora,* mutants are available which show specific alterations in the photoinduction process and in the carotenoid biosynthetic pathway.[11-13] These mutants allow genetic and biochemical analysis of the photoreceptor(s) and the signal transduction chain.

In algae like *Chlorella* and *Scenedesmus,* photoregulatory mutants have been isolated which show a strict light dependency in certain steps of carotenoid synthesis.[7,14,15] They enable the investigation of the light reaction involved without the disturbing dark synthesis of carotenoids normally found in algae. These mutants are induced either by X-irradiation or by chemical treatments as described by Bishop.[16]

Direct Measurement of Light-Induced Carotenoid Synthesis by Difference Spectroscopy

If carotenoid precursors accumulated in the dark show a clear difference in the absorption spectrum compared with that of carotenoid pigments being synthesized during subsequent illumination, the light-dependent carotenoid formation can be followed directly by spectroscopy.

[10] R. W. Harding and W. Shropshire, *Annu. Rev. Plant Physiol.* **31,** 217 (1980).
[11] R. W. Harding and R. V. Turner, *Plant Physiol.* **68,** 745 (1981).
[12] E. Cerda-Olmedo, *Annu. Rev. Microbiol.* **31,** 535 (1977).
[13] F. Degli Innocenti and V. E. A. Russo, *in* "Blue Light Effects in Biological Systems" (H. Senger, ed.), p. 213. Springer-Verlag, Berlin, 1984.
[14] R. Powls and G. Britton, *Arch. Microbiol.* **115,** 175 (1977).
[15] H. Senger and G. Strassberger, *in* "Chloroplast Development" (G. Akoyunoglou, ed.), p. 367. Elsevier, Amsterdam, 1978.
[16] N. I. Bishop, *in* "Methods in Chloroplast Molecular Biology" (M. Edelman, R. B. Hallick, and N.-H. Chua, eds.), p. 51. Elsevier, Amsterdam, 1982.

This is, for example, the case in the light-induced transformation of cis-lycopene to all-*trans*-lycopene observed in mutant C-6D of the alga *Scenedesmus obliquus*.

The absorption spectrum of the cis isomer prolycopene (Fig. 1a) with one maximum at 438.5 nm differs drastically from that of the all-*trans*-lycopene (Fig. 1a') showing three maxima at 440.7, 466.1, and 497.4 nm. To measure the synthesis of all-*trans*-lycopene, which is the photoregulated step in carotenogenesis in mutant C-6D of *Scenedesmus*,[8] the increase in absorption at the long wavelength maximum is plotted during the illumination. For this purpose all following reactions, such as cyclization and oxidations, must be inhibited. Otherwise the trans isomer would be transformed directly to carotenes and xanthophylls. As inhibitor $5 \times 10^{-3} M$ nicotine is added prior to illumination. The measurement can be performed with suspensions of living algae. In these cells a shift to longer wavelengths in the lycopene absorption occurs. Therefore the maximal difference in the absorption of *cis*- and *trans*-lycopene is not at 500 nm as suspected from the spectra recorded in hexane (Fig. 1) but at 520 nm.

Figure 2 shows the optical system in the spectrophotometer. The sample is divided into two aliquots which both contain $5 \times 10^{-3} M$ nicotine and placed in cuvettes at the reference and sample position of a double-beam spectrophotometer. The measuring beam of 521 nm passes through both cuvettes and an interference filter (filter B) (DIL 521, Schott, Mainz,

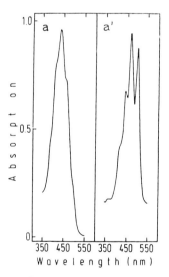

FIG. 1. Absorption spectra of (a) prolycopene (7Z, 9Z, 7'Z, 9'Z) and (a') all-*trans*-lycopene. Spectra were recorded in hexane.

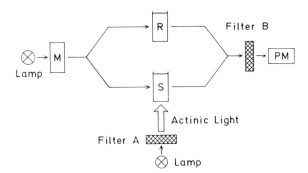

FIG. 2. Optical system for measurement of light-induced absorption changes. M, Monochromator (supplies light of 521 nm); R, reference cuvette; S, sample cuvette; PM, photomultiplier; filter A, actinic light; filter B, DIL 521 (Schott).

Germany) which protects the photomultiplier (PM) against the exciting light. While the reference is kept in the dark, the sample position is cross-illuminated with light of various wavelengths and intensities (filter A). The difference in absorption between reference and sample is then detected during the irradiation.

Figure 3 shows such plots obtained after irradiation of cells of *Scenedesmus* mutant C-6D with narrow blue (400 nm) and red (654 and 706.2 nm) and broad-band red (red Plexiglas filter) light. The slope ($\Delta E/h$) of the tracings equals the rate of light-dependent increase in all-*trans*-lycopene. Using this method direct analysis of the photoregulated step in the carotenogenic pathway is possible.

Action Spectroscopy

The question of what kind of photoreceptor is involved in photoregulation of carotenoid biosynthesis can best be solved via action spectroscopy. Action spectra show the effectiveness of distinct wavelengths in the observed photoregulated process. They therefore reflect the absorption spectrum of the photoreceptor, but one has to consider factors which might influence the shape of the action spectrum, for example, screening by different pigments, fluorescence, or inhomogeneous distribution of the pigments within the sample. For a detailed review of the theory of action spectroscopy including such disturbing effects, see Hartmann.[17]

The first step in the determination of an action spectrum is the measurement of photon fluence curves. Examples of such photon fluence plots

---

[17] K. M. Hartmann, in "Biophysik" (W. Hoppe, W. Lehmann, H. Markl, and H. Ziegler, eds.), p. 197. Springer-Verlag, Berlin, 1977.

[17] CONTROL OF CAROTENOID SYNTHESIS BY LIGHT 181

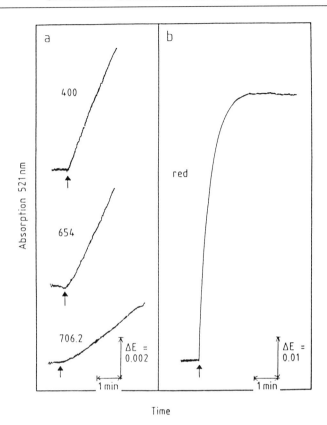

FIG. 3. Kinetics of light-induced increase in absorption of whole cells of *Scenedesmus obliquus* mutant C-6D at 521 nm in the presence of 5×10^{-3} M nicotine. Actinic light of 400, 654, and 706.2 nm (Schott filters) had a photon fluence rate of 12 μmol m^{-2} sec^{-1} (a). Red light of wavelengths above 620 nm had a fluence rate of 110 W m^{-2} (b). Arrows indicate the onset of illumination.

are shown in Fig. 4 for the above-mentioned photoisomerization of lycopene. The reaction (change in the absorption at 521 nm per time) is measured at various photon fluence rates of wavelengths 654 and 706.2 nm. Both plots are linear over a wide range of photon fluence rates. After performing photon fluence plots for all wavelengths applied in the action spectrum, one photon fluence rate is chosen which is in the linear part of all plots. In the actual action spectrum the reaction at this photon fluence rate is plotted against the wavelength.

Figure 5 shows the resulting action spectrum for the photoisomerization of lycopene. Here 20 different wavelengths were investigated. Because the absorption change was detected at 521 nm, no wavelengths between

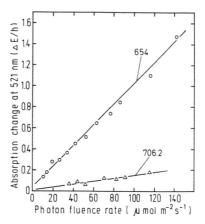

FIG. 4. Fluence response curves of light-induced absorption change at 521 nm. Various photon fluence rates of light of 654 and 706.2 nm (Schott filters) were used.

470 and 580 nm, which would interfere with the measurement, could be tested. The main maxima in the blue (440 nm) and red (670 nm) parts of the spectrum clearly reveal the involvement of chlorophyll molecules known to absorb at these wavelengths.

Photoreceptors Involved in Carotenogenesis

Table I summarizes some characteristics of photoreceptors which have been reported to be involved in carotenogenesis. The most prominent type of photoreceptor playing a role in carotenoid synthesis is the so-called blue light receptor(s). The action spectra show one maximum in the UVA and further peaks in the blue part of the spectrum. The chemical nature of this type of photoreceptor is still a matter of debate. Candidates are flavins, pterins, and carotenoids.[18]

In green algae it was shown that chlorophyll is involved in photoregulation of carotenoid synthesis[7,8] (see also Fig. 5). This is indicated by a blue (440 nm) and a red (670 nm) maximum. In wheat leaves action spectra for the induction of carotenoid synthesis showed red (around 650 nm) and blue (around 450 nm) maxima which resemble the absorption spectrum of protochlorophyll(ide).[19] The authors concluded that carotenogenesis in this plant may depend on the light absorption by protochlorophyll(ide).

[18] H. Senger, ed., "Blue Light Responses: Phenomena and Occurrence in Plants and Microorganisms," Vol. 1. CRC Press, Boca Raton, Florida, 1987.
[19] T. Ogawa, Y. Inoue, M. Kitajima, and K. Shibata, *Photochem. Photobiol.* **18,** 229 (1973).

TABLE I
Characteristics of Photoreceptors Involved in Carotenogenesis

Photoreceptor	Characteristics	Organisms	Refs.
Blue light receptor(s)	Maximum at 370–380 nm, 3 peaks or shoulders at 400–500 nm, >500 nm ineffective	*Fusarium aquaeductuum, Mycobacterium* sp., *Neurospora crassa*	a–c
Chlorophyll	Red light near 670 nm, blue light near 440 nm	*Scenedesmus obliquus, Chlorella vulgaris*	d,e
Protochlorophyll(ide)	Red light near 650 nm, blue light near 450 nm	*Triticum aestivum*	f
Porphyrins	Maximum at 400 nm, several smaller peaks at 450–670 nm	*Mycobacterium marinum, Myxococcus xanthus*	g,h
Phytochrome	Red light (660 nm) is effective, can be reversed by far-red (730 nm)	*Sorghum vulgare, Verticillium agaricinum*	i,j
UV light receptor(s)	Maximum at 370 nm, only very weak blue light effects	*Verticillium agaricinum*	k

^aW. Rau, *Planta* **72**, 14 (1967).
^bC. D. Howes and P. P. Batra, *Arch. Biochem. Biophys.* **137**, 175 (1970).
^cE. DeFabo, R. W. Harding, and W. Shropshire, *Plant Physiol.* **57**, 440 (1976).
^dK. Humbeck, *Planta* **182**, 204 (1990).
^eH. Claes, *Photochem. Photobiol.* **5**, 515 (1966).
^fT. Ogawa, Y. Inoue, M. Kitajima, and K. Shibata, *Photochem. Photobiol.* **18**, 229 (1973).
^gP. P. Batra and H. C. Rilling, *Arch. Biochem. Biophys.* **107**, 485 (1964).
^hR. P. Burchard and S. B. Hendricks, *J. Bacteriol.* **97**, 1165 (1969).
ⁱR. Oelmüller and H. Mohr, *Planta* **164**, 390 (1985).
^jL. R. G. Valadon, M. Osman, and R. S. Mummery, *Photochem. Photobiol.* **29**, 605 (1979).
^kM. Osman and L. R. G. Valadon, *Microbios* **18**, 229 (1977).

Action spectra of carotenoid formation in *Mycobacterium marinum*[20] and *Myxococcus xanthus*[21] exhibit one major maximum at 400 nm and several additional peaks between 450 and 670 nm. From the shape of these spectra and by the comparison with absorption spectra of porphyrin-containing fractions, it seems very likely that a porphyrin is the acting photoreceptor.

There are some reports showing that phytochrome might also have an

[20] P. P. Batra and H. C. Rilling, *Arch. Biochem. Biophys.* **107**, 485 (1964).
[21] R. P. Burchard and S. B. Hendricks, *J. Bacteriol.* **97**, 1165 (1969).

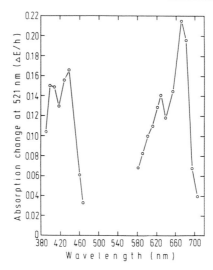

FIG. 5. Action spectrum of light-induced absorption change at 521 nm. Actinic light of the different wavelengths always had a photon fluence rate of 12 μmol m^{-2} sec^{-1}.

effect on carotenoid synthesis.[22,23] The first criterion for the identification of phytochrome action is the demonstration of a red/far-red reversibility in the response. Valadon and co-workers[23] demonstrated that red light induction of carotenogenesis in the fungus *Verticillium agaricinum* is reversed by far-red light. This result might reflect the action of phytochrome in the induction of carotenogenesis in *Verticillium agaricinum*.

Furthermore, some reports demonstrated maximum effects on the photoinduction of carotenogenesis after continuous irradiation with light of the near-UV region (peak at 370 nm), whereas blue light (380–525 nm) produced only very weak effects.[24] Although the spectrum is similar to that of a porphyrin-type photoreceptor, a yet unknown photoreceptor (UV receptor) was suggested as no porphyrins had been identified.

The data summarized in Table I allow the identification of photoreceptor(s) involved in certain steps of the carotenogenic pathway in a specific organism.

Acknowledgments

This chapter is dedicated to our teacher, Professor Horst Senger, on the occasion of his 60th birthday.

[22] R. Oelmüller and H. Mohr, *Planta* **164**, 390 (1985).
[23] L. R. G. Valadon, M. Osman, and R. S. Mummery, *Photochem. Photobiol.* **29**, 605 (1979).
[24] M. Osman and L. R. G. Valadon, *Microbios* **18**, 229 (1977).

[18] Functions of Carotenoids in Photosynthesis

By RICHARD J. COGDELL and ALASTAIR T. GARDINER

Introduction

Carotenoids are essential for the survival of photosynthetic organisms. Indeed, there are no wild-type photosynthetic organisms which exist in the absence of carotenoids, and inhibitors of carotenogenesis are effective herbicides (e.g., metaflurazon). Carotenoids have been shown to have two major functions in photosynthesis. They act as photoprotective agents, preventing the harmful photodynamic reaction, and as accessory light-harvesting pigments, extending the spectral range over which light drives photosynthesis.[1] The first of these two functions depends on the triplet states of the carotenoid, while the second depends on their excited singlet states. However, before we discuss this in more detail and describe how these two functions can be demonstrated experimentally, it is important to consider how carotenoids are organized and arranged within the photosynthetic apparatus.

Most common carotenoids found in photosynthetic cells and organelles are rather hydrophobic molecules and are typically located within the photosynthetic membranes. However, they are not usually freely mobile within the lipid interior of these membranes, but rather are noncovalently bound to the photosynthetic reaction centers and the light-harvesting complexes. It is important, therefore, to emphasize the role of these proteins. Carotenoids are photochemically quite active molecules and can participate in a range of reactions; which of these reactivities are expressed in vivo is determined, largely, by which apoprotein they are bound to.

Figure 1 compares the energy levels of the singlet and triplet states of β-carotene, chlorophyll a, bacteriochlorophyll a, and oxygen. The most striking feature depicted in Fig. 1 is that on the one hand the carotenoid has an energetic $^1B_{2u}^*$ excited singlet state, while on the other hand it has a rather low-lying first excited triplet state. The functional significance of this is that the excited singlet state is energetically able to donate energy to the first excited singlet state of either the chlorophyll or the bacteriochlorophyll (its light-harvesting function), while the low-lying triplet state is able to quench triplet chlorophyll and bacteriochlorophyll, as well as singlet oxygen (its photoprotective function). In this short chapter it is not possible to discuss whether the singlet-singlet energy transfer from the carotenoid to the chlorophylls occurs from the symmetry-allowed $^1B_{2u}^*$ state or the

[1] D. Siefermann-Harms, *Biochim. Biophys. Acta* **811**, 325 (1985).

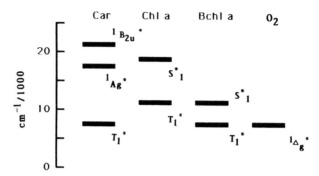

FIG. 1. Schematic representation of the singlet and triplet energies of carotenoids (data for β-carotene), chlorophyll a, bacteriochlorophyll a, and oxygen. For more details, see Cogdell and Frank.[2]

lower lying symmetry-forbidden $^1Ag^*$ singlet state. Readers interested in this point should consult the paper by Cogdell and Frank.[2]

Although carotenoids function just as well in plants and algae as in photosynthetic bacteria, the antenna system of the purple bacteria is an ideal model system in which to demonstrate their functions, because they are spectrally so well defined and have a much simpler organization than those of plants. In the rest of this chapter we describe how to isolate and purify bacterial antenna complexes, and then how to use them to demonstrate the two major functions of carotenoids.

Isolation of Bacterial Antenna Complexes

The preparation of bacterial antenna complexes begins with the solubilization of the photosynthetic membranes. The type and concentration of detergent chosen for this depends on the species of bacteria being used. The detergent must be strong enough to solubilize the complexes but not so harsh that they are denatured. We have found that lauryldimethylamine N-oxide (LDAO) is the most generally applicable detergent. Typically the membranes are adjusted to a standard concentration by diluting them to an absorbance of 50 cm^{-1} at their bacteriochlorophyll near-infrared (NIR) wavelength maximum. Then a suitable amount of detergent is added, usually 1% (w/v). It is easy to see if this causes denaturation. When the complexes are denatured the bacteriochlorophyll a (Bchl$a$) absorption(s) in the NIR shifts to the blue and absorbs at 770 nm. If this happens, try either a lower LDAO concentration or a different detergent.

[2] R. J. Cogdell and H. A. Frank, *Biochim. Biophys. Acta* **895**, 63 (1987).

Once the complexes have been solubilized they must be purified. We currently use two main methods depending on the species.

Sucrose Density Gradient Centrifugation

Centrifugation through sucrose density gradients often works the best, but it can be time consuming if a large quantity of the antenna complex is required. The sucrose concentrations required vary depending on the detergent used; here we describe the conditions that work with LDAO.

Once the photosynthetic membranes have been solubilized they are given a brief low-speed centrifugation (10,000 g, 10 min, 4°), to remove any unsolubilized material. They are then diluted to give a final LDAO concentration of 0.2-0.3% (w/v) and layered on top of the sucrose gradients. The gradients are made up in 28-ml centrifuge tubes in 6-ml steps of 1.2, 0.5, 0.3, and 0.2 M sucrose. The sucrose solutions are prepared in a suitable buffer (typically 20 mM Tris-HCl, pH 8.0) in the presence of 0.2% (w/v) LDAO. The gradients are then centrifuged at 180,000 g (4°) overnight (~ 16 hr).

Figure 2 shows a typical sucrose gradient, in this case using solubilized

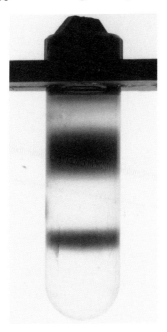

FIG. 2. Typical fractionation of membranes from *Rhodopseudomonas palustris* by sucrose gradient centrifugation. The top band is the B800-850 complex, and the bottom band is the RC-antenna "core" complex.

membranes from *Rhodopseudomonas palustris*. Two bands are seen. The upper band is the major, variable antenna complex, in this case B800–850, and the lower band is the reaction center (RC)–antenna conjugate that represents the "core" of the bacterial photosynthetic unit. The two bands can be conveniently removed with a Pasteur pipette. Figure 3 shows the NIR absorption spectra of the two bands.

Ion-Exchange Chromatography

In our hands the most useful chromatographic material has been Whatman (Clifton, NJ) DE-52 (dimethylaminoethyl-cellulose), which is equilibrated in 20 mM Tris-HCl, pH 8.0, and then poured into a suitably sintered glass column. The solubilized complex is diluted to an LDAO concentration of 0.2% and then loaded onto the column, in 20 mM Tris-HCl, pH 8.0. The column is usually loaded to about one-half of its capacity so as to leave room for chromatography.

Once the column has been loaded the antenna complexes are eluted by raising the concentration of sodium chloride, again in the same Tris buffer supplemented with 0.1% (w/v) LDAO. Usually it is best to use a stepwise elution such as 50, 100, 150, 200, and 300 mM sodium chloride. These columns are quick and easy to run, and the complexes, being brightly colored, are readily seen by eye. This method is best if large quantities of a given complex are required. In most cases the RC–antenna conjugate elutes first, followed by the B800–850 or B800–820 complexes.

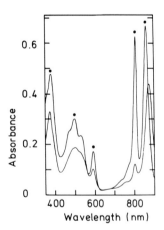

FIG. 3. Absorption spectra of the two bands removed from the gradient shown in Fig. 2. Top band, marked peaks; bottom band, unmarked peaks.

Once they have been prepared most purple bacterial antenna complexes will survive at $-20°$ for months.

Measurement of Singlet–Singlet Energy Transfer from Carotenoids to Bacteriochlorophyll

Experimentally the demonstration of singlet-singlet energy transfer from carotenoids to bacteriochlorophyll, and the determination of the efficiency of this process, involves producing a fluorescence action spectrum. In the absence of a reaction center (i.e., an energy sink), a significant proportion of the light absorbed by an antenna complex will be reemitted as fluorescence. In the case of the B800–850 antenna complex isolated from *Rb. sphaeroides* strain 2.4.1 (Fig. 4A) this fluorescence is mainly emitted from the long-wavelength bacteriochlorophyll absorption band at 850 nm. Any pigments which can transfer energy to bacteriochlorophylls will sensitize this fluorescence, and so the amount of fluorescence can be used as an assay for this energy transfer. Most experiments on this type of measurement use commercial spectrofluorimeters, and so in this chapter we describe only the important practical details, not the construction of the apparatus.

The first task is to determine the fluorescence emission spectrum of the sample so as to know where to monitor the fluorescence during the production of the action spectrum (Fig. 4B). Typically the sample is excited with broad band blue light, and the emission spectrum in the red is determined. Care must be taken so that the sample is sufficiently dilute so that self-absorption does not distort the emission spectrum. Usually this means making sure that the sample has an optical density of less than 0.2 cm^{-1} at the long-wavelength absorption maximum.

The next step in the process is to measure the absorption spectrum of the antenna sample through the region of the spectrum over which the excitation spectrum is to be recorded. The efficiency of the energy transfer is calculated by comparing the absorption spectrum with the excitation spectrum. It must be pointed out here, though, that the absorption spectrum needs to be plotted as fractional absorption ($\equiv$ %transmission), that is, as a linear scale. Often this is forgotten, and a log scale (optical density) is then compared with a linear scale (the fluorescence excitation spectrum), which is of course a basic error.

Finally the fluorescence excitation spectrum is determined. The fluorescence is measured at the peak of the emission spectrum. Typically the emission spectrum of chlorophylls and bacteriochlorophylls is quite broad so that rather wide bandwidths can be used on the emission monochromator (i.e., 10–15 nm). This will improve the size of the signal detected and

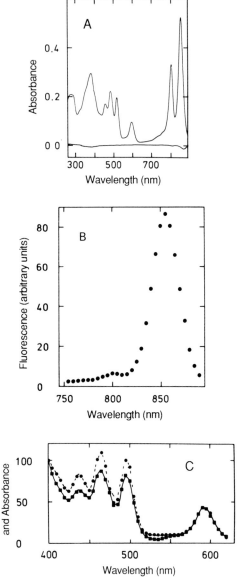

FIG. 4. Absorption, fluorescence emission, and excitation spectra of the B800–850 complex from *Rhodobacter sphaeroides* strain 2.4.1 (A) Absorption spectrum. (B) Fluorescence emission spectrum. The antenna sample was excited with broad band blue light, and the fluorescence was detected after being passed through an RG715 cutoff filter. The detection monochromator was set at a bandwidth of 2 nm. (C) Normalized absorption (●) and fluorescence excitation (■) spectra of the B800–850 complex. The two spectra were normalized at 590 nm. The efficiency of the energy transfer in this case is approximately 95%.

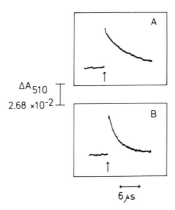

FIG. 5. Kinetics of carotenoid triplet formation induced by a Q-switched ruby laser flash in the B800–850 complex from *Rb. sphaeroides* strain GA. (A) Anaerobic conditions; (B) in air (i.e., $+O_2$).

so give a better signal-to-noise ratio. The excitation spectrum is then determined by exciting the sample with equal numbers of photons at each wavelength throughout the visible region (usually 400–650 nm). On the excitation side satisfactory spectral resolution can be obtained with bandwidths of 1–2 nm.

Once the excitation spectrum has been measured it must be normalized to the absorption spectrum. This is done by making the two spectra coincide at 590 nm, where absorption is due only to bacteriochlorophyll. The logic for this is as follows. One assumes that exciting directly into the bacteriochlorophyll represents an energy transfer efficiency of 100%. Making the two spectra equal at this wavelength then allows the absorption spectrum and the excitation spectrum to be directly compared. If they overlap then the energy transfer efficiency is 100%; if the excitation spectrum is one-half the magnitude of the absorption spectrum the efficiency is 50%. This procedure works well in bacteria where the carotenoids and bacteriochlorophylls are well separated spectrally (e.g., see Fig. 4C). In plants and algae this is not true, and the absorptions of the carotenoids and chlorophylls overlap. This makes exact quantification less clear.

It is now also possible to use picosecond flash photolysis to investigate the kinetics of this singlet–singlet energy transfer. However, a discussion of this is beyond the scope of this chapter. Readers interested in this should consult Refs. 3 and 4.

[3] M. R. Wasielewski, D. M. Tiede, and H. A. Frank, in "Ultrafast Phenomena V" (G. R. Fleming and A. R. Siegman, eds.) p. 388. Springer-Verlag, Berlin, 1986.
[4] T. Gillbro, R. J. Cogdell, and V. Sundstrum, *FEBS Lett.* **235**, 169 (1988).

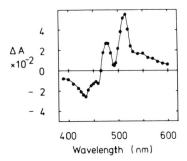

FIG. 6. Difference spectrum of the carotenoid triplet state formed in Fig. 5. The extent of the change immediately following laser excitation is plotted against the wavelength.

Measurement of Triplet–Triplet Energy Transfer from Bacteriochlorophyll to Carotenoid

Triplet-triplet energy transfer from bacteriochlorophyll to the carotenoid occurs in the nanosecond time domain, and then the carotenoid triplet state lasts for a few microseconds (Fig. 5A). Measurement of this process therefore requires flash photolysis. Most experimental setups for this are home built, and good practical descriptions of how to do this are available.[5,6] Typically the samples are excited with a Q-switched laser (usually in the red), and the carotenoid triplet states are detected in the visible region.

The difference spectrum shows bleaching in the region of the carotenoid, ground state absorption bands, and the appearance of one or two new peaks to the red of these (Fig. 6). The decay of the carotenoid triplets is accelerated by the presence of oxygen (this is also shown in Fig. 5B). The quenching of bacteriochlorophyll triplet states by carotenoid depends on the energy level of the carotenoid triplet. It must be lower than that of the bacteriochlorophyll triplet. The energy level of a carotenoid depends on the number of its double bonds. In practice this means more than seven double bonds.[7]

Carotenoids will also quench singlet oxygen. Singlet oxygen is very difficult to monitor *in vivo*. The only direct way to do this is to look at the luminescence of singlet oxygen which occurs at 1260 nm. In aqueous

---

[5] T. G. Monger, R. J. Cogdell, and W. W. Parson, *Biochim. Biophys. Acta* **449**, 136 (1976).

[6] J. McVie, R. S. Sinclair, and T. G. Truscott, *J. Chem. Soc. Faraday Trans. II* **74**, 1870 (1978).

[7] R. Bensasson, E. J. Landard, and B. Maudinas, *Photochem. Photobiol.* **23**, 189 (1976).

samples, however, the yield of singlet oxygen luminescence is low, and its lifetime is short. The effect of carotenoids on singlet oxygen can be demonstrated *in vitro* in organic solvents (e.g., see Ref. 8).

Acknowledgments

A. T. G. has been supported by a grant from the Science and Engineering Research Council.

[8] C. F. Borland, R. J. Cogdell, E. J. Land, and T. G. Truscott, *J. Photochem. Photobiol.* **3,** 237 (1989).

[19] Retinoic Acid Synthesis from β-Carotene *in Vitro*

By JOSEPH L. NAPOLI

Introduction

Retinoic acid supports vitamin A-dependent differentiation through modulating gene transcription.[1] Although the specific enzymes involved physiologically in retinoic acid synthesis have not been characterized in detail, it is now clear that both retinol and β-carotene serve as initial substrates for retinoic acid synthesis and at least three, possibly four, pathways of retinoic acid synthesis exist, at least *in vitro*.[2–7] One pathway starts with the conversion of free retinol to retinal and is catalyzed by a cytosolic NAD-dependent dehydrogenase(s) that is distinct from human class I and class II alcohol dehydrogenases.[2–4] A second pathway uses holo-CRBP as substrate to synthesize retinal and is catalyzed by an NADP-dependent microsomal dehydrogenase(s).[5] In both cases the rate-limiting step is the dehydrogenation of retinol to retinal, followed by an irreversible and rapid dehydrogenation of retinal to retinoic acid. A third pathway, the conversion of β-carotene to retinoic acid,[6] is the subject of this chapter. This report describes high-performance liquid chromatography (HPLC) assays to study the conversion of β-carotene to retinoids. For additional

[1] R. Lotan, *Prog. Clin. Biol. Res.* **259,** 261 (1988).
[2] J. L. Napoli and K. R. Race, *Arch. Biochem. Biophys.* **255,** 95 (1987).
[3] K. C. Posch, W. J. Enright, and J. L. Napoli, *Arch. Biochem. Biophys.* **274,** 171 (1989).
[4] J. L. Napoli, *J. Biol. Chem.* **261,** 13592 (1986).
[5] K. C. Posch, M. H. E. M. Boerman, R. D. Burns, and J. L. Napoli, *Biochemistry* **30,** 6224 (1991).
[6] J. L. Napoli and K. R. Race, *J. Biol. Chem.* **263,** 17372 (1988).
[7] J. L. Napoli and K. R. Race, *Biochim. Biophys. Acta* **1034,** 228 (1990).

background and detail concerning the assay of retinoids, readers are referred to two previous chapters in this series.[8,9]

Purification and Handling of β-Carotene

The laboratory is kept under gold lighting to minimize isomerization of carotenoids and retinoids. Commercial β-carotene and retinoids require purification before use, especially if detection of picomole amounts of products is to be meaningful. Both β-carotene and retinol contain several percentages of a peak that comigrates with retinal on HPLC, as well as other polar contaminants. β-Carotene can be purified by a number of convenient methods. One is to elute it from a Whatman (Clifton, NJ) ODS-2 column (1 × 25 cm) in 48 ml with a mobile phase of tetrahydrofuran/methanol (1:3, v/v). An alternative and more simple method is to use a disposable prepacked column filled with a polar HPLC-grade resin such as microparticulate silica gel. β-Carotene is easily eluted from such columns with hexane, whereas polar contaminants are not.

To purify by the latter method, a saturated solution of β-carotene in 5 ml of hexane is prepared, (this removes polar contaminants that are not hexane soluble) and applied to a Silica Bond Elut column (Analytichem International). Fractions of 1–2 ml are collected. The elution can be monitored crudely by visual inspection of color, but it is prudent to take UV spectra of the eluting fractions and monitor the ratio of 450 nm/340 nm, which indicates the proportion of all-*trans*-β-carotene to various cis isomers of β-carotene.[10] The purified β-carotene is usually used immediately, but it can be stored under an inert gas (nitrogen or argon) at $-20°$ in brown vials in acetone/2-propanol/hexane (3:2:5, v/v). Shortly before use, the storage solvent is evaporated under a gentle stream of nitrogen, and the substrates are dissolved in dimethyl sulfoxide (DMSO). The DMSO solutions of substrate should be prepared no sooner than 1 to 2 hr before use.

High-Performance Liquid Chromatography

Retinoids are analyzed with an automated HPLC system including an automatic sample injector, a tunable UV absorbance detector, and an integrator that is based on a normal-phase HPLC column (Du Pont Zorbax-Sil Reliance-3 cartridges, 3-μm beads, 0.6 × 4 cm). Reversed-

---

[8] J. L. Napoli, this series, Vol. 123, p. 112.
[9] J. L. Napoli, this series, Vol. 189, p. 470.
[10] K. Tsukida, K. Saiki, T. Takii, and Y. Koyama, *J. Chromatogr.* **245**, 359 (1982).

phase systems, especially those that elute retinoic acid near the void volume, should be avoided if specific quantitation of retinoic acid is desired, because of their lower resolution and relative lack of selectivity. Retinoic acid, retinol, and retinal are detected at 340, 325, and 370 nm, respectively.

Retinoic acid is analyzed with a mobile phase of 0.1% (v/v) acetic acid in 1,2-dichloroethane/hexane (1:4, v/v) (Fig. 1) or 0.35% acetic acid in 1,2-dichloroethane/hexane (1:9, v/v), usually at a flow rate of 2 ml/min. The exact proportions of solvents in the mobile phase should not be considered as inflexible; the solvents are adjusted periodically to respond to differences in solvent batches, relative humidity, and specific columns. The following guidelines are used; acetic acid suppresses ionization of retinoic acid and is a determinant of k' (elution volume); 1,2-dichloroethane

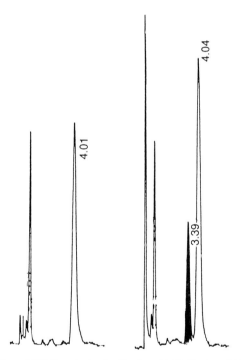

FIG. 1. Normal-phase HPLC analysis of retinoic acid. Retinoic acid (shaded peak) was eluted isocractically from a Du Pont Zorbax-Sil cartridge (0.6 × 4 cm) at 2 ml/min with 0.1% acetic acid in 1,2-dichloroethane/hexane (1:4, v/v) in 6.8 ml (~3.4 min). The peak eluting in 8 ml (~4 min) is the internal standard TIMOTA. β-Carotene 10 μM was incubated for 30 min at pH 6.5 with 3 mM NAD in the absence of protein (*left*, no detectable retinoic acid) or with 0.25 mg of rat liver cytosolic protein (*right*, 35 pmol of retinoic acid, shaded peak). The detector was set at 340 nm.

changes the α characteristics (selectivity) of the mobile phase. Increasing the acetic acid decreases the elution volume, and increasing the 1,2-dichloroethane increases the resolution of all-*trans*-retinoic acid from its isomers, such as 13-*cis*-retinoic acid, and decreases the resolution between all-*trans*-retinoic acid and the internal standard.

Two mobile phases can be used for normal-phase HPLC based on either tetrahydrofuran/hexane or acetone/hexane. The tetrahydrofuran system is exemplified in Fig. 2. β-Carotene elutes in the void volume of this system. In the acetone-based system, retinol elutes in 9.5 ml and retinal elutes in 3.6 ml (acetone/hexane, 1:19, v/v) at a flow rate of 2 ml/min. β-Carotene elutes in less than 2 ml, but it interferes with the retinal peak. To separate retinal from β-carotene, a gradient system can be used. The initial mobile phase is acetone/hexane (1:49, v/v) eluted at 2 ml/min for 5 min. At 5 min a linear gradient is started to reach a final mobile phase of acetone/hexane (1:24, v/v) in 1 min. β-Carotene elutes in less than 3 ml. Retinal elutes in 5.6 ml, and retinol elutes, after final conditions are established, in 21 ml. In experiments with β-carotene substrate concentrations greater than 20 μM, excess β-carotene is removed before HPLC analysis to reduce interference with retinoid peaks by concentrating the first hexane phase to 2 ml and applying the concentrate to a small silica column (Waters Sep-Pak, Milford, MA). The carotenoid(s) are eluted with 2 ml of acetone/hexane (7.5:92.5, v/v). Retinol is eluted with 2 ml of acetone.[6]

Incubations

Incubations with homogenates or subcellular fractions are done in test tubes capable of holding at least 8 ml and are performed at 37° in a buffer containing 150 mM KCl, 2 mM dithiothreitol, and 20 mM HEPES adjusted to the appropriate pH, in a total volume of 1 ml. To initiate the reaction, substrate is added in 10 μl of dimethyl sulfoxide. Controls consist of incubations done in the presence of buffer without an added source of enzymatic activity and/or with boiled protein. The pH optimum seems to

---

FIG. 2. Normal-phase HPLC analysis of retinol and retinal. A Du Pont Zorbax-Sil cartridge (0.6 × 4 cm) was eluted at 2 ml/min with a linear gradient (curve 6 on Waters 600E System Controller) from 1% tetrahydrofuran/hexane (solvent A) to 15% solvent A and 85% hexane. Retinal and retinol (shaded peaks) eluted at 4.95 (9.9 ml) and 8 min (16 ml), respectively. The three plots show (from left to right) increasing amounts of retinal (generated from retinol); namely, 4, 12, and 38 pmol, respectively. In this case the detector was set at 370 nm to optimize sensitivity for retinal. If both retinol and retinal are to be measured, sensitivity for retinol can be enhanced by changing the wavelength to 325 nm after 6 min.

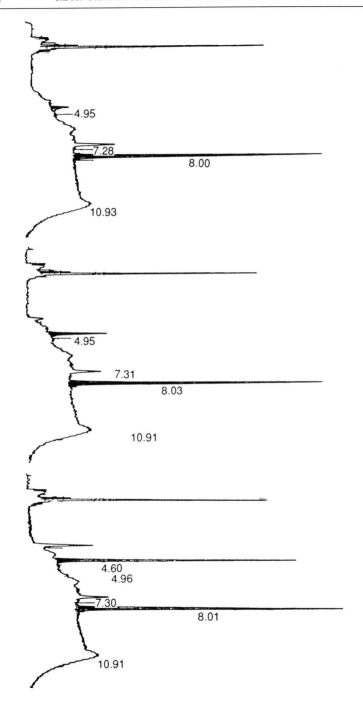

be 7 for retinol production, but may vary for retinoic acid production (Ref. 6 and Fig. 3) and should be determined for each source of protein. Linearity of rate with protein and time depends on the source of enzyme activity, and should also be determined for each specific case. Pyridine nucleotide cofactors are usually not required to observe retinoid production from β-carotene, but they influence the amount of metabolites and their proportions relative to each other. If retinoic acid is the object of the study, it is prudent to add NAD to incubations, at least during preliminary trials (Table I).

The incubations are quenched with sufficient 0.025 N KOH in ethanol to raise the pH above 12 (for incubations done at pH 7.0 in a 1-ml incubation volume, 1 ml of the ethanolic KOH solution is added), and the internal standard for retinoic acid [all-trans-7-(1,1,3,3-tetramethyl-5-indanyl)-3-methylocta-2,4,6-trienoic acid (TIMOTA)] is added in ethanol (10–100 μl). Neutral retinoids and β-carotene are extracted from the alkaline aqueous phase with 5 ml of hexane (first hexane by vortexing 5 min with a multitube vortexer. The tubes are centrifuged for 2 to 5 min to effect clean separation of the phases. The alkaline aqueous phase is then acidified to a pH below 2 with 4 N HCl (75 μl is sufficient for the 1-ml incubation volume done at pH 7). Retinoic acid and the other acidic products are extracted by vigorous vortexing with a second volume of hexane (second hexane). The concentration of base must be kept low, because a final

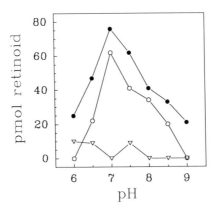

FIG. 3. Effects of pH on the conversion of β-carotene to retinoids by rat liver cytosol. Incubations were conducted for 30 min with 0.25 mg of protein, 10 μM β-carotene, and 3 mM NAD. The retinoids measured were retinoic acid (●), retinol (○), and retinal (△). Data are represented as net picomoles above those observed in the "no protein" controls. Neither retinoic acid nor retinol was observed in the absence of protein.

TABLE I
EFFECTS OF COFACTORS ON CONVERSION OF
β-CAROTENE TO RETINOIDS BY RAT LIVER
CYTOSOL[a]

Cofactor	Product (pmol)		
	Retinoic acid	Retinal	Retinol
None	37 ± 4	23 ± 2	61 ± 6
3 mM NAD	110 ± 4	0	51 ± 4
3 mM NADH	57 ± 1	0	59 ± 4

[a] Incubations were done at pH 7.5 for 30 min with 0.25 mg of rat liver cytosolic protein and 10 μM β-carotene. Data are the means ± S.E. of triplicates. No retinoids were detected when the incubations were conducted under the same conditions but without protein.

concentration of 0.2 N KOH or NaOH, for example, causes the conversion of all-*trans*-retinoic acid to geometric isomers at 37°, including 13-*cis*-retinoic acid.[8,9]

The analytes may be stored in hexane overnight at −20°. Samples are not stored in mobile phases containing either acid or chlorinated hydrocarbons. Immediately prior to HPLC analysis, the solvent is evaporated with a gentle stream of nitrogen and the residue is dissolved in 0.1 ml of mobile phase.

Detergents

Detergents are not required to detect the synthesis of retinoic acid from β-carotene with homogenates or subcellular fractions. In fact, detergents seem to have an adverse effect, especially with low concentrations (20 μM or less) of β-carotene as substrate. For example, the generation of retinoids from β-carotene by liver cytosol was completely inhibited by the following: 0.02 and 0.1% HS-15; 0.02, 0.075, 0.1, 0.15, and 0.3% 1-*S*-octyl-β-D-thioglucopyranoside. Inhibition by 0.3% CHAPS or CHAPSO was 62%.

Pathway of Retinoic Acid Synthesis from β-Carotene

Retinal has been reported as a major initial product of β-carotene metabolism *in vitro*, which is then converted to retinol overwhelmingly, if

TABLE II
EFFECTS AND RETINOIC ACID PRODUCED FROM β-CAROTENE, RETINAL, OR MIXTURE OF β-CAROTENE AND RETINAL BY RAT INTESTINAL CYTOSOL[a]

Substrate	Product (pmol)	
	Retinol	Retinoic acid
15 μM β-Carotene	75 ± 4	52 ± 3
0.5 μM Retinal	0	38 ± 3
15 μM β-Carotene + 0.5 μM retinal	76 ± 3	76 ± 5

[a] Incubations were done at pH 7.5 for 30 min with 0.5 mg of rat intestinal cytosolic protein. Data are the means ± S.E. of 5 replicates.

not exclusively.[11,12] Although this is consistent with observations *in vivo* of β-carotene metabolism (i.e., retinol, not retinoic acid, is the major product *in vivo*),[13] it is inconsistent with *in vitro* work showing that retinoic acid is the major *in vitro* product of the metabolism of free retinal.[6] The *in vivo/in vitro* discrepancy may be the result of disrupting, during preparation of subcellular fractions, protein–protein complexes that direct the product(s) of β-carotene metabolism down specific pathways. Alternatively the dilution of enzymes and/or cofactors by homogenization and subcellar fractionation could affect the relative proportions of products. Comparison of reaction pathways *in vitro* does not necessarily (in fact, do not usually) represent what occurs in intact cells, where protein concentrations are much higher and substrate concentrations are much lower than those normally used for *in vitro* assays conducted under standard Michaelis–Menten conditions.

As far as the mechanism of the production of retinoids from β-carotene is concerned, the difference between the products of β-carotene and retinal metabolism *in vitro* is inconsistent with an exclusive "central cleavage" mechanism. Two types of experiments illustrate this point and indicate that β-carotene conversion to retinoids *in vitro* is not a simple case of generating retinal and releasing it into solution to diffuse to enzymes that convert it to retinol and retinoic acid. The first type is based on the observation that the relative amounts of retinol and retinoic acid generated

[11] J. A. Olson and O. Hayaishi, *Proc. Natl. Acad. Sci. U.S.A.* **54**, 1364 (1965).
[12] M. R. Lakshman, I. Mychovsky, and M. Attlesey, *Proc. Natl. Acad. Sci. U.S.A.* **86**, 9124 (1989).
[13] N. I. Krinsky, M. M. Mathews-Roth, S. Welankiwar, P. K. Sehgal, N. C. G. Lausen, and M. Russett, *J. Nutr.* **120**, 81 (1990).

from β-carotene are different than those generated from retinal. With retinal as substrate, the major product *in vitro* is retinoic acid. In the absence of cofactor, little or no retinol is detected. In the presence of NAD with retinal as substrate, 10- to 100-fold more retinoic acid than retinol will usually be observed. With β-carotene as substrate *in vitro,* retinol is detected in the absence of cofactor, and 1.5-- to 2-fold more retinol is observed than retinoic acid. In the presence of NAD, retinoic acid is the major product, but it is only 1.5- to 2-fold greater than the retinol (Ref. 6 and Table I). This phenomenon is illustrated in Table II.

Other data that illustrate this point are shown in Figs. 4 and 5. The effects of 4-methylpyrazole (4-MP) and 4'-(9-acridinylamino) methane-sulfon-*m*-anisidide(*m*-AMSA) on the production of retinol and retinoic acid depend on whether the substrate is retinal or β-carotene. 4-MP causes 50% inhibition of β-carotene conversion to retinoic acid at 0.27 mM, but 5.5 mM is required to inhibit retinoic synthesis from retinal 50%, a 20-fold difference. Retinol synthesis from retinal is not inhibited by 4-MP, but retinol synthesis from β-carotene is. *m*-AMSA causes 50% inhibition of retinoic acid and retinol synthesis from β-carotene at approximately 6 μM, but it does not inhibit retinoic acid or retinol synthesis from retinal at concentrations up to 100 μM.

Clearly, the metabolism of retinal in solution is different from the metabolism of β-carotene. These data suggest that *in vitro* mechanisms

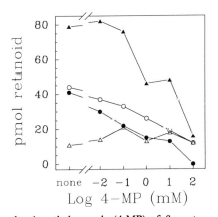

FIG. 4. Inhibition by 4-methylpyrazole (4-MP) of β-carotene or retinal conversion to retinol and retinoic acid by rat intestinal cytosol. Incubations were done for 30 min at pH 7.0 with 2 mM NAD, 0.5 mg of protein, and 10 μM β-carotene or 1 μM retinal. The retinoids measured were retinoic acid from retinal (▲), retinoic acid from β-carotene (●), retinol from retinal (△), and retinol from β-carotene (○).

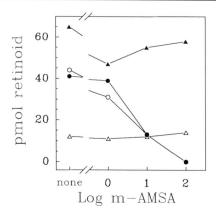

FIG. 5. Inhibition of β-carotene or retinal conversion to retinol and retinoic acid in rat intestinal cytosol by m-AMSA [4'-(9-acridinylamino)methanesulfon-m-anisidide]. Incubation conditions are described in the legend to Fig. 3. The retinoids measured were retinoic acid from retinal (▲), retinoic acid from β-carotene (●), retinol from retinal (△), and retinol from β-carotene (○). The x axis is the log of the m-AMSA concentration (μM).

must be operating instead of, or in addition to, the central cleavage mechanism, and therefore provide support for "excentric cleavage."

Acknowledgments

The technical assistance of Gregory Salerno is gratefully acknowledged. This work was supported by National Institutes of Health Grants DK36870 and DK38885 and U.S. Department of Agriculture Grant 87-CRCR-1-2332.

[20] β-Carotene Modulation of Delayed Light Emission from Aggregated Chlorophyll

By SHAN YUAN YANG

Introduction

Delayed light emission (DLE) of chlorophyll (Chl) *in vivo* was discovered accidentally by Strehler and Arnold in an attempt to use the firefly luminescence method to measure ATP production by algae.[1] DLE was also

[1] B. L. Strehler and W. Arnold, *J. Gen. Physiol.* **34**, 809 (1951).

observed with leaves[1] and chloroplasts[2] and with photosynthetic bacteria.[3] It is related to the partial process of photosynthesis. Thus DLE at different time scales (nanoseconds, milliseconds, and seconds) has been used in the study of primary photochemical reactions, electron transfer, photophosphorylation, etc.

Delayed light emission has been observed in a suspension of aggregated Chl.[4] It was shown that the intensity and decay rate of DLE from aggregated Chl depend on several factors, such as the intensity and duration of illumination, the presence of electron donors and acceptors, and the presence of deaggregating reagents. Of special interest is the fact that not only do unilluminated pheophytin, Chl$b$, and carotenoids have effects on the DLE from aggregated Chl$a$, but the illuminated pigments can also induce DLE from unilluminated aggregated Chl$a$, although no DLE can be detected after illumination.[5-7] These phenomena can be used to probe energy storage and transfer between aggregated Chl and other photosynthetic pigments.

In this chapter, procedures for the modulation by β-carotene of the DLE from aggregated Chl and detection of energy storage and transfer between carotene and Chl are given.

Materials and Methods

Petroleum Ether and Hexane. In DLE measurements to avoid the interference by compounds which may emit light or absorb DLE, petroleum ether (A.R., 60–90°) or hexane (99%) is refluxed with sulfuric acid several times, until the sulfuric acid layer no longer turns yellow. The solvent is then shaken with 0.1 N KMnO$_4$, 10% H$_2$SO$_4$ solution, dried with CaCl$_2$, and distilled.

Chlorophyll a and Its Aggregates. Chl$a$ is prepared from an acetone extract of fresh spinach leaves using a microcrystalline cellulose chromatographic column. The Chl eluate is dried and dissolved in warm, water-saturated petroleum ether. The solution is kept at 5° overnight or longer in the dark, to form aggregated Chl$a$. Formation of aggregates is indicated by

---

[2] B. L. Strehler, *Arch. Biochem. Biophys.* **34**, 239 (1951).
[3] W. Arnold and J. Thompson, *J. Gen. Physiol.* **39**, 311 (1956).
[4] S. Y. Yang, *Lumin. Display Devices* **6**, 140 (1985).
[5] S. Y. Yang, *Chin. Sci. Bull.* **33**, 1215 (1988).
[6] S. Y. Yang, *Zhiwu Shengli Xuebao (Acta Phytophysiologica Sin.)* **15**, 83 (1989).
[7] S. Y. Yang and S. S. Brody, *Z. Naturforsch. C: Brosci.* **44C**, 132 (1990).
[8] W. Bertsch, J. West, and R. Hill, *Biochim. Biophys. Acta* **172**, 525 (1969); W. Bertsch, J. West, and R. Hill, in "Application of Chlorophyll Fluorescence" (H. L. Lichtenthaler, ed.). Kluwer Academic Publ., Dordrecht, The Netherlands, 1988.

the solution turning turbid and the color changing from blue to green. The aggregated Chl, which may be in the form of microcrystals, is collected by centrifugation and suspended in petroleum ether or hexane with a vortex mixer. To reduce the concentration of Chl monomer as far as possible, the suspension is again centrifuged and suspended in petroleum ether or hexane. All procedures are performed in dim light.

all-trans-β-Carotene. all-*trans*-β-Carotene is obtained from Merck Chemical Co. and recrystallized from benzene and methanol, or further purified by chromatography.

Apparatus for Measuring Delayed Light Emission from Chlorophyll a

The instrumentation constructed for DLE measurement is shown schematically in Fig. 1. The light source for illumination is a 12-V 75-W projection lamp. A 5-cm layer of 0.1% $CuSO_4$ solution is used to remove heat. The sample is put into the syringe and illuminated. It is then injected into a control cuvette or a cuvette which contains aggregated Chl in the dark chamber. The time required for injection is about 0.8 sec. Because the DLE from aggregated Chl is very faint, the cuvette is put as close against the window of the photomultiplier as possible. The photomultiplier is sensitive to red light. The photocurrent is amplified and recorded. The response time of the system is about 0.5 sec.

Determinations of Energy Transfer between Illuminated β-Carotene and Aggregated Chlorophyll

When a solution of β-carotene, pheophytin, or Chl*b* is illuminated for 30 sec and then injected into the cuvette in a dark chamber, no DLE is

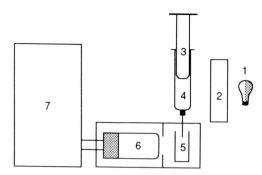

FIG. 1. Schematic illumination setup for measuring the delayed light emission from chlorophyll aggregates. (1) Light source; (2) heat filter; (3) syringe; (4) suspension of aggregated Chl; (5) cuvette, 10 × 5 × 45 mm; (6) photomultiplier tube; and (7) amplifier and recorder.

detected in the absence of Chl aggregates. In the presence of aggregated Chl$a$, however, a DLE signal is detected (Fig. 2), which indicates that an energy storage reaction has occurred and produced an energized intermediate. When the energized intermediate interacts with aggregated Chl, the energy can be transferred to the aggregated Chl and released as DLE of Chl.

We do not know what kind of reaction has taken place on illumination of the carotene molecule (pheophytin, Chl$a$, and Chl$b$), nor how carotene turns into an energized state, that is, whether it changes its configuration or forms an active intermediate, etc. For convenience's sake, we call a substance that causes DLE of Chl an energized state (ES).

Characteristics of Delayed Light Emission from Chlorophyll Induced by Energized State

By injecting solutions of carotene having various concentrations, illuminated at different light intensities, and for different durations into cuvettes containing aggregated Chl in the dark chamber, a group of DLE curves is obtained. These curves reflect the concentration of ES and the reaction rate between ES and aggregated Chl, and the areas under the curves represent the relative amounts of energy stored by carotene under illumination. The fact that the DLE of Chl has a much lower intensity than fluorescence, but has a much longer duration (up to several minutes), indicates that the energy stored in ES is considerable.

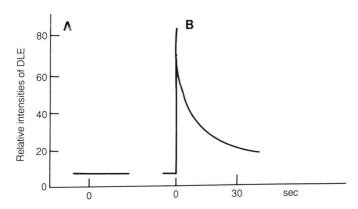

FIG. 2. Light emission of chlorophyll a aggregates induced by preilluminated β-carotene. One milliliter of a 2 μg/ml β-carotene–petroleum ether solution after 30 sec of illumination was injected into a cuvette containing (A) 0.5 ml petroleum ether or (B) 0.5 ml of a 10 μg/ml aggregated Chl$a$–petroleum ether solution.

Life-time of Energized State of Carotene

The amount of the ES of carotene is indicated by the intensity of the DLE from aggregated Chl and is a function of intensity and time of illumination. A maximum is reached at about 1 min of preillumination with white light (~ 1.6 nmol quanta m^{-2} Sec^{-1}).

When preilluminated carotene solution is placed in the dark for various time intervals, and then injected into cuvettes containing aggregated Chl, the intensity of the DLE decreases as the dark interval lengthens (Fig. 3). After a dark interval of about 2 min, the intensity of DLE decreases to about one-half of that with a zero dark interval. This shows that the ES of carotene decays along a nonradiative pathway in the dark.

From Fig. 3 it can be observed that the decay of ES of carotene is not a first-order process. The first half-life of the ES is about 2 min, and the next half-life is about 10 min. On the basis of the fact that the decay rate of the ES is many orders of magnitude lower than that of the singlet or triplet state, the possibility of involvement of any singlet or triplet state of carotene can certainly be eliminated.

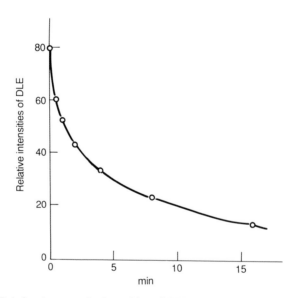

FIG. 3. Relation between the intensities of DLE and duration of dark intervals after illumination of carotene. Illuminated β-carotene was injected, after various dark intervals, into a cuvette containing Chl$a$ aggregates.

Energized State Migration between Chlorophyll and Carotene

Preillumination of Chl aggregates results in DLE.[5] On adding acetone to Chl aggregates following preillumination, a "burst" of light emission appears, and the DLE is terminated (Fig. 4A). Acetone acts to convert the aggregated Chl to its monomeric form which does not exhibit DLE. An energized state which gives rise to DLE exists only in aggregated Chl. When the aggregated Chl is deaggregated into monomeric Chl, which contains no ES, the energy of ES is released as a burst of emitted light.

No "burst" of light is observed when acetone is added to a mixture of Chl aggregates in the presence of carotene in any of the following combinations: aggregated Chl plus preilluminated carotene, preilluminated aggregated Chl plus carotene, or preilluminated aggregated Chl plus preilluminated carotene. In all these cases, the addition of acetone results in an abrupt drop in DLE (Fig. 4B-D). Phenomenologically, it appears that in

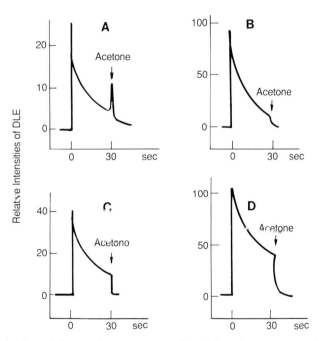

FIG. 4. Light emission of β-carotene contacting with Chl$a$ aggregates and after Chl$a$ aggregates were deaggregated. (A) Illuminated Chl$a$ aggregates only; (B) illuminated β-carotene contacting with aggregated Chl$a$; (C) illuminated Chl$a$ aggregates contacting with β-carotene; (D) β-carotene mixed with Chl$a$ aggregates and illuminated. After 30 sec the solution was injected with 0.5 ml acetone, and the Chl$a$ aggregates were deaggregated. Relative intensities were in the same scale. Final concentrations: Chl, 10 μg/ml; carotene, 2 μg/ml.

the presence of β-carotene all ES in aggregated Chl will migrate to carotene, on deaggregation by acetone, so that no "burst" of light is observed.

These "burst" phenomena suggest that carotene might draw the energy from the excited Chl and bring the Chl down to the ground state, thus protecting the Chl from photooxidation or other reactions.

Acknowledgments

I am grateful to Prof. H. C. Yin and T. D. Wang for critical reading of the manuscript.

[21] Relation of Cis–Trans Isomers of Carotenoids to Developmental Processes

By H. J. Nelis, P. Sorgeloos, and A. P. De Leenheer

Introduction

Natural carotenoids mainly occur in their most stable stereochemical configuration, namely, the all-trans form.[1] If cis isomers are present in biological materials such as fruits and vegetables, they have often been formed artifactually from the all-trans form, mainly as a result of processing.[2-4] This trans to cis isomerization is considered an unwanted phenomenon, particularly when provitamin A carotenoids are involved, because the cis isomers possess a lower biological potency.[5] The small amounts of *cis*-β-carotene in human blood probably originate from the diet,[6,7] although formation from all-*trans*-β-carotene in the body itself has also been suggested.[6]

However, the genuine presence of *cis*-carotenoids has been documented in a number of organisms, including higher plants,[8] photosynthetic

---

[1] B. C. L. Weedon, *in* "Carotenoids" (O. Isler, ed.), p. 267. Birkhaüser, Basel, Switzerland, 1971.
[2] J. P. Sweeney and A. C. Marsh, *J. Assoc. Off. Anal. Chem.* **53**, 937 (1970).
[3] J. P. Sweeney and A. C. Marsh, *J. Am. Diet. Assoc.* **59**, 238 (1971).
[4] F. W. Quackenbush, *J. Liq. Chromatogr.* **10**, 643 (1987).
[5] K. L. Simpson, *Proc. Nutr. Soc.* **42**, 7 (1983).
[6] A. L. Sowell, D. L. Huff, E. W. Gunter, and W. J. Driskell, *J. Chromatogr.* **431**, 424 (1988).
[7] W. Gray Rushin, G. L. Catignani, and S. J. Schwartz, *Clin. Chem.* **36**, 1986 (1990).
[8] G. Tóth and J. Szabolcs, *Phytochemistry* **20**, 2411 (1981).

bacteria,[9] algae,[10] and the brine shrimp *Artemia*.[11] The *cis*-canthaxanthins demonstrated in the latter organism were found to display a more characteristic biological fate than their all-trans counterpart.[11,12] This observation led us to hypothesize about a biological function for these unusual carotenoids in connection with development.[12,13] The aim of this chapter is to describe the methods that permit the quantitation of *cis*-canthaxanthins, besides the all-*trans*-canthaxanthin and their metabolic precursors, in different stages of the *Artemia* life cycle, so as to relate their occurrence, fate, and tissue distribution to developmental processes.

Biological Samples

Artemia cysts are available from the collection at the Artemia Reference Center/Laboratory for Aquaculture at the State University of Ghent, Ghent, Belgium. Cysts can be analyzed either as received (air dry) or after dehydration, full hydration, or decapsulation. Dehydrated cysts are prepared by drying 10-mg samples in an oven at 40° for 24 hr, followed by cooling in a desiccator. Hydration is achieved by soaking dehydrated cysts in water for 5 hr at room temperature using small plastic filter cups (Beckman, Fullerton, CA). Decapsulation is carried out according to the method of Bruggeman *et al.*[14] Nauplii are produced by incubating cysts under standard conditions in seawater (35 g of NaCl/liter).[14] Pigmented animals at various developmental stages are obtained by transferring nauplii to 2-liter aquariums and feeding them twice a day on living *Dunaliella tertiolecta* algae under standard culture conditions.[15]

---

[9] M. Lutz, I. Agalidis, G. Hervo, R. J. Cogdell, and F. Reiss-Husson, *Biochim. Biophys. Acta* **503**, 287 (1978).

[10] L. J. Borowitzka and M. A. Borowitzka, *in* "Biotechnology of Vitamins, Pigments and Growth Factors" (E. J. Vandamme, ed.), p. 15. Elsevier Applied Science, London, 1989.

[11] H. J. C. F. Nelis, P. Lavens, L. Moens, P. Sorgeloos, J. A. Jonckheere, G. R. Criel, and A. P. De Leenheer, *J. Biol. Chem.* **259**, 6063 (1984).

[12] H. J. C. F. Nelis, P. Lavens, M. M. Z. Van Steenberge, P. Sorgeloos, G. R. Criel, and A. P. De Leenheer, *J. Lipid Res.* **29**, 491 (1988).

[13] H. J. Nelis, P. Lavens, L. Moens, P. Sorgeloos, and A. P. De Leenheer, *in* "Biochemistry and Cell Biology of *Artemia*" (T. H. MacRae, J. C. Bagshaw, and A. H. Warner, eds.), p. 159. CRC Press, Boca Raton, Florida, 1989.

[14] E. Bruggeman, P. Sorgeloos, and P. Vanhaecke, *in* "The Brine Shrimp *Artemia*" (G. Persoone, P. Sorgeloos, O. Roels, and E. Jaspers, eds.) Vol. 3, p. 261. Universa Press, Wetteren, Belgium, 1980.

[15] P. Sorgeloos, E. Bossuyt, P. Lavens, P. Léger, P. Vanhaecke, and D. Versichele, *in* "CRC Handbook of Mariculture" (J. P. McVey, ed.), Vol. 1, p. 71. CRC Press, Boca Raton, Florida, 1983.

Standard Compounds

all-*trans*-Canthaxanthin, echinenone, β-apo-8'-carotenal, and β-apo-8'-carotenoic acid ethyl ester were gifts from Hoffmann-LaRoche (Basel, Switzerland). β-Carotene was purchased from Fluka (Buchs, Switzerland). *cis*-Canthaxanthin (collective term for a mixture of one predominating and two minor mono-*cis*-canthaxanthins, substantially free from all-*trans*-canthaxanthin) is prepared by stereomutation from all-*trans*-canthaxanthin, as described previously.[11] The mixture of *cis*-canthaxanthins (both the natural ones in *Artemia* and the synthetic ones) will further be referred to as "*cis*-canthaxanthin."

Liquid Chromatographic Instrumentation

Any isocratic liquid chromatograph equipped with a variable-wavelength detector can be used. In our laboratory the instrument consists of a Varian 5020 pump (Varian Associates, Palo Alto, CA), a Valco N60 valve injector (Valco, Houston, TX) equipped with a 100-μl loop, an HP 1040A multichannel photodiode array detector connected to an HP 9121 dual-disk drive, an HP 7470A plotter, and an HP 85 computer (all from Hewlett-Packard, Palo Alto, CA), and an SP 4100 integrator (Spectra-Physics, San Jose, CA). Although for the quantitative work a variable-wavelength detector without scanning capability will suffice, the use of a photodiode array detector is strongly recommended to permit on-line peak identity confirmation based on absorption spectra.[16] The detector wavelength used to monitor the signal is 470 nm.

Liquid Chromatographic Conditions

For reversed-phase chromatography a 5-μm Zorbax ODS column (15 × 0.46 cm) (Du Pont, Wilmington, DE) is used. The mobile phase consists of a mixture of acetonitrile–methanol–dichloromethane (40:50:10, v/v, containing 0.15% triethylamine), and the flow rate is 1 ml/min. In our initial studies,[11] the chromatographic conditions were slightly different in that a 25 × 0.46 cm 5 μm Zorbax ODS column was used, eluted with acetonitrile–methanol–dichloromethane (5:3:2, v/v). However, superior resolution between all-*trans*- and *cis*-canthaxanthin is obtained with the former eluent, which has become standard. Normal-phase separations are carried out on a 10-μm CP-Spher Si column (25 × 0.46 cm) (Chrompack, Middelburg, The Netherlands), eluted with dichloromethane–2-propanol (99.4:0.6, v/v) at a flow rate of 1 ml/min.

[16] A. P. De Leenheer and H. J. Nelis, this series, Vol. 213 [22].

Conditions for preparative HPLC are as follows: column, 8-μm ROSIL silica (50 × 0.9 cm) (RSL, Eke, Belgium); mobile phase, dichloromethane–2-propanol, 99.5:0.5, v/v; flow rate, 6 ml/min.

Identification of cis-Canthaxanthin in Artemia

The initial identification of cis-canthaxanthin in Artemia cysts was based on absorption spectrometry (HPLC–photodiode array detection), mass spectrometry, and comparison with authentic cis-canthaxanthin.[11] For the preparative isolation of cis-canthaxanthin, 25 g of cysts are first homogenized in 150 ml of the pH 7.6 buffer described below (see section on analysis of subcellular fractions). After centrifugation at 1500 g (5 min, ambient temperature), the precipitate (mainly yolk platelets) is homogenized in 30 ml of a mixture of acetonitrile–methanol–dichloromethane–formic acid (48:30:20:2, v/v), using a Potter–Elvehjem device. A subsequently developed superior extraction procedure based on deoxycholate–methanol, used for small samples (see section on analysis of cysts), can likewise be scaled up for the preparative isolation of cis-canthaxanthin from yolk platelets. To the extract is added 20 ml of water, and the pigments are reextracted with 50 ml of dichloromethane. The extract is dried over anhydrous sodium sulfate, concentrated to 1 ml, and subjected to preparative HPLC. Three fractions corresponding to mono-cis-canthaxanthins are collected, concentrated under reduced pressure, and subjected to direct inlet mass spectrometry (70 eV). Similarly, authentic cis-canthaxanthins can be purified from a stereomutation mixture using preparative HPLC. "Routine" identification of cis-canthaxanthin in quantitative procedures relies on comparison with authentic cis-canthaxanthin (agreement between retention times and absorption spectra recorded on-line with the aid of the photodiode array detector).

Protein Determination

Proteins are determined in the solid residue remaining after extraction of the carotenoids from Artemia samples (cysts, nauplii, later developmental stages, and dissected tissues). Aqueous samples (i.e., suspensions of subcellular fractions) are directly analyzed for protein content after solubilization. Specifically, the whole solid extraction residue or a 0.2- to 0.5-ml aliquot of a subcellular fraction is solubilized with 10 ml of 3.3% (w/v) sodium dodecyl sulfate (SDS) in 0.27 M NaOH. The solution is quantitatively transferred to a 50-ml volumetric flask and brought to volume with distilled water. A 0.5- to 1.0-ml sample is then used for the protein determination according to the method of Peterson,[17] except for the deoxycholate/trichloroacetic acid step which is omitted. The contents of SDS and NaOH in reagent A are reduced so as to compensate for the amounts already present in the solubilization mixture.

Quantitation

Quantitation of all-*trans*-canthaxanthin, *cis*-canthaxanthin, echinenone, and β-carotene is based on the determination of peak height ratios [compound of interest versus internal standard (β-apo-8'-carotenal or β-apo-8'-carotenoic acid ethyl ester)]. Standardization is carried out with the authentic substances. The exact concentrations of standard stock solutions in dichloromethane are determined spectrophotometrically with calculations based on the respective extinction coefficients. The experimentally determined molar absorption coefficients of the standard *cis*-canthaxanthin and all-*trans*-canthaxanthin in acetonitrile–methanol–dichloromethane (5:3:2, v/v) are 80,000 and 125,000 M^{-1} cm^{-1}, respectively. Carotenoid levels in the biological matrix are expressed as micrograms per milligram of protein.

Sample Preparation

Analysis of Artemia Cysts

Two to ten milligrams of cysts are pretreated in a Potter–Elvehjem apparatus with 150 μl of 5% sodium deoxycholate and further homogenized and extracted with 2 ml of methanol. After centrifugation and removal of the supernatant, this step is repeated once more. The internal standard (β-apo-8'-carotenal or β-apo-8'-carotenoic acid ethyl ester) is added to the combined supernatants, and an 100-μl aliquot is injected. In some instances a slightly hazy solution may be obtained after centrifugation. The addition of 1.5 ml acetonitrile will readily clear up the solution.

Analysis of Artemia Nauplii

Freshly hatched nauplii are separated from empty cyst shells and unhatched cysts in a separatory funnel. They are collected on a 120-μm sieve. Most developmental experiments in our studies were carried out with *Artemia* from the San Francisco Bay strain (SFB 522). Hatching for this strain can usually be observed after 15 hr, but sufficient nauplii can only be collected after 16–17 hr (20% hatching). The extraction procedure is the

[17] G. L. Peterson, *Anal. Biochem.* **83**, 346 (1977).

same as for the cysts except that all volumes of reagents are reduced (100 μl of 5% sodium deoxycholate, 1.5 ml of methanol, and 1 ml of acetonitrile), in case very small samples have to be analyzed.

Analysis of Later Developmental Stages of Artemia

For the analysis of *Artemia* at various developmental stages, including mature (sexually active) animals, the procedure is the same as for nauplii. Volume reduction of reagents is mostly desirable as often only a few animals (5–10) may be available.

Analysis of Dissected Tissues from Artemia

Mature female *Artemia* can be dissected under a binocular although this procedure requires some skill and training. Organs (gut, shell gland, ovaries) and other body sections (oocytes and hemolymph) can be removed and separately analyzed along with the remaining carcass. The isolated tissues of 10–20 animals are pooled to give samples of sufficient size. The above procedure for nauplii can be further miniaturized while maintaining the ratio between the reagents.

Analysis of Subcellular Fractions from Artemia Cysts

Hydrated cysts can be subjected to a crude subcellular fractionation procedure, mainly for the purpose of isolating the yolk platelets. About 10 g of cysts are homogenized in 40 ml of a pH 7.6 buffer (containing 30 mM Tris, 70 mM KCl, 9 mM MgCl$_2$, 150 mM sucrose, and 1 mM 2-mercaptoethanol) with the aid of a mortar. Following differential centrifugation at 5600 g (10 min), 15,000 g (10 min), and 30,000 g (10 min) four fractions are obtained. The first one mainly contains yolk platelets, while the second is enriched in mitochondria. On top of the postmitochondrial supernatant (PMS) (fraction 3) a deeply orange floating lipid layer is present. Yolk platelets can be disrupted by continuous stirring of a suspension in buffered sodium chloride (2 M in 50 mM Tris, pH 9), as described by de Chaffoy de Courcelles and Kondo,[18] to yield the lipovitellin and an insoluble fraction enriched in membranes. In our procedure a 10 times higher ratio of extraction buffer to protein is used compared to the original procedure because this gives higher extraction efficiencies.

Solid subcellular fractions (e.g., yolk platelets) are analyzed as cysts. However, aqueous samples (e.g., PMS and lipovitellin) require an additional step to remove the water. To this end a 200- to 500-μl aliquot of the sample is mixed with 150 μl of 5% sodium deoxycholate and 3 ml of methanol. After centrifugation (1500 g, 5 min, ambient temperature) and isolation of the supernatant, the solid residue is once more extracted in the same way. Water (2 ml) is added to the combined supernatants, and the mixture is applied on top of a Bond-Elut C_{18} cartridge (500 mg, 6 ml, Analytichem), preconditioned successively with methanol and 70% methanol. The sample is forced through the cartridge by applying gentle pressure (rubber bulb); the cartridge is then washed with 70% methanol (2 ml) and eluted with 5 ml of the reversed-phase chromatographic solvent (acetonitrile–methanol–dichloromethane, 5:4:1, v/v, containing 0.15% triethylamine). The internal standard is added, and an 100-μl aliquot is injected. Alternatively, to remove water the carotenoids can be extracted in diethyl ether after treatment of the aqueous subcellular fraction with sodium deoxycholate–methanol. Five milliliters of diethyl ether is added, the organic phase is isolated and dried over anhydrous sodium sulfate, and the residue is redissolved in the chromatographic solvent (1–5 ml, depending on the sensitivity required).

Comments

Demonstration of cis-*Canthaxanthins in Artemia Cysts*

For a long time all-*trans*-canthaxanthin and its metabolic precursors echinenone and β-carotene have been considered to be the major carotenoids in *Artemia*.[19–22] The application of nonaqueous reversed-phase chromatography in conjunction with photodiode array detection to encysted embryos of this organism led to the discovery of genuine *cis*-canthaxanthin[11] (Fig. 1), which had already been recognized but not identified in 1969 by Hata and Hata.[23] By normal-phase chromatography on silica the *cis*-canthaxanthin fraction can be further differentiated into one major and two minor mono-*cis*-canthaxanthins (Fig. 2). These were tentatively identified as the 13,9- and 15-*cis* isomers, respectively,[11] on the basis of their absorption spectra, comparison with synthetic *cis*-canthaxanthins, and direct inlet mass spectrometry.

The authentic character of *cis*-canthaxanthin is supported by a number of observations. Supplementation of cysts with all-*trans*-canthaxanthin

[19] N. I. Krinsky, *Comp. Biochem. Physiol.* **16**, 181 (1965).
[20] W.-J. Hsu, C. O. Chichester, and B. H. Davies, *Comp. Biochem. Physiol.* **32**, 69 (1970).
[21] B. H. Davies, W.-J. Hsu, and C. O. Chichester, *Comp. Biochem. Physiol.* **33**, 601 (1970).
[22] T. Soejima, T. Katayama, and K. L. Simpson, *in* "The Brine Shrimp *Artemia*" (G. Persoone, P. Sorgeloos, O. Roels, and E. Jaspers, eds.), Vol. 2, p. 613. Universa Press, Wetteren, Belgium, 1980.
[23] M. Hata and M. Hata, *Comp. Biochem. Physiol.* **29**, 985 (1969).

FIG. 1. Chromatogram of an extract of Reference *Artemia* Cysts obtained by reversed-phase chromatography. Column, 5-µm Zorbax ODS (25 × 0.46 cm); mobile phase, acetonitrile–methanol–dichloromethane (5:3:2, v/v); flow rate, 1 ml/min; detection, 470 nm. Peak 1, all-*trans*-Canthaxanthin; peak 2, *cis*-canthaxanthin (mixture of mono-*cis*-canthaxanthins).

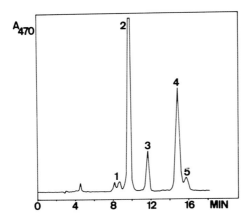

FIG. 2. Chromatogram of an extract of *Artemia* cysts (Macau) obtained by normal-phase chromatography. Column, 10-µm CP-Spher Si (25 × 0.46 cm); mobile phase, dichloromethane–2-propanol (99.4:0.6, v/v); flow rate, 1 ml/min; detection, 470 nm. Peak 1, Di-*cis*-canthaxanthins; peak 2, all-*trans*-canthaxanthin; peak 3, 9-*cis*-canthaxanthin; peak 4, 13-*cis*-canthaxanthin; peak 5, 15-*cis*-canthaxanthin. All identities except for all-*trans*-canthaxanthin are tentative.

before homogenization does not lead to the formation of amounts of *cis*-canthaxanthin above the amount initially present in the cysts. On storage of extracts, no trans to cis conversion is observed. On the contrary, as the cis/trans ratio in extracts of cysts significantly exceeds the equilibrium ratio found in stereomutation mixtures, a progressive cis to trans back-isomerization takes place, a process which is greatly accelerated by light and heat. Finally, the observed selective occurrence/distribution of *cis*-canthaxanthins in particular tissues and developmental stages, to be described below, provides strong evidence suggesting their nonartifactual character, because all samples were analyzed under similar conditions.

Carotenoid Profiles of Major Developmental Stages of Artemia

Figure 3 shows the carotenoid profiles of *Artemia* decapsulated cysts, nauplii hatched from these cysts, and growing animals (preadults) fed on *Dunaliella tertiolecta* algae. The former two mainly contain a mixture of all-*trans*- and *cis*-canthaxanthins, but the cis/trans ratio in cysts is considerably higher than that in nauplii. Echinenone, the metabolic precursor of canthaxanthin, occurs in very low amounts. In older animals, more echinenone as well as β-carotene and a number of xanthophylls, including lutein, are present. Except for echinenone all these compounds are sup-

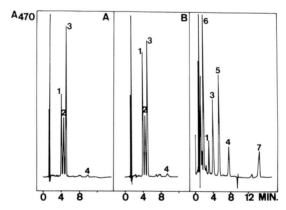

FIG. 3. Carotenoid patterns of major developmental stages of *Artemia*. (A) Chromatogram of an extract of decapsulated cysts (SFB 522); (B) chromatogram of an extract of 11-hr starved nauplii hatched from these cysts; (C) chromatogram of an extract of immature growing animals (preadults) fed on *Dunaliella tertiolecta*. Column, 5-μm Zorbax ODS (15 × 0.46 cm); mobile phase, methanol–acetonitrile–dichloromethane (5:4:1, v/v), containing 0.15% triethylamine; flow rate, 1 ml/min; detection 470 nm. Peak 1, all-*trans*-Canthaxanthin; peak 2, *cis*-canthaxanthin; peak 3, β-apo-8'-carotenal (internal standard); peak 4, echinenone; β-apo-8'-carotenoic acid ethyl ester (internal standard); peak 6, lutein; peak 7, β-carotene.

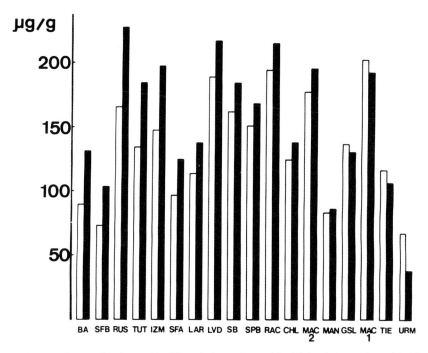

FIG. 4. all-*trans*-Canthaxanthin (□) and *cis*-canthaxanthin (■) levels (μg/g dry weight) in dehydrated cysts of different geographical origin. Abbreviations are explained in Table I.

plied by the algal diet, as *Artemia* is incapable of biosynthesizing carotenoids *de novo*. *cis*-Canthaxanthin, unlike the all-trans form, is no longer detected in preadult animals.

Fate of cis- and all-trans-Canthaxanthin In Artemia Development

The quantities of all-*trans*- and *cis*-canthaxanthin in fully dehydrated cysts is strongly strain dependent[24] (Fig. 4 and Table I). Hydration of cysts consistently leads to an increase in the *cis*-canthaxanthin level, irrespective of the strain[24] (Fig. 5 and Table I). However, this trans to cis isomerization is reversible on subsequent dehydration, and the "cis/trans cycle" can be repetitively reproduced (Fig. 6).

The progress of the all-*trans*- to *cis*-canthaxanthin conversion as a function of hydration time is illustrated in Fig. 7. For Reference *Artemia*

[24] H. J. Nelis, P. Lavens, P. Sorgeloos, M. Van Steenberge, and A. P. De Leenheer, in "*Artemia* Research and its Applications" (W. Decleir, L. Moens, H. Slegers, E. Jaspers, and P. Sorgeloos, eds.), Vol. 2, p. 99. Universa Press, Wetteren, Belgium, 1987.

TABLE I
GEOGRAPHICAL STRAINS OF Artemia [a]

Genus, species and strain	Origin	Abbreviation	Batch no.	Year
Artemia franciscana				
Manaure	Colombia	MAN	2[b]	1983
Macau	Brasil	MAC1	200[b]	1978
Macau	Brasil	MAC2	971051	1979
Chaplin Lake	Canada	CHL	241[b]	1979
San Francisco Bay	United States	SFB	288–2596	1976
San Pablo Bay	United States	SPB	1628	1978
Great Salt Lake	United States	GSL	285[b]	1978
Reference *Artemia* Cysts	—	RAC	—	1979
Artemia parthenogenetica				
Shark Bay	Australia	SB	299[b]	1980
Tientsin	China	TIE	242[b]	1978
Lavalduc	France	LVD	—	1980
Tuticorin	India	TUT	215[b]	1978
Izmir	Turkey	IZM	209[b]	1979
Altai-Siberia	Russia	RUS	—	1983
Artemia persimilis				
Buenos Aires	Argentina	BA	223[b]	1979
Artemia urmiana				
Lake Urmia	Iran	URM	338[b]	1982
Artemia tunisiana				
Larnaca	Cyprus	LAR	305[b]	1980
Sfax	Tunisia	SFA	362[b]	1983

[a] (See Figs. 4 and 5.)
[b] Numbers refer to the collection at the Artemia Reference Center.

Cysts[25] the cis/trans peak height ratio (C/T) increased from 1.02 (full dehydration) to 1.28 (30 min of hydration), and reached a plateau $(C/T\ 1.50)$ after 3 hr. In the course of further preemergence embryonic development the all-*trans*- and *cis*-canthaxanthin levels (hence also the cis/trans ratio) changed very little (Fig. 8). Over a period of 15 hr the average concentrations of both forms per milligram of protein were 1.14 ± 0.07 $\mu g/mg$ ($n = 13$, all-trans) and 1.12 ± 0.06 $\mu g/mg$ of protein ($n = 13$, cis).

Early Naupliar Development. After hatching of cysts there is a progressive decrease in *cis*-canthaxanthin and a concomitant increase in all-*trans*-canthaxanthin (Fig. 8). At the same time, the total canthaxanthin content (all-trans plus cis) per milligram of protein shows little variation (2.20 ±

[25] P. Sorgeloos, *Mar. Ecol. Prog. Ser.* **3**, 363 (1980).

CAROTENOID ISOMERS IN *Artemia* DEVELOPMENT

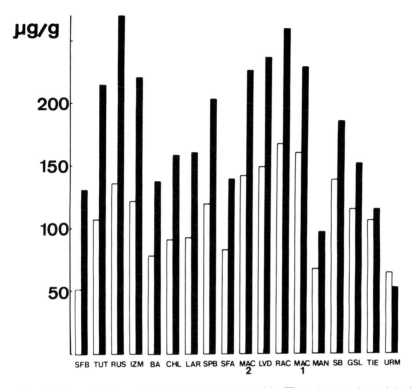

FIG. 5. all-*trans*-Canthaxanthin (□) and *cis*-canthaxanthin (■) levels (μg/g dry weight) in hydrated cysts of different geographical origin. Abbreviations are explained in Table I.

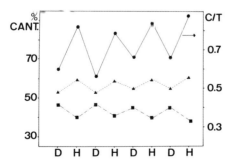

FIG. 6. Influence of repeated dehydration (40°) (D) and hydration (20°) (H) on the relative distribution of *cis*- and all-*trans*-canthaxanthin in Reference *Artemia* Cysts. Upper trace (●) shows peak height ratios (C/T, cis/trans ratio); middle trace (▲) shows percent *cis*- and lower trace (■) percent all-*trans*-canthaxanthin in the extract.

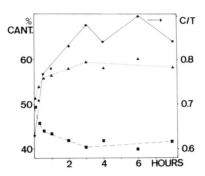

FIG. 7. Influence of hydration time on relative distribution of *cis*- and all-*trans*-canthaxanthin in Reference *Artemia* Cysts. Upper trace (●) shows peak height ratios (C/T, cis/trans); middle trace (▲) shows percent *cis*- and lower trace (■) percent all-*trans*-canthaxanthin in the extract.

0.14 µg/mg, $n = 8$). This picture was confirmed by repeated comparative analysis of 4- and 24-hr-old nauplii ($n = 10$ in both groups). During this 20-hr period the *cis*-canthaxanthin level dropped from 0.97 ± 0.03 to 0.69 ± 0.02 µg/mg of protein, whereas the all-*trans*-canthaxanthin concentration increased from 1.25 ± 0.05 to 1.65 ± 0.04 µg/mg. The total mean canthaxanthin levels in both groups did not differ by more than 5% (2.23 ± 0.08 and 2.34 ± 0.06 µg/mg, respectively).

Later Developmental Stages. Once the older nauplii start feeding on microalgae, the increase in their biomass is accompanied by a decrease in absolute canthaxanthin levels (Fig. 9 and Table II). *cis*-Canthaxanthin, unlike the all-trans isomer, becomes undetectable in immature (preadult) animals. However, it reappears in reproductively active females but re-

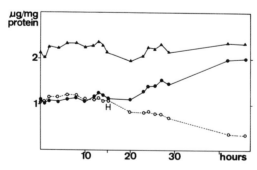

FIG. 8. Concentration (µg/mg of protein) of all-*trans*-canthaxanthin (●), *cis*-canthaxanthin (○), and total canthaxanthin (▲) in the course of early *Artemia* development (SFB 522). H indicates the time of hatching.

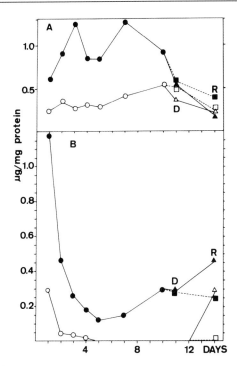

FIG. 9. Levels of β-carotene, echinenone, and all-*trans*- and *cis*-canthaxanthin during the life cycle of *Artemia* (SFB) fed on *Dunaliella tertiolecta*. (A) β-Carotene in undifferentiated animals (●), males (■), and females (▲); echinenone in undifferentiated animals (○), males (□), and females (△). (B) all-*trans*-Canthaxanthin in undifferentiated animals (●), males (■), and females (▲); *cis*-canthaxanthin in undifferentiated animals (○), males (□), and females (△). D indicates the point at which differentiated immature males and females can be distinguished and separated; R indicates reproductively active animals.

mains negligible or virtually absent in mature males (Fig. 9 and Table II). Ongoing metabolic transformation of ingested β-carotene to ketocarotenoids is reflected by the appearance of echinenone, the monoketo precursor of canthaxanthin (Fig. 9 and Table II). Maximum canthaxanthin biosynthesis in females is clearly situated at an advanced stage of the reproductive cycle (Table II). Both *cis*- and all-*trans*-canthaxanthin levels in females at an early stage (D_1) of the cycle are indeed significantly lower than those at the later stages D_2 and D_3. However, while the concentration of *cis*-canthaxanthin increases more than 4-fold over this period, that of the all-*trans* form does so only by a factor 1.7. At the same time, the echinenone levels gradually decrease.

TABLE II
CAROTENOID LEVELS IN MATURE ANIMALS IN COURSE OF TWO SUCCESSIVE REPRODUCTIVE CYCLES

Number of reproductive cycle	n	Carotenoid level[a] (μg/mg of protein)		
		all-*trans*-Canthaxanthin	*cis*-Canthaxanthin	Echinenone
A. Reproductively inactive males				
0	2	0.28 ± 0.02	N.D.[b]	0.57 ± 0.01
B. Reproductively active males				
1	2	0.19 ± 0.04	0.02 (—)	0.25 ± 0.06
2	2	0.21 ± 0.04	0.02 ± 0.01	0.26 ± 0.01
Mean	4	0.20 ± 0.03	0.02 ± 0.01	0.26 ± 0.03
C. Reproductively inactive females				
0	2	0.28 ± 0.03	N.D.[b]	0.54 ± 0.07
D. Reproductively active females				
D_1 Early stage: oocytes in ovaries				
1	3	0.26 ± 0.02	0.06 ± 0.03	0.58 ± 0.05
2[c]	3	0.33 ± 0.07	0.08 ± 0.03	0.49 ± 0.09
Mean	6	0.30 ± 0.06	0.07 ± 0.03	0.54 ± 0.08
D_2 Intermediate stage: unfertilized eggs in oviduct				
1	2	0.44 (—)	0.24 ± 0.02	0.44 ± 0.05
2[c]	3	0.48 ± 0.10	0.27 ± 0.05	0.29 ± 0.05
Mean	5	0.46 ± 0.07	0.26 ± 0.04	0.35 ± 0.09
D_3 Late stage: developing eggs in uterus[d]				
1	2	0.51 ± 0.01	0.29 ± 0.03	0.39 ± 0.04
2	3	0.49 ± 0.03	0.33 ± 0.05	0.25 ± 0.02
Mean	5	0.50 ± 0.03	0.31 ± 0.04	0.31 ± 0.08

[a] Mean = S.D.
[b] Not detectable.
[c] No offspring from the first cycle remained in the uterus.
[d] No vitellogenesis in the ovaries.

Organ and Tissue Distribution of Carotenoids in Artemia. Figure 10 shows the distribution of all-*trans*-, *cis*-canthaxanthin, and their metabolic precursors echinenone and β-carotene in dissected organs and tissues of vitellogenic and non-vitellogenic animals. The highest abundance of *cis*-canthaxanthin is observed in the ovaries of vitellogenic females and in the oocytes of both groups of animals. all-*trans*-Canthaxanthin predominates in the carcass and, unlike the cis isomer, is also present in the gut. Concentrations of both isomers in the hemolymph are comparable. Echinenone is mainly found in the carcass and the gut, and β-carotene is nearly exclusively confined to the latter. Hence concentrations of β-carotene and echinenone serve only as indicators of carotenoid ingestion and metabolic transformation, respectively.

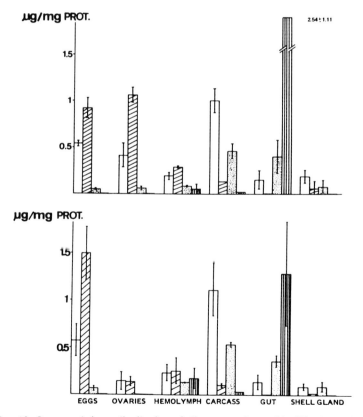

FIG. 10. Organ and tissue distribution of all-*trans*-canthaxanthin (□), *cis*-canthaxanthin (▨), echinenone (▣), and β-carotene (▥) in vitellogenic (top) and nonvitellogenic (bottom) female *Artemia* (SFB).

TABLE III
QUANTITATIVE ANALYSIS OF SUBCELLULAR FRACTIONS PREPARED FROM *Artemia* SFB 288/2606 AND GSL Z-627 CYSTS

	Canthaxanthin concentration (μg/mg of protein)			
Subcellular fraction	all-*trans*-Canthaxanthin	*cis*-Canthaxanthin	Total canthaxanthin	cis/trans ratio
A. Analysis of fractions without Bond-Elut cleanup procedure[a] (SFB 288/2606)				
Total cysts	0.35	0.53	0.88	1.51
Total homogenate	0.33	0.46	0.79	1.39
PMS	0.47	0.32	0.79	0.68
Yolk platelets	0.24	0.86	1.10	3.58
Insoluble fraction	0.27	0.59	0.86	2.19
Lipovitellin	0.20	0.83	1.03	4.15
B. Analysis with solid-phase extraction procedure[b] (GSL Z-627)				
Total homogenate[c]	0.64 (0.03)	1.07 (0.095)	1.71 (0.12)	1.67
PMS[d]	0.31 (—)	0.17 (—)	0.48 (—)	0.55
Yolk platelets[d]	0.78 (0.028)	1.28 (0.035)	2.60 (0.007)	2.33

[a] Solid fractions were analyzed directly; aqueous suspensions were subjected to extraction with diethyl ether after treatment with sodium deoxycholate–methanol.
[b] All fractions were analyzed as suspensions. Aliquots were taken for canthaxanthin and protein determination.
[c] Figures in parentheses are S.D. ($n = 3$).
[d] Figures in parentheses are S.D. ($n = 2$).

Subcellular Localization. Within the cysts, the yolk platelets are highly enriched in *cis*-canthaxanthin, as opposed to the postmitochondrial supernatant (PMS), which mainly contains the all-*trans* form (Table III). The predominance of *cis*-canthaxanthin relative to the all-trans form in the lipovitellin is even more obvious (Table III).

Interpretation of Results

Crustaceans are well known to mobilize carotenoids in their gonads and eggs when they prepare for reproduction.[26] In general, the carotenoids occur in association with proteins.[27] However, apart from cleavage of these carotenoproteins,[27] no changes in the fate of carotenoids have been demonstrated in connection with development.[13] To the best of our knowledge *Artemia* is the first example of a crustacean in which a given carotenoid (*cis*-canthaxanthin) does display a highly characteristic biological fate in relation to specific developmental stages. The compound lies at the basis of sexual dimorphism, in that it is abundant in reproductively active females

[26] B. M. Gilchrist and W. L. Lee, *Comp. Biochem. Physiol.* **42B**, 263 (1972).
[27] D. F. Cheesman, W. L. Lee, and P. F. Zagalsky, *Biol. Rev.* **42**, 131 (1967).

but not in males. Within the females, the reproductive organs and oocytes are strongly enriched in *cis*-canthaxanthin, compared to the rest of the body (carcass, gut). This is unlike the case for the all-trans form, which is more or less uniformily distributed throughout the whole animal.

The higher affinity of ovaries and oocytes for *cis*-canthaxanthin relative to the all-trans form may suggest a role in connection with the process of reproduction itself. Alternatively, mobilization of the compound in the gonads could merely have the purpose of transferring it to the cysts, where the actual function would then be exerted. Most likely the encysted embryos are indeed equipped with *cis*-canthaxanthin not to serve larval requirements, because the compound starts to disappear at a fast rate immediately after hatching. The fact that at the same time the total (cis plus all-trans) canthaxanthin concentration per milligram of protein remains virtually constant indicates that *cis*-canthaxanthin is (re)converted to the (already present) all-trans form and not to some unknown metabolite. Since in the absence of feeding activity no new proteins are formed, the catabolism of "total" canthaxanthin apparently keeps equal pace with the disappearance of proteins, namely, yolk consumption.

The accumulation of *cis*-canthaxanthin in the yolk platelets (and in the lipovitellin) may point to a protective role. It has been suggested that in crustaceans carotenoprotein formation would temporarily prevent the digestion of proteins by the embryo.[27] Of both canthaxanthin isomers, the cis form may possess the correct stereochemical conformation required to put a "lock" on certain proteins. This lock would have to be removed after hatching to make the protein available for consumption by the nauplius. *cis*-Canthaxanthin, once dissociated from the protein, would then be spontaneously converted to the thermodynamically more favorable all-trans form. The gradual disappearance of *cis*-canthaxanthin observed in the nauplii would thus reflect the progressive deblocking of the yolk protein in the course of its digestion by the animal. In this hypothesis, the reversible cis to trans conversion in cysts would have no special significance but would only reflect conformational changes in the (caroteno) protein (hence also in the carotenoid moiety itself) in relation to dehydration/hydration. However, conclusive evidence indicating that *cis*-canthaxanthin protects lipovitellin from proteolytic breakdown, for example, is still lacking. Since 1988 when this study was terminated, there has indeed been no followup to the work on *cis*-canthaxanthin in *Artemia*.

Acknowledgments

H.J.N. and P.S. are Research Associates of the Belgian Foundation for Scientific Research (National Fonds Voor Wetenschappelijk Onderzoek). This work was supported by Fonds voor Kollektief Fundamenteel Onderzoek Contract 2.0012.82 and by Fonds voor Geneeskundig Wetenschappelijk Onderzoek Contracts 3.0011.81 and 3.0048.86.

[22] In Vitro Biological Methods for Determination of Carotenoid Activity

By JOEL L. SCHWARTZ

Introduction

Carotenoids are synthesized in plants and microorganisms from acetylcoenyzme A by a series of condensation reactions.[1] In plant photosynthesis chlorophyll *a* is the most important pigment. It is through the light-harvesting capacities of the chlorophyll complex that light is absorbed and there is a transfer of singlet excitation to the energy centers of the plant.[2] Chlorophylls *a* and *b* are associated with other pigments, such as carotenoids. The carotenoids are divided into unsubstituted carotenes and xanthophylls.[3] β-Carotene functions as the main accessory light-harvesting pigment in chlorophyll *a*, transferring its excitation energy to chlorophyll *a* by an exchange mechanism.[4] β-Carotene will also accept triplet excitation energy, perhaps playing a indirect role in electron transfer. The xanthophylls lutein, neoxanthin, and violaxanthin have also been shown to play a role in plant photoreactivity.[5] Lutein has been shown to transfer excitation energy to chlorophyll *a* and could act with β-carotene to deactivate chlorophyll triplet states.[6] Carotenoids have unusually large electric field effects, producing large photocurrents, but it is the protein complex of the carotenoids that produce permanent dipole moments between the ground state and the achieved excited states.[7] These data may cause us to view the biological effects observed in animal cells with carotenoids as a possible manifestation of the electron transfer capacities of these molecules in plants. These results also provide additional support for caution when using an assay system that requires the release of electrons from one quantum level to another (e.g., thymidine incorporation, proliferation assay).

In general studies have been directed to the biological effects of carotenoids in the prevention and inhibition of neoplastic growth, enhance-

[1] T. W. Goodwin, *in* "The Biochemistry of the Carotenoids" (T. W. Goodwin, ed.), Vol. 1, p. 33. Chapman & Hall, New York, 1980.
[2] T. W. Goodwin, *in* "The Biochemistry of the Carotenoids" (T. W. Goodwin, ed.), Vol. 1, p. 77. Chapman & Hall, New York, 1980.
[3] T. W. Goodwin, *in* "The Biochemistry of the Carotenoids" (T. W. Goodwin, ed.), Vol. 1, p. 193. Chapman Hall, New York, 1980.
[4] K. R. Naqvi, *Photochem. Photobiol.* **31**, 523 (1980).
[5] G. F. W. Searle and J. S. C. Wessels, *Biochim. Biophys. Acta* **504**, 84 (1978).
[6] D. Siefermann-Harms and H. Ninnemann, *Photochem. Photobiol.* **35**, 719 (1982).
[7] D. S. Gottfried, M. A. Steffen, and S. G. Boxer, *Science* **251**, 662 (1991).

ment of immune responses, and prevention of chronic physiologic diseases such as atherosclerosis. This chapter deals with the biological assays currently available for *in vitro* analysis of cell biological effects of carotenoids, with a particular concentration on neoplastic cells and lymphoid cells. We focus on these cell types because of our experience. To our knowledge there are other reports of *in vitro* assays investigating the absorption of β-carotene through adenocarcinoma and intestinal mucosal cells, but these are not discussed here.

The primary focus for the study of the biological function of carotenoids has been their ability to capture free radicals and function as antioxidants. This concept, although sufficient to explain some of the biological effects of carotenoids in some microenvironments, is not comprehensive enough to explain other results. To understand the relevance of *in vitro* assays we first briefly describe some of the *in vivo* studies which have been performed. In many of the *in vivo* studies *in vitro* segments are associated, especially in investigations concerning immune responsiveness following carotenoid treatment. Therefore we later describe these assays in greater detail.

In vivo β-carotene treatment of mice with mammary carcinoma or adenocarcinoma, rats with colon carcinoma, and hamsters with salivary and oral carcinomas has resulted in the inhibition of tumor growth.[8-12] β-Carotene has been shown to inhibit either initiation or promotional phases in oral carcinogenesis as well as to prevent oral carcinoma development.[13,14] β-Carotene therefore appears to induce a change in the growth pattern of tumor cells at an early stage in their development. Further studies, especially *in vitro* analysis of the fundamental processes involved in tumor suppression, are needed.

A potential mechanism for this inhibitory activity may be found in changes in the oxygen state of the tumor cells. For example, the molecule glutathione is required in cells to regulate generation of oxygen free radicals and thiol-reactive molecules.[15] Our studies have shown that γ-glutamyltranspeptidase (GGT, γ-glutamyltransferase), the enzyme involved with the formation of glutathione in cells, was increased during oral carcinogenesis in the hamster but was significantly reduced following β-

---

[8] E. Seifter, G. Rettura, and J. Padawer, *J. Natl. Cancer Inst.* **21**, 409 (1984).
[9] G. Rettura, F. Stratford, S. M. Levenson, and E. Seifter, *J. Natl. Cancer. Inst.* **69**, 73 (1982).
[10] B. S. Alam, S. Q. Alam, J. C. J. Weir, and W. A. Gelson, *Nutr. Cancer* **6**, 4 (1984).
[11] J. L. Schwartz, D. Suda, and G. Light, *Biochem. Biophys. Res. Commun.* **130**, 1130 (1986).
[12] J. L. Schwartz, E. Flynn, and G. Shklar, *Ann. N.Y. Acad. Sci.* **598**, 92 (1990).
[13] J. L. Schwartz, *Adv. Med. Biol.* **262**, 227 (1988).
[14] J. L. Schwartz, D. Sloane, and G. Shklar, *Tumor Biol.* **10**, 297 (1989).
[15] W. H. Habig, M. J. Pabst, and W. B. Jokoby, *J. Biol. Chem.* **249**, 7130 (1974).

carotene administration, possibly indicating a change in oxygen state in forming neoplastic cells.[16]

Seifter et al. examined the short-term effects on mice and found an increase in thymic weight, stimulation of allograft rejection, and inhibition of virally induced tumor growth.[17,18] In rat studies using dietary β-carotene and another carotenoid, canthaxanthin, Bendich and Shapiro[18a] indicated that T- and B-lymphocyte proliferation was enhanced. The immune enhancing effects were concluded to be independent of the provitamin A activity of the β-carotene. In a tumor model system using a Meth A fibrosarcoma, Tomita et al. demonstrated that adoptive immune-reactive cytotoxic T lymphocytes which presumably had contacted β-carotene *in vivo* could augment specific tumor growth.[19] In an extension of this study, Tomita et al. treated mice with carotenoids, β-carotene, canthaxanthin, or astaxanthin, and showed that cytotoxic lymphocytes from lymph nodes mixed with tumor cells could control the growth of the tumor cells.[20] Schwartz also demonstrated that sites of carcinoma growth inhibition show the localization of cytotoxic lymphocytes and macrophages.[13] Importantly, β-carotene in combination with other modalities such as radiation or chemotherapy has resulted in the increased survival of mice with mammary carcinoma.[21]

In an unrelated area, Prince et al. have demonstrated that an emulsion of β-carotene produced a maintainable serum level in rabbits and accumulated in atherosclerotic plaques in these rabbits. The plaques were eliminated using a laser with a visible light frequency directed at the β-carotene-containing plaques both in rabbits and humans.[22,23]

The *in vivo* results have provided sufficient reason to pursue the mechanisms of action for the carotenoids. To accomplish this task there has been a development of *in vitro* methods to answer fundamental questions of cell biological activity.

---

[16] D. Suda, J. L. Schwartz, and G. Shklar, *Eur. J. Cancer Clin. Oncol.* **23**, 43 (1987).

[17] E. Seifter, G. Rettura, and S. M. Levenson, in "Quality of Foods and Beverages: Recent Development in Chemistry and Technology" (G. Charalambous and G. Inglett, eds.). Academic Press, New York, 1981.

[18] E. Seifter, G. Rettura, J. Padawer, and S. M. Levenson, *J. Natl. Cancer Inst.* **68**, 835 (1982).

[18a] A. Bendich and S. S. Shapiro, *J. Nutr.* **116**, 2259 (1986).

[19] Y. Tomita, K. Himeno, K. Nomoto, H. Endo, and T. Hirohata, *J. Natl. Cancer Inst.* **78**, 679 (1987).

[20] Y. Tomita, K. Himeno, K. Nomoto, H. Endo, and T. Hirohata, Abstract, 8th International Symposium in Carotenoids in Boston, 1987.

[21] E. Seifter, G. Rettura, J. Padawer, et al., *J. Natl. Cancer Inst.* **73**, 1167 (1984).

[22] M. R. Prince, S. K. Frisoli, et al., *J. Cardiovasc. Pharmacol.* **17**, 343 (1991).

[23] T. G. Truscott, *J. Photochem. Photobiol.* **6**, 359 (1990).

A. *Methods for Treatment of Whole Cells in Vitro*

To our knowledge the *in vitro* assays below have used only the following carotenoids: β-carotene, canthaxanthin, and α-carotene. Although there may be additional methods employed for the solubilization of carotenoids, the methods described here appear to be the most common and usable. The choice of method to be used in a particular study should also be based on the method employed for quantitation. For example, if [^{3}H]thymidine incorporation is required, the removal of undissolved crystals becomes very important because the carotenoids will quench the counts per minute obtained in the scintillation counter for a given sample. Further discussion of this characteristic is given below.

1. Serial Dilution of Stock Solution of Carotenoid Medium with or without Alcohol. Apparently the most common method for carotenoid treatment of cells is solubilization of the carotenoids with a dilution of a stock medium.[19,24,25] The medium, usually containing 10% fetal calf serum, is mixed with the carotenoid, usually β-carotene, at a final concentration of $10^{-3} - 10^{-8}$ M. If ethanol is added the final concentration should not be greater than 0.1%, and the solution should be stored for extended periods under nitrogen or argon gas at $-20°$. For shorter periods of time the gas is not required. In a slight modification of this method, Rhodes[25] states, that prior to use, crystals of the carotenoid β-carotene should be dissolved in chloroform. The technique is as follows: "5 μl of chloroform containing 18 μg of β-carotene was added at the center of the interface to 9 ml of serum-free medium in a glass tube and mixed by intermittent vortex agitation until all droplets of chloroform had disappeared (about 10–15 min); 1 ml of fetal calf serum was then added to give a final concentration of 3 $\times$ 10^{-6} M β-carotene."

We have used the above technique many times; it is simple and fast. Its major disadvantages are that we cannot achieve the complete or nearly complete dissolution of carotenoid crystals. This results in a poor distribution of the carotenoid. Therefore, some aliquots contain higher concentrations of the carotenoid than others. In assays such as those demonstrated by Stich and Dunn[24] observing micronuclei appearance, or in determinations of protein, RNA, or DNA alterations, in which processing of the cellular contents eliminates the undissolved crystals, this method would be acceptable.

In addition, Bertram, and colleagues recently demonstrated that tetrahydrofuran could be used to solubilize β-carotene for *in vitro* transforma-

[24] H. F. Stich and B. P. Dunn, *Int. J. Cancer* **38**, 713 (1986).
[25] J. Rhodes, *J. Natl. Cancer Inst.* **70**, 833 (1983).

tion assays. Solubilization with tetrahydrofuran appears to offer a simple, highly effective, and nontoxic means of delivering carotenoids of interest to cell. The carotenoid and tetrahydrofuran is stored in the dark at $-35°$ under N_2 and dispersed using a hypodermic through sealed neoprene septum. Care should be taken not to introduce air into the vials. Final carotenoid concentration is 10^{-5} M with 0.5% tetrahydrofuran with the carotenoid introduced rapidly into the cell culture medium.[25a]

2. *Beadlets of Carotenoids.* Beadlets (water-soluble)[26,27] of canthaxanthin and/or β-carotene can be obtained from Hoffmann-LaRoche (Nutley, NJ). They contain 0.2% (w/w) of the carotenoid, but they also contain butylated hydroxytoluene (BHT). Therefore, an additional control for the BHT should be used. A control beadlet is also available from Hoffmann-LaRoche on request. We have found the beadlets to have considerably enhanced solubilization of the carotenoids. The only disadvantage is the addition of another agent, BHT, an antioxidant. This could be a problem in assays with carotenoids in studies that focus on the oxygen state of the cells.

3. *Emulsions to Dissolve Carotenoids.* Murakoshi et al.[22,28] in their *in vitro* studies have used an emulsion with 0.5% sucrose ester P-1570 (Mitsubishi-Kasei Food Co., Tokyo), 1.0% Sansoft 8000 (Taiyo Co., Tokyo), 0.2% L-ascorbyl stearate, and 4.0% peanut oil. They state that the vehicle control did not alter the proliferation of the cells being tested. To treat cells the emulsion containing $5-20$ μM of carotenoid (α- or β-carotene) was mixed with 99.5% ethanol, and 10 μl of this sample emulsion was added to the culture medium [Dulbecco's modified Eagle's medium, (DMEM)]. Another emulsion, which is under patent consideration, can be obtained from Cardiospectrum (MA). This phospholipid emulsion does not contain the above ingredients. Unfortunately, this emulsion has only been used *in vivo*.

The development of a homogeneous emulsion would have as its probable *in vitro* advantage an increase in cellular uptake and distribution. The presence of emulsifing agents in this vehicle may effect conjugations of the carotenoids to proteins of cellular structures. In addition there may be a problem with stability of the components of the emulsion, as well as changes in osmotic tensions in the target cells, creating swelling or shrinkage.

[25a] J. S. Bertram, A. Pung, M. Churley, et al., *Carcinogenesis* **11**, 671 (1991).
[26] A. Bendich, *Ann. N.Y. Acad. Sci.* **587**, 168 (1990).
[27] A. Pung, J. E. Rundhaug, C. N. Yoshizawa, and J. S. Bertram, *Carcinogenesis* **9**, 1533 (1988).
[28] M. Murskoshi, J. Takayasu, O. Kimura, et al., *J. Natl. Cancer Inst.* **81**, 1649 (1989).

4. *Liposomes Encapsulating Carotenoids.* Watson et al.[29] reported that liposomes of canthaxanthin (from 10^{-8} up to 10^{-4} M) were used to inhibit the growth of various tumors (JB/MS melanoma, B16F10 melanoma, and PYB6 fibrosarcoma). In another study Schwartz et al.[30] described micellar constructs that were formed by using a probe sonicator placed into a solution of β-carotene (or canthaxanthin) and medium (DMEM) plus 10% fetal bovine serum. Before placement into the medium carotenoids were dissolved in dimethoxyethylene. The dimethoxyethylene was evaporated and the remaining slurry was placed into the medium above. This solution (300-70 μM) was then added to *in vitro* cultures of human lung (SK-MES) and oral carcinoma cells (SCC-25). In a previous study, Schwartz et al. formed liposomes from phosphatidycholine, phosphatidylserine, and phosphatidylethanolamine, in a ratio of 1:1:1 (a mixture of unilamellar and multilamellar vesicles, 70-300 nm). These liposomes were used to encapsulate β-carotene to form a solubilized carotenoid vehicle.[12]

Liposomes present the problem of variable size. This can be corrected, to a great degree, with a liposome maker (Liposomat, MM Developments, Ottawa, ON). In our experience, there is also a problem with stability. When making our liposomes we were required to form the liposome usually 1 day before use. Advantages to the liposome treatment are a more rapid uptake of the carotenoid and the increased possibility to observe and quantitate the distribution and localization of the carotenoid-liposome inside the cells using electron micrographs or by labeling of the liposome with hydrophobic radiolabeled linkers (Zynaxis, Malvern, PA). In addition, the liposome method allows for the more precise quantitation of the amount of carotenoid added to the culture. Before incorporation of the carotenoid into the liposome the amount of carotenoid can be assessed by high-performance liquid chromatography (HPLC), and after the formation of micellar or liposome constructs an aliquot of the material can be quantitated for carotenoid quantity.

5. *Carotenoid Conjugation to Membranes.* Carotenoids can be combined with membrane structures such as those from microsomes by pelleting microsomal membranes at 105,000 g and gently homogenizing the pellet with the carotenoid in 10 μl of chloroform, using a Potter-Elvehjem homogenizer.[31] The microsomes, resuspended in buffer, can demonstrate carotenoid incorporation by HPLC analysis or through cytochrome *P*-450 reductase activity. This method allows for the conjugation of carotenoids to cellular structures; unfortunately, the arrangement may be artifactual.

[29] R. R. Watson, D. S. Huang, and M. C. Lopez, *FEBS Lett.* **929**, 3252a (1991).
[30] J. L. Schwartz, E. Flynn, D. Trickler, and G. Shklar, *Nutr. Cancer* **16**, 107 (1991).
[31] G. E. Vile and C. C. Winterbourn, *FEBS Lett* **238**, 353 (1988).

II. Effects of Carotenoids on Proliferation of Cells in Culture

A. Quenching Characteristics of Carotenoids

Figure 1 demonstrates that as the concentration of the carotenoid, β-carotene, was increased from 0 to 560 μM and mixed with 1 μCi of [^{3}H]thymidine (the amount of [^{3}H]thymidine placed in wells for proliferation assays) there was as much as a 97.7% reduction in the number of counts per minute. This quenching was not seen with α-tocopherol. In a second study we compared the [^{3}H]thymidine counts per minute, obtained from a human lymphoma cell line, HUT 78, to the counts obtained from HUT 78 cells by treating the cells with β-carotene (70 μM) right before harvesting (Fig. 2). In addition, we compared β-carotene incubation (2 hr) and subsequent solubilization with dimethoxyethylene (100 μl added to glass disks containing the cells and undissolved β-carotene) to the identical treatment of HUT 78 cells with β-carotene, followed by a separation of the undissolved crystals from the cells.

Ficoll-Hypaque Gradient Centrifugation Technique for Undissolved Carotenoid Crystals and Cells. Separation involves placing the cells and β-carotene crystals over Ficoll-Hypaque (Pharmacia, Piscataway, NJ). The Ficoll-Hypaque was used because it offers through density gradient centrif-

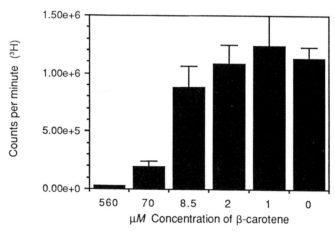

FIG. 1. β-Carotene quenching of [^{3}H]thymidine. β-Carotene in concentrations of 0–560 μM was mixed with 1 μCi of [^{3}H]thymidine. In addition 1 ml of Biofluor scintillation fluid was added to triplicate tubes. Counts were then obtained from a scintillation counter. It appears that undissolved crystals of β-carotene above 8.5 μM of carotenoid significantly reduced the number of observable counts. β-Carotenes from various commercial sources gave similar results.

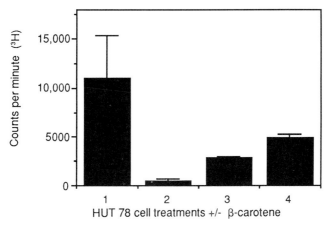

FIG. 2. Quenching of [^{3}H]thymidine counts with β-carotene in HUT 78 cells. The human lymphoma cell line HUT 78 (10^5 cells/well) was cultured in vitro in a [^{3}H]thymidine proliferation assay without β-carotene (1), with β-carotene (70 μM) (2) (2 hr), without β-carotene but with β-carotene (70 μM) added to the cultures before harvest (2 hr) and dissolved with dimethoxyethylene (1 ml) in each scintillation tube containing the harvested cells (3), and with β-carotene (70 μM, 2 hr) but with the crystals separated from the cells by Ficoll-Hypaque and the cells then replated and incubated with [^{3}H]thymidine before harvesting and counting (4). All conditions were performed in triplicate, and a mean and standard deviation of counts were obtained. The results indicated that dissolving or separating the crystals of carotenoid from the tumor target cells provided a clear indication of the decreased proliferation of these lymphoma cells in the presence of carotenoid.

ugation the ability to separate even noncellular materials. Routinely, we use 3 ml of Ficoll-Hypaque for 10^6 cells, which are layered over the Ficoll-Hypaque slowly as the tube is placed at an acute angle. The tube containing the cells and β-carotene crystals are then centrifuged (Sorvall RT6000B) at 2000 rpm (H-1000B rotor, bucket r = 18.67, 1303 RCF value) for 10 min, with the brake off. The viable cells are found in a white layer above the Ficoll-Hypaque; with a slow rotational movement and suction from a Pasteur pipette, the cells are removed, washed twice in the required medium, and then resuspended and counted in a hemocytometer, using trypan blue dye exclusion (0.25%). Viable cells (10^5 cells/well) are then incubated 24 hr with 1 μCi of [^{3}H]thymidine per well, harvested onto glass disks using a PHD Cell Harvester (Cambridge, MA), and counted in a β scintillation counter (Beckman).

This process has been used for human and animal lymphoid cells (T, B, NK, LAK lymphocytes and monocytes) and epidermal and neuroectoderm-derived cells (normal breast, keratinocytes, melanocytes, and malignant carcinomas and sarcomas). We have also separated canthaxanthin and α-carotene crystals in this manner. These carotenoids require repeating the above process with Ficoll-Hypaque at least twice to remove the crystals. Following this procedure we have observed only one or two crystals in 10 wells of a 96-well plate, a reduction in crystals of approximately 95%. This procedure not only allows the investigator to perform assays with radioisotopes without the worry of quenching, but viability assays with tetrazolium salts (e.g., MTT) can also be accomplished without the concern for abnormally high absorbance readings. The major disadvantage to this technique is the amount of time required and the very small number of remaining crystals. Unless otherwise mentioned this technique was used with all studies described below.

B. *Nonneoplastic Cells*

1. Animal Lymphocytes and Macrophages. In general, the process of analysis of immune cells from animals treated with carotenoids has evolved into two paths. The first involves carotenoid treatment of the animal, then the removal of the immunologic tissue, such as the spleen, lymph node, or the exudate from the peritoneum. The other path is to remove the immune cells from the animal's spleen, lymph nodes, peripheral blood, peritoneal exudate, or tissue, then treat them with carotenoids *in vitro*. The methods to accomplish the isolation and purification of various cell types have been well documented and will not be given here.[32] In the past the removal of crystals of carotenoids has been difficult, but the methods provided above should reduce this previously encountered difficulty.

Bendich and Shapiro[33] fed rats 2 g/kg (0.2%) β-carotene, canthaxanthin, or a basal diet for up to 66 weeks. The *in vitro* immune responses of splenocytes to T- and B-lymphocyte mitogens were then determined to be consistently enhanced in the β-carotene or canthaxanthin group. Shklar and Schwartz[14,34] also demonstrated enhanced cytotoxicity by T lymphocytes and macrophages examined *in vitro*.

The process which produced these results was thought to be due to the antioxidant character of the carotenoids. For example, β-carotene has been

---

[32] M. E. Kanof and P. Smith, in "Current Protocols in Immunology" (J. E. Coligan, A. M. Kruisbeek, D. H. Margulies, E. M. Shevach, and W. Strober, eds.), p. 7.05. National Institutes of Health, Bethesda, Maryland, 1991.
[33] A. Bendich and S. S. Shapiro, *J. Nutr.* **116**, 2259 (1986).
[34] G. Shklar and J. L. Schwartz, *Eur. J. Cancer Clin. Oncol.* **24**, 839 (1988).

shown to have a great capacity to trap superoxide anion[35] and perhaps other free radicals. Free radicals, such as hydroxyl radicals, are used by immune effectors such as monocytes to damage foreign organisms.[36] Unfortunately, these oxygen-reactive substances can also damage the immune effector cells[37] and bystander naive immune cells. The process of either immune cell recognition or response to antigen (processing or activation) is very complex. Therefore, the assessment of the oxidative state in relation to a myriad of others factors [autocrine or paracrine regulation of growth factors, cytokine/lymphokine production, histocompatibility antigens (class I and/or class II), and other minor cell surface epitopes] must be carefully considered.

Our laboratory has recently begun this kind of analysis. In general, the preliminary analysis of cytotoxic T lymphocytes and macrophages from the hamster demonstrated an enhanced cytokine production of tumor necrosis factor (TNF) and cytotoxicity to a similar carcinomatous tumor target as seen *in vivo*.[14] Further assays have disclosed that these immune cells will exhibit more stress protein on their surface (70 and/or 90 kDa) (data not shown). These proteins can be expressed by cells undergoing changes in their oxidative state.[38] The gene for the 70-kDa stress protein is linked to the histocompatibility gene complex as well as the gene for TNF.[39] The function of these proteins may be a signal for self- and non-self-recognition.[40] For example, another stress protein (65 kDa) has been linked to the development of autoimmune disorders, such as arthritis and thyroiditis.[40,41] We have also observed in animals treated orally with β-carotene and associated with the inhibition of oral carcinogenesis the presence of lymphocytes and macrophages in the inflammatory infiltrate. These immune effectors stress proteins while in contact with dysplastic or malignant cells. In the tumor control, there was a marked decrease in cell surface expression of the stress proteins, even though there were lymphocytes exhibiting stress proteins in close proximity to the invasive carcinoma (data not shown). Further studies are indicated to relate the observations of tumor recognition by the immune cells to the stress proteins and their level

---

[35] A. Bendich, *Adv. Exp. Med. Biol.* **262**, 35 (1990).
[36] R. M. Bobior, *Clin. Invest.* **73**, 599 (1984).
[37] G. F. Weber, *J. Clin. Chem. Clin. Biochem.* **28**, 569 (1990).
[38] H. R. B. Pelham, *Trends Genet.* **1**, 31 (1985).
[39] H. R. B. Pelham, *Cell (Cambridge, Mass.)* **46**, 959 (1989).
[40] I. J. Benjamin, B. Kroger, and R. S. Williams, *Proc. Natl. Acad. Sci. U.S.A.* **87**, 6263 (1990).
[41] I. R. Cohen, *Annu. Rev. Immunol.* **9**, 567 (1991).

of oxygen-protective enzymes (e.g., superoxide dismutase, glutathione S-transferase).

Among other *in vitro* studies that utilized *in vivo* carotenoid or β-carotene treatment of animals were studies by Tomita *et al.*[19] and Tengerdy *et al.*[42] As stated above, Tomita *et al.* treated BALB/c mice with β-carotene and inoculated s.c. 10 syngeneic BALB/c Meth A fibrosarcoma cells. The tumor cells were rejected following subsequent inoculation. Another syngeneic tumor Meth 1 fibrosarcoma was not rejected by these mice, indicating that the rejection was possibly due to the cellular immune recognition of unique tumor epitopes. The cytotoxic T lymphocytes derived from the β-carotene-treated mice appeared to be the primary immune population involved in rejection of the tumor, as determined by a Winn assay system. Tengerdy *et al.* treated chickens with β-carotene then investigated the humoral response to *Escherichia coli*.[43] They found that β-carotene was not effective against *E. coli* infection but became effective if the carotenoid was combined with vitamin E.

Pung *et al.*[27] and Rundhaug *et al.*[44] have demonstrated that the chemically and physically induced transformation of 10T1/2 cells can be inhibited by β-carotene (beadlet form) and that the inhibition of transformation was a carotenoid effect and not a retinoid response. In contrast, a report by Wamer *et al.* showed that in A31-1 melanoma cells β-carotene was converted to vitamin A after 5 days of *in vitro* culture.[45] Using a mouse T-helper lymphocyte cell clone, Ar5, we observed that β-carotene treatment (2 hr, 1.2 μM) in combination with concanavalin A (Con A; 2 μg/ml) or interleukin 2 (IL-2; 1000 U/ml, 24 hr) resulted in an enhanced proliferation of the T cells (Fig. 3).

2. *Human Lymphocytes and Monocytes.* Studies by Rhodes *et al.*[25,46] have led the way in using human monocytes *in vitro*. They examined the effects of retinol (vitamin A) and β-carotene (provitamin A) on the human interferon action on produced monocytes. β-Carotene and vitamin A had opposite responses by the monocytes toward human interferon activity. Vitamin A inhibited the stimulatory action of interferon on the monocyte membrane function, whereas, this inhibition was reversed by β-carotene. In addition, β-carotene increased slightly the interferon system. Using supernatant derived from a human superficial urinary bladder carcinoma that contained usually high levels of retinoid, it was demonstrated that the

[42] R. P. Tengerdy, *Ann. N.Y. Acad. Sci.* **587**, 24 (1990).
[43] R. P. Tengerdy and J. C. Brown, *Poult. Sci.* **56**, 957 (1990).
[44] J. E. Rundhaug, A. Pung, C. M. Read, and J. S. Bertram, *Carcinogenesis* **9**, 1541 (1988).
[45] W. G. Warmer, R. R. Wei, and V. P. Dunkel, *Am. Assoc. Cancer Res.* **31**, 141a (1990).
[46] J. Rhodes, P. Stokes, and P. Abrams, *Cancer Immunol. Immunother.* **16**, 189 (1984).

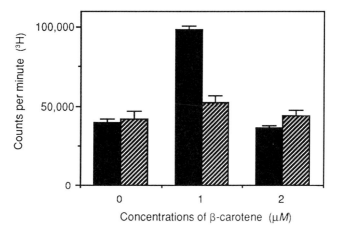

FIG. 3. Proliferation of Ar5 cells (Th1 clone). Cells of the normal mouse T-helper clone (Ar5) were incubated in a [^{3}H]thymidine proliferation assay with β-carotene at 8.5 μM (1) or 2.0 μM (2) for 2 hr. The mouse clone requires either IL-2 (■) or mitogen (Con A, ▨) stimulation to proliferate. β-Carotene was added to stimulated cultures and was observed to increase their proliferation at the 8.5 μM concentration (1), but it had little effect on proliferation at 2.0 μM (2). All conditions were performed in triplicate and provided a mean and standard deviation for counts from each condition.

supernatant was inhibitory toward the monocyte membrane function. This inhibition was reversed with β-carotene.

In vitro studies using human peripheral blood T lymphocytes and monocytes have been conducted. For the purposes of this discussion we focus on the techniques involved. Human peripheral blood lymphocytes were obtained from the blood bank service of the Dana Farber Cancer Institute (Boston, MA). The whole blood is obtained from volunteer donors, and we are provided the white blood cell, buffy coat, of the peripheral blood. The cells are placed over Ficoll-Hypaque as described above, in as many tubes as required. This is a well-documented method for separating white blood cells and other blood types. Mononuclear cells and platelets collect on top of the Ficoll-Hypaque because they have a lower density. Red blood cells and granulocytes collect at the bottom, for they have a higher density. Platelets are separated from the mononuclear cells by subsequent washing or by centrifugation through a fetal calf serum cushion gradient which allows the movement of mononuclear cells but not platelets. The mononuclear cells are purified by adherence as previously described,[32] and the T-cell population is obtained by depleting the nonadherent population of IgG- and IgM-expressing B lymphocytes. This is accomplished using goat anti-human IgG and IgM antibodies and complement as previously documented.[32] Once the B lymphocytes are eliminated, the T-cell population is isolated and purified from the LAK or NK cell populations using a panning technique.[32] This technique utilizes antibodies to CD4$^+$ and CD8$^+$ cells (Becton Dickinson, San Jose, CA); mouse anti-human, 1:300). The LAK cells are formed by incubation with 1000 U of recombinant human IL-2 per milliliter for 10^6 cells/ml (Cellular Products, Buffalo, NY) for 3 to 5 days. To obtain natural killer (NK) cells, they are isolated prior to IL-2 activation using a nylon wool column followed by rosetting with sheep red blood cells.[32]

Human peripheral T lymphocytes were incubated with β-carotene and/or canthaxanthin for 2 hr (8.2 μM). The cells were then separated from undissolved crystals as described above. The isolated and purified human peripheral blood monocytes were plated into wells of a 96-well plate (10^5) as described above. To these wells the human peripheral blood lymphocytes were added (99% viable, determined by trypan blue dye exclusion) in a ratio of 20:1. In addition we added tetanus toxoid (1 mg/ml), and the wells were then incubated for 5 days (37°, 20% CO_2). On the fifth day 1 μCi/well of [^{3}H]thymidine was added. The wells were incubated for another 24 hr, then harvested as described above to determine proliferation of the T lymphocytes. The results indicated a comparative increase in proliferation of the T cells treated with β-carotene or canthaxanthin. This response could have resulted from increased recognition of tetanus antigen (Fig. 4).

NK and T lymphocytes were treated with β-carotene or canthaxanthin (8 μM, 2 hr), then incubated with antiserum to stress proteins (70 and 90 kDa) and fixed in 2.0% paraformaldehyde. The cellular level of these proteins was determined by fluoresceinated activated cell sorter (FACS) analysis. The results showed these lymphocytes express a 90-kDa stress protein (10.2%) and the 70 kDa stress protein (11.7%). In contrast, the lymphoma K562, β-carotene treatment increased cell surface level of the 90 kD protein (53%), whereas canthaxanthin increased the 90-kDa cell surface level to 63%. There was no increase of the cell surface 70-kDa protein, but there was a reduction with β-carotene (5.0%) and canthaxanthin (29.0%). In a preliminary study we incubated LAK cells with the erythroleukemia tumor target (K562, ^{51}Cr radiolabeled). The tumor target was initially incubated with β-carotene or canthaxanthin (8 μM, 2 hr). The LAK cells demonstrated a increased decreased of the tumor target. In addition, we could reverse this tumor target lysis by preincubating the tumor target with antisera to the 70- or 90-kDa stress proteins (data not shown). Importantly, at the concentration of β-carotene used to increase T or LAK cell activity, the proliferation of the lymphoma HUT 78 or the

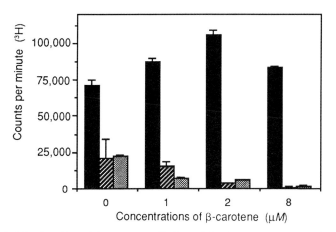

FIG. 4. Selective effect of β-carotene on [^{3}H]thymidine incorporation. In a [^{3}H]thymidine proliferation assay, human peripheral T cells were incubated with peripheral blood monocytes (20:1) and tetanus toxoid (1 mg/ml) in identical triplicate wells. The T cells were incubated with β-carotene in three concentrations (1.0, 2.0, and 8.0 μM) for 2 hr prior to their addition to the monocyte wells. Crystals of undissolved β-carotene were removed by Ficoll-Hypaque. The T cells were determined to be 99% viable by trypan blue dye exclusion (0.25%). The results indicated that there was an increase in the proliferation of the T cells in 24 hr with β-carotene treatment, perhaps indicating increased recognition of tetanus antigen. In contrast, the human lymphoma cell line HUT 78 and the leukemia cell line U973 (10^5 cells/well), treated in an identical manner with β-carotene (1.0, 2.0, and 8.0 μM), demonstrated a dose-dependent decrease in proliferation, indicating a selective effect on growth of lymphoid type cells with β-carotene treatment. (■) T + Mon, (▨) HUT 78, (□) U937.

monocytic leukemia cell line U937 could not be inhibited (Fig. 4). Some of the treated leukemia and lymphoma cells were replated and cultured for another 3 days. On the third day they were placed into serum-free medium overnight. The supernatant that was removed was ultrafiltered and concentrated (Centripep, Amicon Danvers, MA). The supernatant (pH 7.4) was then added to HUT 78 lymphoma cells, and a thymidine proliferation assay was performed. A neat or 1:2 dilution of the supernatant reduced the proliferation of the lymphoma target cells (Fig. 5).

In a preliminary study Leslie and Dubey (1982) added β- or α-carotene to human lymphocytes and found an increase in lysis of tumor targets mediated by NK cells.[46a]

3. Human Nonneoplastic Cells. In three separate studies β-carotene has demonstrated the selective ability to inhibit the growth of squamous cell

---

[46a] C. A. Leslie and D. P. Dubey, *Fed. Proc.* **41**, 331 (1982).

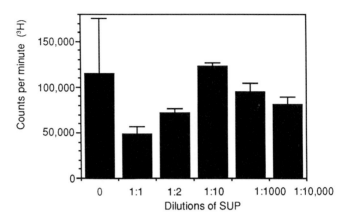

FIG. 5. Inhibition of HUT 78 with treated supernatant. In a continuation of the HUT 78 proliferation assay, some HUT 78 cells (10^6 cells) which had been treated with β-carotene as indicated in Fig. 4 were replated and allowed to grow normally in vitro for 3 days. The cells were then resuspended in medium without serum and incubatd for another 24 hr. The cells were then pelleted, and the supernatant was ultrafiltered and concentrated (2 times) using Centriprep tubes. The supernatant in dilutions of 1:1, 1:2, 1:10, 1:1000, and 1:10,000 was added (100 μl) to identical triplicate wells containing HUT 78 cells, which were then given [^{3}H]thymidine. Proliferation of the HUT 78 cells that received the 1:1 and the 1:2 dilutions was markedly reduced. This assay indicated that β-carotene treatment altered the state of the HUT 78 cells inducing the secretion of a possible protein that caused the inhibition of a secondary group of HUT 78 cells.

carcinoma but not normal keratinocytes.[47,48,48a] The keratinocytes were obtained from human foreskin or mammilliplasties (Clonetics, San Diego, CA). When comparing the proliferation of normal cells to carcinoma cells, the differentiated keratinocytes from mammilliplasties must be used. The results indicated that carotenoids (70 or 300 μM) produced no inhibition of normal keratinocytes.[47,48] In an identical set of studies using human breast cells and melanocytes, the carotenoids (70 μM) incubated with these cells (2 hr) either increased the proliferation of the normal cells or had no significant effect on their growth.[48] We also observed that, particularly with the breast cells, the cells appeared to form adenomatous patterns of growth, while the melanocytes appeared to generate more melanin (L-dopa stain), and keratinocytes more keratin (Shklar stain), perhaps indicating an in-

[47] J. L. Schwartz, R. P. Singh, B. Teicher, et al., Biophys. Biochem. Res. Commun. **169**, 941 (1990).
[48] J. L. Schwartz and G. Shklar, J. Oral Maxillo. Facial Surg. **50**, 367 (1992).
[48a] J. L. Schwartz, B. A. Teicher, and J. M. Stringham, Cancer Res. **33**, 44 (1992).

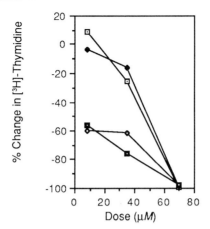

FIG. 6. Proliferation of HCPC-1 cells. In a [^{3}H]thymidine proliferation assay the hamster oral carcinoma cell line HCPC-1 was incubated with various doses of β-carotene (10, 30, and 70 μM) for 2 (□), 6 (◆), 12 (■), and 24 (◇) hr. The percent change in the number of counts obtained from identical triplicate wells was determined as (experimental − control)/control × 100, where a negative value indicated a decrease in the proliferation counts below control (0%). The results shown indicated that with increasing dose and time there was less proliferation of the hamster carcinoma cells with β-carotene treatment.

crease in differentiation (data not shown).[48] Further studies are required to substantiate these observations.

C. Neoplastic Cells

1. Animal Neoplastic Cells. Carotenoids have been incubated with mouse neoplastic cells (JB/MS melanoma, B16F10 melanoma, PYB6 fibrosarcoma, and 10T1/2 fibroblasts)[25a,29] and hamster oral carcinoma.[58] Canthaxanthin (10^{-4}–10^{-8} M) placed into liposomes, ethanol, or dimethyl sulfoxide (DMSO) was shown to delay the growth of these tumor cells after 72 hr of incubation. No inhibitory effect on growth was found for nontransformed NIH 3T3 cells. In the case of the hamster cell line HCPC-1, an oral carcinoma, the carotenoid (β-carotene, 10–70 μM) with time (2–24 hr) reduced the growth of cells (Fig. 6).

2. Human Neoplastic Cells

a. Proliferation effects. To our knowledge there are only a few studies available that have investigated the *in vitro* effects of carotenoids on human tumor cells. Nishino *et al.*[49] reported that α- and β-carotenes in an

---

[49] H. Nishino, J. Takayasu, T. Hasegawa, *et al., 4th Int. Congr. Cell Biol. Montreal* 1.2.40a (1988).

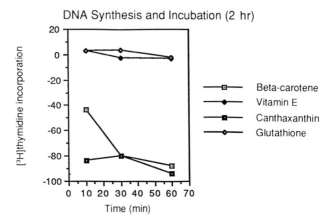
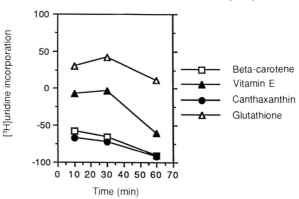
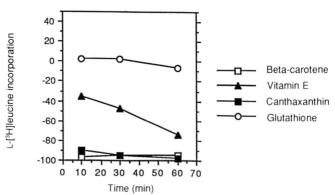

emulsion (see Section I,A,3) at 2, 5, or 20 μM inhibited the growth of various human malignant tumor cells, such as GOTO (neuroblastoma),[50] PANC-1 (pancreatic cancer), A172 (glioblastoma), and HGC-27 cells (gastric carcinoma). We have reported that canthaxanthin and β-carotene (70 μM) could inhibit SCC-25, SQ-38 (oral carcinomas), ZR75, MCF-7 (breast carcinomas), SK-MES (lung carcinoma), and A375 cells (malignant melanoma) after 6 or 24 hr of incubation.[48] It should be recalled that the determination of cell growth inhibition was observed after the carotenoids were removed. [Indicated above (Section II,B,1).] Therefore the growth inhibition was a result of a profound metabolic change in the tumor cells.

b. *Changes in RNA, DNA, and protein synthesis.* Deng et al. have shown sister chromatid exchanges (SCE) to be increased when the carcinogen MNNG was added to cultures of V_{79} cells under conditions free of the enzyme system to convert β-carotene to vitamin A. Aflatoxin B_1-induced SCE was, in contrast, inhibited by β-carotene. The increase in SCE could be suppressed when β-carotene was combined with α-tocopherol but not alone at low concentrations (2–5 μM).[51] Stich and Dunn used CHO cells treated with genotoxic agents to induce micronuclei *in vitro*.[24] This induction was inhibited with β-carotene and substantiated *in vivo*, human responses.[52,53]

β-Carotene may have a direct inhibitory effect on carcinogens such as 7,12-dimethylbenzanthracene (DMBA). In a preliminary report β-carotene has been observed to inhibit the formation of the diol epoxide derived from the carcinogen. Perhaps this process may be occurring in *in vivo* during inhibition studies.[11-14]

In a recent study, we investigated the response by the breast carcinoma cell line MCF-7 after treatment with β-carotene, canthaxanthin, α-

[50] H. Nishino, J. Takayasu, T. Hasegawa, *et al., Kyoto Igakkai Zasshi (J. Kyoto Prefect. Univ. Med.)* **97**, 1097 (1988).
[51] D. Deng, H. Guogeng, *et al., Zhinghua Zhonglui Zazhi* **10**(2), 89 (1988).
[52] H. F. Stich, M. P. Rosin, A. P. Horaby, *et al., Int. J. Cancer* **42**, 195 (1988).
[53] H. F. Stich, *in* "Mechanisms in Tobacco Carcinogenesis," p. 99. Cold Spring Harbor Laboratory, Cold Spring Harbor, New York, 1986.

FIG. 7. The human breast carcinoma MCF-7 was treated with β-carotene, vitamin E, canthaxanthin, or reduced glutathione (70 μM) for 2 hr, then incubated with either [³H]thymidine to determine DNA incorporation, [³H]uridine to determine RNA incorporation, or [³H]leucine (1 μCi/well) to ascertain incorporation into newly synthesized protein. The incorporation was observed at intervals of 10, 30, and 60 min. The carotenoids consistently demonstrated a decrease in the percent counts. The results indicate that the process of cell change induced by the carotenoids is relatively rapid and persistent.

tocopherol, and glutathione (70 μM) for 2 hr. The relative changes in amounts of DNA were observed through [^{3}H]thymidine incorporation. RNA levels were determined through [^{3}H]uridine and protein synthesis was ascertained with the incorporation of [^{3}H]leucine (1 μCi/well). We have also demonstrated that β-carotene enhanced the expression of newly synthesized proteins.[47,48] The results indicated that the carotenoids reduced the incorporation of [^{3}H]thymidine within 10 min, which continued to be reduced to an even greater extent by 60 min (80% reduction compared to untreated control cells). This general pattern was also observed for uridine and leucine incorporation. The results indicated that carotenoids, at least in this cell line, could directly alter the metabolic state of a tumor target (Fig. 7). For a further discussion of the changes in protein synthesis, see Section III.

  c. *Alteration in tumor cell viability.* We have previously quantitated the decrease in viability of tumor targets (SCC-25 and SK-MES) following incubation (5 hr) with β-carotene and canthaxanthin (70 and 300 μM) using an MTT assay.[47] To determine if this change in viability in SCC-25 cells was due to the accumulation in the cell of a retinoid, or vitamin A derivative, we incubated tumor cells with β-carotene, canthaxanthin, vita-

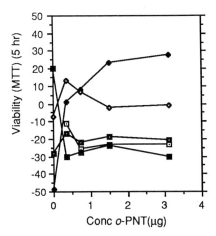

FIG. 8. Effect of o-PNT on SCC-25 viability. A viability assay (MTT) with the human oral carcinoma cell line SCC-25 was performed. The assay was dependent on the precipitation of a tetrazolium salt complex resulting from the action of the mitochondrial enzyme succinate dehydrogenase. We assessed the percent change in viability (see Fig. 6) resulting from the treatment with o-phenanthroline (o-PNT; □, 0.5, 1.5, and 3.0 µg), with or without β-carotene (♦), vitamin E (■), canthaxanthin (◇), or reduced glutathione (■) (70 μM) after 5 hr of incubation. The results indicated that, with the addition of o-PNT, to the carotenoids, especially β-carotene, we observed an increased level of the mitochondrial enzyme succinate dehydrogenase.

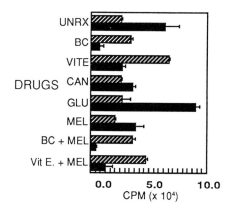

FIG. 9. A [^{3}H]thymidine incorporation assay with normal human keratinocytes (▨) and the oral carcinoma cell line SCC-25 (■) was performed. The cells (10^5 cells/well) were incubated (2 hr) with β-carotene (BC), vitamin E (VITE), canthaxanthin (CAN), reduced glutathione (GLU), melphalan (MEL), β-carotene plus melphalan (BC + MEL), and vitamin E plus Melphalan (VitE + MEL). All the drugs except melphalan were given at a concentration of 70 μM; melphalan was at 100 μM. The actions of the carotenoids and vitamin E were inhibitory toward the proliferation of SCC-25 cells. In addition, β-carotene vitamin E plus melphalan and produced an increased antitumor growth response, without producing a similar result in the normal cells.

min E, reduced glutathione (70 μM) and the Zn^{2+} or Fe^{2+} chelator o-phenanthroline (o-PNT; 0.5–3.0 μg) for 5 hr. This inhibitor has been used to inhibit 15′,15′-dioxygenase reductase, which converts β-carotene to vitamin A in the intestine.[54] We observed that the viability of the SCC-25 cells was not reduced but increased as we increased the concentration of o-PNT and maintained the concentration of β-carotene. In contrast, the increasing concentration of o-PNT resulted in no change in the viability of SCC-25 cells when the carotenoid canthaxanthin was incubated with the cells. However, the viability of SCC-25 cells with vitamin E, glutathione, a single treatment, was also decreased (Fig. 8).

The HPLC analysis of these cells following treatment with β-carotene did not show any retinoid in the cells, only β-carotene. The response we observed may be a result of the chelating capacity of o-PNT. Alterations in the metal ions may effect the electron transport system, indirectly altering catalase activity, changing the level of peroxides in the cells, or effecting the binding of various proteins with zinc-binding sites.

d. Suppression of tumor growth with combinations of chemotherapy. We have recently reported that when β-carotene (50 μM) was combined with

[54] J. A. Olson, *J. Nutr.* **119**, 105 (1989).

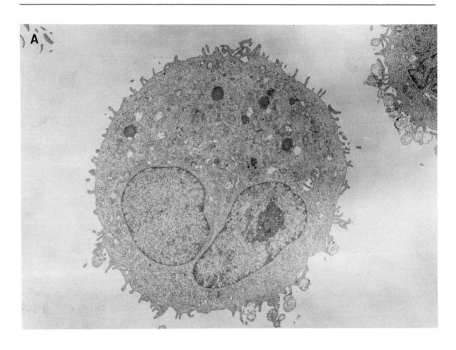

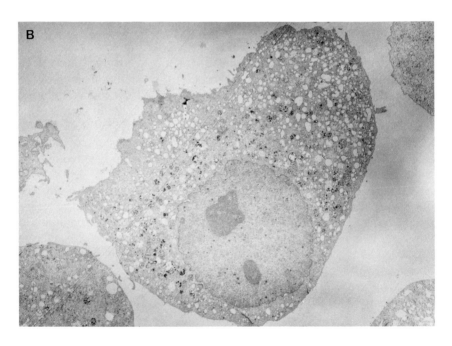

the chemotherapy agents cisplantin or melphalan (50 or 100 μM) we observed a marked reduction in the clonogenic growth of the SCC-25 tumor cells. In the case of the alkylating agent melphalan, there was a 1000-fold increase in the antitumor activity of this agent.[55,56] Melphalan is a chemotherapy agent that can suppress tumor growth by reducing the level of nonprotein sulfhydryls.[56]

To substantiate the clonogenic effect and to determine if this response was selective, we performed a thymidine incorporation assay using the SCC-25 cells and their normal counterpart, human epidermal keratinocytes. The results indicated that β-carotene and melphalan did not inhibit the proliferation of the keratinocytes but significantly reduced the proliferation of the SCC-25 cells (Fig. 9). These studies tend to confirm the observations of Seifter and Rettura, that the combination of chemotherapy, cyclophosphamide, and β-carotene increased the survival of tumor-bearing mice.[18,21]

e. Ultrastructural changes. Using ultrastructural techniques previously employed,[14] we treated SCC-25 cells with β-carotene (70 μM). After 2 hr, the cells were treated with trypsin plus EDTA (0.25%), washed as described above to separate crystals from the cells, and pelleted. The pellet was then fixed as previously described.[14] Electron micrographs showed the loss of pseudopodia and a loss of organelles and microfilaments. Specifically, there was an increased vacuolation of the β-carotene-treated cells, with a swelling of mitochondria and endoplasmic reticulum (Fig. 10).

D. Cell Cycle Effects

The mechanism for carotenoid inhibition of tumor cell growth is unknown, although there are indications from the results of β- plus α-

[55] J. Tanaka, J. L. Schwartz, T. S. Herman, B. A. Teicher, *Cancer Res. Assoc.* **32**, 373a (1991).
[56] J. L. Schwartz, J. Tanka, V. Khandakar, *et al., Cancer Chemother. Pharmacol.* **29**, 207 (1991).

FIG. 10. Cells of the human oral cancer cell line SCC-25 (10^6 cells) were treated with β-carotene (70 μM) for 2 hr, then removed and washed as previously indicated. The tumor cells, untreated or treated, were pelleted, and electron micrographic sections (10 Å) were made. (a) Untreated carcinoma cell exhibiting pseudopodia and membrane blebs on its surface. In the cytoplasm there are myelin bodies present and the usual distribution of mitochondria, Golgi, endoplasmic reticulum, and microfilaments. Magnification: ×4050. (b) β-Carotene-treated carcinoma cell showing a loss of pseudopodia and the irregular configuration of the cell. The cytoplasmic contents appear to contain vacuoles and electron-dense precipitations (β-carotene?). In addition, organelles such as the mitochondria and endoplasmic reticulum became swollen and vacuolated in appearance and there was noticeable chromatin clumping in the nucleus. Magnification: ×2430.

carotene treatment of GOTO cells that cells accumulate in G_0-G_1 while decreasing the expression of the protooncogene n-*myc*.[28] In recent studies with SCC-25 cells and keratinocytes, we analyzed the cell cycles following treatment with β-carotene (70 μM, 2 hr). The cells were treated with propidium iodide in a sodium citrate lysing buffer containing Triton X-100. The cell cycle was analyzed using a FACS and the Cell Fit Program (Coulter, Hialeah, FL). The SCC-25 cells tend to accumulate at the G_2-M phase of the cell cycle. If the cells were depleted of serum and then serum plus medium added with β-carotene (70 μM), the cells accumulated at the G_0-G_1 phase of the cell cycle (Fig. 11). Other tumor cell lines such as MCF-7 or A375 tend to accumulate at G_0-G_1 of the cell cycle, without serum starvation. We have recently shown that as β-carotene inhibited the carcinogen DMBA induction of oral carcinogenesis there was a reduction in the number of cells in synthesis (S) phase.[48a]

III. Effects of Carotenoids on Activity and Synthesis of Cellular Proteins

A. Oxidative State

We reported that the inhibition of the clonogenic activity of SCC-25 cells was directly altered if the oxygen concentration in which the cells were grown, was changed.[55,56] Specifically, we found that if the cells were placed in an oxygen-poor environment and incubated for 1 or 6 hr with β-carotene (5–100 μM) there was a significant decrease in the inhibitory activity of β-carotene toward SCC-25 growth (Fig. 12).

Investigating the level of protective enzyme systems in the β-carotene-treated tumor target we found that there was a significant reduction in superoxide dismutase (SOD) and glutathione S-transferase (GST). In addition, the level of nonprotein sulfhydryls was also depressed.[55,56] Notable was the lack of reduction of these constituents in SCC-25 cells treated with canthaxanthin, vitamin E, or reduced glutathione. In another assay, we treated the SCC-25 cells with β-carotene (70 μM, 2 hr) alone or in combination with 125 U of superoxide dismutase (Cu-Zn) (Sigma, St. Louis, MO) (5 hr), then added 1 μCi/well of [^{3}H]thymidine overnight and assayed for proliferation and/or the level of SOD as previously described.[56] The results indicated that the addition of exogenous SOD could reverse the decrease in proliferation of SCC-25 cells following treatment with β-carotene (Table I).

These results as well as the determinations by Burton and Ingold[57] that

---

[57] G. W. Burton and K. U. Ingold, *Science* **224**, 569 (1984).

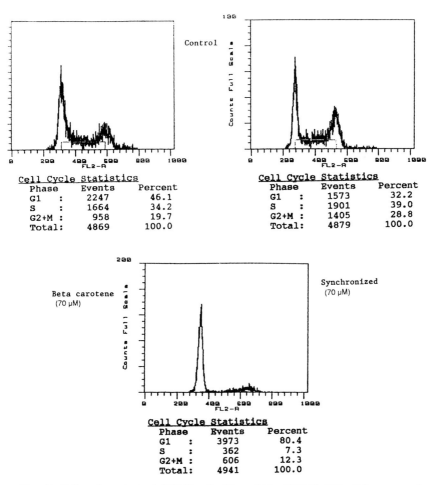

FIG. 11. Cell cycle analysis of SCC-25 cells. The cell line 3CC-25 (10^6 cells) was treated with β-carotene (70 μM) for 2 hr. The cells, untreated and treated, were then stained and lysed with propidium iodide in a sodium citrate buffer containing Triton X-100. The phases of the cell cycle were then determined by FACS analysis of 5000 cells. In the untreated population the highest percentage of cells were in G_1, followed by S (synthesis), then G_2 plus M (mitosis). In the β-carotene-treated cultures the cell population was approximately equally distributed, resulting in an increase in the percentage of cells in G_2 plus M compared to controls. If the tumor cells were first deprived of serum for 24 hr, then β-carotene (70 μM) was added back with medium containing serum (synchronized), the cells treated with β-carotene remained predominantly in G_1. The control cells were virtually identical to the previous controls shown.

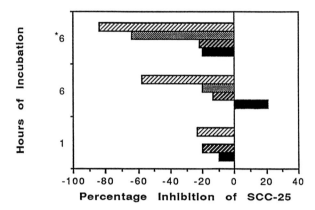

FIG. 12. Effect of hypoxia and β-carotene on SCC-25 cell growth. The tumor cell line SCC-25 (5 × 10⁶ cells/flask) was placed into an incubator with 5.0% carbon dioxide and 95.0% oxygen (oxic conditions, 37°) and treated with β-carotene (5–100 μM) for 6 hr. Hypoxic conditions were achieved by placing the cell into a flask containing 95% nitrogen gas and 5.0% oxygen for 1 or 6 hr. After treatment the cells were replated, and colonies were counted 5 days later. There was observed an increase in the number of colonies formed and therefore a decrease in the perent inhibition by β-carotene for the cells treated in an oxygen-poor environment. The percent inhibition was obtained as (no. of experimental colonies − no. of control colonies)/no. of control colonies × 100. Asterisk marks the oxic state. β-Carotene concentrations: ▨, 100 μM, ■, 50 μM, ▨, 10 μM, ■, 5 μM.

β-carotene would act as an antioxidant in an environment with a low partial pressure of oxygen but could function as a prooxidant at a higher partial pressure cause us to hypothesize that β-carotene induces an oxidative stress in the tumor targets. This oxidative stress results in the expression of stress proteins which could alter the metabolic state of the tumor cell.[57a]

B. Stress Proteins

The evidence for the increased expression of the stress proteins following β-carotene treatment is as follows: immunoflourescent staining (Fig. 13) quantitated the level of stress proteins (70 and 90 kDa) in human and hamster oral carcinoma cells by FACS (data not shown). We have also observed the expression of these proteins in a 10% polyacrylamide gels electrophoresed in the presence of sodium dodecyl sulfate (SDS-PAGE) (Fig. 14). In addition, Western immunoblotting and immunoprecipitation provided further evidence for the increased expression of these proteins in

[57a] J. L. Schwartz, G. Shklar, and D. Trickler, *J. Dent. Res.* **71**, 286 (1992).

TABLE I
PERCENTAGE CHANGE IN SUPEROXIDE DISMUTASE RELATED TO THYMIDINE
PROLIFERATION

Agent	SOD[a] (%)	+SOD[b] (%)	[^{3}H]Thymidine[c] (%)	[^{3}H]Thymidine + SOD[d] (%)
β-Carotene	−48.6	24.0	−44.0	−7.0
Vitamin E	15.1	9.7	4.0	5.0
Reduced glutathione	14.5	11.7	12.0	17.0
Canthaxanthin	−17.8	19.2	−46.0	−28.7

[a] The percent change in the level of SOD present in treated tumor cells (units/ml/mg of protein). The percent level of SOD present without exogenous SOD as determined by the percent difference [(expt − control)/control × 100].
[b] The level of SOD present with the addition of 125 U of CuZn SOD for 10^5 cells [(exogenous SOD − endogenous SOD)/endogenous SOD × 100].
[c] The percent change in [^{3}H]thymidine in treated tumor cells. The percent change in [^{3}H]thymidine without the addition of SOD [(expt − control)/control × 100].
[d] The percent change in [^{3}H]thymidine with the addition of SOD as indicated in footnote b using the above formulas.

the supernatant and treated tumor cells (breast, oral carcinomas, and malignant melanoma) lysates[57b] (data not shown). Notable has been the lack of increased expression of these stress proteins in nonmalignant keratinocytes or melanocytes.[48] The possible significance of the enhanced expression of stress proteins may be found in their ability to alter the folding and configuration of other proteins. In addition, the increases in synthesis of mRNAs for these proteins becomes an almost exclusive function of the triggered cell. This results in the reduced translation of preexisting mRNAs coding for other cellular polypeptides of protooncogene or oncogene origin.[59] Specifically, we have observed decreases in the proteins derived from protooncogene *ras* and mutated *ras* at the valine-12 position (data not shown). In addition, other growth related proteins have been observed to be decreased, such as epidermal growth factor receptor, transforming growth factor α, and mutant p53663.

C. Protooncogene Expression

As indicated above, Murskoshi *et al.* showed that there was a decrease in the expression of the protooncogene n-*myc* in association with the

[57b] S. Toma, E. C. Albanese, R. Palumbo *et al., Anti-Cancer Drug Des.* **2**, 581 (1991).
[58] J. L. Schwartz and D. Z. Antoniades, submitted for publication.
[59] D. Halevy, D. Michlovitz, and M. Oren, *Science* **250**, 113 (1990).

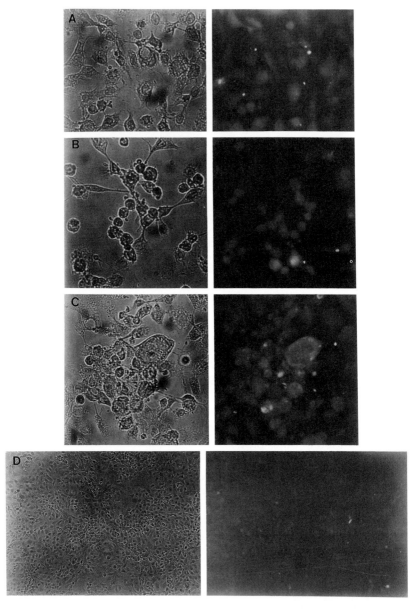

FIG. 13. The cell line HCPC-1 was grown to 70% confluence, and the cells were then treated with β-carotene (70 μM) for 2 hr. The cells were stained with a primary mouse monoclonal antibody which recognizes wild-type p53 (A) (1:200), a primary mouse monoclonal antibody which recognizes the 70-kDa stress protein (B) (1:200), and a primary mouse monoclonal antibody to the 90-kDa stress protein (C) (1:200). A second fluoresceinated goat anti-mouse antibody was used to visualize the primary antibody. Untreated tumor cells demonstrated only a faint reaction with the antibodies (D). Magnifications: A–C, ×292; D, ×146.

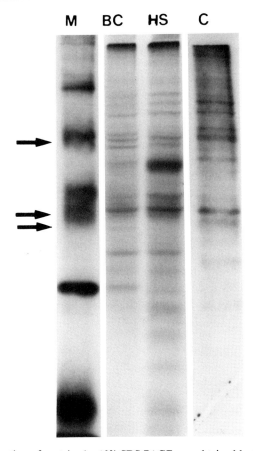

FIG. 14. Separation of proteins by 10% SDS-PAGE was obtained by lysing SCC-25 cells after treatment with β-carotene (70 μM) for 2 hr. Lanes are as follows: M, markers; BC, β-carotene; HS, heat shock, a control for stress protein expression; C, untreated tumor cells. Arrows indicate the presence of proteins expressed by the treated SCC-25 cells but not the untreated controls. Many of the same proteins designated by the arrows are also seen in the heat-shock control lysate.

GOTO cells arresting in G_0-G_1. We have found using slot-blot Northern analysis that the mRNA level for c-*myb* was increased slightly above the actin protein control in a population of SCC-25 cells arrested in G_2 and M (data not shown). We have also observed that there may an increase in the expression of the associated tumor-suppressive protein, namely wild-type p53, in SCC-25 cells (Fig. 13). More definitive studies are required to form an accurate assessment of these changes in oncoprotein expression. One such assay is the chloramphenicol acyltransferase assay (CAT). Using the

CAT assay COS (monkey kidney cells) cells are transfected with the nucleotide sequence for the mutated *ras* (valine-12) or mutated p53. We can then investigate the possible direct relationship of the stress protein promotor 70B or the stress protein on the expression of mutated *ras* and p53 expression.

D. Fibronectin Production

We have observed that the incubation of tumor cells (carcinomas, melanoma) with β-carotene (70 μM) for 5–6 hr resulted in their lifting from the wells or plates. Canthaxanthin also produces this effect, but to a much lesser degree. Therefore, in a preliminary study varying dose and time, we treated the breast carcinoma cell line MCF-7 with carotenoid and assessed the level of fibronectin on the tumor cell surface using FACS analysis. Fibronectin was chosen because it is a common extracellular marix protein, which may be used by the tumor cells to adhere to the plastic. Figure 15 shows the results after 2 hr of treatment with β-carotene (70 μM). There was a decrease in the cellular fibronectin level with β-carotene treatment (Fig. 15). Further *in vitro* analysis is required to conclude if a reduction in extracellular proteins such as fibronectin is respon-

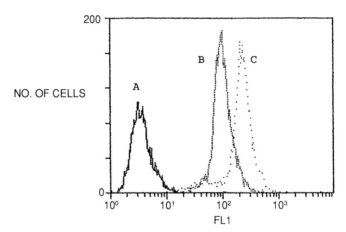

FIG. 15. FACS analysis of fibronectin. The tumor cell line MCF-7, a breast carcinoma, was treated with β-carotene (70 μM) for 2 hr. (A) MCF-7 cells were treated but were not incubated with a primary antibody, only the second fluorescent goat anti-mouse antibody (1:500). (B) Positive staining of MCF-7 cells after treatment with β-carotene and incubation with the primary antibody mouse anti-human fibronectin (1:100). (C) Level of staining for fibronectin by the untreated MCF-7 cells. The results indicated that β-carotene reduced the level of fibronectin on these breast carcinoma cells.

sible for the previous *in vitro* observations and if these proteins are associated with the metastatic potential of tumors *in vivo*.

E. Other Possible Actions

The studies noted above appear to point to a profound metabolic change in tumor cells. Phosphorylation assays have indicated that there may be five or six major protein differences between carotenoid-treated tumor cells and the untreated controls. Preliminary analysis of protein kinase C in SCC-25 cells has shown increases in that protein following carotenoid treatment. Another direction for further investigation should be studies into the differentiation of cells. The molecular and cell biological effects of differentiation relating to the apoptosis of the tumor cells should be considered. The studies presented here clearly indicate that more investigation will be required to form an accurate determination of the biological activities of the carotenoids. Many of the studies that must be performed for the carotenoids have been accomplished for the retinoids, which can be used as an experimental guide. With the considerable reduction of the problem with undissolved crystal precipitates, the future for *in vitro* analysis looks promising.

IV. Concluding Remarks

The sensitivity of most of the tests described above required only small quantities of material. This development allows the investigator to obtain a complete dose-response curve with less than 1 mg of carotenoid. The proliferation assay (thymidine incorporation) is the most sensitive assay to assess carotenoid function. Specificity of the tests is critical in determining importance. That is why in many of the above assays we have included both nonmalignant and malignant cells in identical protocols, performed at the same time, under the identical conditions.

Precision is critical in knowing the end point of the assay. It must be reproducible, quantitative, and become modified in a dose-responsive manner. Many of the assays provided above fulfill these criteria, but a continual reassessment of the parameters of each assay must be forthcoming. The use of serum-free medium will help to focus on the selective expression of protein induced by carotenoid treatment.

The ease of performing the above assays has been considered. Even though many of the assays require many hours to complete, the results can be obtained within days of setup. With time the *in vitro* results should parallel the *in vivo* results, and we have found evidence already in our studies with the hamster oral carcinoma tumor model. We have observed the increased expression of stress proteins *in vivo* with β-carotene, while there was a decrease expression of stress proteins in the untreated oral carcinomas, a pattern seen *in vitro* with the hamster oral cancer cell line HCPC-1.

The techniques now available should open up many new avenues for research and provide a greater understanding of the possible clinical relevance of carotenoids for the treatment of immunologic disorders, cardiovascular abnormalities, and neoplastic disorders.

Acknowledgments

This work is supported in part by Cancer Prevention Technologies and the Milheim Foundation.

[23] Enzymatic Conversion of all-*trans*-β-Carotene to Retinal

By M. R. LAKSHMAN and CHITUA OKOH

Introduction

Carotenoids are the sole biological precursors of retinoids. β-Carotene is the major precursor of this vitamin. Dietary retinoids in humans are derived primarily from the intake of carotenoids. β-Carotene cleavage (BCC) enzyme catalyzes the conversion of all-*trans*-β-carotene to 2 mol of retinal.[1-3] In addition to β-carotene, several other carotenoids and apocarotenoids are cleaved to yield retinal,[4] and a rough correlation exists between the rate of retinal formation and the biological activity of a given carotenoid.[5] We previously showed that BCC enzyme is distributed in a variety of herbivorous and carnivorous species, with the notable exception of the cat in which BCC enzyme is virtually absent.[6] Generally it is

---

[1] DeW. S. Goodman and H. S. Huang, *Science* **149**, 879 (1989).
[2] J. A. Olson and O. Hayaishi, *Proc. Natl. Acad. Sci. U.S.A.* **54**, 1364 (1965).
[3] DeW. S. Goodman, H. S. Huang, and T. Shiratori, *J. Biol. Chem.* **241**, 1929 (1966).
[4] M. R. Lakshman, J. L. Pope, and J. A. Olson, *Biochem. Biophys. Res. Commun.* **33**, 347 (1968).
[5] J. A. Olson and M. R. Lakshman, in "Fat Soluble Vitamins" (H. F. Deluca and J. W. Suttie, eds.), p. 213. University of Wisconsin Press, Madison, 1969.
[6] M. R. Lakshman, H. Chansang, and J. A. Olson, *J. Lipid Res.* **13**, 477 (1972).

more abundant in herbivores than in carnivores. More recent studies[7] have shown the existence of carotenoid cleavage enzyme activity where the site of attack on the carotene molecule was other than the central 15,15'-ethylenic bond, resulting in the formation of various β-carotenals. Nonetheless, it is clear that the BCC enzyme may be the major, if not the sole, enzyme responsible for the conversion of carotenoids to retinoids. Thus, BCC enzyme plays an important role in vitamin A nutrition and could be the sole provider of vitamin A in individuals that are dependent exclusively on dietary carotenoids for their vitamin A requirements.

Recent findings indicate that the serum concentration of β-carotene in the cord blood of term and preterm infants is one-eighth the concentration in the maternal serum.[8] Breast-feeding replenishes the plasma β-carotene levels of the infant to normal within 4–6 days because of high β-carotene content of the colostrum.[8] Furthermore, low birth-weight infants have lower plasma vitamin A levels than the normal birth-weight infants.[9-11] Whereas the median liver vitamin A concentration of healthy adults in the United States is around 100 μg/g,[12] newborn infants have only 20 μg/g.[13,14] Thus, the vitamin A requirements of neonates would be met by the conversion of β-carotene to vitamin A, provided the BCC enzyme is present in the neonatal tissues. Thus, BCC enzyme may play an important role in the vitamin A requirements of the neonates. BCC enzyme is known to be distributed in the intestine, liver, and corpus luteum, the activity being highest in the intestine.[1-5]

Assay Method

Principle. Using the intestinal mucosa from rabbit and human neonates as the source of the β-carotene cleavage (BCC) enzyme activity, it will be demonstrated conclusively in this chapter that retinal is, in fact, the product of BCC enzyme using (1) a novel method of synthesizing the *O*-ethyl oxime of the enzymatic product as its derivative and subsequent separa-

---

[7] X.-D. Wong, G.-W. Tang, J. G. Fox, N. I. Krinsky, and R. M. Russel, *Arch. Biochem. Biophys.* **285**, 8 (1991).
[8] E. M. Ostrea, Jr., J. E. Balun, R. Winkler, and T. Porter, *Am. J. Obstet. Gynecol.* **154**, 1014 (1986).
[9] R. B. Brandt, D. G. Mueller, J. R. Schroeder, K. E. Guyer, B. V. Kirkpatrick, N. E. Hutcher, and F. E. Ehrlich, *J. Pediatr. (St. Louis)* **92**, 101 (1978).
[10] J. P. Shenai, F. Chytil, A. Jhaveri, and M. T. Stahlman, *J. Pediatr. (St. Louis)* **99**, 302 (1981).
[11] J. Navarro, M. B. Causse, N. Desquilbet, F. Herve, and D. Lalemand, *J. Pediatr. Gastroenterol. Nutr.* **3**, 744 (1984).
[12] N. Raica, Jr., J. Scott, L. Lowry, and H. E. Sauberlich, *Am. J. Clin. Nutr.* **25**, 291 (1972).
[13] T. Moore, "Vitamin A." Elsevier, Amsterdam, 1957.
[14] N. Montreewasuwat and J. A. Olson, *Am. J. Clin. Nutr.* **32**, 601 (1979).

tion, identification, and quantitation by high-performance liquid chromatography (HPLC), (2) unequivocal identification of the retinal (O-ethyl) oxime derivative by its characteristic absorption spectrum, mass spectrum, and its repeated crystallization to constant specific activity after it was enzymatically formed from β-[^{14}C]carotene, and finally (3) enzymatic reduction of retinal to retinol and its subsequent esterification to retinyl palmitate with their characteristic absorption spectra and distinctive elution profiles on HPLC.

Animals and Other Supplies. New Zealand White rabbits (2–3 kg) are procured from National Institutes of Health (Bethesda, MD); male Wistar-Furth albino rats (190–220 g) are from Charles River Breeding Laboratories (Wilmington, MA). The animals are maintained on their normal diet for at least 2 weeks before experimentation. All chemicals and reagents are of analytical or ultrapure grade. All organic solvents are of HPLC grade and are routinely filtered through a 0.45-μm filter before use. All the solvents used for extraction have 50 mg/liter of butylated hydroxytoluene (BHT).

Neonatal Tissues. Autopsy samples of 12 premature infants were obtained by Dr. Lois Johnson, Pennsylvania Hospital (Philadelphia, PA). The proximal portion of each small intestine is removed and stored at $-70°$ until assayed for BCC enzyme activity.

Isolation of Enzyme. The procedure is essentially according to our earlier publication.[15] Briefly, each intestinal segment is washed with 0.154 M ice-cold saline. All the subsequent procedures are carried out at 4° unless otherwise stated. The intestine is slit open longitudinally, and the mucosa is scraped into a beaker containing the homogenizing buffer [0.1 M potassium phosphate buffer, pH 7.8, containing 1 mM dithiothreitol (DTT)]. The mucosa is homogenized with 5 volumes of the homogenizing buffer, and the homogenate is centrifuged at 100,000 g for 1 hr. The supernatant solution is subjected to 0–60% ammonium sulfate saturation and centrifuged at 16,000 g for 15 min. The pellet is dissolved in the homogenizing buffer to give a final protein concentration of approximately 10 mg/ml and stabilized by adding 1 mM reduced glutathione (GSH). The procedure followed for the isolation of the enzyme from the placentas is identical to that for the intestine. Each enzyme fraction is assayed for β-carotene cleavage activity as described below. The protein in various fractions is determined according to Lowry et al.[16]

Enzyme Assay. The standard assay mixture is made up as follows: 100

[15] M. R. Lakshman, I. Mychkovsky, and M. Attlesey, *Proc. Natl. Acad. Sci. U.S.A.* **86,** 9124 (1989).
[16] O. H. Lowry, N. J. Rosebrough, A. L. Farr, and R. J. Randall, *J. Biol. Chem.* **193,** 265 (1951).

nmol β-carotene in 0.1 ml benzene is mixed with 180 μl of 1/10 diluted Tween 20 in water, and the benzene is removed by a gentle stream of nitrogen. To this substrate are added the following components at the indicated final concentrations: potassium phosphate buffer, pH 7.8, 100 mM; GSH, 1 mM; ferrous sulfate, 1 mM; nicotinamide, 15 mM; and the intestinal enzyme fraction, approximately 7 mg; the final volume is always made up to 2 ml with water. Blank tubes have either no enzyme preparation or an equivalent amount of boiled enzyme fraction (boiled at 100° for 5 min). After incubation at 37° for 60 min in a shaking water bath (50 excursions/min) under F40 gold fluorescent light (Ace Electric Co., Washington, D.C.), the reaction is stopped by adding 2 ml methanol. The O-ethyl oxime derivative is prepared essentially according to Van Kuijk et al.[17] Briefly, to the incubation mixture, 100 μl of 0.1 M O-ethylhydroxylamine hydrochloride in 0.1 M potassium phosphate buffer, pH 6.5, and 100 μl methanol containing cholesterol (50 μg/ml) are added. After 10 min at 25°, 6 ml of water is added, and the whole reaction mixture is thoroughly extracted with three 10-ml portions of light petroleum. The lipid extracts are combined, evaporated to near dryness under a gentle stream of nitrogen, and the residue finally redissolved in 1 ml of methanol.

Chromatographic Analysis. All HPLC analyses are carried out using a Gilson HPLC automated system (Gilson Inst, Middleton, WI) equipped with a Kratos Model 783 variable wavelength detector (ABI Analytical, Ramsey, NJ). A 50-μl aliquot of the final methanol extract is subjected to HPLC on a 4.6 × 25 cm Vydac TP (reversed-phase C_{18} column, 9% carbon load, particle size 5 μm) with a linear gradient solvent system of methanol–water containing 0.5% ammonium acetate as the mobile phase at a flow rate of 1 ml/min. The linear gradient is increased over a 10-min period from 92:8 to a 98:2 (v/v) methanol–water system. Under these conditions authentic retinal (O-ethyl) oxime has a retention time of 8.1 min while retinol, retinal, and β-carotene have retention times of 4.8, 5.3, and 16.0 min, respectively. The recovery of added retinal as its O-ethyl oxime is 95 ± 3% ($n = 4$) under these conditions.

Mass Spectrum of O-Ethyl Oxime Derivative of Enzymatic Product and Crystallization of [^{14}C]Retinal (O-Ethyl) Oxime Formed from β-[^{14}C]Carotene on Incubation with Enzyme. Twenty microcuries of β-[15,15'-^{14}C]carotene (specific activity 21 μCi/μmol) is mixed with 1.5 μmol of cold β-carotene (both from Hoffmann-LaRoche, Nutley, NJ) and chromatographed on a 1 × 30 cm column of 10% (v/w) water-deactivated neutral alumina (30 g; Brockman Grade I; Sigma, St. Louis, MO). The fraction that elutes with 1% (v/v) acetone in petroleum ether is collected and shows

---

[17] F. J. G. M. Van Kuijk, G. J. Handleman, and E. A. Dratz, *J. Chromatogr.* **348**, 241 (1985).

the characteristic absorption spectrum of β-carotene. The recovery of β-carotene based on radioactivity and extinction coefficient routinely amounts to 70% of the starting material.

The purified β-carotene is used in the following experiment, which is carried out just like the standard assay described above except that β-[15,15'-^{14}C]carotene [specific activity 17,600 disintegrations/min (dpm)/nmol] is added instead of nonradioactive β-carotene as the substrate. At the end of 60 min of incubation, 10 such assay mixtures are combined and treated with O-ethylhydroxylamine hydrochloride as described above. The final lipid extract is subjected to thin-layer chromatography on silica gel G (250 μm thick) using 1.5% (v/v) acetone in isooctane as the developing solvent. The derivative of the product, retinal (O-ethyl) oxime (R_f 0.43), is completely separated from the substrate, β-carotene (R_f 0.83). The product band is scraped off the thin-layer plate and extracted thoroughly with acetone. One-tenth of the total amount of this extract is evaporated under a gentle stream of nitrogen, redissolved in 10 μl of ethanol, and analyzed for its mass spectrum in a LKB Model 9000 mass spectrometer (Bromma, Sweden). The conditions are 70 eV by direct insertion probe, source temperature 270°, and ionization current 20 μA. The other nine-tenths of the extract is combined with 87 mg of authentic nonradioactive retinal (O-ethyl) oxime in 2 ml ethanol and allowed to crystallize at $-80°$. The orange-yellow needle crystals are redissolved in ethanol and recrystallized from ethanol 2 more times. The ^{14}C specific activity (dpm/mg) of the crystalline retinal (O-ethyl) oxime is determined at each crystallization step.

Enzymatic Reduction and Esterification of Retinal Formed from β-Carotene

Reduction. The lipid extract of the standard incubation mixture of β-carotene with the BCC enzyme of the rabbit intestine is incubated for 1 hr at 37° with 250 mM NADH$_2$, 1 mM GSH, and 1 mg of horse liver alcohol dehydrogenase (Sigma) in a 2-ml final volume of 0.1 M potassium phosphate buffer, pH 6.0. The reaction is stopped by adding 2 ml ethanol followed by extraction with light petroleum. The lipid extract is tested for its absorption spectrum and for its HPLC profile.

Esterification. The enzymatically formed retinol fraction from the above experiment is evaporated under nitrogen and redissolved in 50 μl of ethanol and incubated with 0.14 mM palmitic acid bound to 0.14 mM fatty acid-free bovine serum albumin, 0.14 mM coenzyme A, and 10 mg of the rat pancreatic acetone powder (source of esterase) in a 0.6-ml final volume of potassium phosphate buffer, pH 7.3. The reaction is stopped by the addition of 1.6 ml ethanol. After adding 1 ml water, the reaction mixture is thoroughly extracted twice with 10-ml portions of light petroleum. After a 5-min centrifugation at 600 g, the light petroleum extracts are combined, evaporated under nitrogen, and finally redissolved in 0.5 ml dichloromethane. An aliquot of this fraction is subjected to HPLC on a ODS column (4.6 mm × 10 cm, 3-μm particle size) using a solvent system of 100% methanol containing 0.5% (w/v) ammonium acetate. The initial flow rate of the solvent is 1 ml/min for the first 3 min, which is increased linearly to 2 ml/min over the next 1 min and thereafter kept constant at 2 ml/min for the subsequent 10 min. Under these conditions authentic retinol and retinyl palmitate completely separate from each other, exhibiting sharp peaks with retention times of 1.97 and 10.01 min, respectively.

Major Findings

It has been unequivocally proven that β-carotene is cleaved by an enzyme from the intestinal mucosa of rabbits, rats, and human neonates and that the product of this reaction is, in fact, retinal. It has been previously observed[18] that there may be accompanying activities in the cytosol fraction of the intestine which could catalyze the further metabolism of retinal to retinol or retinoic acid. Furthermore, the crude cytosol fraction from the rat intestine, kidney, testes, and liver has also been shown to convert β-carotene to retinoic acid.[19] In view of these observations, we decided to fractionate the cytosol fraction of the rabbit intestinal mucosa as described above. The BCC enzyme activity was localized in the 45–60% acetone pellet fraction of the initial 0–60% ammonium sulfate pellet fraction of the cytosol. More importantly, this fraction was devoid of either retinal reductase or retinal oxidase activity (data not shown). Since our aim was to test whether retinal was the product of β-carotene cleavage, we used the 45–60% acetone pellet fraction of the intestinal mucosa throughout this study.

The HPLC analysis of the lipid extract from the incubation of β-carotene with the rabbit intestinal enzyme preparation (Fig. 1A) showed a peak with a retention time of 7.99 min corresponding to authentic retinal (O-ethyl) oxime with a retention time of 8.1 min (Fig. 1B). In contrast, as shown in Fig. 1C, the extract from the boiled enzyme incubation mixture failed to show any HPLC peaks corresponding to retinal (O-ethyl) oxime or retinal or retinol. Furthermore, the mass spectrum of the O-ethyl oxime derivative of the enzymatic product (Fig. 2A) was identical to that (Fig. 2B) of authentic retinal (O-ethyl) oxime (m/z 327, 45%; m^+ and m/z 282, 100%, methoxy). These results conclusively prove the identity of the enzymatic product as retinal.

[18] H. Singh and H. R. Cama, *Biochim. Biophys. Acta* **370**, 49 (1974).
[19] J. L. Napoli and K. R. Race, *J. Biol. Chem.* **263**, 17372 (1988).

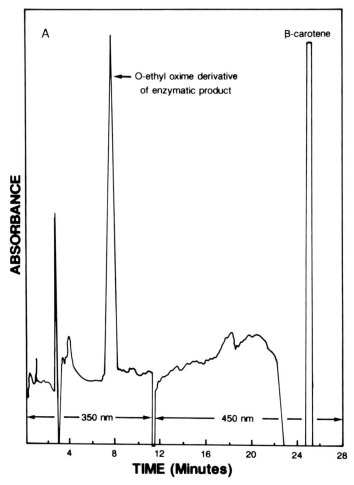

FIG. 1. HPLC profiles of the lipid extracts from the standard assay mixtures from (A) rabbit BCC enzyme, (B) authentic retinal (O-ethyl) oxime, and (C) boiled rabbit BCC enzyme. The experimental details and the HPLC conditions are given in the text. Detection wavelength, 360 nm.

Table I summarizes the recovery of radioactivity in crystalline retinal (O-ethyl) oxime formed after the lipid extract from the incubation of β-[15,15'-^{14}C]carotene with the partially purified (45–60% acetone pellet fraction) BCC enzyme fraction from rabbit intestine. Nearly 87% of the original radioactivity in the final ethanol solution containing the concentrate of retinal (O-ethyl) oxime was recovered in its crystalline form with a

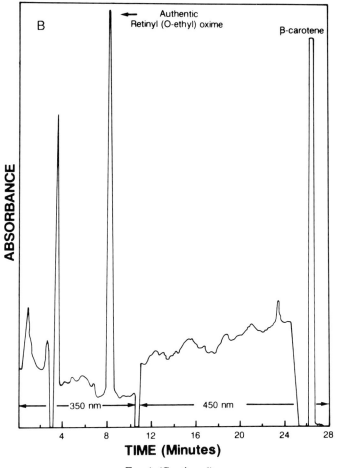

FIG. 1. (Continued)

concomitant recovery of the product (first crystallization). It is clear that even after three crystallizations the specific activity of the isolated retinal (O-ethyl) oxime remained constant (1343 dpm/mg). This is further evidence for the identity of the enzymatic product to be retinal.

The enzymatic product had an absorption spectrum in light petroleum identical to that of authentic retinal. Furthermore, it gave rise to retinol with an absorption spectrum in light petroleum similar to that of authentic retinol, with its absorption maximum at 326 nm when incubated with $NADH_2$ and horse liver alcohol dehydrogenase. This enzymatically formed retinol, on HPLC, had a retention time of 1.96 min, identical to

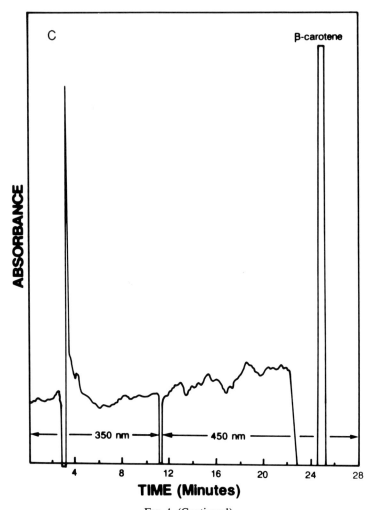

FIG. 1. (Continued)

that of authentic retinol. Its identity was further confirmed by its ready conversion to retinyl palmitate on incubation with palmitic acid, coenzyme A, and rat pancreatic esterase as evidenced by the appearance of a HPLC peak corresponding to authentic retinyl palmitate with a retention time of 10.0 min.

In view of all the above evidence, it is unequivocally proven that retinal is, indeed, the enzymatic product of β-carotene cleavage by the intestinal mucosal enzyme, and, therefore, the original findings of Goodman et al.[1]

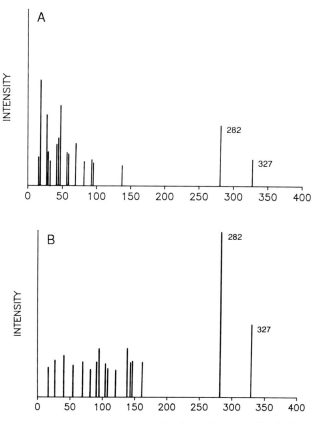

FIG. 2. Mass spectrum of (A) enzymatically formed retinal (O-ethyl) oxime and (B) authentic crystalline retinal (O-ethyl) oxime.

and Olson and Hayaishi[2] are confirmed. Based on the amount of retinal (O-ethyl) oxime measured by HPLC in a typical assay, the BCC enzyme activity of the 45–60% acetone pellet fraction of the rabbit intestinal mucosa was calculated to be close to 1 nmol retinal formed/mg protein/hr.

These studies were extended to humans, and the following results were obtained. Twelve premature infants, ten of whom died soon after birth and two at ages 2 and 4 months, and an adult were the subjects of this study. Postconceptual age at the time of death ranged from 22 to 37 weeks, and the time between death and autopsy ranged from 1 to 52 hr. HPLC profiles of the enzymatic product from one of the neonatal intestinal enzyme incubations showed the characteristic retinal (O-ethyl) oxime peak, unlike

TABLE I
COCRYSTALLIZATION OF ENZYMATICALLY FORMED [^{14}C]RETINAL AS ITS O-ETHYL OXIME WITH AUTHENTIC NONRADIOACTIVE RETINAL (O-ETHYL) OXIME[a]

Stage of crystallization	Volume of mother liquor (ml)	Amount of oxime (mg)	Total radioactivity (dpm)	Specific radioactivity (dpm/mg)
Before crystallization	2	87	117,000	1345
First crystallization	2	75	102,000	1360
Second crystallization	4	46	61,040	1327
Third crystallization	4	43	57,600	1340

[a] The enzymatic product formed from β-[^{14}C]carotene was converted to its O-ethyl oxime derivative as described in the text and mixed with 87 mg of authentic nonradioactive retinal (O-ethyl) oxime in 2 ml ethanol and allowed to crystallize at $-80°$. The crystals were redissolved in the indicated volumes of ethanol and the crystallization process repeated through the third crystallization. Based on the total radioactivity and the amount of retinal (O-ethyl) oxime (assuming a E_{1cm} value of 1800 at 363 nm in ethanol) recovered at each stage of crystallization, the specific radioactivity (dpm/mg) at each crystallization step was calculated.

the extract from the boiled enzyme incubation. This result clearly shows that the BCC enzyme does exist in the neonatal intestine and that its product of action on β-carotene is retinal.

The results of the measured intestinal BCC enzyme activity in the autopsy samples of 12 neonates showed that BCC enzyme activity ranged widely from 3.3 to 1210 pmol/mg/hr. Except for one isolated sample which had an activity of 1.2 nmol retinal formed/mg protein/hr, all others showed relatively low activities compared to the rabbit intestinal activity of 1 nmol/mg/hr.[15] It must be pointed out that the time lag between the time of death and the time of storage of the intestinal tissue at $-70°$ varied between 1 and 52 hr. Significantly, the BCC activity in a fresh biopsy sample of an adult intestine was found to be 0.35 nmol/mg/hr.

To verify whether the BCC activity was influenced by the storage conditions of the intestinal tissue, its activity was measured in rats as a function of time after death and temperature at which the carcass was stored. The results showed that as much as 80% of the original activity of the fresh intestine is lost on storage of the dead animal for 8 hr at $25°$ followed by storage at $4°$ for 16 hr. These are normally the prevailing conditions in most hospitals for humans from the time of death to their storage in the morgue. Thus, the observed activities in the human autopsy samples may be underestimated, presumably because of marked loss of enzyme activity from the time of death to the time of assay. Therefore the true activity of the enzyme in neonates can be assessed only after the extent of the loss of its activity on storage of the human samples can be accurately measured.

Precautions

Because of the relative instabilities of the BCC enzyme and carotenoids, one has to take extreme precautions in assaying this enzyme activity. Some of the precautions to be taken are listed below. (1) BCC enzyme is very sensitive to poisoning by heavy metals because it is a sulfhydryl enzyme. Therefore, the enzyme must be stored in the presence of sulfhydryl-protecting agents such as reduced glutathione or dithiothreitol. Furthermore, all chemicals and reagents including water must be of high purity and free of heavy metals. (2) The enzyme becomes more labile as it is purified. It is also unstable to freezing and thawing. However, the enzyme can be stored, preferably, at $-70°$ as a 0-50% ammonium sulfate pellet fraction containing 1 mM reduced glutathione without appreciable loss of activity over a 1-month period.[15] (3) Because β-carotene and other carotenoids are unstable in the presence of light, oxygen, and heavy metals,[20,21] the substrates should be freshly purified in the presence of antioxidants such as BHT and used immediately. (4) Preparation of the carotenoid substrate as a micellar solution requires careful attention. Many detergents are inhibitory at higher concentrations, and many of them vary in their purity from batch to batch. Therefore, one has to test each batch of the detergent used before finding the most suitable one. In our studies, Tween 20 worked very well. (5) All organic solvents used in the extraction and HPLC analyses must be free of peroxides, and the use of 50 mg/liter BHT in all these solvents is strongly recommended.

Enzyme Properties

The BCC enzyme from various species exhibits similar properties.[5,6,22] It is localized in the cytosolic fraction, requires molecular oxygen, and yields retinal as the sole identifiable product. The K_m for β-carotene is 2-10 μM, and the V_{max} for β-carotene ranges from 0.003 to 1.7 nmol retinal formed/mg/hr at 37°, depending on the species. The pH optimum is between 7.5 and 8.5. Various iron chelators such as α,α'-dipyridyl,

[20] O. Isler, ed. "Carotenoids." Birkhauser Verlag, Basel, 1971.
[21] J. A. Olson, in "Handbook of Vitamins" (L. J. Machlin, ed.), p. 1. Dekker, New York, 1984.
[22] J. A. Olson, in "Biosynthesis of Isoprenoid Compounds" (J. W. Porter and S. P. Spugeon, eds.), Vol. 2, p. 371. Wiley, New York, 1983.

o-phenanthroline, and possibly EDTA inhibit the enzyme. Similarly, sulfhydryl agents such as N-ethylmaleimide, iodoacetamide, p-chloromercuribenzoate, and heavy metal ions strongly inhibit the enzyme activity. Reduced glutathione or DTT protect the enzyme, while ferrous ions enhance its activity. Nicotinamide may also protect the enzyme. The enzyme shows fairly broad substrate specificity. Thus, apart from β-carotene, the following carotenoids are cleaved at reasonably appreciable rate by the BCC enzyme: 3,4,3′,4′-tetrahydro-β-carotene, 5,6-epoxy-β-carotene, 5,6-epoxy-β,ϵ-carotene, β,ϵ-carotene, 3,4-dehydro-β,ψ-caroten-16-al, 10′-β-apocarotenal, and 10′-β-apocarotenol.[4,18,22]

Summary

Enzymatic conversion of all-*trans*-β-carotene to retinal by a partially purified enzyme from rabbit, rat, and human neonatal intestinal mucosa has been demonstrated. The enzymatic product was characterized based on the following evidence. First, the product gave rise to its O-ethyl oxime by treatment with O-ethylhydroxylamine with an absorption maximum at 363 nm in ethanol characteristic of authentic retinal (O-ethyl) oxime. High-performance liquid chromatography of this derivative yielded a sharp peak with a retention time of 7.99 min, corresponding to the authentic compound. The enzyme blank and boiled enzyme blank failed to show any significant HPLC peaks corresponding to retinal (O-ethyl) oxime or retinal or retinol. Second, the mass spectrum of the O-ethyl oxime of the enzymatic product was identical to that of authentic retinal (O-ethyl) oxime (m/z 327, 45%; m^+ and m/z 282, 100%, methoxy). Third, the ^{14}C radioactivity persisted to constant specific activity even after repeated crystallization of the retinal (O-ethyl) oxime isolated from the enzyme reaction with purified β-[^{14}C]carotene. Fourth, the enzymatic product exhibited an absorption maximum at 370 nm in light petroleum characteristic of authentic retinal. Furthermore, it was reduced by horse liver alcohol dehydrogenase to retinol with an absorption maximum at 326 nm in light petroleum. This retinol was enzymatically esterified to retinyl palmitate by rat pancreatic esterase with a retention time of 10 min on HPLC, corresponding to authentic retinyl palmitate. Thus, the enzymatic product of β-carotene cleavage by the partially purified intestinal enzyme has been unequivocally confirmed to be retinal.

Similarly, enzymatic conversion of all-*trans*-β-carotene to retinal by an intestinal mucosal enzyme from autopsy samples of human neonates has also been demonstrated. Based on the observed activities among intestinal samples from 12 premature infants, the BCC enzyme activity ranged from 3.3 to 1210 pmol/mg mucosal protein/hr. However, the observed activities in the human autopsy samples may be markedly underestimated, presumably because of marked loss of enzyme activity from the time of death to the time of assay. Therefore, the true activity of the enzyme can be assessed only after the extent of the loss of its activity on storage of the human samples can be accurately measured. Nonetheless, the demonstration of BCC enzyme activity in human neonates shows that β-carotene may be an important source of vitamin A nutrition during gestation.

Acknowledgments

The authors wish to thank Dr. Lois H. Johnson, Department of Pediatrics, Pennsylvania Hospital (Philadelphia, PA), and Dr. Hemmige N. Bhagavan, Hoffmann-LaRoche, Inc. (Nutley, NJ), for help in the collection of the human samples. The authors are grateful to Drs. Henry M. Fales and William E. Comstock, Division of Chemistry, National Institutes of Health, for the mass spectrometric analyses and identification of retinal (O-ethyl) oxime in this study. This work was supported by a grant from National Cancer Institute (CA 39999) and by Hoffmann-LaRoche, Inc.

[24] Methods for Investigating Photoregulated Carotenogenesis

By MIKIRO TADA

Introduction

The biosynthesis of carotenoid pigments is photoregulated in a wide range of organisms, and this photoregulation has been studied by many workers as a typical example of a biosynthetic process photoregulated *in vivo*. In some nonphotosynthetic microorganisms, in which carotenogenesis in the dark is either nil or very slight, short exposure to light followed by a return to darkness can considerably stimulate carotenogenesis. Studies for elucidating the biochemistry of this photoresponse have been conducted mainly with *Mycobacterium* sp.,[1] *Mycobacterium marinum*,[2,3] *Neurospora crassa*,[4,5] *Fusarium aquaeductuum*,[6] and *Rhodotorula min-*

[1] H. C. Rilling, *Biochim. Biophys. Acta* **60**, 548 (1962).
[2] P. P. Batra and H. C. Rilling, *Arch. Biochem. Biophys.* **107**, 485 (1964).
[3] P. P. Batra, *J. Biol. Chem.* **242**, 5630 (1967).
[4] M. Zalokar, *Arch. Biochem. Biophys.* **56**, 318 (1955).

uta.[7] The focus of this chapter is the mechanism by which light regulates the *de novo* synthesis of carotenoids in the red yeast, *Rhodotorula minuta*.

Convenient Method for Determining Carotenoid Content

To elucidate the biochemistry of photoregulated carotenogenesis, it is necessary to obtain extensive data on the relationship between carotenoid production and light, and it is preferable to analyze these data stoichiometrically. A convenient and accurate method for the determination of carotenoid content is required. The conventional method usually involves measurement of carotenoids after disruption and extraction of the cells, and hence is laborious and requires a large amount of cells. Accordingly, a new method was devised to measure the cell carotenoid content without cell disruption and extraction.[7]

As will be shown later, the amount of carotenoid produced in *R. minuta* cells depends on the light intensity during incubation. Cells grown under different light intensities are washed and suspended in water (2.7 mg dry weight/ml). The suspensions are placed in a spectrophotometer, and the absorption spectra are read by reflectance against the air phase as a blank. The absorption spectra thus obtained show a flat baseline and clear absorption maxima at 503 and 537 nm and an absorption minimum at 522 nm. A linear relationship exists between the difference [$\Delta A(503 - 522$ nm)] read from the absorption spectrum and the carotenoid content determined by the ordinary extraction procedure. Therefore, this newly devised method is employed for subsequent quantitative determinations of cell carotenoid content.

Carotenoid Production under Continuous Illumination

No effect of light has been found on the rate of cell multiplication or consumption of glucose and amino acids by *R. minuta*. However, the carotenoid content of cells cultured under illumination differs markedly from those cultured in the dark; namely, it is 140 μg/g dry cells with culture under 4000 lux fluorescent light for 120 hr but 10–15 μg/g dry cells with culture in the dark.

To elucidate the dependence of carotenoid production on illumination, a fluorescent lamp is used as the light source. Cells used in this experiment had been incubated previously for 48 hr at 26° in the dark. The results

---

[5] R. W. Harding, P. C. Huang, and H. K. Mitchell, *Arch. Biochem. Biophys.* **129**, 696 (1969).
[6] W. Rau, *Planta* **72**, 14 (1967).
[7] M. Tada and M. Shiroishi, *Plant Cell Physiol.* **23**, 541 (1982).

summarized in Fig. 1 show that the carotenoid content of the cells depends on the light intensity during the incubation. The cells are incubated at 26° under light of various intensities, and carotenoid contents are determined during incubation. The increase in carotenoid content is linear at all intensities tested until 70 hr, but the rate of increase depends on the light intensity. The relationship between the rate of increase [Vc ($\mu g\ g^{-1}\ hr^{-1}$)] and the light intensity [I (erg $cm^{-2}\ sec^{-1}$)] is expressed by the empirical equation $Vc = 0.74 \log I - 1.46$. The plateau level of carotenoid content reached within 75 hr of incubation also depends on the light intensity. The relationship between the plateau level of carotenoid content [C ($\mu g\ g^{-1}$)] and the light intensity is expressed by the empirical equation $C = 52 \log I - 81$. The carotenoid content of cells incubated for 72 hr under a light intensity of 5.1×10^3 erg $cm^{-2}\ sec^{-1}$ further increased when the cells were incubated at 4.8×10^4 erg $cm^{-2}\ sec^{-1}$. The rate of increase after exposure to the higher intensity is essentially the same as that observed with cells that had been incubated under the higher intensity from the beginning. Such phenomena indicate that the yeast *R. minuta* is able to recognize the intensity of light and to reflect the recognition in both the rate of carotenoid synthesis and the quantity of carotenoid synthesized.

Effect of Temperature on Carotenoid Production

Cells grown in the dark are suspended in water (5 mg/ml).[8] The cell suspension is placed in a glass container 15 cm high with an internal diameter of 2.5 cm. The glass container is fixed in a glass vessel filled with

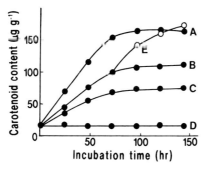

FIG. 1. Dependence of the increase in cell carotenoid content on light intensity. The cells were incubated at 26° under light of various intensities: 4.8×10^4 (A), 5.1×10^3 (B), 1.1×10^3 (C), and 0 erg $cm^{-2}\ sec^{-1}$ (D). A portion of cells that had been incubated under illumination at 5.1×10^3 erg $cm^{-2}\ sec^{-1}$ was exposed to 4.8×10^4 erg $cm^{-2}\ sec^{-1}$ after 72 hr (E).

antifreeze liquid adjusted to $-1.5 \pm 0.5°$ and exposed to light at an intensity of 1×10^5 erg cm^{-2} sec^{-1} for 4 hr using a slide projector equipped with a blue filter. During the period of illumination, air or nitrogen gas is bubbled into the cell suspension to provide uniform illumination and to prevent freezing. The cells are harvested immediately after illumination, resuspended in growth medium (2.5 mg/ml), and incubated for 15 hr in the dark at 26°, after which the cell carotenoid content is determined.

The carotenoid content of cells does not increase during the illumination at $-1.5°$. However, when the cells that have been exposed to light are incubated in the dark, the cell carotenoid content increases. These results suggest that the promotion of carotenoid biosynthesis in *R. minuta* by light occurs in two phases: the first involves a photochemical reaction independent of temperature (light reaction process), and the second involves a biochemical reaction dependent on temperature but independent of light (dark reaction process). Rilling,[9] using *Mycobacterium* sp., studied the induced capability of carotenoid synthesis acquired by exposure to light and called it "memory," and he proposed that this memory might be a photochemical product playing an important role as inducer of carotenogenesis.

Carotenoid Production in the Dark Reaction Process

Cells are exposed to light for various periods at $-1.5°$, then incubated in the dark, and the time course of carotenoid production is studied in detail. As shown in Fig. 2, there is a lag period of approximately 2 hr before

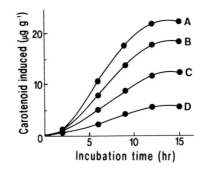

FIG. 2. Effect of illumination time in the light process on the rate of carotenoid production and the quantity of carotenoid produced during the dark process. Cells grown in the dark were exposed to light at an intensity of 1×10^5 erg cm^{-2} sec^{-1} at $-1.5°$ for various periods: 480 (A), 120 (B), 10 (C), and 5 min (D). Immediately after exposure to light, the cells were incubated at 26° in the dark and were harvested at intervals to determine the amount of carotenoids.

the accumulation of carotenoid in the cells becomes detectable. Thereafter, the carotenoid content increases linearly up to 10 hr, followed by a more gradual increase; carotenoid synthesis is completed within about 12 hr regardless of the duration of exposure to light. However, the quantity and rate of carotenoid production increase as the duration of exposure is prolonged. In other words, the light dose applied in the light reaction process regulates not only the amount of carotenoid produced, but also the rate of carotenoid production during the dark reaction process.

The relationship between the light dose [D (erg cm^{-2})] and the carotenoid production [C' (μg g^{-1})] during the dark incubation could be expressed by the empirical equation $C' = 9.1 \log D - 62.5$. Furthermore, the relationship between the light dose and the rate of carotenoid production [Vc' (μg g^{-1} hr^{-1})] determined on the basis of the slopes in Fig. 2 could be expressed by the empirical equation $Vc' = 0.81 \log D - 5.6$. This kind of relationship is also observed for carotenoid production in cells growing under continuous illumination, but in this case the light intensity played a major role in controlling the quantity and rate of carotenoid production.

Effect of Oxygen on Photoinduction of Carotenoid Production

The influence of oxygen on carotenoid production in *R. minuta* has been studied.[8] Pure (99.999%) nitrogen gas is passed through an aqueous suspension of cells for 1 hr before their exposure to light and during the illumination. When cells illuminated under such anaerobic conditions are incubated under aerobic conditions in the dark, they produce almost the same amount of carotenoid as those illuminated under aerobic conditions. When the dark incubation at 26° is carried out under anaerobic conditions, no carotenoid production occurs even if illumination is carried out under aerobic conditions. These results prove that the photochemical reaction in photoinductive carotenogenesis in *R. minuta* is oxygen-independent, but the dark process is completely dependent on oxygen. Howes *et al.*[10] stated that the oxygen requirement for the photochemical reaction in photoinduction of carotenogenesis is a universal phenomenon. However, the results of Rau[11] and the present experiments indicate that the oxygen requirements for the photochemical reaction may vary from species to species.

[8] M. Tada and M. Shiroishi, *Plant Cell Physiol.* **23**, 549 (1982).
[9] H. C. Rilling, *Biochim. Biophys. Acta* **79**, 464 (1964).
[10] C. D. Howes, P. P. Batra, and C. F. Blakely, *Biochim. Biophys. Acta* **189**, 298 (1969).
[11] W. Rau, *Planta* **84**, 30 (1969).

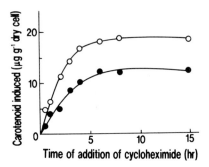

FIG. 3. Effect of cycloheximide on carotenoid production in the dark reaction process. Cells were exposed to light at 1×10^5 erg cm^{-2} sec^{-1} at $-1.5°$ for either 120 (○) or 30 min (●) and then subjected to dark incubation at 26°. At each incubation time, cycloheximide was added to the cultures to a concentration 5×10^{-4} M, and the dark incubation was continued. The cells were harvested at 15 hr after the beginning of the dark incubation to determine the level of carotenoid production.

Effect of Cycloheximide on Carotenoid Production in the Dark

Cycloheximide (CHI),[12] an inhibitor of protein synthesis in eukaryotes, if added immediately after exposure of the organisms to light, blocks carotenoid biosynthesis in *N. crassa*,[13] *F. aquaeductuum*,[14] and *Verticillium agaricinum*.[15] Thus, illumination is thought to induce the synthesis *de novo* of one or more enzymes in the carotenoid biosynthetic pathway. The effect of CHI on carotenoid production in the dark reaction process in photoinductive carotenogenesis in *R. minuta* has been studied (Fig. 3).

Two groups of cells are exposed to different light doses and incubated at 26° in the dark to allow them to synthesize carotenoid. CHI is added to the incubation medium to give a concentration of 5×10^{-4} M immediately after or at a defined time after the beginning of the dark incubation. The carotenoid content of the cells is determined 15 hr after the beginning of the incubation. Addition of CHI immediately after the beginning of incubation causes complete inhibition of carotenoid production. The magnitude of inhibition, however, becomes smaller as the time of CHI addition is delayed, being practically zero when CHI is added 5 hr after the onset of the dark incubation. The same tendency is observed with the other group of cells exposed to a lower light dose in the light reaction process.

[12] M. Tada and M. Shiroishi, *Plant Cell Physiol.* **23**, 567 (1982).
[13] R. W. Harding and H. K. Michell, *Arch. Biochem. Biophys.* **128**, 814 (1968).
[14] E. Bindl, W. Lang, and W. Rau, *Planta* **94**, 156 (1970).
[15] R. S. Mummery and L. R. G. Valadon, *Physiol. Plant.* **28**, 254 (1973).

These results indicate that carotenogenic enzymes in *R. minuta* are synthesized *de novo* within 5 hr of the dark incubation at 26°. Since the level of carotenoids produced most likely reflects the amount of enzymes synthesized, it is reasonable to conclude that the light dose given to cells in the light reaction process regulates the rate of synthesis and the amount of carotenogenic enzymes synthesized in the dark reaction process.

Photoinductive carotenogenesis in *N. crassa*[16] and *F. aquaeductuum*[17] is also inhibited by inhibitors of nucleic acid synthesis. In *V. agaricinum*,[15] however, these chemicals do not inhibit carotenoid production. The effect of an inhibitor of nucleic acid synthesis on carotenoid production in *R. minuta* is studied by using 5-fluorouracil, which is expected to inhibit RNA synthesis.[12] However, this chemical does not inhibit the carotenoid production in the dark reaction process, even at the high concentration of 10^{-2} M, which leads to greater than 80% inhibition of cell multiplication. On the basis of these results, the photoregulation of synthesis of carotenogenic enzymes in *R. minuta* is considered to be controlled at the level of translation.

Photoinduction of 3-Hydroxy-3-methylglutaryl-CoA Reductase

It is well known that 3-hydroxy-3-methylglutaryl-CoA (HMG-CoA) reductase (mevalonate:NADP$^+$ oxidoreductase, EC 1.1.1.34) plays an extremely important role in isoprenoid biosynthesis and that mevalonate is a common precursor of isoprenoid compounds.[18] The activity of HMG-CoA reductase in cell-free extracts taken from *R. minuta* cells incubated for 50 hr under continuous illumination is measured according to the procedures of Sugino *et al.*[19] Activity in cells incubated in the dark ranges from 1.8 to 2.0 nmol mg-N^{-1} min^{-1}. Activities in cells under illumination with five levels of light intensity, namely, 5×10^2, 1×10^3, 5×10^3, 1×10^4, and 5×10^4 erg cm^{-2} sec^{-1}, however, are 2.60, 2.89, 3.35, 3.36, and 4.03 nmol mg-N^{-1} min^{-1}, respectively. Thus, the HMG-CoA reductase activity [A (nmol mg-N^{-1} min^{-1})] increases with the increase in light intensity [I (erg cm^{-2} sec^{-1})], the relationship being expressed by the empirical equation $A = 0.72 \log I + 0.80$. This equation is similar to the equation which showed the relationship between the rate of carotenoid production in cells grown under continuous illumination and the intensity of light given. The similarity of the two equations is strong evidence that HMG-CoA reductase activity is involved in photoregulated carotenoid biosynthesis in *R. minuta*.

---

[16] R. E. Subden and C. Bobowski, *Experientia* **29**, 965 (1973).
[17] W. Rau, *Pure Appl. Chem.* **47**, 237 (1976).
[18] M. Tada and M. Shiroishi, *Plant Cell Physiol.* **23**, 615 (1982).
[19] M. Sugino, H. Okamatsu, and T. Idei, *Agric. Biol. Chem.* **42**, 2009 (1978).

The activity of HMG-CoA reductase in the light reaction process and the dark reaction process may also be investigated separately by regulating the temperature. The HMG-CoA reductase activity in cells does not increase during illumination at $-1.5°$. However, when illuminated cells are incubated in the dark at $26°$, the activity in the cells increases. In addition, the increase in HMG-CoA reductase activity in the dark reaction process is inhibited completely by the addition of cycloheximide at the beginning of this process. The inhibitory effect of cycloheximide, however, gradually decreases with delay of the addition, and it is essentially nil when CHI is added 4 hr after the beginning of the dark incubation.

The time course of HMG-CoA reductase activity in cells during dark incubation at $26°$ after illumination at $-1.5°$ is shown in Fig. 4. The activity of HMG-CoA reductase in the cells immediately after illumination is about the same as that in cells before illumination, but it begins to increase after the start of dark incubation. The first 3 hr produces a marked linear increase, which ends after 4 hr. During the next 4 hr, a high level of activity is maintained, but from 8 hr after the beginning of dark incubation, the activity decreases linearly to the value for cells grown in the dark. These results clearly indicate that the synthesize of HMG-CoA reductase in *R. minuta* is induced by light, that the synthesis of the enzyme is completed in about 4 hr, and that the turnover number for this light-induced enzyme is high.

Comparison with the results for cells exposed to different light doses suggests that the light dose given to cells in the light reaction process regulates both the maximum activity and the rate of increase in activity.

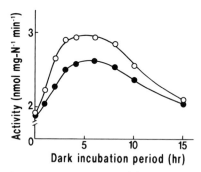

FIG. 4. Time course of HMG-CoA reductase activity *in vivo* during dark incubation after exposure to light. Cells first incubated in the dark for 48 hr were exposed to light of an intensity of 1×10^5 (O) or 5×10^4 erg cm^{-2} sec^{-1} (●) for 150 min at $-1.5°$, and then they were incubated in the dark at $26°$.

TABLE I
Relationship between Light Dose Given to Cells and HMG-CoA Reductase Activity[a]

Light dose (erg cm^{-2})	Light intensity (erg cm^{-2} sec^{-1})	Illumination time (hr)	HMG-CoA reductase activity (nmol mg-N^{-1} min^{-1})
1.8 × 10^8	5 × 10^4	1	2.25
	1 × 10^4	5	2.31
3.6 × 10^8	1 × 10^5	1	2.47
	5 × 10^4	2	2.45
	1 × 10^4	10	2.51
7.2 × 10^8	1 × 10^5	2	2.77
	5 × 10^4	4	2.68

[a] Cells were incubated in the dark at 26° for 5 hr after exposure to light at $-1.5°$.

Actually, the relationship between the activity $[A]$ and the light dose $[D]$ could be expressed by an empirical equation, $A = 0.08 \log D - 4.35$. The empirical equation resembles the equation which expresses the relationship between the rate of carotenoid production in the dark reaction process and the light dose given to cells in the light reaction process. The similarity of the two equations is evidence that the rate of carotenoid production in the dark reaction process is controlled by the amount of HMG-CoA reductase which, in turn, is under the control of the light dose given to cells during the light reaction process. It was also confirmed that this photoresponse follows the Roscoe–Bunsen reciprocity law (Table I). This result also indicates that the amount of photochemical product(s) regulates both the amount and the rate of synthesis of HMG-CoA reductase in the dark reaction process. Thus, the results of a study with a cell-free system provide an effective explanation for the photoregulation of carotenogenesis in *R. minuta* and are in good agreement with the results of *in vivo* experiments.

Presence of Other Photoinducible Enzymes

Which part of the pathway is regulated by light can be determined by comparing the levels of the intermediates and related compounds in dark-grown versus illuminated cells. To elucidate whether photoregulative enzymes other than HMG-CoA reductase exist in the carotenogenic pathway in *R. minuta*, the contents of phytoene, ubiquinone, and ergosterol in cells incubated under various light intensities have been determined.[20,21] It is found that the production of phytoene and ubiquinone is clearly regulated by the intensity of light in a manner similar to the regulation of carotenoid production, but ergosterol production is not influenced by light. All carotenoid, ergosterol, and isoprenoid side chains in ubiquinone are synthesized by the well-established terpenoid pathways, and mevalonate is a common precursor of isoprenoid compounds. These facts suggest that more than one enzyme, the synthesis of which is regulated by light, may be involved in the carotenogenic pathway beyond mevalonate.

Mevinolin, a highly specific competitive inhibitor of HMG-CoA reductase, has been used in such an investigation.[22] In the experiment, a black light (Toshiba FL-20S-BLB) is used as a light source, and mevinolin is purchased from Sigma Chemical Co. (St. Louis, MO). Cells illuminated for four different periods of time at $-1.5°$ are transferred to growth media that contains various concentration of mevinolin, and then cultures are incubated at $26°$ in the dark for 15 hr for subsequent measurement of the amount of carotenoid produced during the dark incubation (Fig. 5).

The addition of mevinolin results in a marked decrease in the photoinduced production of carotenoid. The curves illustrating the relationship between the production of carotenoid and the concentration of mevinolin are hyperbolic, indicating that mevinolin acts as a competitive inhibitor. It is noteworthy that the relationship between the inhibition ratio and the concentration of mevinolin is approximately identical, regardless of the light dose given to the cells, at concentrations of mevinolin below 30 μM. In cases of competitive inhibition, the inhibition ratio is generally given as the ratio of concentrations of substrate and inhibitor. The curves shown in the inset to Fig. 5 indicate, therefore, that the level of HMG-CoA in the *R. minuta* cells is maintained at a constant value regardless of light dose to the cells. Consequently, it appears that the enzymes involved in the formation of HMG-CoA are not subject to photoregulation.

Illuminated cells are incubated in growth medium that contains various concentrations of mevalonate and 100 μM mevinolin, as required for the complete inhibition of carotenoid production in these cells. Although the photoinduced production of carotenoids does not occur in the absence of mevalonate, when mevalonate is added to the medium the production of carotenoids increases, depending on the concentration of mevalonate, and

---

[20] M. Tada, M. Shiroishi, K. Hasegawa, T. Suzuki, and K. Iwai, *Plant Cell Physiol.* **23**, 607 (1982).

[21] M. Tada, *Plant Cell Physiol.* **30**, 1193 (1989).

[22] M. Tada, M. Tsubouchi, K. Matsuo, H. Takimoto, Y. Kimura, and S. Takagi, *Plant Cell Physiol.* **31**, 319 (1990).

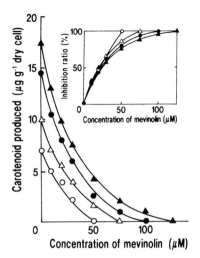

FIG. 5. Effect of mevinolin on the photoinduced production of carotenoids. Cells grown in the dark were illuminated with a light intensity of 2.4×10^4 erg cm^{-2} sec^{-1} at $-1.5°$ for various periods of time: 5 (○), 10 (△), 30 (●), and 60 min (▲). Immediately after illumination, the cells were incubated at 26° for 15 hr in the dark in growth media that contained various concentrations of mevinolin, and the amounts of carotenoids produced during the dark incubation were determined. The inset shows the relationship between the concentration of mevinolin and the inhibition ratio.

reaches the plateau level at 3 mM. The quantities of carotenoids produced at plateau levels are approximately equal to those produced in the absence of mevinolin. These results seem to imply that an enzyme(s) regulated by light is involved in the process of carotenoid formation beyond mevalonate.

To determine whether the photoregulation of the enzyme postulated above involves activation of a preexistent enzyme protein or induction of synthesis of the enzyme *de novo,* the effect of cycloheximide on the photoinduced production of carotenoids has been examined. Illuminated cells are subjected to dark incubation in growth medium that contains both 100 μM mevinolin and 3 mM mevalonate. Addition of cycloheximide immediately after the beginning of incubation causes a complete inhibition of carotenoid production. However, the extent of inhibition becomes smaller as the timing of the addition of cycloheximide is delayed, and inhibition is practically nonexistent when cycloheximide is added 5 hr after the beginning of the dark incubation. These results reveal that the postulated enzyme is synthesized *de novo* as a result of photoinduction, and that the synthesis of the enzyme is complete within about 5 hr. These phenomena resemble to a great extent the photoinductive synthesis of HMG-CoA reductase. It is suggested, therefore, that the synthesis of the postulated enzyme is regulated by a mechanism similar to that involved in regulation of the synthesis of HMG-CoA reductase.

It appears from these results that one or more photoinducible enzymes, such as HMG-CoA reductase, may be present in the carotenogenic pathway beyond mevalonate. However, identification and characterization of the photoinducible enzyme(s) have yet to be achieved.

Action Spectrum for Photoinduced Carotenogenesis

Action spectroscopy is a useful, nondestructive method for the characterization of a photoreceptor involved in mediation of a photoresponse.[23] The determination of an accurate action spectrum usually provides an indication of the chemical nature of the substance that is acting as a photoreceptor.

Cells of *R. minuta* grown in the dark are suspended in sterilized water at a concentration of 0.5 mg/ml. The cell suspension is placed in an optical quartz cuvette with a 1-cm light path after being bubbled with nitrogen gas for 30 min to expel dissolved oxygen. The cuvette is capped with a Teflon plug, to prevent access by air. Monochromatic light is obtained with an Okazaki Large Spectrograph.[24] The rate of incidence of photons is adjusted by setting up an appropriate neutral filter in front of the cuvette and is measured at the surface of the cuvette. In this experiment, the term fluence (photon cm^{-2}) refers to the average incident photon density to which individual cells are exposed in the cuvette, and the defined fluence is calculated by considering the decrease in photon density caused by absorption and scattering during passage across the 1-cm light path.[25] Cells exposed to monochromatic light are immediately collected by suspending in fresh medium. They are then incubated in the dark at 26° for 14 hr to allow the induced carotenogenesis to proceed.

The source of error in the determination of fluence–response curves is the variability in the sensitivity of cells to light stimuli.[26] It has been previously confirmed that the sensitivity of *R. minuta* cells to light does not change when cells are incubated for more than 6 hr under anaerobic

[23] M. Tada, M. Watanabe, and Y. Tada, *Plant Cell Physiol.* **31**, 241 (1990).
[24] M. Watanabe, M. Furuya, Y. Miyoshi, Y. Inoue, I. Iwahashi, and K. Matsumoto, *Photochem. Photobiol.* **36**, 491 (1982).
[25] J. Jagger, *in* "The Science of Photobiology" (K. C. Smith, ed.), p. 1. Plenum/Rosetta, New York, 1977.
[26] W. Shropshire, Jr., *in* "Phytochrome" (K. Mitrakos and W. Shropshire, Jr., eds.), p. 161. Academic Press, London, 1972.

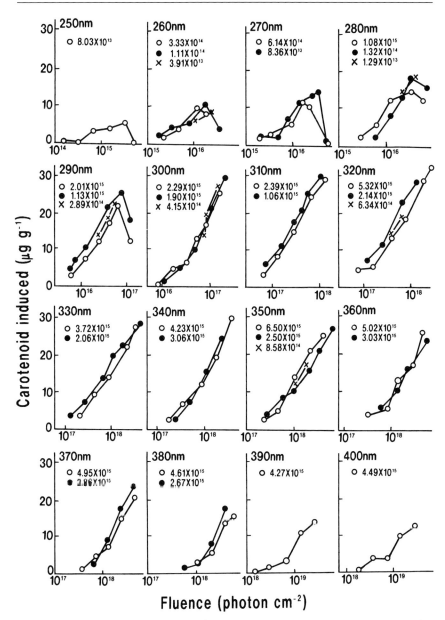

FIG. 6. Fluence–response curves at various wavelengths over the range 250 to 400 nm for the induction of carotenogenesis. The numbers below the wavelength represent the fluence rate (photons cm^{-2} sec^{-1}) used. Over the range from 260 to 380 nm, more than two fluence–response curves were measured to confirm reciprocity at different fluence rates.

conditions at 20°. Thus, irradiation is carried out under anaerobic conditions and at 20°. The confirmation of reciprocity is also indispensable for an accurate action spectrum.[26]

Fluence–response curves for 16 varieties of monochromatic light at wavelengths from 250 to 400 nm are shown in Fig. 6. At all wavelengths tested, the quantity of carotenoid induced increases linearly with increasing fluence, although irradiation of higher fluence, at wavelengths below 290 nm, inhibits the biosynthesis of carotenoids. It is apparent that the slopes of the fluence–response curves at all wavelengths tested, with the exception of 250 nm, are almost identical. At wavelengths in the range of 260 to 380 nm, various fluence rates (photons cm^{-2} sec^{-1}) are used to determine fluence–response curves, and these results are shown together in Fig. 6. The result that, even with different fluence rates, the relationship between the fluence and response holds constant at all wavelengths tested indicates that the Roscoe–Bunsen reciprocity law is valid at each wavelength.

The number of photons required to induce a carotenoid production of 10 μg/g is determined for each wavelength from the fluence–response curves. In the case of illumination at 250 nm and at wavelengths in the range from 410 to 450 nm, values are estimated by extrapolation of straight lines on the graphs. The reciprocal of the number of photons determined in this manner is designated as the relative quantum effective-

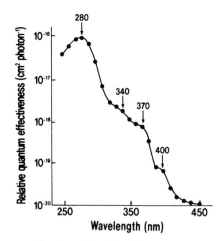

FIG. 7. Action spectrum for photoinduction of carotenogenesis in *Rhodotorula minuta*. The relative quantum effectiveness on the logarithmic vertical axis is expressed in terms of the reciprocal of the fluence required for photoinduction of production of carotenoids to a level of 10 μg/g.

ness. Thus, an action spectrum is obtained by plotting the relative quantum effectiveness as a function of wavelength over the range from 250 to 450 nm. As shown in Fig. 7, the action spectrum for photoinduction of carotenogenesis in *R. minuta* displays an extremely large peak at 280 nm and shoulders at 340, 370, and 400 nm. The relative quantum effectiveness at these wavelengths is 8.5×10^{-17}, 1.4×10^{-18}, 7.2×10^{-19}, and 6.3×10^{-20}, respectively. The fact that the slopes of the fluence–response curves at all wavelengths from 260 to 400 nm are almost identical suggests that the one peak and three shoulders in the action spectrum are generated by the same molecule. It is possible that the large peak in the UV region is due to protein, and that the three shoulders in the near-UV region are attributable to the prosthetic moiety of a complex of protein and pigment. It can be presumed, therefore, that a new type of chromoprotein plays a major role as a photoreceptor for photoinduced carotenogenesis in *Rhodotorula minuta*.

[25] Photoinduction of Carotenoid Biosynthesis

By JAVIER ÁVALOS, EDUARDO R. BEJARANO, and E. CERDÁ-OLMEDO

Introduction

The biosynthesis of carotenoids is almost universally regulated by light.[1,2] A partial explanation is provided by the role of carotenoids in photosynthesis. The regulation of carotenogenesis is an aspect of the regulation of photosynthesis in plants, algae, and many bacteria and will be shunned in this chapter, which is restricted to nonphotosynthetic bacteria and fungi. The photoprotective role of the carotenes could be a more universal basis for the stimulation of carotenogenesis by light, but does not explain the partial inhibition of carotenogenesis by light in some microorganisms, for example, the yeast *Phaffia rhodozima*[3] and the mold *Blakeslea trispora*.[4]

---

[1] W. Rau, in "Biosynthesis of Isoprenoid Compounds" (J. W. Porter and S. L. Spurgeon, eds.), p. 123. Wiley, New York, 1983.
[2] W. Rau and E. L. Schrott, in "The Blue Light Responses: Phenomena and Occurrence in Plants and Microorganisms" (H. Senger, ed.), Vol. 1, p. 43, CRC Press, Boca Raton, Florida, 1987.
[3] G.-H. An and E. A. Johnson, *Antonie van Leeuwenhoek* **57**, 191 (1990).
[4] R. P. Sutter, *J. Gen. Microbiol.* **64**, 215 (1970).

Light is a source of information for many organisms that do not use it as a source of energy. Carotenogenesis is one of the aspects of life that are modulated by light in these organisms (Table I).[5-24] In some cases no carotenoids are made in the dark; most often small amounts of carotenoids are made in the dark, but much more in the light. Light governs other diverse phenomena, such as circadian rhythms, tropisms, taxis, morphogenesis, and biochemical activities.

Response

The variable to be measured in photocarotenogenesis is the rate of carotenoid biosynthesis following illumination. If carotenoids are assumed to be stable, one can measure accumulation instead of synthesis. The validity of this assumption can be tested by a pulse–chase labeling experiment.[25]

Simple and reliable carotenoid analyses are essential for progress in this field. Accurate analyses involve extraction, chromatographic separation, identification, and quantification of individual carotenoids.[26-28] Concen-

---

[5] A. Martínez-Laborda, J. M. Balsalobre, M. Fontes, and F. J. Murillo, *Mol. Gen. Genet.* **223**, 205 (1990).
[6] R. P. Burchard and S. B. Hendricks, *J. Bacteriol,* **97**, 1165 (1969).
[7] P. P. Batra, R. M. Gleason, and J. Jenkins, *Biochim. Biophys. Acta* **177**, 124 (1969).
[8] P. P. Batra and H. C. Rilling, *Arch. Biochem. Biophys.* **107**, 485 (1964).
[9] F. Kato, Y. Koyama, S. Muto, and S. Yamagishi, *Chem. Pharm. Bull.* **29**, 1674 (1981).
[10] C. D. Howes and P. P. Batra, *Arch. Biochem. Biophys.* **137**, 175 (1970).
[11] H. C. Rilling, *Biochim. Biophys. Acta* **79**, 464 (1964).
[12] O. B. Weeks, F. K. Saleh, M. Wirahadikusumah, and R. A. Berry, *Pure Appl. Chem.* **35**, 63 (1973).
[13] M. Tada and M. Shiroishi, *Plant Cell Physiol.* **23**, 541 (1982).
[14] M. Tada, M. Watanabe, and Y. Tada, *Plant Cell Physiol.* **31**, 241 (1990).
[15] K. Bergman, A. P. Eslava, and E. Cerdá-Olmedo, *Mol. Gen. Genet.* **123**, 1 (1973).
[16] E. R. Bejarano, J. Avalos, E. D. Lipson, and E. Cerdá-Olmedo, *Planta* **183**, 1 (1991).
[17] M. El-Jack, A. Mackenzie, and P. M. Bramley, *Planta* **174**, 59 (1988).
[18] M. Zalokar, *Arch. Biochem. Biophys.* **56**, 318 (1955).
[19] E. C. De Fabo, R. W. Harding, and W. Shropshire, *Plant Physiol.* **57**, 440 (1976).
[20] W. Rau, *Planta* **72**, 14 (1967).
[21] J. Avalos and E. Cerdá-Olmedo, *Curr. Genet.* **11**, 505 (1987).
[22] J. Avalos and E. L. Schrott, *FEMS Lett.* **66**, 295 (1990).
[23] L. R. G. Valadon and S. Mummery, *Microbios* **4**, 227 (1971).
[24] K. C. Hsiao and L. O. Björn, *Physiol. Plant,* **54**, 235 (1982).
[25] F. J. Murillo, S. Torres-Martínez, C. M. G. Aragón, and E. Cerdá-Olmedo, *Eur. J. Biochem.* **119**, 511 (1981).
[26] G. Britton, this series, Vol. 111, p. 113.

TABLE I
PHOTOINDUCTION OF CAROTENOIDS IN BACTERIA AND FUNGI

Organism	Main carotenoid	Refs.[a]
Bacteria		
Myxococcus xanthus	Myxobactone	5, **6**
Mycobacterium marinum	β-Carotene	7, **8**
Mycobacterium smegmatis	Complex mixture	9
Mycobacterium sp.	—	10, 11
Flavobacterium dehydrogenans	Complex mixture	12
Yeast		
Rhodotorula minuta	Torulene	13, **14**
Fungi		
Phycomyces blakesleeanus	β-Carotene	15, **16**
Aspergillus giganteus	β-Carotene	17
Neurospora crassa	Neurosporaxanthin	**18, 19**
Fusarium aquaeductuum	Neurosporaxanthin	20
Gibberella fujikuroi (*Fusarium moniliforme*)	Neurosporaxanthin	21, 22
Verticillium agaricinum	Torulene	23, 24

[a] References with action spectra are shown in boldface type.

trations are usually referred to the volume or the dry weight of the sample. These methods are unnecessarily time-consuming in studies of photoinduction that require many determinations.

Chromatography is superfluous for organisms with simple carotenoid compositions: *Phycomyces blakesleeanus* and *Aspergillus giganteus* produce nearly pure β-carotene on solid media. In other cases a computer may replace chromatography: multicomponent analysis of the absorption spectrum of an extract[29] yields individual quantitative determinations of the carotenoids present in the extract. Precise absorption spectra of the pure carotenoids present in the mixture are needed.

Appropriate conditions must be allowed for the response. Carotenoid biosynthesis depends on such environmental factors as temperature, pH, and composition of the growth medium. As a surprising example,[30] photocarotenogenesis in *Neurospora crassa* is particularly efficient when irradiation and response occur at 6°.

---

[27] M. Ruddat and O. H. Will, this series, Vol. 111, p. 189.
[28] This series, Vol. 213, Section II.
[29] P. Jochum and E. L. Schrott, *Anal. Chim. Acta* **157**, 211 (1984).
[30] R. W. Harding, *Plant Physiol.* **54**, 142 (1974).

Biomass Absorbance as Estimate of Carotenoid Content

Mycelial absorbance provided a convenient estimate of carotenoid content in the study of photocarotenogenesis in *Neurospora*[18,19] and *Phycomyces*.[16,31] The procedure to measure β-carotene in *Phycomyces* mycelia will be described; it can be readily modified to measure other carotenoids. Mycelial pads are detached from the agar and mounted upside down on glass. Light passed through a heat-absorbing filter and an interference filter (452 nm, near the β-carotene absorption maximum, or 651 nm, not absorbed by β-carotene) is delivered to the mycelium through an optical fiber. The optical density of the mycelium is defined as the logarithm of the ratio of the picoammeter readings when the mycelium is either left out or placed in the optical path. The difference D between the optical densities at 452 and 651 nm is obtained for each mycelium. The photoinduced β-carotene absorbance is the difference between the D values of light-exposed and unexposed cultures.

This procedure is justified[16] by chemical analyses of β-carotene content and by the observation that in many carotene-free mutant mycelia OD_{452} equals OD_{651} plus a constant (more exactly, $OD_{452} = 0.987\ OD_{651} + 0.209$, correlation coefficient 0.990).

Stimulus

Culture and Illumination Conditions

Knowledge of the life-style of the organism is required to find out optimal, or at least suitable, conditions to elicit the response. Cultures to be illuminated should be homogeneous and thin enough to keep shading at an acceptable level. Shaken liquid cultures with low titers of unicellular microorganisms are close to these ideals. Agar cultures are preferable in the case of fungi that flocculate in broth; the thickness of the mycelial pad can be controlled by modifying the nutritional strength of the medium.

Stimulation and response may occur under normal growing conditions or after transfer of the cells to a phosphate–glucose solution (see Fig. 1). Oxygen is a critical factor: its presence during illumination may be partially[32,33] or absolutely[34,35] required, depending on the organism; illumination in oxygen may be more efficient than illumination in air.[36] Overheat-

---

[31] M. Jayaram, D. Presti, and M. Delbrück, *Exp. Mycol.* **3**, 42 (1979).
[32] W. Rau, *Planta* **59**, 123 (1962).
[33] W. Rau, *Planta* **84**, 30 (1969).
[34] C. D. Howes, P. P. Batra, and C. F. Blakeley, *Biochim. Biophys. Acta* **189**, 298 (1969).
[35] M. Tada and M. Shiroishi, *Plant Cell Physiol.* **23**, 549 (1982).

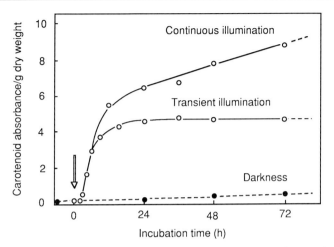

FIG. 1. Time course of carotenoid accumulation in *Fusarium aquaeductuum* in the dark, after transient illumination, and under continuous illumination. Cells grown to exponential phase in submerged culture in the dark were suspended in KH_2PO_4 (16.5 mM) and glucose (2%, w/v) and kept for 2 hr in the dark before illumination. The arrow indicates the period of the light pulse and the beginning of continuous illumination. Aeration was maintained throughout. (Redrawn from Ref. 1.)

ing is a potential risk at high irradiances. Temperature should be kept under control during illumination (e.g., by means of a thermoregulated bath).

Photoinduction in microorganisms cannot be assumed to be constant in space and time. Competence to respond to light depends on the phase of the life cycle and the age of the culture. For example, *Phycomyces* mycelia show a maximal competence at the end of the growth phase, before the onset of sporangiophore formation.[16] As a rule, aging decreases photoinduction.

Light Sources

Light for photoinduction of carotenogenesis is generated and measured as in other photobiological experiments.[37-39] Different light sources differ in emission spectra, intensity, stability, and lifetime.

[36] R. W. Harding, P. C. Huang, and H. K. Mitchell, *Arch. Biochem. Biophys.* **129**, 696 (1969).
[37] K. M. Hartmann, in "Biophysics" (W. Hoppe, W. Lohmann, H. Markl, and H. Ziegler, eds.) p. 115. Springer-Verlag, Berlin, 1983 (German edition, 1982).
[38] M. G. Holmes, in "Techniques in Photomorphogenesis" (H. Smith and M. G. Holmes, eds.), pp. 43 and 81. Academic Press, Orlando, Florida, 1984.
[39] P. Galland and E. D. Lipson, in *"Phycomyces"* (E. Cerdá-Olmedo and E. D. Lipson, eds.), p. 375. Cold Spring Harbor Laboratory, Cold Spring Harbor, New York, 1987.

Incandescent (tungsten) lamps and fluorescent tubes are inexpensive and durable. Halogen lamps do not last as long but are more powerful and stable; slide projectors equipped with halogen lamps are a good choice to obtain high irradiances over small surfaces. For higher irradiances, xenon or mercury arc lamps may be used.

For monochromatic irradiation, interference filters of the desired maximal wavelength (Schott, Mainz, Germany, or Duryea, PA; Balzers, Liechtenstein) are combined with a bright light source. Monochromatic light from spectrophotometers is too dim for most uses, but more powerful diffraction-grating monochromators are available.

Infrared radiation overheats the samples, particularly when using incandescent lamps and very powerful lamps of other kinds. It should be excluded by using an appropriate filter (e.g., KG-1, Schott); for blue light illumination a copper sulfate solution (1%, w/v) is convenient. If the response is limited to blue light, manipulations in the dark may be carried out under red safelight.

There is considerable confusion in the use of physical terms in photobiology. Intensity should be restricted to the energy output of a source per unit time; the term, however, is often used in a casual way for the more correct terms of irradiance or energy fluence rate, the light energy per unit surface and unit time ($W\ m^{-2}$). Exposure or fluence ($J\ m^{-2}$) is the product of irradiance and time. The corresponding quantum expressions are photon irradiance (mol $m^{-2}\ sec^{-1}$) and photon exposure (mol m^{-2}).

Relation between Stimulus and Response

A simple way to study photocarotenogenesis is to compare the concentrations of carotenoids present in cultures grown continuously in the light and in the dark. A stimulus–response curve is obtained by repeating the experiments under various irradiances. Transient illuminations are more informative and should be preferred to continuous illuminations. Transient illuminations are measured in terms of exposure; different exposures are obtained by varying the irradiance, the duration of the pulse, or both. One must establish first the validity of the Bunsen–Roscoe law of reciprocity: the results must depend on exposure and not on the particular values of irradiance and time used in each experiment.

Variations in irradiance may be achieved through variations in the output of the light source or its distance to the culture, and by intercalation of neutral density filters or beam splitters between the source and the culture. Variable transformers allow variations in voltage, and thus in the output of incandescent and other kinds of lamps; attention must be given to the modification of light quality (emission spectrum). Cultures may be placed at different distances from a point light source in a long, dark room to obtain irradiances that vary inversely with the square of the distance. Neutral density filters of defined optical density may be purchased (Schott). Gray plastic sheets (Plexiglas G, Röhm, Darmstadt, Germany, or Röhm and Haas, Philadelphia, PA) are less uniform in transmission but are available in large sizes. Beam splitters reflect part of the light and transmit the rest. In the "threshold box"[15] a collimated light beam goes through a set of beam splitters to illuminate many samples with different irradiances.

Time Course of Carotenoid Photoinduction

The time course of carotenoid accumulation depends on whether the stimulus is transient or continuous (see Fig. 1). The short lag between stimulus and response reflects the time needed for signal transduction, synthesis or activation of enzymes, and biosynthesis of colored carotenoids. Transient illumination results in a transient activation of carotenoid synthesis; under continuous illuminations an elevated accumulation rate can be achieved for a long time.

A second light pulse may result in a second response. In *Fusarium* the second response reproduces the time course of the first.[40] In *Neurospora* a light pulse causes a temporary insensitivity to a second light pulse.[41]

Stimulus–Response Curves

Each stimulus (a scalar value) corresponds to a time course of the response (a function). To obtain a stimulus–response curve, the response is measured as a scalar value taken from the time course. Following the transient illumination in Fig. 1, one could take the slope (rate of carotenoid synthesis) or the final carotenoid accumulation. A common alternative is to take the carotenoid accumulated after a certain time.

The threshold is an outstanding parameter of the stimulus–response relationship. Defined as the smallest irradiance or exposure that gives a significant response, the value of the threshold depends on the ability to discriminate response from noise. Alternatively, an explicit algebraic function may be fitted to the stimulus–response relationship, and the threshold defined and calculated from this function (see Fig. 2).

The stimulus–response relationship may have several components,

---

[40] W. Rau and A. Rau-Hund, *Planta* **136,** 49 (1977).
[41] E. L. Schrott, *Planta* **151,** 371 (1981).

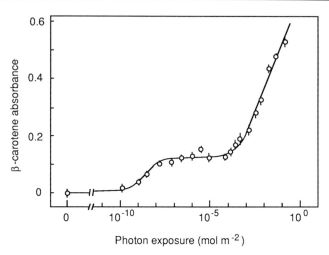

FIG. 2. Stimulus–response curve for photocarotenogenesis in *Phycomyces blakesleeanus*. Photoinduced β-carotene absorbance was measured in mycelia grown for 72 hr in the dark, illuminated for 11 hr with various irradiances of monochromatic radiation (wavelength 469 nm), and incubated for a further 12 hr in the dark. The data (triplicate determinations in three independent experiments) were fitted by computer to the empirical function $A = ax/(x + b) + c \log(x/d) + [c^2 - \log(x/d)^2 + k^2]^{1/2}$, where A is the photoinduced β-carotene absorbance and x is the exposure. The first term, $ax/(x + b)$, is a sigmoidal function when A is plotted in terms of the logarithm of the exposure. The other terms represent a hyperbola when A is plotted as a function of the logarithm of the exposure. The computer program estimates the parameters a, b, c, d, and k, and their standard errors. For the sigmoidal component the threshold can be defined as b/e^2, the intersection of the horizontal axis with the tangent of the curve at the inflection point. For the hyperbolic component, the threshold can be defined as d, the intersection of the horizontal and slant asymptotes. (Redrawn from Ref. 16.)

each with its own threshold (Fig. 2). Photocarotenogenesis is often a two-step response,[31,42] geared to very different exposure intervals.

Theoretically there are no upper limits to the size of the stimulus, but practical limits are set by the maximal output of the light sources and the induction of cellular damage. A saturation of the response at high exposures has been observed in some cases (*Mycobacterium*,[10] *Neurospora*[18]) but not in others at the same or higher exposures (*Fusarium*,[20] *Rhodotorula*,[35] *Phycomyces*[16]).

The response to high exposures may be underestimated if the carotenoids are significantly destroyed by light. It is not easy to correct for this

---

[42] E. L. Schrott, *Planta* **150**, 174 (1980).

effect, because the extent of photolysis varies with the carotenoid and its molecular environment. The problem does not arise when the inducing wavelengths are not absorbed by the carotenoids, for example, in mutants that accumulate colorless carotenes, such as phytoene.

Action Spectra

Action spectra represent the wavelength dependence of photoresponses. Concordance of the action spectrum of a photoresponse with the absorption spectrum of a pigment present in the cell unveils the pigment as a candidate photoreceptor for the response. For an excellent description of action spectroscopy and its applications, see Ref. 37.

For each wavelength there is a corresponding stimulus–response curve. To obtain an action spectrum, the effectiveness of each wavelength must be defined as a value extracted from the stimulus–response curve. Common choices are the reciprocal of the threshold, the reciprocal ($1/b$, in Fig. 2) of the exposure at the inflection point, or the reciprocal of the exposure that produces an arbitrary predetermined response. In the case of Fig. 2, each of the two components of the response has its own action spectrum.

Action spectra have been determined for photocarotenogenesis in many bacteria and fungi (Table I). In most cases, the main peaks lie between 350 and 500 nm, as befits a blue light-mediated response.

Chemical Modification of Photocarotenogenesis

Many chemicals induce carotenogenesis in the dark. This is the case, for example, of antimycin A in *Mycobacterium*,[7] retinol, trisporic acids, and many aromatics in *Phycomyces*,[43,44] and *p*-hydroxymercuribenzoate in *Fusarium*.[45] It is difficult to establish the relation of activation by a chemical with photoinduction. If their mechanisms of action are independent, when the response is saturated by one of them, the other should have no additional effect. Photoinduction seems to be independent of activation by *p*-hydroxymercuribenzoate in *Fusarium*,[45] by antimycin A in *Mycobacterium*,[7] and by trisporic acids in *Phycomyces*.[46]

---

[43] E. Cerdá-Olmedo, in *"Phycomyces"* (E. Cerdá-Olmedo and E. D. Lipson, eds.), p. 199. Cold Spring Harbor Laboratory, Cold Spring Harbor, New York, 1987.
[44] E. R. Bejarano, F. Parra, F. J. Murillo, and E. Cerdá-Olmedo, *Arch. Microbiol.* **150**, 209 (1988).
[45] R. R. Theimer and W. Rau, *Planta* **106**, 331 (1972).
[46] N. S. Govind and E. Cerdá-Olmedo, *J. Gen. Microbiol.* **132**, 2775 (1986).

Incorporation of photoreceptor analogs with altered absorption spectra could lead to changes in the action spectrum. Thus, the addition of roseoflavin, an analog of riboflavin, to *Phycomyces* cultures modified the action spectrum for phototropism,[47] an indication that riboflavin is a photoreceptor *in vivo*.

Photosensitizing dyes, such as methylene blue and toluidine blue, act as artificial photoreceptors in *Fusarium aquaeductuum*[48] and *F. oxysporum*[49]; red light induces carotenogenesis in the presence of those dyes but not in their absence. This suggests photooxidation as the mechanism of action of the natural photoreceptor. Application of reducing agents points to the same conclusion: dithionite and hydroxylamine inhibit photoinduction in *F. aquaeductuum*.[50]

Inhibitors of protein synthesis may indicate whether photoinduction involves *de novo* synthesis of enzymes. A positive answer is suggested by the use of chloramphenicol in bacteria[7,11] and cycloheximide in fungi.[17,30,51,52] Similar results are obtained with other protein inhibitors, such as blasticidin S, and amino acid analogs, such as DL-*p*-fluorophenylalanine or 4-methyl-DL-tryptophan.[30] Additional information is given by the use of inhibitors of RNA synthesis, such as actinomycin D, which may indicate whether light affects transcription or translation.

Competitive enzyme inhibitors may provide information about the effect of light on a specific reaction. Mevinolin, a competitive inhibitor of 3-hydroxy-3-methylglutaryl-CoA reductase, decreases photoinduction in *Rhodotorula*[53] in such a way that light is believed to increase the amount of enzyme, but not its activity.

Genetic Alteration of Photoinduction

Mutants with altered photoinduction are isolated and studied with methods that are not very different from those used for mutants of other phenotypes. For the methods related to the carotene mutants of *Phycomyces*, see Ref. 54.

[47] M. K. Otto, M. Jayaram, R. H. Hamilton, and M. Delbrück, *Proc. Natl. Acad. Sci.* **78**, 266 (1981).
[48] J. Lang-Feulner and W. Rau, *Photochem. Photobiol.* **21**, 179 (1975).
[49] N. V. Guzhova, O. Y. Varik, L. B. Rubin, and G. Y. Fraikin, *Mikol. Fitopatol. (U.S.S.R.)* **11**, 467 (1977).
[50] R. R. Theimer and W. Rau, *Planta* **92**, 129 (1970).
[51] R. W. Harding and H. K. Mitchell, *Arch. Biochem. Biophys.* **128**, 814 (1968).
[52] W. Rau, I. Lindemann, and A. Rau-Hund, *Planta* **80**, 309 (1968).
[53] M. Tada, M. Tsubouchi, K. Matsuo, H. Takimoto, Y. Kimura, and S. Takagi, *Plant Cell Physiol.* **31**, 319 (1990).
[54] E. Cerdá-Olmedo, this series, Vol. 110, p. 220.

The *pic* mutants of *Phycomyces* are specifically deficient in photoinduction.[55] Constitutive photoinduction is difficult to distinguish from changes in other regulations of carotenogenesis in mutants that make as much carotenoids in the dark as in the light. Such mutants have been isolated in *Fusarium*,[56] *Gibberella*,[21] and *Phycomyces*.[16]

Other mutants exhibit defects in photoinduction and in other forms of regulation. Such is the case with some *mad* mutants of *Phycomyces*[55,57] and the *wc* mutants of *Neurospora*.[58] These mutants affect common steps in the detection and transduction of signals for different responses.

A refined genetic analysis of carotenoid photoinduction has been carried out in *Myxococcus*.[5,59-61] Two DNA regions contain structural genes for carotenogenesis. Insertion of a reporter *lac* gene into a structural gene[59] showed that expression is induced through a light-driven mechanism in which the loci *carA* and *carR* are involved. These are defined by constitutive mutants that make carotenoids both in the dark and in the light. Gene *carR*, unlinked to the structural genes, codes for a general negative regulator, probably a transmembrane protein that could be a photoreceptor or a signal transducer; *carA* is a *cis*-acting element necessary for the transcription of the contiguous structural genes.

Biochemistry

Photoinduction of carotenogenesis does not occur *in vitro*, but extracts from light-grown cells are often more carotenogenic than those from dark-grown cells.[62-64] In the case of *A. giganteus*, which makes no carotenoids in the dark, cell-free extracts from dark-grown mycelia are not carotenogenic at all.[65] Early steps in the pathway have been shown to be photoinduced in

---

[55] I. López-Díaz and F. Cerdá-Olmedo, *Planta* **150**, 134 (1980).
[56] R. R. Theimer and W. Rau, *Biochim. Biophys. Acta* **177**, 180 (1969).
[57] L. M. Corrochano and E. Cerdá-Olmedo, *J. Photochem. Photobiol.* **B6**, 325 (1990).
[58] F. Degli Innocenti and V. E. A. Russo, in "Blue Light Effects in Biological Systems" (H. Senger, ed.), p. 213. Springer-Verlag, Berlin, 1984.
[59] J. M. Balsalobre, R. M. Ruiz-Vázquez, and F. J. Murillo, *Proc. Natl. Acad. Sci. U.S.A.* **84**, 2359 (1987).
[60] A. Martínez-Laborda, M. Elías, R. Ruiz-Vázquez, and F. J. Murillo, *Mol. Gen. Genet.* **205**, 107 (1986).
[61] A. Martínez-Laborda and F. J. Murillo, *Genetics* **122**, 481 (1989).
[62] J. Avalos, D. Nelki, A. Mackenzie, and P. M. Bramley, *Biochim. Biophys. Acta* **966**, 257 (1988).
[63] U. Mitzka-Schnabel and W. Rau, *Phytochemistry* **20**, 63 (1981).
[64] L. M. Salgado, J. Avalos, E. R. Bejarano, and E. Cerdá-Olmedo, *Phytochemistry* **30**, 2587 (1991).
[65] M. El-Jack, P. M. Bramley, and A. Mackenzie, *Phytochemistry* **26**, 2525 (1987).

several organisms.[66-68] These observations suggest that light induces gene expression, but they do not rule out a possible effect of light on enzyme activity.

Illumination induces the synthesis of new mRNA in *Fusarium*[69] and activates transcription of mRNA for at least five new polypeptides in *Neurospora*.[70] Direct evidence for photoinduced transcription comes from the cloning of structural genes for carotenogenesis in *Neurospora*. Light causes a 15-fold increase in the transcription of *al-3*, probably responsible for geranylgeranyl pyrophosphate synthase,[71] and a 70-fold increase in the transcription of *al-1*, encoding phytoene dehydrogenase.[72]

Complementation of available mutants after transformation with gene libraries should lead to the cloning and characterization of other genes responsible for photoinduction and other forms of regulation of carotenogenesis.

[66] J. H. Johnson, B. C. Reed, and H. C. Rilling, *J. Biol. Chem.* **249**, 402 (1974).
[67] M. Tada and M. Shiroishi, *Plant Cell Physiol.* **23**, 615 (1982).
[68] R. W. Harding and R. V. Turner, *Plant Physiol.* **68**, 745 (1981).
[69] E. L. Schrott and W. Rau, *Planta* **136**, 45 (1977).
[70] J. A. A. Chambers, K. Hinkelammert, and V. E. A. Russo, *EMBO J.* **4**, 3649 (1985).
[71] M. A. Nelson, G. Morelli, A. Carattoli, N. Romano, and G. Macino, *Mol. Cell. Biol.* **9**, 1271 (1989).
[72] T. J. Schmidhauser, F. R. Lauter, V. E. A. Russo, and C. Yanofsky, *Mol. Cell. Biol.* **10**, 5064 (1990).

Section II

Molecular and Cell Biology

A. Genetics
Articles 26 through 28

B. Biosynthesis
Articles 29 through 38

[26] Evolutionary Conservation and Structural Similarities of Carotenoid Biosynthesis Gene Products from Photosynthetic and Nonphotosynthetic Organisms

By GREGORY A. ARMSTRONG, BHUPINDER S. HUNDLE, and JOHN E. HEARST

Introduction

Carotenoids have attracted considerable interest as a major class of natural pigments with important roles in photosynthesis, nutrition, and photooxidative protection.[1] The biochemical purification and characterization of carotenoid biosynthesis enzymes have proved difficult tasks, thus hindering direct structural studies of these proteins.[2] Recent developments in the molecular description of carotenoid biosynthesis genes from photosynthetic bacteria *(Rhodobacter capsulatus)*, nonphotosynthetic bacteria (*Erwinia herbicola* and *Erwinia uredovora*), and fungi *(Neurospora crassa)*, however, offer powerful indirect approaches for predicting structural properties of the gene products. The nucleotide sequencing of carotenoid biosynthesis genes from *R. capsulatus (crtA, crtB, crtC, crtD, crtE, crtF, crtI, crtK)*,[3] *E. herbicola*[4,5] and *E. uredovora (crtB, crtE, crtI, crtX, crtY, crtZ)*,[6] and *N. crassa (al-1)*[7] provides a wealth of information to be analyzed. In the absence of the purification to homogeneity of the corresponding carotenoid biosynthesis enzymes, predictions made on the basis of deduced amino acid sequences serve as a starting point for the characterization of these proteins.

In this chapter we describe methods useful for the comparison and analysis of the deduced amino acid sequences of carotenoid biosynthesis enzymes from these organisms. We also discuss the significance of sequence similarities between these proteins and other gene products from

---

[1] G. Britton, "The Biochemistry of Natural Pigments." Cambridge Univ. Press, Cambridge, 1983.
[2] P. D. Fraser and P. M. Bramley, this volume [33].
[3] G. A. Armstrong, M. Alberti, F. Leach, and J. E. Hearst, *Mol. Gen. Genet.* **216,** 254 (1989).
[4] G. A. Armstrong, M. Alberti, and J. E. Hearst, *Proc. Natl. Acad. Sci. U.S.A.* **87,** 9975 (1990).
[5] B. S. Hundle, M. Alberti, V. Nievelstein, P. Beyer, G. A. Armstrong, D. Burke, H. Kleinig, and J. E. Hearst, unpublished data (1991).
[6] N. Misawa, M. Nakagawa, K. Kobayashi, S. Yamano, Y. Izawa, K. Nakamura, and K. Harashima, *J. Bacteriol.* **172,** 6704 (1990).
[7] T. J. Schmidhauser, F. R. Lauter, V. E. A. Russo, and C. Yanofsky, *Mol. Cell. Biol.* **10,** 5064 (1990).

computer databases in terms of the biosynthetic reactions catalyzed and cofactor requirements.

The hydrocarbon backbone of C_{40} carotenoids arises through a series of prenyl transfer reactions.[8] Three molecules of the C_5 building block isopentenyl pyrophosphate (IPP) are added successively to allylic substrates in $1'-4$ condensations, starting with dimethylallyl pyrophosphate (DMAPP; C_5), the allylic isomer of IPP. These consecutive reactions yield the products geranyl pyrophosphate (GPP; C_{10}), farnesyl pyrophosphate (FPP; C_{15}), and geranylgeranyl pyrophosphate (GGPP; C_{20}), which are precursors for many branches of isoprenoid metabolism. In the first reactions unique to carotenoid biosynthesis, two molecules of GGPP undergo a $1'-2-3$ condensation yielding prephytoene pyrophosphate (PPPP; C_{40}), followed by the loss of pyrophosphate and insertion of a double bond to yield phytoene. These early reactions leading to the formation and dehydrogenation of the first C_{40} carotenoid phytoene are common to most carotenogenic organisms, although later biosynthetic reactions diverge.[9]

The specific biosynthetic pathways operating in *R. capsulatus*,[10] in *Escherichia coli* expressing cloned *Erwinia crt* genes,[6,11] and in *N. crassa*[12] have been summarized. Three genetic loci necessary for the synthesis of phytoene from GGPP via PPPP *(crtB, crtE)*[10] and phytoene and dehydrogenation *(crtI)*[13] have been identified in *R. capsulatus*. Cognate genes have been identified in *Erwinia* by comparisons of deduced sequence similarities between the *R. capsulatus* and *Erwinia* gene products and by mutational analyses.[4-6] The *N. crassa crtI* homolog, *al-1*, has been identified by functional complementation and by the predicted sequence similarity of Al-1 to *R. capsulatus* CrtI and CrtD.[7,12] *In vitro* precursor accumulation in cell extracts from *R. capsulatus* mutants implicates CrtB in the condensation of two molecules of GGPP to PPPP and CrtE in the conversion of PPPP to phytoene.[10] CrtI/Al-1 probably perform multiple dehydrogenations, converting phytoene to neurosporene in *R. capsulatus* (or further to lycopene and 3,4-dehydrolycopene in *N. crassa*), while CrtD mediates the specialized dehydrogenation of methoxyneurosporene to spheroidene observed in some photosynthetic bacteria.[12] CrtY directs the two-step con-

---

[8] H. Kleinig, *Annu. Rev. Plant Physiol. Plant Mol. Biol.* **40**, 39 (1989).
[9] T. W. Goodwin, this volume [29].
[10] G. A. Armstrong, A. Schmidt, G. Sandmann, and J. E. Hearst, *J. Biol. Chem.* **265**, 8329 (1990).
[11] B. S. Hundle, P. Beyer, H. Kleinig, G. Englert, and J. E. Hearst, *Photochem. Photobiol.* **54**, 89 (1991).
[12] G. E. Bartley, T. J. Schmidhauser, C. Yanofsky, and P. A. Scolnik, *J. Biol. Chem.* **265**, 16020 (1990).
[13] G. Giuliano, D. Pollock, H. Stapp, and P. A. Scolnik, *Mol. Gen. Genet.* **213**, 78 (1988).

version of the acyclic carotenoid lycopene to bicyclic β-carotene in *Erwinia*.[5,6]

Methods and Applications

Procedures for Alignment of Deduced Amino Acid Sequences

Many of the sequence comparisons described here were recognized during searches of the NBRF and SWISS-PROT protein sequence databases using the FASTA amino acid sequence alignment program provided in the Genetics Computer Group (GCG) software package (version 6.2; June, 1990).[14] FASTA, described in detail elsewhere,[15] compares a test sequence against a databank of other sequences, ultimately yielding an alignment *(opt)* score in which deletions and insertions of amino acids have been considered to yield an optimal alignment. The pairwise FASTA amino acid sequence alignments were refined by maximizing the normalized alignment scores (NAS).[16] To determine the NAS, the number of identities between the two sequences are multiplied by 10 (20 in the case of cysteines). The number of gaps multiplied by 25 is subtracted from this sum. This result is divided by the average length of the two sequences and multiplied by 100, yielding the NAS. The maximal NAS method offers the advantage that sequence alignments of related proteins can be readily optimized without the use of a computer. Although the results obtained by the FASTA and NAS alignments are similar, FASTA tends to overlook regions of limited identity at the ends of aligned sequences and also yields different results in regions of low identity containing gaps. NAS refinements of the FASTA sequence alignments yielded the results shown in Figs. 1, 2, and 4. Percent identities for these pairwise comparisons, calculated using the shorter of the two sequences, are given in Table I.

Sequence Alignment of CrtB, CrtE, and Other Related Protein Sequences

At the time of the nucleotide sequence determination of the *R. capsulatus crtB* and *crtE* genes, no similarity was observed between the deduced protein sequences and other proteins described in the literature.[3] Sequence similarities were subsequently found between *R. capsulatus* CrtB and deduced protein sequences from the *E. herbicola*[4] and *E. uredovora*[6] *crt* genes by inspection. In addition, *R. capsulatus* and *E. herbicola* CrtB were shown, using the FASTA program, to resemble the protein product of a

---
[14] J. Devereux, P. Haeberli, and O. Smithies, *Nucleic Acids Res.* **12**, 387 (1984).
[15] W. R. Pearson, this series, Vol. 183, p. 63.
[16] R. F. Doolittle, this series, Vol. 183, p. 99.

TABLE I
ANALYSIS OF AMINO ACID SEQUENCE RELATIONSHIPS

Sequence comparison[a]	Length of sequence[b]	Number of identities[c]	Number of gaps[c]	Percent identity[b,c] (%)	FASTA opt score[d]
T-pTOM5 / Eu-CrtB	296	75	4	25.3	330
T-pTOM5 / Eh-CrtB	309				318
T-pTOM5 / Rc-CrtB	339				298
Eu-CrtB / Eh-CrtB	296	191	—	64.5	1061
Eu-CrtB / Rc-CrtB	296				433
Eh-CrtB / Rc-CrtB	309	101	5	32.7	409
Ec/IspA / Eu-CrtE	299	100	6	33.4	335
Ec-IspA / Eh-CrtE	299				311
Ec-IspA / Rc-CrtE	289				303
Ec-IspA / Bs-GerC3	299				230
Ec-IspA / Cp-CrtE	299				197
Ec-IspA / Ec-ORFX	131				125
Ec-IspA / Y-HPS	299				101
Eu-CrtE / Eh-CrtE	302	160	—	53.0	837
Eu-CrtE / Rc-CrtE	289				275
Eu-CrtE / Bs-GerC3	302				191
Eu-CrtE / Cp-CrtE	302				214
Eu-CrtE / Ec-ORFX	131				61
Ec-IspA / Y-HPS	299				118
Eh-CrtE / Rc-CrtE	289	88	10	30.4	283
Eh-CrtE / Bs-GerC3	307				178
Eh-CrtE / Cp-CrtE	307				173
Eh-CrtE / EC-ORFX	131				60
Eh-CrtE / Y-HPS	307				135
Rc-CrtE / Bs-GerC3	289	74	6	25.6	202
Rc-CrtE / Cp-CrtE	289				191
Rc-CrtE / Ec-ORFX	131				78
Rc-CrtE / Y-HPS	289				114
Bs-GerC3 / Cp-CrtE	323	106	2	32.8	520
Bs-GerC3 / Ec-ORFX	131				196
Bs-GerC3 / Y-HPS	348				259
Cp-CrtE / Ec-ORFX	131	43	1	32.8	193
Cp-CrtE / Y-HPS	323				354
Rc-CrtD / Eu-CrtI	492	128	9	26.0	397
Rc-CrtD / Eh-CrtI	492				440
Rc-CrtD / Rc-CrtI	494				426
Rc-CrtD / Nc-Al-1	494				298
Eu-CrtI / Eh-CrtI	492	375	—	76.2	2087
Eu-CrtI / Rc-CrtI	492				1069
Eu-CrtI / Nc-Al-1	492				708

TABLE I (continued)
ANALYSIS OF AMINO ACID SEQUENCE RELATIONSHIPS

Sequence comparison[a]	Length of sequence[b]	Number of identities[c]	Number of gaps[c]	Percent identity[b,c] (%)	FASTA opt score[d]
Eh-CrtI / Rc-CrtI	492	205	4	41.7	1091
Eh-CrtI / Nc-Al-1	492				591
Rc-CrtI / Nc-Al-1	524	167	8	31.9	682

[a] Abbreviations used here and in Fig. 1–5: T, tomato *(Lycopersicon esculentum)*; Eu, *E. uredovora*; Eh, *E. herbicola*; Rc, *R. capsulatus*; Ec, *E. coli*; Bs, *B. subtilis*; Cp, *C. paradoxa*; Y, yeast *(Saccharomyces cerevisiae)*; Nc, *N. crassa*.
[b] Using the shorter of the two compared sequences.
[c] Alignments shown in Fig. 1–5.
[d] Magnitude of the FASTA *opt* score increases with the length of the compared sequences at a fixed value of sequence identity.

tomato cDNA (pTOM5) (Fig. 1),[4] differentially expressed during fruit ripening.[17] Table I lists the FASTA *opt* alignment scores determined for pairwise comparisons of CrtB/pTOM5 sequences and shows that the identity between the pTOM5 protein and *E. uredovora* CrtE is 25.3%. FASTA alignment values between the pTOM5 protein and all three CrtB proteins are similar. Because the degree of identity in these comparisons lies close to the 25% rule-of-thumb threshold for probable homology between two sequences,[16] it is instructive to observe that many of the identities with pTOM5 occur in regions most highly conserved in CrtB proteins from photosynthetic and nonphotosynthetic bacteria. Arguments for a relationship between pTOM5 and carotenoid biosynthesis, such as lycopene accumulation in the ripening tomatoes in which pTOM5 is differentially expressed and the similarity of the molecular weights of pTOM5 and the purified bifunctional phytoene synthase from red pepper chromoplasts,[18] have been detailed elsewhere.[4] The identity between CrtB proteins from photosynthetic and nonphotosynthetic bacteria is 32.7% (Table I), indicating homology between the prokaryotic sequences.

In the case of CrtE, in addition to published sequence comparisons between the *R. capsulatus* and *Erwinia* proteins,[4,6] several groups have identified related protein sequences in other organisms. A gene product

[17] J. Ray, C. Bird, M. Maunders, D. Grierson, and W. Schuch, *Nucleic Acids Res.* **15**, 10587 (1987).
[18] B. Camara, this volume [32].

```
T-pTOM5   MSVALLMVVSPCDVSNGTSFMESVREGNRFFDSSRHRNLVSNERINRGGGKQTNNGRKFSVRSAILATPSGERTMTSEQMVYDVVLRQAALVKRQLRSTNELEVKPDIPGNLGLLSEAYDRCGEVCAE   130
Eu-CrtB                                                                                                                                MAV     3
Eh-CrtB                                                                                                                       MSQPPLLDHATQTMAN    16
                                                                                                                                   **
Rc-CrtB                                                                                                                       MIAEADMEVCRELIRT    16
                                                                                                                                   **

T-pTOM5   YAKTFNLGTMLMTPERRRAIWAIYVMCRRTDELVGPNASY-ITPAALDRMENRLEDV------FNGRPFDMLDGALSDTVSNFPVDIQPFR---DMIEGMRMDLRKSRYKNFDELYLYCYVVAGTVGLMS   251
                                        *             *                                           *     *    *       *   *  * **
Eu-CrtB   GSKSFATASKIFDAKTRRSVLMLYAWCRHCDDVIDDQTIGFQARQPALQTPEQRLMQLEMKTRQAYAGSQMHEPAFAAFQEVAMAHDIAPAYAFDHLEGFAMDVREAQYSQLDDTLRYCYHVAGVVGLMM   133
          ***********
Eh-CrtB   GSKSFATAAKLFDPATRRSVLMLYTWCRHCDDVIDDQTHGFASEAAAEEATQRLARLRTLTLAAFEGAEMQDDPAFAAFQEVALTHGITPRMALDHLDGFAMDVAQTRYVTFEDTLRYCYHVAGVVGLMM   146
          ***********
Rc-CrtB   GSYSFHAASRVLPARVRDPALALYAFCRVADDEVDEVGAPR-DKAAAVLKLGDRLEDI-----YAGRPRNAPSDRAFAAVVEEFEMPRELIPEAL---LEGFAWDAEGRWYHTLSDVQAYSARVAAAVGAMM   138

                              II
T-pTOM5   VPIMGIAPESKATTESVYNAALALGIANQLTNILRDVGEDARRGRVYLPQDELAQAGLSDEDIFAGRVTDK-WRIFMKKQIHRRARKFFDEAEKGVTELSSASRFPVWASLVLVYRKILDEIEANDYNFTK   380
           * ** *** ****  *** * * * ** ** ** *****    *   **** * ***      *    * *   **
Eu-CrtB   AQIMGVRD------NATLDRACDLGLAFQLTNIARDIVDDAHGRCYLPASWLEHEGLNKENYAAPENRQA-LSRIARRLVQEAEPYYLSATAGLAGLPLRSAWAIATAKQVYRKIGVKVEQGGQQAWDQ   256
          * ** *          * *** **** ** ** * ***  *  *      *          * *****   * *    ***  * **   **   * *****  ** ********* ***   ** *
Eh-CrtB   ARVMGVRD------ERVLDRACDLGLAFQLTNIARDIIDDAAIDRCYLPAEWLQDAGITPENYAARENRAA-LARVAERIIDAAEPYYISSQAGLHDLPPRCAWAIATARSVYREIGIKVKAAGGSAWDR   269
            * *                 *** *** ** **       *       *      * ** *    *    *             **    *  * *  **** *      ** **      **
Rc-CrtB   CVLMRVRN------PDALARACDLGLAMQMSNIARDVGEDARAGRLFLPTDWMVEEGIDPQAFLADPQPTKGIRRVTERILNRADRLYWRAATGVRLPFDCRPGIMAAGKIYAAIGAEVAKAKYDNITR   262

T-pTOM5   RAYVSKSKQVD----CITYCICKISCASYKNASLQR                                                                                              412
Eu-CrtB   RQSTTTPEKIT----LLLAASGQALTSRMRAHPPRPAHLWQRPL                                                                                      296
Eh-CrtB   RQHTSKGEKIA----MLMAAPGQVIRAKTTRVTPRPAGLWQRPV                                                                                      309
          *
Rc-CrtB   RAHTTKGRKLWLVANSAMSATATSMLPLSPRVHAKPEPEVAHLVDAAAHRNLHPERSEVLLISALMALKARDRGLAMD                                                    339
```

FIG. 1. Sequence alignments of CrtB proteins and a related tomato gene product. Dashes indicate gaps, identical residues are shown by asterisks, and amino acid positions are numbered at the right. A conserved domain (overlined) is described in the text and in Fig. 3. Database accession numbers for nucleotide sequences (G, Genbank; E, EMBL) and protein sequences (S, SWISS-PROT) are as follows: Eu-CrtB, D90087(G); Eh-CrtB, M38423(G); Rc-CrtB, X52291(E); T-pTOM5, Pto5$Lyces(S).

encoded in the *Cyanophora paradoxa* cyanelle genome has been observed to resemble *R. capsulatus* CrtE and has been proposed to play an analogous role in carotenoid biosynthesis.[19] A *Bacillus subtilis* gene of unknown function *(gerC3)*[20] and a partial open reading frame (ORF) from *E. coli* (ORF X)[21] were also found to exhibit sequence similarity to *R. capsulatus* CrtE,[20] as was the *E. coli ispA* gene encoding FPP synthase (FPS).[22] While preparing this chapter we also observed that yeast hexaprenyl pyrophosphate synthase (HPS)[23] displayed significant sequence similarity to CrtE and to the related proteins. These findings, summarized in the FASTA *opt* alignment scores (Table I) and in the refined sequence alignments (Fig. 2), are intriguing as (1) *B. subtilis* and *E. coli* do not normally synthesize carotenoids and (2) *E. coli* IspA, yeast HPS, and CrtE perform different isoprenoid biosynthetic reactions. For the alignments shown in Fig. 2, apart from the highly conserved *Erwinia* CrtE proteins (53% identity), the pairwise comparisons yield overall sequence identities of 25.6 to 32.8% (Table I), indicating probable homology.[16]

CrtB and CrtE Share Domains Found in Prenyltransferases

Sequences conserved between CrtE (the putative phytoene synthase) and related proteins could form domains required for the binding of related but nonidentical isoprenoid pyrophosphate substrates. The first evidence to support this theory comes from the recent observation that yeast HPS and eukaryotic FPS enzymes contain three highly conserved domains (I, II, III).[23] Domains I and II include the sequence (I,L,V)XDDXXD given in single-letter amino acid code where X can be any amino acid. These authors also observed that domain II occurs in the yeast IPP transferase, encoded by the *MOD5* gene.

A survey of CrtE and related proteins, using the FIND program in the GCG software package to search for core DDXXD sequences, revealed the conservation of domains I and II (Fig. 3). For *E. coli* IspA and the three CrtE sequences, it was necessary to allow insertion of additional amino acids in the new consensus to obtain optimal alignment in domain I (Fig. 3A; 8 of 25 residues are absolutely conserved). For domain II, residues at 6 of 13 positions are strongly conserved (Fig. 3B). Furthermore, inspection

---

[19] C. B. Michalowski, W. Löffelhardt, and H. J. Bohnert, *J. Biol. Chem.* **266**, 11866 (1991).
[20] D. J. Henner, EMBL database secession number M80245 (1992).
[21] Y.-L. Choi, T. Nishida, M. Kawamukai, R. Utsumi, H. Sakai, and T. Komano, *J. Bacteriol.* **171**, 5222 (1989).
[22] S. Fujisaki, H. Hara, Y. Nishimura, K. Horiuchi, and T. Nishino, *J. Biochem. (Tokyo)* **108**, 995 (1990).
[23] M. N. Ashby and P. A. Edwards, *J. Biol. Chem.* **265**, 13157 (1990).

```
Ec-IspA                    MDFPQQLEACVKQANQALSRFIAPLPFQNTPVVETMQY--GALLGGKRLRPFLVATGHMF---G--VSTNLDAPAA-AVECIHAYSLIHDDL      86
Eu-CrtM                    MTVCAKKHVHLTRDAAEQLLADIDRRLDQI------LPVEGERDVGAAMRE--GALAPGKRIRPMLLLTARDL--GCAVSHDGLLDLAC-AVEMVHAASLILDDM    96
Eh-CrtM                    MVSGSKAGVSPHREIEVMRQSIDDHLAGL------LPETDSQDIVSLAMRE--GVMAPGKRIRPLIMLLAARDL--RYQGSMPTLLDLAC-AVELTHTASLMLDDM    95
Rc-CrtM                    MSLDKRIESA------LVKAL------SPEALGESPPLLAALPYGVFPGGARIRPTI--LVSVAL--ACGDDCPAVTDAAAVALEIMHCASLVHDDL           82
Bs-GerC3   MLNIIRLLAESLPRISDGNENTDVWNDMKFKMAYSFLNDDID------VIERE---LEQTVRSDYPLLSEAGLHLLQAGGKRIRPVF--VLLSG--MFGDYDINKIKYVAVTLEMIHMASLVHDDV   114
Cp-CrtM                    MASITNILAPVENELD------LLTKN---LKKLVGSGHPILSAASEHLFSASGKRPRPAI--VLLISKATMENEIITSRHRRLAEITEIIHTASLVHDDI    90

                                                                                                                                           I
Ec-IspA    PAMDDDDLRRGLPTCHVKFGEANAILAGDALQTLAFSILSDADMPEVSDRDRISMISELASASGIAGMCGGALDLDAEGKHVPLDALERIHRHKTGALIRAAVRLGALSAGDKGRRALPVLDKYAESIG   216
Eu-CrtM    PCMDDAKLRRGRPTIHSHYGEHVAILAAVALLSKAFGVIADADGLTPLAKNR--AVSELSNAIGMQGLVQGQFKDLSEGDKPRSAEALMTNHFKTSTLFCASMQMASIVANA--SSEARDCLHRFSLDLG   223
Eh-CrtM    PCMDNAELRRGQPTTHKKFGESVAILASVGLLSKAFGLIAATGDLPGERRAQ--AVNELSTAVGVQGLVLVGQFPRDLNDAALDRTPDAILSTNHLKTGILFSAMLQIVAIASAS--SPSTRETHAFALDFG   222
Rc-CrtM    PAFDNADIRRGKPSLHKAYNEPLAVLAGDSLLIRGFEVLADVGAVNPDRALK--LISKLGQLSGARGGICAGGAWESESKVD-----LAAYHQAKTGALFIAATQMGAIAAGY----EAEPWFDLGMRIG   201
Bs-GerC3   --IDDAELRRGKPTIKAKMDNRIAMYTGDYMLAGSLEMMTRINEPKAHRILS--QTIEVCL-GEIEQIKDKYNMEQNLRTY-----LRRIKR-KTALLIAVSCQLGAIASGA-DEKIHKALYWFGYYVG   232
Cp-CrtM    --LDESDVRRGIPTVHSDFGTKIAILAGDFLFAQSSWYLANLESLEVVKLIS--KVITDFAE-GEIRRGLNQFKVDLITEEY----LEKSFY-KTASILAASSKAAALLSHV-DLTVANDLYNYGRHLG   208
Ec-ORFX                                                                                                             IPDLQDYGRYLG         12

Ec-IspA    LAFQVGQDDILDVVGDTATIGKRQGADQQLG---KSTYPALLGLE-----------QARKKARDLIDDARQSLKQLAEQSLDTSALEALADYIIQRNK                                299
Eu-CrtM    QAFQLLDDLIDGMTDT---GKDSN--QDAG--KSTLVNLLGPR------------AVEERLRQHLQLASEHLSAACQHGHATQHFIQAWFDKKLAAVS                                302
Eh-CrtM    QAFQLIDDLIDRDHPET---GKDRN--KDAG--KSTLVNRLGAD------------AARQKLREHIDSADKHLTFACPQGGAIRQFMHLWFGHHLADWSPVMKIA                          307
Rc-CrtM    SAFQIADDLKDALMSAEAMGKPAG--QDIANERPNAVKTMGIE------------GARKHLQDVLAGAI--ASIPSCP-GEAKLAQMVQLYAHKIMDIPASAERG                          289
Bs-GerC3   MSYQIIDDILDFTSTEEELGKPVG---GDLLQGNVTLPVLYALK---NPALKNQLKLINSETTQEQLEPII-EEIKKTD-AIEASMAVSEMYLQKAFQKINTLPRGRARSSLAAIAKYIGKRKF       348
Cp-CrtM    LAFQIVDDILDFTSSTEELGKPSC---SDLKKGNLITAPVLFALE---QNSELIPLIQRQFSEPKDEYTL-QIVEETK-AIEKTRELAMEHAQVAIQCLENLPFSSSKEALKLITKYVLERLY       323
Ec-ORFX    TAFQLIDDLLDYNADGEQLGKNVG---DDLNEGKPTLPLLHAMHHGTPEQAQMIRTAIEQGNGRHLLEPVL-EAMNACG--SLEWTRQREAEEADKAIAALQVLPDTPWREALIGLAHIAVQRDR       131
```

FIG. 2. Sequence alignments of CrtE proteins and related gene products. Protein sequences are presented as in Fig. 1. Conserved domains (overlined) are described in the text and in Fig. 3. Database accession numbers for nucleotide sequences (G, GenBank; E, EMBL) and protein sequences (N, NBRF) are as follows: Eu-CrtE, D90087(G); Eh-CrtE, M38424(G); Ec-IspA, D00694(G); Cp-CrtE, M61174(G); Ec-ORFX, PV0010(N); Rc-CrtE, X52291(E). Ec-ORFX was reported as a partial sequence and is numbered here from the first amino acid shown.[21] The sequence of the *gerC3* gene encoding Bs-GerC3 has been submitted to GenBank.[20]

A

Domain I

```
                    s  D D       D
Ec-IspA      73  V E C I H A Y S L I H D D L P A M D D D D L R R G   97
Eu-CrtE      83  V E M V H A A S L I L D D M P C M D D A K L R R G  107
Eh-CrtE      82  V E L T H T A S L M L D D M P C M D N A E L R R G  106
Rc-CrtE      69  L E L M H C A S L V H D D L P A F D N A D I R R G   93
Bs-GerC3    101  L E M I H M A S L V H D D V - - I D D A E L R R G  123
Cp-CrtE      77  T E I I H T A S L V H D D I - - L D E S D V R R G   99
Y-HPS       182  V E M I H T A S L L H D D V - - I D H S D T R R G  204
R-FPS        92  V E L L Q A F F L V L D D I - - M D S S Y T R R G  114
H-FPS        92  V E L L Q A F F L V A D D I - - M D S S L T R R G  114
Y-FPS        89  I E L L Q A Y F L V A D D M - - M D K S I T R R G  111
Consensus        E         L     D D       D             R R G
```

B

Domain II

```
                    s  D D     D
Ec-IspA     215  I G L A F Q V Q D D I L D  227
Eu-CrtE     222  L G Q A F Q L L D D L T D  234
Eh-CrtE     221  F G Q A F Q L L D D L R D  233
Rc-CrtE     200  I G S A F Q I A D D L K D  212
Bs-GerC3    231  V G M S Y Q I I D D I L D  243
Cp-CrtE     207  L G L A F Q I V D D I L D  219
Ec-ORFX      11  L G T A F Q L I D D L L D   23
Y-HPS       356  L G I C F Q L V D D M L D  368
R-FPS       235  M G E F F Q I Q D D Y L D  247
H-FPS       235  M G E F F Q I Q D D Y L D  247
Y-FPS       232  L G E Y F Q I Q D D Y L D  244
Y-MOD5      210  P E P L F Q R L D D R V D  222
Consensus           G     F Q     D D     D

T-pTOM5     275  L G I A N Q L T N R L R D  287
Eu-CrtB     151  L G L A F Q L T N I A R D  163
Eh-CrtB     164  L G L A F Q L T N I A R D  176
Rc-CrtB     156  L G L A M Q M S N I A R D  168
```

FIG. 3. (A, B) CrtE, CrtB, and related proteins contain conserved domains defined in prenyltransferases. Farnesyl pyrophosphate synthase (FPS), hexaprenyl pyrophosphate synthase (HPS), and isopentenyl pyrophosphate transferase (MOD5) protein sequences not presented in Figs. 1 and 2 are from yeast (Y), rat (R), and human (H). The sequence used to define domains I and II (shown above in **A** and **B**; s, small or hydrophobic residue = I, L, V) was identified by aligning the FPS and HPS proteins.[23] Database accession numbers for nucleotide sequences (G, GenBank) and protein sequences (N, NBRF) are as follows: Y-HPS, J05547(G); R-FPS, A27772(N), A34713(N), and B34713(N); H-FPS, A33415(N) and A35726(N); Y-FPS, A34441(N); Y-MOD5, A26717(N). The positions of each sequence within the respective proteins are indicated at the left and right, with dashes indicating gaps. The new consensus shown indicates residues conserved in all 10 sequences **(A)** or at least 11 of 12 sequences **(B)**. Matches with the new consensus are marked in bold type. A region similar to domain II was also found in the CrtB and pTOM5 proteins **(B)**.

of the CrtB (the putative PPPP synthase) and tomato pTOM5 proteins revealed domain II-like sequences lacking two of the aspartate residues from the new consensus. None of the CrtE or related proteins (Fig. 2) demonstrate significant overall sequence similarity to the CrtB proteins (data not shown). Our results thus redefine the (I,L,V)XDDXXD consensus originally defined for domains I and II in eukaryotic FPS and yeast HPS enzymes,[23] to EXXXXXXLXXDDX$_{2-4}$DXXXXRRG for domain I and GXXFQXXDDXXD for domain II, and suggest the occurrence of a looser domain II consensus in the CrtB and pTOM5 proteins (GXXXQXXXXXXD).

How can these data be rationalized in terms of the diverse biosynthetic reactions catalyzed by these enzymes? CrtB, as the putative PPPP synthase, catalyzes the 1'-2-3 condensation of two molecules of GGPP, while CrtE, as the putative phytoene synthase, binds PPPP and converts it to phytoene. The FPS and yeast HPS enzymes mediate the 1'-4 condensations of allylic isoprenoid pyrophosphates with IPP, while yeast MOD5 donates allylic DMAPP to a tRNA substrate. It has been proposed that the aspartates of domains I and II form salt bridges between the FPS and HPS proteins and magnesium salts of their pyrophosphate substrates.[23] Such a hypothesis would be consistent with the participation of domains I and II in CrtE and domain II-like sequences in CrtB in isoprenoid pyrophosphate binding.

The conservation of domains I and II in CrtE, in prenyltransferases (*E. coli* IspA, FPS, yeast HPS, MOD5), and in proteins of unknown function from noncarotenogenic prokaryotes (*E. coli* ORFX and *B. subtilis* GerC3) suggests functions for the last group in biosynthetic reactions involving at least one isoprenoid pyrophosphate substrate (e.g., quinone biosynthesis, dolichol biosynthesis, tRNA modification[8,24]). A thorough examination of the FASTA *opt* scores and sequences for CrtE and related proteins (Table I, Fig. 2) reveals the rough division of these sequences into two groups. The CrtE proteins and *E. coli* IspA form one cluster, while *B. subtilis* GerC3, *E. coli* ORFX, *C. paradoxa* CrtE, and yeast HPS form a second cluster. This grouping raises the possibility that *C. paradoxa* CrtE, proposed to be the cyanelle CrtE homolog, may rather be involved in a noncarotenogenic isoprenoid biosynthetic reaction. The eukaryotic FPS enzymes demonstrate no significant similarity outside of domains I and II to the proteins aligned in Fig. 3 (data not shown) and are hence not included in Table I.

The unique nature of domains I and II was demonstrated by searching for the consensus sequences shown in Fig. 3 among all proteins in the NBRF database (release 42.0, 3/90) using the FIND program. For domain I, three searches were performed to allow for the observed spacing varia-

[24] M. M. Sherman, L. A. Petersen, and C. D. Poulter, *J. Bacteriol.* **171**, 3619 (1989).

tions within the aspartate cluster. In a total of 27,711 sequences no exact matches were found excluding the proteins already described. Therefore, these motifs define fingerprints unique to proteins which bind isoprenoid pyrophosphate substrates. These examples clearly illustrate how a judicious comparison of protein sequences, always within the framework of their biological roles, can provide a starting point for suggesting enzymatic function and structural features.

Sequence Alignment of Carotenoid Dehydrogenases (CrtI, CrtD, Al-1)

For the carotenoid dehydrogenases, the determination of the deduced protein sequences for *R. capsulatus* CrtI and *R. capsulatus* CrtD revealed that these proteins share two highly conserved domains, one N-terminal, the other C-terminal.[3] These domains were recognized by comparing the two sequences using a homology matrix program provided in a sequence analysis software package.[25] This search was motivated by the reasoning that both proteins had been proposed to participate in analogous dehydrogenations of phytoene for *R. capsulatus* CrtI and methoxyneurosporene for *R. capsulatus* CrtD. The predicted amino acid sequence of *R. capsulatus* CrtI was subsequently revised to begin at an upstream ATG codon, adding 33 new N-terminal amino acids, when it was realized that new alignments of *R. capsulatus* CrtI with *N. crassa* Al-1,[12] with *R. capsulatus* CrtD,[10] or with *E. herbicola* CrtI[4] allowed a substantial increase in sequence similarity. Carotenoid dehydrogenase function for *N. crassa* Al-1 and *R. capsulatus* CrtI is detailed elsewhere in this volume.[26] CrtI homologs in *E. herbicola*[4,5] and *E. uredovora*[6] were later identified by inspection of sequence similarities and by mutational analyses.

FASTA sequence alignments between the carotenoid dehydrogenases are detailed in Fig. 4 and Table I. The alignments confirm the N- and C-terminal regions of sequence similarity originally noted between *R. capsulatus* CrtI and *R. capsulatus* CrtD.[3] Table I indicates that *R. capsulatus* CrtI, *E. herbicola* CrtI, and *E. uredovora* CrtI are closely related. *Neurospora crassa* Al-1 is, as the only eukaryotic carotenoid dehydrogenase, more distantly related. *Rhodobacter capsulatus* CrtD, as the only dehydrogenase not using phytoene as its primary substrate, is also distantly related to both the prokaryotic CrtI subgroup and to *N. crassa* Al-1. That the carotenoid dehydrogenases constitute an evolutionarily conserved class of proteins is revealed by the sequence identities ranging from 26.0 to 76.2% (Table I). The carotenoid dehydrogenases also show a higher degree of conservation than either the CrtB or CrtE proteins [compare identities

[25] J. Pustell and F. Kafatos, *Nucleic Acids Res.* **10**, 51 (1982).
[26] G. E. Bartley, A. Kumle, P. Beyer, and P. A. Scolnik, this volume [34].

βαβ ADP-binding fold

```
Rc-CrtD   MRSETDVVVIGARMGLAAAIGAAAAGLRVTVEAGDAPGGKARAV-PTPGPADTGPTVLTMRHVLDALFAACGTRAE-EHLTLIPRLARHFWPDGSSLDLFTD-TEANIEAIRAFAGDKEAA    123
Eu-CrtI        MKPTTVIGAGFGGLALAIRLQAAGIPVLLLEQRDKPGGRAVVY-EDQGFTFDAGPTVITDPSAIEELFALAGKQLK-EYVELLPVTPFYRLCMESGKVFNYDNDQTRLEAQIQQFNPRDVEG-   120
Eh-CrtI        MKKTVVIGAGFGGLALAIRLQAGIPTVLLEQRDKPGGRAYVW-HDQGFTFDAGPTVITDPTALEALFTLAGRRME-DYVRLLPVKPFYRLCWESGKTLDYANDSAELEAQITQFNPRDVEG-  120
Rc-CrtI   MSKNTEGMGRAVVIGAGLGLGLAAAMRLGAKGYKVTVVDRLDRPGGRSSI-TKGGHRFDLGPTVTVPDRLRELWADCGRDFD-KDVSLVPMEPFYTIDFPDGEKYTAYGDDAKVKAEVARISFGDVEG-   127
Nc-Al-1   MAETQRPRSAIIVGAGAGGIAVAARLAKAGVDVTVLEKNDFTGGRCSLIHTKAGYREFDQGPSLLLLPGLFRETFEDLGTTLEQEDVELLQCFPNYN IWFSDGKRFSPTTDNAMKVEIEKWEGPD--G-  126

Rc-CrtD   AFRRFDHLTGLWEAFHRSVIA--------APKPDLWRIAAATVTRPQLWPALRPGLTMRDLLAHHFKDPRLAQLFGRYATVVGGRPGATPAVLSLIWQAEVQ-GVWAIREGMHGVAAALARVAEKGVRFH    246
Eu-CrtI   -YQRFLDYSRAVFKEGYLKLGT--------VPFLSFERDMLRAAPQLAKLQAWRSVISKVASYI------EDEHLRQAFSFHSLLVGGNPFATSIVTLIHALERENGVWFPRGGTGALVQGMIKLFQDLGGEVV   239
Eh-CrtI   -YRRFLAYSQAVFQEGYLRLGS--------VPFLSFERDMLRAGPQLLKLQAWQSVYQSVSRFI------EDEHLRQAFSFHSLLVGGNPFTTSIYTLIHALERENGVWFPEGGTGALVNGMVKLFTDLGGEIE   239
Rc-CrtI   -FRHFMHDAKARYEFGYENLGR--------KPMSKLNDLIKVLPTEGWLRADRSVGHAKKMV-----KDDHLRFALSFHPLFIGGDPHVTSMYILVSQLEKKFGVHYAIGGVQAIADAMAKVITDQGEMR     246
Nc-Al-1   -FRRYLSWLAEGHQHYETSLRHVLHRNFKSIELLADPRLVVTLMALHPFESIWHRAGRYF----KTDRMQRVFTFATMVMGMSFPDAPATYSLLQYSELAEGIWYPRGGFHKVLDALVKIGERMGVKYR     251

Rc-CrtD   YGAKAKRIVR--------KEGRVTAVEIETGVSIPCGACIFNGDPGALRDGLLG----DAARASMEKSPRPAPSLSAWVWAFGATP---IGVDLAHHNVFFTADP-ELEFGPIGAGEMPEEPTLYICAQDREM   363
Eu-CrtI   LNARVSHMET--------TGNKIEAVHLEDGRRFLTQAVASNADVHTYRDLLSQ-HPAAVKQSNKLQTKRMSNSLFVLYFGLNH------HHDQLAHHTVCFGPRYRELIDEIFNHDGLAEDFSLYLHAPCVTD   359
Eh-CrtI   LNARVEELVV--------ADNRVSQVRLADGRIFDTDAVASNADVVNTYKKLLGH-HPVGQKRAAALERKSMSNSLFVLYFGLNQ----PHSQLAHHTICFGPRYRELIDEIFTGSALADFSLYLHSPCVTD   359
Rc-CrtI   LNTEVDEILV--------SRDGKATGIRLMDGTELPAQVVVSNADGHTYKRLLRN-RDRWRWTDEKIDKKRWSMGLFVWYFGCTKGTAKMWKDVGHHTVVVGPRKEHVQDIEIKGELAEDMSLYIVRPSVTD   370
Nc-Al-1   LNTGVSQVLTDGGKNGKKPKATGVQLENGEVLNADLVVVNADLVYTYNNLLPKEIGGIKKYANKLINNRKASCSSISFYWSLSG---MAKELETHNIFLAELYKESFDAIFERQALPDDDFSFYHVPSRVD   378

Rc-CrtD   QAPVPEIERFEIINMGPAGHQPFP-----------QEEAQCRARTFPMLAAMGLIFSPDPE-TRALITPALLSRRFPGSLGAIVGGSPEGTL-ATFRRPLARTGLKGLYLLAGGGTHPGAGVPMALTSGTHAARA    484
Eu-CrtI   SSLAPEGCGSYYVLAPVPHLGTANL----DMTVEGPKLRDRIFAYLEQHYMPGLRSQLVTHRMEFTPDFRDQLNAYHGSAFSVEPVLTQSAWFRPHNRDKTITNLYLVGAGTHPGAGIPGVGISSAKATAGL    486
Eh-CrtI   PSLAPPGCASFYVLAPVPHLGNAPL----DWAQEGPKLRDRIFDYLEERYMPGLRSQLVTQRIFTPADFHDTLDAHLGSAFSIEPLLTQSAMFRPHNRDSDIANLYLVGAGTHPGAGIPGVVASAKATASL    486
Rc-CrtI   PTAAPKGDDTFYVLSPVPNLGFDNGV--DWSVEAEKYKAKVLKVIEERLLPGVAEKITEEVVFTPETFRDRYLSPLGAGFSLEPRILQSAWFRPHNASEEVDGLYLVGAGTHPGAGVPSVIGSGELVAQ-    497
Nc-Al-1   PSAAPPDRAVIALVPVGHLLQNGQPELDNPTLVSKARAGVLATIQARTGLSLSPLITEEIVNTPYTWETKFNLSKGAILGLAHDFNVLAFRPRRTKAGQMDNAYFVGASTHPGTGVPIVLAGAKITAE-    507

Rc-CrtD   LLADRISAAK    494
Eu-CrtI   MLEDLI        492
Eh-CrtI   MIEDLQ        492
Rc-CrtI   MIPDAPKPETPAAAAPKARTPRAKAAQ    524
Nc-Al-1   QILEETFPKNTKVPWTTNEERNSERMRKEMDEKITEEGIIMRSNSKFPGRGSDAFEGAMEVVNLLSQRAFPLIVALMGVLYFLLFVR    595
```

between *E. herbicola* and *E. uredovora* for CrtI (76.2%), CrtB (64.5%), and CrtE (53.0%)].

FAD/NAD(P)-Binding Domains in Carotenoid Dehydrogenases and Cyclases

Further FASTA searches comparing carotenoid dehydrogenase sequences against other proteins in the databases exposed the conservation of the N-terminal region with a variety of eukaryotic and prokaryotic proteins known to bind FAD and/or NAD(P) cofactors[4] and with a fingerprint defined for ADP-binding folds with a $\beta\alpha\beta$ structure. Similar observations were also made by other researchers.[12] Figure 5 shows the alignments of the N termini of the carotenoid dehydrogenases to each other and to the ADP-binding fingerprint.[27,28] We define a new consensus for carotenoid dehydrogenase putative ADP-binding folds on the basis of these alignments.

The deduced sequences of the *Erwinia* CrtY (lycopene cyclase) proteins, identified by mutational analyses, have recently become available.[5,6] The two proteins are 57.6% identical,[5] including a conserved N-terminal region reminiscent of the dehydrogenase ADP-binding fold (Fig. 5). Many key residues in the dehydrogenase consensus and the fingerprint for ADP-binding folds[27,28] are conserved in the cyclases. Deviations from the fingerprint occur at the third conserved glycine, which is replaced by asparagine, and in the size of the variable turn region between the α helix and the second β sheet. Interestingly, however, a revisitation of the structure of ADP-binding folds for enzymes containing both FAD and NAD(P) binding sites indicates that conservation of the third glycine in the fingerprint is not as strict as originally proposed.[29]

What is the biochemical evidence for FAD/NAD(P) cofactor requirements for carotenoid dehydrogenation and lycopene cyclization? Although the N-terminal sequence alignments (Fig. 5) are suggestive, dehydrogenation and cyclization cofactor requirements have not yet been established in

[27] R. K. Wierenga, P. Terpstra, and W. G. J. Hol, *J. Mol. Biol.* **187**, 101 (1986).
[28] G. Eggink, H. Engel, G. Vriend, P. Terpstra, and B. Witholt, *J. Mol. Biol.* **212**, 135 (1990).
[29] J. H. McKie and K. T. Douglas, *FEBS Lett.* **279**, 5 (1991).

FIG. 4. Sequence alignments of carotenoid dehydrogenases. Protein sequences are labeled as in Fig. 1. The overlined region indicates a highly conserved putative ADP-binding fold for FAD/NAD(P) cofactors (see Fig. 5). Database accession numbers for nucleotide sequences (G, GenBank; E, EMBL) are as follows: Eu-CrtI, D90087(G); Eh-CrtI, M38423(G); Nc-Al-1, M33867(G); Rc-CrtD and Rc-CrtI, X52291(E).

```
              β β β β β β    α α α α α α α α α α α α α α α      β β β β β β
              b s   s G G    G         G            s   s        s   s   a
Eu-CrtI    3  P T V V I G A G F G G L A L A I R L Q A A G - - I P V L L L E Q R D K P G G  38
Eh-CrtI    3  K T V V I G A G F G G L A L A I R L Q A A G - - I P T V L L E Q R D K P G G  38
Rc-CrtI   10  R A V V I G A G L G G L A A A M R L G A K G - - Y K V T V V D R L D R P G G  45
Rc-CrtD    6  D V V V I G A R M G G L A A A I G A A A A G - - L R V T V V E A G D A P G G  41
Nc-Al-1    9  S A I I V G A G A G G I A V A A R L A K A G - - V D V T V L E K N D F T G G  44
Consensus       V V I G A G   G G L A   R L   A A G       V       E   D   P G G

Eu-CrtY    6  D L I L V G A G L A N G L I A L R L Q Q Q Q P D M R I L L I D A A P Q A G G  43
Eh-CrtY    3  D L I L V G G G L A N G L I A W R L R Q R Y P Q L N L L L I E A G E Q P G G  40
```

FIG. 5. Carotenoid dehydrogenases (CrtI, CrtD, Al-1) and cyclases (CrtY) contain putative ADP-binding βαβ folds for FAD or NAD(P) cofactors. Protein sequences are presented as in Fig. 3. Conserved N-terminal domains are compared to a fingerprint for an ADP-binding fold (shown above: G, glycine; b, basic or hydrophilic residue; s, small or hydrophobic residue; a, acidic residue; the glycines and the acidic residue are strictly conserved).[27,28] Arg-13 of Rc-CrtD (underlined) may represent the site of a point mutation as the *crtD223* mutant allele of the *crtD* gene was sequenced.[3,4] The consensus shown indicates residues conserved in at least four of the five dehydrogenases. The gene encoding Eu-CrtY appears in the GenBank database under accession number D90087. The sequence of the *E. herbicola crtY* gene and the full deduced protein sequence of Eh-CrtY will be reported elsewhere.[5]

prokaryotes. The daffodil chromoplast system has been used to study cofactors needed for the eukaryotic dehydrogenation and cyclization reactions, however. This work indicates a requirement for NADPH in the conversion of lycopene to β-carotene, involving an isomerization and two cyclization reactions,[30,31] although no direct requirement for FAD/NAD(P) cofactors was observed for the dehydrogenation.[32] In addition to the ADP-binding fold, the conservation between carotenoid dehydrogenases and FAD-binding disulfide oxidoreductases of residues involved in FAD binding and protein dimerization has been proposed.[12,26] On the other hand, we find that additional residues involved in FAD binding in other dehydrogenases[28] are not obviously conserved in the carotenoid dehydrogenases. Future biochemical experiments to explore cofactor requirements for prokaryotic carotenoid cyclases and dehydrogenases should prove instructive.

While molecular genetics cannot directly answer biochemical questions, the two fields complement each other. Specific predictions made on the basis of deduced structural features in proteins can be tested by biochemists, while biochemical properties of proteins can be dissected genetically on the molecular level. The study of carotenoid biosynthesis enzymes provides a perfect case in point.

[30] P. Beyer and H. Kleinig, this series, Vol. 213 [8].
[31] P. Beyer, U. Kröncke, and V. Nievelstein, *J. Biol. Chem.* **266**, 17072 (1991).
[32] P. Beyer, M. Mayer, and H. Kleinig, *Eur. J. Biochem.* **184**, 141 (1989).

Acknowledgments

We thank Dr. D. J. Henner (Genentech) and Dr. H. J. Bohnert (University of Arizona) for communicating protein sequences and sequence alignments prior to publication. This chapter is based on work supported by a National Science Foundation Graduate Fellowship (G.A.), by a National Institute of Environment and Health postdoctoral training Grant No. 2T32 ES07075-11 (B.H.), by National Institutes of Health Grant GM 30786 (J.H.), and by the Office of Basic Energy Sciences, Biological Energy Division, U.S. Department of Energy, under Contract DE-ACO30-76SF00098 (J.H.).

NOTE ADDED IN PROOF. Recent data indicate that CrtB and CrtE homologs function in the conversion of geranylgeranyl pyrophosphate to phytoene and in the synthesis of geranylgeranyl pyrophosphate, respectively.

[27] Cloning of Carotenoid Biosynthetic Genes from Maize

By BRENT BUCKNER and DONALD S. ROBERTSON

Introduction

Several genes have been cloned from maize by a strategy commonly referred to as transposon tagging.[1-14] For this cloning strategy, a transposon is used to induce a mutation in a gene of interest. The transposon, which should have previously been cloned, can then be used as a hybrid-

---

[1] N. V. Federoff, D. B. Furtek, and O. E. Nelson, Jr., *Proc. Natl. Acad. Sci. U.S.A.* **81**, 3825 (1984).
[2] C. O'Reilly, N. S. Shepherd, A. Pereira, Z. Schwarz-Sommer, I. Bertram, D. S. Robertson, P. A. Peterson, and H. Saedler, *EMBO J.* **41**, 877 (1985).
[3] K. C. Cone, F. A. Burr, and B. Burr, *Proc. Natl. Acad. Sci. U.S.A.* **83**, 9631 (1986).
[4] U. Wienand, U. Weydemann, U. Niesbach-Kloesgen, P. A. Peterson, and H. Saedler, *Mol. Gen. Genet.* **203**, 202 (1986).
[5] M. McLaughlin and V. Walbot, *Genetics* **117**, 771 (1987).
[6] N. Theres, T. Scheele, and P. Starlinger, *Mol. Gen. Genet.* **209**, 193 (1987).
[7] R. J. Schmidt, F. A. Burr, and B. Burr, *Science* **238**, 960 (1987).
[8] M. Motto, M. Maddaloni, G. Ponziani, M. Brembilla, R. Martta, N. Di Fonzo, C. Soave, R. Thompson, and F. Salamini, *Mol. Gen. Genet.* **212**, 488 (1988).
[9] S. L. Dellaporta, I. Greenblatt, J. L. Kermicle, J. B. Hicks, and S. R. Wessler, in "Chromosome Structure and Function: Impact of New Concepts, 18th Stadler Genetics Symposium" (J. P. Gustafson and R. Appels, eds.), p. 263. Plenum, New York, 1988.
[10] D. R. McCarty, C. B. Carson, P. S. Stinard, and D. S. Robertson, *Plant Cell* **1**, 523 (1989).
[11] R. A. Martienssen, A. Barkan, M. Freeling, and W. C. Taylor, *EMBO J.* **8**, 1633 (1989).
[12] S. Hake, E. Vollbrecht, and M. Freeling, *EMBO J.* **8**, 15 (1989).
[13] C. Lechelt, T. Peterson, A. Laird, J. Chen, S. L. Dellaporta, E. Dennis, J. W. Peacock, and P. Starlinger, *Mol. Gen. Genet.* **219**, 225 (1989).
[14] B. Buckner, T. L. Kelson, and D. S. Robertson, *Plant Cell* **2**, 867 (1990).

ization probe to identify and clone a transposon-hybridizing DNA fragment that cosegregates with the mutant phenotype. This DNA fragment is cloned so that it includes some of the DNA that is adjacent to the transposon. This adjacent DNA can be demonstrated to be part of or the entire gene of interest and can then be used as a hybridization probe to clone the standard nonmutant allele of the gene.

Several genes involved in carotenoid biosynthesis have been identified in maize.[15] Mutants of these genes exhibit white to pale-yellow kernel and typically albino seedling phenotypes. Biochemical analysis of these mutants has shown that they accumulate intermediates in the carotenoid biosynthetic pathway.[15,16] Therefore, these mutants are probably deficient in one or more of the enzymes responsible for biosynthesis of carotenoids. Transposon-induced mutant alleles of several carotenoid biosynthetic genes have been produced in maize.[17] In addition, other maize genes have been characterized that have a tissue-specific pattern of carotenoid expression (e.g., white kernel, green seedling) and, therefore, may be genes involved in the regulation of carotenoid biosynthesis. One such gene, the $y1$ gene, has been cloned using a transposon tagging and cloning strategy.[14] In this chapter we discuss the strategy and methods used to isolate carotenoid biosynthetic genes in maize. These strategies and methods can be applied to any plant containing an endogenous or introduced transposon system.

Tagging

The first step in a transposon tagging strategy is to use a transposon to induce a mutation in a gene of interest. If the gene can be propagated as a homozygous recessive nonlethal allele, the following crossing strategy can be employed to produce a transposon-induced mutation of the gene. Plants that are homozygous for a dominant wild-type allele of the gene of interest and carry copies of the transposon are pollinated by plants homozygous recessive for the same gene. The resulting kernels are all expected to be heterozygous and display the wild-type phenotype for the target gene of interest. However, if a transposon inserts in the target gene some progeny will be expected to be mutant. These mutants should be propagated and allele-tested to the gene of interest to confirm that the observed mutant phenotype is caused by a mutation at this gene. By using Robertson's

---

[15] D. S. Robertson, I. C. Anderson, and M. D. Bachmann, in "Maize Breeding and Genetics" (D. B. Walden, ed.), p. 461. Wiley, New York, 1978.

[16] F. Fong, J. D. Smith, and D. E. Koehler, in "Third International Symposium on Pre-Harvest Sprouting in Cereals" (J. E. Krueger and D. E. LaBerge, eds.), p. 188. Westview Press, Boulder, Colorado, 1983.

[17] D. S. Robertson, unpublished results, 1984.

Mutator transposon system as a tag, this strategy has been used to generate several hundred independently induced transposon-tagged alleles of the *y1* locus of maize.[18]

Many maize carotenoid biosynthetic mutants are lethal when the gene is homozygous.[15] Transposon-induced alleles of these genes can also be produced. Yellow-kerneled plants that contain an active transposon system are crossed to another yellow-kerneled line that does not contain that transposon system. The resulting F_1 plants are self-pollinated, and recessive carotenoid mutants are scored in the F_2 progeny as kernel (pale-yellow or white) or seedling mutants (albino). As a control both of the parents of the F_1 plants are self-pollinated to demonstrate that any mutation found in the F_2 generation is not a result of a mutation (perhaps non-transposon-induced) which was already present in a heterozygous condition in one of the parent plants. This cross will only be possible if two viable ears can be produced on each parent plant. The wild-type sibling kernels on the F_2 ear segregating for the mutant are used to propagate the mutant allele since approximately two-thirds of them should carry the transposon-induced mutant allele. Selfing these plants will identify those that carry the mutant allele. These plants are also outcrossed to available carotenoid mutants to test for allelism. Utilizing this method several different carotenoid-deficient mutants have been generated in maize.[17]

Identification of Transposon-Tagged DNA Fragment

Rationale of Method

Once a transposon-induced mutation of a gene has been produced, the cloned transposon can be used as a hybridization probe to determine if a particular transposon-hybridizing DNA restriction fragment can be correlated with the gene of interest. This method requires that the transposon-induced mutation be segregating in a population of plants and that the plants with the mutant allele can be easily identified and separated from the wild-type siblings. If the mutant allele is a recessive nonlethal, a population of sibling plants should be produced by crossing a plant that is heterozygous for a standard dominant allele and the transposon-induced allele to a plant that is homozygous for a standard recessive allele of this gene. The wild-type and mutant progeny of this test cross will segregate in a 1 : 1 ratio. DNA is isolated from the wild-type and mutant sibling plants and is digested with a restriction endonuclease that (1) does not cleave within the transposon presumed to be in the gene and (2) produces ends

[18] D. S. Robertson, *Mol. Gen. Genet.* **200**, 9 (1985).

which are compatible for cloning in bacteriophage λ vectors. The digested DNA is then electrophoresed, blotted to hybridization membranes, and hybridized with radioactively labeled transposon DNA. It is expected that the transposon-mutated gene will cosegregate as a transposon-hybridizing DNA fragment exclusively with the mutant plants (Fig. 1). Two transposon-hybridizing DNA fragments cosegregate with the mutant phenotype in Fig. 1. The shorter fragment was cloned and used to show that the longer fragment resulted from a partial digestion of the DNA due to methylation.[14] Therefore, a single transposon-mutated sequence can result in more than one cosegregating fragment when using methylation-sensitive restriction endonucleases. Since the DNA blot analysis may not resolve all transposon-hybridizing fragments, it may be necessary to perform this analysis with several different restriction endonucleases before a cosegregating fragment is detected.

If the mutant allele conditions a recessive seedling lethal trait, an alternative method to identify a transposon-hybridizing DNA fragment that associates with the mutant phenotype must be used. A large number of plants, all heterozygous for the same transposon-induced allele, can be analyzed by DNA blot analysis. It is expected that very few transposon-

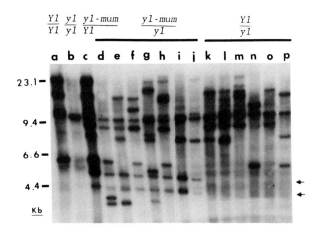

FIG. 1. Identification of transposon-hybridizing restriction endonuclease fragments that cosegregate with the mutant genotype. A plant carrying a transposon-induced mutant allele of the *y1* gene *(y1-mum/Y1)* was crossed to a homozygous *y1* plant. The resulting ear segregated in a 1:1 ratio for white *(y1-mum/y1)* and yellow *(Y1/y1)* kernels. *Bam*HI-digested DNA from sibling plants derived from white and yellow kernels was analyzed by DNA blot hybridization using Robertson's *Mu3* transposon as a hybridization probe. The genotype of the plants was as follows: lane a, *Y1/Y1;* b, *y1/y1;* c, *y1-mum/Y1;* d to j, *y1-mum/y1;* k to p, *Y1/y1.* Arrows indicate the *Mu3*-hybridizing fragments that cosegregate with the *y1-mum* genotype.[14] (Reprinted from Ref. 14 by permission of the American Society of Plant Physiologists.)

hybridizing DNA fragments will be common to all of these plants. These fragments are candidates for being the transposon-tagged gene of interest. This approach has been used successfully by Martienssen et al.[11]

Restriction Endonuclease Digestion

Occasionally we have found that maize and other higher plant DNAs are not digested to completion when treated with restriction endonucleases as recommended by the manufacturer. To obtain complete digests we routinely perform our digestions of 5–10 μg of DNA in a volume of 300 μl of the manufacturer's recommended reaction buffer and 50 units of restriction endonuclease. The restriction endonuclease is added in 25-unit aliquots at time 0 and at 1.5 hr of a 3-hr incubation. The digestion is made 0.25 M in sodium acetate (pH 5.2), and the DNA is precipitated by adding 2.5 volumes of 95% ethanol and incubating at $-20°$ for 0.5–1 hr. The DNA precipitate is collected by centrifugation for 10 min in a microcentrifuge, the ethanol is removed, and the DNA pellet is air-dried for 5 min and then resuspended in 20 μl of TE (10 mM Tris-HCl, 1 mM EDTA, pH 8.0).

Electrophoresis

The digested DNA is electrophoresed on $25 \times 20 \times 0.5$ cm 0.85% agarose gels made in TAE (40 mM Tris, 20 mM sodium acetate, 1 mM EDTA, pH 8.0). Electrophoresis is carried out for approximately 20 hr at 20–30 V until the bromphenol blue indicator dye has moved approximately 20 cm. After electrophoresis the gel is stained in 0.5 μg/ml ethidium bromide for 30 min, illuminated with 300 nm UV light (Foto UV 440 DNA transilluminator, Fotodyne Inc., New Berlin, WI), and photographed with a Polaroid MP-4 camera using type 55 positive/negative film and a Kodak (Rochester, NY) No. 9 Wratten gelatin filter. Prior to photographing the gel a hole is made in each of the ethidium bromide-stained DNA bands of standard length markers using a glass pipette and aspiration to remove the plug of agarose. The holes are used as a template to mark the position of the molecular length markers on the membrane onto which the DNA will be transferred.

DNA Transfer

The DNA is transferred by capillary action onto hybridization membranes [such as Genetran (Plasco, Inc., Woburn, MA) or Zetaprobe (Bio-Rad, Richmond, CA)] by the method of Southern with minor modifica-

---

[19] E. M. Southern, *J. Mol. Biol.* **98**, 503 (1975).

tions.[19] The gel is rinsed twice in 0.5 M NaOH, 3.0 M NaCl for 20 min, once in distilled water for 15 sec, and then twice in 0.5 M Tris-HCl, pH 7.0, 3.0 M NaCl for 30 min. All rinses are done at room temperature with gentle shaking. Wicking paper (3MM, Whatman, Clifton, NJ) is placed on the upper surface of the gel, and a platform with the same dimensions as the gel is placed on the wicking paper. The platform is inverted so that the gel now rests on top of the wicking paper and platform. The gel is covered with a hybridization membrane that has been presoaked in water for 10 min followed by a rinse in 20× SSC (SSC is 0.15 M NaCl, 15 mM trisodium citrate) for 60 sec. The hybridization membrane is covered with three pieces of 3MM filter paper, an approximately 6-cm stack of paper towels, a plexiglass sheet, and approximately 500 g of weight. The blotting setup is suspended over a tray containing 20× SSC so that the two opposite edges of the wicking paper are immersed in the solution. Transfer is for 12–16 hr. After blotting all but the last sheet of blotting paper is removed and the remainder inverted so that the wicking paper is on the upper surface. The wicking paper is removed, and the hybridization membrane is marked through the holes in the gel using a ballpoint pen. The hybridization membrane is then washed in 2× SSC at room temperature for 20 min with gentle shaking and dried in an oven at 80° under reduced pressure for 2–2.5 hr.

Hybridization of DNA Gel Blots

The membrane to which DNA has been transferred is prehybridized in 15 ml of 50% formamide, 5× Denhardt's solution [1× Denhardt's solution is 0.02% Bovine serum albumin (BSA), 0.02% Ficoll, and 0.02% polyvinylpyrrolidone],[20] 5× SSC, 0.1% sodium dodecyl sulfate (SDS) and 250 µg/ml sheared, denatured salmon sperm DNA at 42° for 3–4 hr. Hybridization is performed at 42° for 18 hr using fresh solution made as above and also containing approximately 2×10^7 counts/min (cpm)/ml ^{32}p-labeled transposon DNA. The DNA is made radioactive by random-primed synthesis as described by the manufacturer of the labeling kit (Boehringer Mannheim, Indianapolis, IN). After hybridization the membrane is rinsed briefly in 2× SSC, 0.5% SDS and then twice for 20 min at room temperature under the same conditions with gentle shaking. The membrane is then rinsed for 30 min and 20 min with gentle shaking in 0.1× SSC, 0.5% SDS at room temperature and 60°, respectively. To identify the transposon-hybridizing DNA the membrane is then exposed to Kodak XAR-5 film using a Du Pont (Wilmington, DE) Lightning Plus intensifying screen for 24 hr to 7 days at −80°.

[20] D. T. Denhart, *Biochem. Biophys. Res. Commun.* **23**, 641 (1966).

Cloning Transposon-Tagged DNA

Isolation of DNA to Be Cloned

Once a transposon-tagged sequence that cosegregates with the gene of interest on a DNA gel blot analysis is identified it can be cloned. Approximately 60 μg of DNA isolated from a plant that contains the transposon-tagged restriction fragment that cosegregates with the mutant phenotype is digested as above using the same type of restriction endonuclease that was used in the cosegregation analysis. Following alcohol precipitation the DNA is resuspended in a total of 500 μl of TE. The DNA is electrophoresed on 0.85% SeaKem GTG agarose (FMC, Rockland, ME) gels prepared and run as above except that the maize DNA digest is loaded into a 150 × 2 × 4 mm well. Wells for the standard molecular length markers are as normal on either side of the large well.

Following electrophoresis the areas of the gel containing the standard molecular length markers are excised and stained in 0.5 μg/ml ethidium bromide for 30 min. The central portion of the gel containing the maize DNA is covered with plastic wrap to prevent drying and is stored at 4°. The molecular length markers are visualized with UV light, and holes are punched in the gel, using a glass pipette and aspiration to remove the plug of agarose. These holes mark the upper and lower areas of the gel known to contain the transposon-tagged restriction fragment that cosegregates with the mutant phenotype. The gel pieces containing the molecular length markers are then placed back beside the central portion of the gel, and the area of the gel containing the maize DNA digest delineated by the holes is excised.

The agarose is placed into dialysis tubing containing TAE. Electroelution is performed at 100 V for 3–5 hr essentially as described by Maniatis et al.[21] The buffer is removed from the dialysis tubing and placed in a microcentrifuge tube. This solution is centrifuged in a microcentrifuge for 2 min to pellet any small gel fragments, and the supernatant is transferred to a new microcentrifuge tube. The supernatant is then made 0.25 mM in sodium acetate (pH 5.2) and precipitated with 2.5 volumes of 95% ethanol. The DNA precipitate is collected by centrifugation for 15 min at 4° in a microcentrifuge. The ethanol is removed, and the DNA precipitate is rinsed in 70% ethanol, air-dried for 5 min, and resuspended in 10–20 μl TE. If any precipitate does not dissolve in the TE, the solution is centrifuged for 2 min in a microcentrifuge, and the resulting DNA solution is removed to a separate tube. Prior to cloning approximately 0.2 μg of the

---

[21] T. Maniatis, E. F. Fritsch, and J. Sambrook, *in* "Molecular Cloning: A Laboratory Manual." Cold Spring Harbor Laboratory, Cold Spring Harbor, New York, 1982.

size-fractionated DNA is analyzed by DNA blot hybridization using the transposon as a hybridization probe. The cosegregation DNA fragment should appear as a sharp band, indicating no nuclease degradation.

Cloning

The size-fractionated DNA is now ready to be cloned in an appropriate phage λ cloning vector. The bacteriophage λ cloning vector used should have a cloning site compatible with the restriction endonuclease that is used to identify the cosegregating fragment and should also accept the length of the DNA fragment to be cloned. The size-fractionated DNA is ligated, packaged in bacteriophage λ packaging extracts (Gigapeck Gold, Stratagene, La Jolla, CA), and plated as recommended by the manufacturer of the packaging extract.

Plaque Lifts

The packaged phage are plated in the appropriate host strain on square phage typing grid dishes (Falcon, Oxnard, CA), containing NZY media. The plates are prepared 2 days in advance of phage plating, and prior to use they are warmed to 37°. To remove excess liquid from the plates any condensation present on the lids is removed and the plates are placed in a tissue culture hood with their lids off and the blower running for 10–15 min. The phage are plated in NZY media containing 0.7% agar at a density of approximately $0.5-1 \times 10^4$ plaques/plate. Plates are incubated at 37° until plaques are clearly evident but not larger than 1 mm in diameter. The plates are then placed at 4° for 1 hr to harden the top agar. Each plate is left in the refrigerator until just prior to plaque lifting. Using well-dried plates and refrigerating them prior to the process of lifting the plaques will help to prevent the top agar from pulling off during the plaque lifts.

Plaque lifts are done by the method of Benton and Davis[22] with minor modifications.[23] An 8×8 cm nitrocellulose filter (Millipore, Bedford, MA) is placed on the top agar surface for 1 min. A felt-tipped permanent marker is used to make three asymmetrically placed spots on the filter and in corresponding locations on the bottom of the phage typing grid dish. The filter is peeled off and immediately placed for 3 min, plaque side up, on three layers of Whatman 3MM filter paper saturated with 1.5 M NaCl, 0.5 M NaOH. During this lysis/denaturation step a second filter is placed

---

[22] W. D. Benton and R. W. Davis, *Science* **196**, 180 (1978).
[23] P. J. Mason and J. G. Williams, *in* "Nucleic Acid Hybridization: A Practical Approach" (B. D. Hames and S. J. Higgins, eds.), p. 113. IRL Press, Washington, D.C., 1987.

on the plate, marked in the same locations as the first filter, and allowed to absorb phage for 3 min. Following lysis/denaturation the filters are transferred to three layers of 3MM filter paper saturated with 1.5 M NaCl, 1 M Tris HCl, pH 7.5, for 5 min and then transferred to 3MM filter paper saturated with 600 mM NaCl, 4 mM EDTA, 80 mM Tris-HCl, pH 7.8, for 2.5 min. The filters are dried at room temperature for 20 min, placed between two sheets of 3MM paper, and baked at 80° under vacuum for 2 hr.

Hybridization and Plaque Purification

As many as 30 filters are prehybridized in 200 ml of 5× Denhardt's solution, 5× SSC, 0.1% SDS, and 250 µg/ml sheared, denatured salmon sperm DNA at 65° for 4 hr with gentle shaking. Hybridization is performed for 18 hr under the same conditions using 125 ml of solution prepared as above but also containing approximately 3 × 10^6 cpm/ml ^{32}P-labeled transposon DNA. Prehybridization and hybridization are carried out in a 11 × 11 cm square-bottomed plastic food container (Rubbermaid, Wooster, OH). With the indicated amounts of prehybridization and hybridization solution and gentle agitation, the filters do not stick to each other or to the container walls. Insertion of positive control filters at various levels in the filter stack indicate that hybridization occurs uniformly throughout the stack. After hybridization the filters are rinsed as described for DNA gel blots.

The replica filters are arranged side by side on 25 × 30 cm fiberboard covered with a single sheet of Whatman 3MM paper. Square filters are easy to orient during identification of positive plaques and are also useful for maximizing the number of filters that can be autoradiographed on any one X-ray film. The center of the 3MM paper is marked with three or four asymmetrically placed radioactive ink dots (~200 cpm/dot). These radioactive dots are used to orient and position the developed X-ray film when locating positive plaques. To prevent radioactive contamination of the board, a small square of laboratory film (Parafilm) should be placed between the fiberboard and the 3MM paper. The filters are then exposed to Kodak XAR-5 film using a Du Pont Lightning Plus intensifying screen for approximately 48 hr at −80°. Radioactive plaques should be evident after 24 hr, but the additional exposure time will help to distinguish positive plaques, which will be well-defined round spots, from artifactual dark spots which are more irregularly shaped. The film can be aligned on top of the filters using the radioactive ink marks. The orientation marks on the filters are marked on the film with a marking pen. The plates with positive plaques are aligned on the film, and the general region that contains a positive plaque can be delineated on the bottom of the plate.

Positive plaques are purified as described by Davis et al,[24] with minor modifications. The appropriate bacteria for phage propagation are grown to an OD of 0.5 at 600 nm in TB broth supplemented with 0.2% maltose and 10 mM MgSO$_4$. The cells are collected by centrifugation and resuspended in 0.5 volume of sterile 10 mM MgSO$_4$. Three hundred microliters of these cells in 3 ml of NZY broth containing 0.7% agar is poured into NZY plates and allowed to cool for 5 min. A sterile toothpick is touched to each plaque in the region containing the positive plaque, and then the toothpick is touched to the bacterial lawn on the newly poured plate. Positive and negative control plaques can also be transferred in this fashion to the bacterial lawn. The plate is then incubated for 16–18 hr at 37°. The plaques are lifted and hybridized as described above. The length of autoradiography is usually 24 hr or less. Positive plaques are excised with a glass pipette and placed into 1 ml of sterile SM (0.1 M NaCl, 8 mM MgSO$_4 \cdot$ H$_2$O, 50 mM Tris-HCl, 0.01% gelatin) with 50 µl of chloroform. The tube is gently vortexed and incubated at 4° for at least 1 hr to allow phage to disperse into the medium. The phage solution is then titered, and plates that contain few plaques (10 to 100) are lifted and hybridized as described above. Positive plaques at this stage are typically pure clones.

Subcloning of DNA Flanking Transposon

Higher plant genomes such as maize contain a large amount of repetitive DNA,[25] and therefore the cloned DNA may contain some repetitive DNA. If a DNA containing repetitive sequences is used as a hybridization probe of a DNA gel blot of maize DNA it will hybridize to a large number of restriction fragments. This will make it difficult or impossible to distinguish the tagged gene of interest. Therefore, a portion of the cloned DNA containing no repetitive DNA should be isolated.

The phage λ clone DNA is isolated,[24] and if the cloned insert is short (<14 kb) it is subcloned[26] into a plasmid vector. If the insert is too large to be subcloned in its entirety into a plasmid vector, the λ DNA is digested with a restriction endonuclease that cleaves the transposon once or not at all and that produces DNA ends which are compatible with cloning in a plasmid vector. The digest is analyzed by DNA blot hybridization as described above using the transposon as a hybridization probe. The restriction fragment(s) that hybridizes to the transposon is subcloned into a plasmid vector and will contain the transposon and maize DNA adjacent to the transposon. Subclones are restriction mapped, and a maize restric-

[24] L. G. Davis, M. D. Dibner, and J. F. Battey, "Basic Methods in Molecular Biology." Elsevier Science, New York, 1986.
[25] S. Hake and V. Walbot, *Chromosoma* **79,** 251 (1980).
[26] J. R. Greene and L. Guarente, this series, Vol. 152, p. 512.

tion fragment adjacent to the transposon is used as a hybridization probe of a DNA gel blot of maize DNA. If it hybridizes to repetitive DNA, other restriction fragments that map closer to the transposon are used in a similar analysis until a fragment that does not contain repetitive DNA is obtained. This DNA is then subcloned into a plasmid vector and is used to confirm that the cloned sequence is a portion of the gene of interest.

Confirmation of Cloned Sequence as Gene of Interest

Several strategies that can be used to try to determine whether the cloned sequence is at least a portion of the gene of interest are described below. When possible it is best to utilize more than one of these methods.

Linkage

Using the cloned DNA adjacent to the transposon as a hybridization probe of a DNA blot analysis of plants segregating for the transposon-induced allele will indicate if the cloned sequence is tightly linked to the gene of interest. The cloned DNA should cosegregate with the mutant phenotype and also hybridize to the nonmutant allele of the gene. If it does not, a crossover has uncoupled the transposon-tagged sequence from the gene of interest, and this is evidence that the cloned sequence is not the gene of interest. By using this analysis a large number of plants can be analyzed; however, this analysis will only demonstrate linkage and will not prove that the sequence is the gene of interest.

RNA Analysis

If the cloned DNA is the gene of interest, it is expected to hybridize to RNA in tissue expressing the gene but not to the RNA from tissue where the gene is not expressed. (It should be noted that this assumption may not be valid if the gene is under posttranscriptional regulation.) We isolate RNA by the method of McCarty,[27] but other methods can be used.[28] The RNA is electrophoresed, blotted to hybridization membranes, and hybridized to cloned DNA adjacent to the transposon by the method of Thomas[29,30] or Alwine *et al.*[31,32]

---

[27] D. R. McCarty, *Maize Gen. Coop. Newslett.* **60,** 61 (1986).
[28] A. Barkan, D. Miles, and W. C. Taylor, *EMBO J.* **5,** 1421 (1986).
[29] P. S. Thomas, *Proc. Natl. Acad. Sci. U.S.A.* **77,** 5201 (1980).
[30] P. S. Thomas, this series, Vol. 100, p. 255.
[31] J. C. Alwine, D. J. Kemp, and G. R. Stark, *Proc. Natl. Acad. Sci. U.S.A.* **74,** 5350 (1977).
[32] J. C. Alwine, D. J. Kemp, B. A. Parker, J. Reiser, J. Renart, G. R. Stark, and G. M. Wahl, this series, Vol. 68, p. 220.

Analysis of Other Transposon-Tagged Alleles

If an independently derived transposon-induced mutation of the gene of interest is available, the demonstration by DNA blot analysis that it contains an insertion equal in length to the transposon presumed to be within the gene is supporting evidence that the cloned sequence is in fact the gene of interest. The cloned DNA is used as a hybridization probe of a DNA blot of DNA from a plant with an independently derived transposon-induced allele of the gene, which has been digested by a restriction endonuclease that does not cleave within the transposon presumed to be in the gene. If the cloned DNA is part of the gene of interest, the length polymorphism associated with the transposon-induced allele will be greater in length than the progenitor allele by an amount equal to that of the transposon. The cloned sequence is not expected to detect a length polymorphism between the transposon-induced and progenitor allele if it is not part of the gene of interest. Additionally, this allele can be cloned, as described above, using the cloned DNA adjacent to the transposon insertion as a hybridization probe. The presence of a transposon within the same region of DNA in this second allele suggests that the cloned sequence is the gene of interest.

Analysis of Revertants

Another method to determine whether the cloned sequence is the gene of interest is to use the cloned sequence as a hybridization probe of a DNA blot of DNA from tissue that has reverted to the wild-type phenotype. Reversions occur when a transposon within a gene transposes out of that sequence, leaving behind a functional gene. Reversion events can occur both germinally and somatically. Germinal revertants can be generated if the mutant phenotype is a homozygous recessive nonlethal by crossing a plant that is homozygous for the transposon-induced allele, or heterozygous for the transposon-induced allele and a standard recessive allele, to a plant that is homozygous for a standard recessive allele. All resulting progeny are expected to exhibit the recessive phenotype; however, if the transposon presumed to be present in the gene of interest transposes out of the locus, the allele may function normally and result in a wild-type phenotype. In maize this would be evident as a wild-type kernel and/or seedling. DNA from the revertant individuals and sibling nonrevertant individuals can be analyzed by DNA blot hybridization analysis. Somatic reversion events can be observed phenotypically in plants that are homozygous recessive for the mutant allele as a sector of wild-type tissue within the mutant background. If the sector is large it can be excised, and the DNA can be isolated from the sector as well as from the surrounding mutant tissue and be analyzed by DNA blot analysis. If the cloned sequence is part of the gene of interest, a length polymorphism indistinguishable from the progenitor allele will be observed in a DNA blot analysis of DNA from germinal or somatic revertant tissue while a length polymorphism associated with the transposon-induced allele is present in DNA isolated from the nonrevertant tissue. These data would demonstrate that the cloned DNA sequence is part of the gene of interest.

Acknowledgments

The authors are grateful to Diane Janick-Buckner for critical reading of the manuscript. This work was supported by a grant from The Rockefeller Foundation's International Program on Rice Biotechnology. The *y1-mum* alleles used in these studies were the result of work supported by National Science Foundation Grants PCM 76-28758, PCM 79-23052, and PCM 83-02214.

[28] Protection by Cloned Carotenoid Genes Expressed in *Escherichia coli* against Phototoxic Molecules Activated by Near-Ultraviolet Light

By R. W. TUVESON and G. SANDMANN

Introduction

The evidence that carotenoids protect plants, animals, and microorganisms against photodynamic action by quenching triplet state photosensitizers, singlet oxygen ($^1O_2$), and perhaps radical oxygen species has been reviewed.[1–3] Photosynthetic bacteria contain carotenoids that serve to protect against photodynamic action probably involving chlorophyll as an endogenous photosynthesizer.[4] In the nonphotosynthetic bacterium *Myxococcus xanthus*, protoporphyrin IX accumulates to high levels when the cells enter stationary phase. Concomitant with the accumulation of protoporphyrin IX, carotenoids are synthesized probably to protect against photodynamic action resulting from protoporphyrin IX serving as an en-

[1] N. I. Krinsky, *in* "Photophysiology—Current Topics" (A. C. Giese, ed.), p. 123. Academic Press, New York and London, 1968.
[2] N. I. Krinsky, *in* "Carotenoids" (O. Isler, H. Gutman, and U. Solms, eds.), p. 669. Birkhauser, Basel and Stuttgart, 1971.
[3] M. M. Mathews-Roth, *Photochem. Photobiol.* **40**, 63 (1984).
[4] M. Griffiths, M. R. Sistrom, and G. Cohen-Bazire, *Nature (London)* **176**, 1211 (1955).

dogenous photosensitizer.[5,6] Epiphytic and phytopathogenic bacteria associated with aboveground plant parts (e.g., *Erwinia herbicola*[7] frequently produce yellow pigments, presumably carotenoids, to protect against photodynamic damage that may result from the absorption of visible light by the chlorophyll of the host plant acting as a photosensitizer. Many fungi produce carotenoids, and in the case of *Neurospora crassa* these pigments have been shown to protect against photodynamic action involving specific exogenous dyes.[8-10] Because *N. crassa* as well as many other fungi are exposed to sunlight in nature, it is plausible that the carotenoids serve to protect against light-induced damage at the wavelengths being absorbed by endogenous photosensitizers such as cytochromes or flavins.[11,12] Normally *Escherichia coli* does not synthesize carotenoids, but Perry et al.[7] were able to clone into *E. coli* a 15.6-kb DNA fragment from an epiphytic bacterium, *E. herbicola,* carrying genes allowing for the synthesis of the yellow pigments characteristic of this species. The carotenoid pigments expressed in *E. coli* strain LE392(pPL376) have been identified as phytoene (15-*cis* isomer), β-carotene (all-*trans,* 9-*cis,* and 15-*cis*), β-cryptoxanthin (3-hydroxy-β-carotene), zeaxanthin (3,3'-dihydroxy-β-carotene), and the corresponding carotenoid glycosides.[13] The cloned DNA fragment specifying the genes controlling the biosynthesis of the carotenoids characteristic of *E. herbicola* and expressed in *E. coli* has been partially sequenced.[14] Based on analogy with DNA sequences from *Rhodobacter capsulatus,* it has been suggested that the sequences derived from *E. herbicola* carry information for three enzymes involved in the early steps in the biosynthesis of carotenoid pigments [prephytoene pyrophosphate synthase (CrtB), phytoene synthase (CrtE), and phytoene dehydrogenase (CrtI)] in addition to other as yet unidentified gene products. The idea that the early genes concerned with carotenoid biosynthesis are highly conserved is supported by the observation that the deduced amino acid sequences for phytoene dehydrogenase from *N. crassa* and *R. capsulatus* coded for by the *al-1* and

[5] R. P. Burchard and M. Dworkin, *J. Bacteriol.* **91**, 535 (1966).
[6] R. P. Burchard, S. A. Gordon, and M. Dworkin, *J. Bacteriol.* **91**, 896 (1966).
[7] K. L. Perry, T. A. Simonitch, K. J. Harrison-Lavoie, and S.-T. Liu, *J. Bacteriol.* **168**, 607 (1986).
[8] T. W. Goodwin, *Prog. Ind. Microbiol.* **11**, 31 (1972).
[9] P. L. Blanc, R. W. Tuveson, and M. L. Sargent, *J. Bacteriol.* **125**, 616 (1976).
[10] S. A. Thomas, M. L. Sargent, and R. W. Tuveson, *Photochem. Photobiol.* **33**, 349 (1981).
[11] R. W. Tuveson and L. J. Sammartano, *Photochem. Photobiol.* **43**, 621 (1986).
[12] R. E. Lloyd, J. L. Rinkenberger, B. A. Hug, and R. W. Tuveson, *Photochem. Photobiol.* **52**, 897 (1990).
[13] G. Sandmann, W. S. Woods, and R. W. Tuveson, *FEMS Microbiol. Lett.* **71**, 77 (1990).
[14] G. A. Armstrong, M. Alberti, and J. E. Hearst, *Proc. Natl. Acad. Sci. U.S.A.* **87**, 9975 (1990).

crtI genes, respectively, are approximately 30% identical.[15] The *E. coli* strains carrying and expressing genes for carotenoid biosynthesis are useful for assessing the protection offered by carotenoids to a variety of photosensitizers as well as near-ultraviolet light (near-UV; 320–400 nm) along.[16,17]

Methods for Using *Escherichia coli* to Assess Protection by Carotenoids

Two *E. coli* strains (HB101; LE392) have been transformed with three plasmids: (1) pHC79, the original plasmid into which the *E. herbicola* carotenoid biosynthetic genes were cloned; (2) pPL376, a derivative of pHC79 carrying genes necessary for carotenoid biosynthesis; and (3) pdel16, a derivative of pPL376 from which a 1.9-kb *Ava*I fragment has been deleted leading to the accumulation of lycopene in *E. coli* strains carrying this plasmid (probably equivalent to the plasmid originally described as pHL545[7]).

The recipient and plasmid-carrying strains are grown in Luria–Bertani (LB) medium containing 10 g tryptone (Difco, Detroit, MI) 5 g yeast extract (Difco), and 10 g NaCl per liter.[18] Cells are grown at 37° with vigorous shaking (300 rpm) in side-arm flasks (Bellco, Vineland, NJ) containing either 10 or 50 ml of LB to which is added 50 μg/ml ampicillin for the plasmid-containing strains. Vigorous shaking is essential since carotenoid biosynthesis is strongly oxygen dependent. Growth is monitored by measuring the change in absorbance with a Klett–Summerson colorimeter equipped with a green filter. Stationary phase cells must be used since the synthesis of carotenoids does not occur until well into stationary phase (usually 2.5 hr after entering the transition from exponential to stationary growth phase). A 5-ml sample of cells is removed, washed 3 times with 67 mM, pH 7.0, phosphate buffer (K-K), diluted in cold buffer (ice bath temperature) to approximately 5.0×10^8 cells/ml, and placed in a 16 × 160 mm test tube with a magnetic stirring bar at the bottom. For strains expected to produce carotenoids, carotenoid synthesis can be recognized by the obvious pigmentation of the bacterial pellets following centrifugation (bright yellow for strains carrying plasmid pPL376; pink for strains carrying plasmid pdel 16). In experiments involving photosensitiz-

---

[15] T. J. Schmidhauser, F. R. Lauter, V. E. A. Russo, and C. Yanofsky, *Mol. Cell. Biol.* **10**, 5064 (1990).
[16] R. W. Tuveson, R. A. Larson, and J. Kagan, *J. Bacteriol.* **170**, 4675 (1988).
[17] R. W. Tuveson, G. R. Wang, T. P. Wang, and J. Kagan, *Photochem. Photobiol.* **52**, 993 (1990).
[18] J. H. Miller, "Experiments in Molecular Genetics." Cold Spring Harbor Laboratory, Cold Spring Harbor, New York, 1972.

ing molecules, the chemicals are dissolved in either 95% ethanol or dimethyl sulfoxide (DMSO) and added to the cell suspension to give the final desired concentration. A 1-ml sample is withdrawn and held in the dark as a check on light-independent toxicity. Viability of cells in this sample is assessed after all manipulations are completed with the near-UV-treated suspension.

The minimal medium (SEM) used to assess viability consists of minimal A medium[18] supplemented with the nutritional requirements of the particular *E. coli* cloning strain (HB101, proline and leucine; LE392, methionine). In experiments involving plasmid-carrying strains, ampicillin (50 μg/ml) is added to the minimal A agar plating medium.

The broad-spectrum near-UV source used for assessing protection by carotenoids is identical to that used in all our previous experiments involving broad-spectrum near-UV.[19] It consists of a bank of four lamps (GE40BLB, integral filter) emitting between 300 and 425 nm, with maximum intensity at 350 nm. The source is maintained in a cold room at 10° to reduce heating effects. Sham near-UV irradiation of foil-wrapped cell suspensions show no inactivation.[19] When measured with a DRC-100 Digital Radiometer equipped with a DIX-365 sensor (Spectroline), the fluence rate ranges from 40 to 43 J sec^{-1} m^{-2} during the course of the experiments described here.

Carotenoid Protection

In previous experiments,[13,16] six *E. coli* strains were used to investigate carotenoid protection from a variety of inactivating agents. In those experiments, the recipient strains, *E. coli* HB101 and LE392, together with the same strains transformed with the cloning vector pHC79, as well as the plasmid (pPL376) into which the carotenoid biosynthetic genes had been cloned, were used.

When HB101 and its plasmid-containing derivatives were treated with far-UV (200-290 nm), the inactivation kinetics were indistinguishable.[16] This result is exactly what might have been expected since the inactivating events produced by far-UV are based on the direct absorption of these wavelengths by DNA leading to the formation of pyrimidine dimers among other minor lesions. Furthermore, since carotenoids are associated with the cell membrane,[1] carotenoids would not be expected to offer protection from far-UV inactivation. These results are consistent with those reported with *N. crassa* in which the inactivation kinetics of wild-type (carotenoid containing) and albino (carotenoidless) conidia following

[19] R. W. Tuveson and R. B. Jonas, *Photochem. Photobiol.* **30**, 667 (1979).

treatment with far-UV were essentially identical.[9] Using the same *E. coli* strains described here, Perry *et al.*[7] reported that the pigmented and nonpigmented strains were equally sensitive to inactivation by UV Since these investigators did not specify the wavelengths investigated, we assume that they were using far-UV, explaining the absence of carotenoid protection.

When *E. coli* strains HB101 and HB101(pHC79) were treated with broad-spectrum near-UV, the inactivation kinetics were essentially identical. The results with the carotenoid-producing strain [HB101(pPL376)] were different, showing apparent protection by carotenoids at higher fluence levels.[16] Apparently carotenoids do offer some protection against the inactivating effects of near-UV, at least at higher fluences. The complex character of the inactivation curve obtained with the carotenoid-producing strain [HB101(pPL376)] reflects the possibility that both DNA and the membrane are important lethal targets for near-UV, as has been suggested.[20,21] These observations are consistent with results obtained using an *N. crassa* carotenoid-producing and a carotenoidless mutant, showing that carotenoids offer only limited protection against the inactivating effects of near-UV alone.[9,10] Perhaps the fact that both *E. coli* and *N. crassa* carotenoids offer limited protection against near-UV inactivation means that carotenoids are capable of quenching only $^1O_2$ or triplet state photosensitizers. Near-UV inactivation is oxygen dependent and apparently generates chiefly superoxide, leading to H_2O_2 and ultimately hydroxyl radical.[22,23] It cannot be excluded that near-UV may also generate a limited amount of $^1O_2$, thus accounting for the limited protection seen when carotenoid-producing cells are treated with near-UV alone. If carotenoids can protect by quenching hydroxyl radical, then it should be possible to demonstrate that the carotenoid-producing strains [HB101(pPL376) or LE392(pPL376)] are protected against inactivation by H_2O_2, which in the presence of Fe(II) leads to hydroxy radical. Using the glucose–glucose oxidase H_2O_2-generating system,[23] we have attempted to demonstrate such protection with the *E. coli* strains described here with results that are uniformly negative. We conclude that under the conditions of this assay, carotenoids do not protect against H_2O_2 and ultimately hydroxyl radical.

The *E. coli* carotenoidless [HB101, HB101(pHC79)] and carotenoid-producing strains were tested with a variety of photosensitizing molecules (phototoxins) activated by near-UV. It is important to note that the maximum near-UV fluence used to activate the phototoxins described below

---

[20] S. H. Moss and K. C. Smith, *Photochem. Photobiol.* **33**, 202 (1981).
[21] A. Eisenstark, *Environ. Mol. Mutagen.* **10**, 317 (1987).
[22] L. J. Sammartano and R. W. Tuveson, *Photochem. Photobiol.* **40**, 607 (1984).
[23] L. J. Sammartano, R. W. Tuveson, and R. Davenport, *J. Bacteriol.* **168**, 13 (1986).

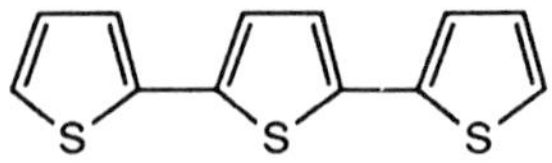

FIG. 1. Chemical structure of α-terthienyl.

results in no inactivation of the three *E. coli* strains when used alone. Psoralen and 8-methoxypsoralen are near-UV-activated molecules that inactivate chiefly by forming cycloadducts to DNA. Since the inactivation by these molecules is not oxygen dependent, it was expected that protection by carotenoids would not occur. This is exactly what was observed.[16]

When the carotenoidless and carotenoid-producing strains were treated with near-UV-activated molecules that exhibited oxygen dependence, the degree of protection offered by carotenoids depended on the particular molecule. For example, carotenoids offered only modest protection against inactivation by near-UV-activated phenylheptatriyne, suggesting that this oxygen-dependent molecule probably generated chiefly superoxide leading to H_2O_2 and, ultimately, hydroxyl radical *in vivo*. However, the fact that some protection is afforded by the presence of carotenoids must mean that some singlet oxygen is also formed *in vivo*. Of all the near-UV-activated oxygen-dependent molecules investigated to date, α-terthienyl (α-T, Fig. 1) represents one in which the available evidence suggests that the chief damaging product is $^1O_2$.[24] The target for attack by α-T plus near-UV seems to be the cell membrane since carotenoids do strongly protect against inactivation by this near-UV-activated molecule and it appears to be incapable of inducing mutations. Because the α-T molecule is extremely hydrophobic, it may simply be incapable of entering cells to attack DNA. This possibility was checked by using purified *Haemophilus influenzae* transforming DNA and pBR322 DNA.[25] Using these DNAs it was shown that near-UV-activated α-T did not destroy *H. influenzae* transforming activity. However, such treatment was capable of converting supercoiled pBR322 DNA to the open circular form. Open circular pBR322 DNA is capable of transforming *E. coli* probably because the cell contains sufficient ligase to repair nicks in the DNA backbone. The fact that *H. influenzae* DNA seemed not to be affected by treatment with α-T plus near-UV may simply reflect the fact that the cell repairs any single-strand DNA nicks during the process of uptake. If α-T does penetrate cells and produces single-strand breaks following treatment with near-UV, such nicks are probably rapidly repaired by ligase and would not influence survival.

[24] K. R. Downum, R. W. E. Hancock, and G. H. N. Towers, *Photochem. Photobiol.* **36**, 517 (1822).
[25] T. P. Wang, J. Kagan, R. W. Tuveson, and G. R. Wang, *Photochem. Photobiol.* **53**, 463 (1991).

The effects of α-T plus near-UV have been reinvestigated using recipient *E. coli* strains HB101 and LE392 carrying plasmids pHC79 (original cloning vector) and pPL376 (carrying genes specifying carotenoid biosynthesis) as well as plasmid pdel 16 (a deletion plasmid resulting in the accumulation of lycopene). The results of these experiments (Fig. 2) confirm our previous results with *E. coli* HB101[16] and LE392[13] and their plasmid-carrying derivatives. As expected, the carotenoid-producing strains (carrying plasmid pPL376) are protected against inactivation by α-T plus near-UV. The strain carrying the plasmid allowing for the accumulation of lycopene (pdel16) is as sensitive to inactivation by α-T plus near-UV as is the recipient strain or the strain carrying the cloning vector (pHC79). This result supports the view that acyclic carotenes probably do not function to quench triplet state photosensitizers or $^1O_2$. The only

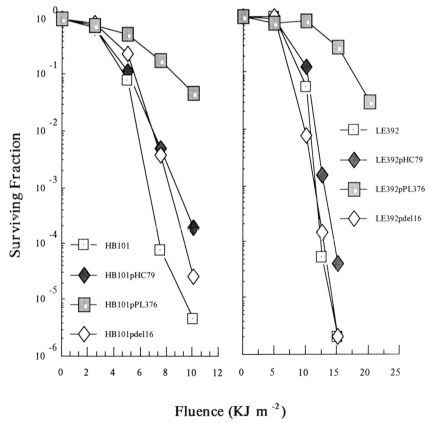

FIG. 2. Near-UV fluence–response curves for various strains of *Escherichia coli* in the presence of α-terthienyl (10 μg/ml).

difference between the experiments involving HB101 and its derivatives and LE392 and its derivatives is that the HB101 strains are about twice as sensitive to inactivation by α-T plus near-UV when compared to the LE392 strains (compare the fluence axes for the two sets of experiments).

Conclusions

The *E. coli* strains described here can be used to determine whether carotenoids offer protection against inactivation by a particular photosensitizer, suggesting that either (1) the triplet state of the photosensitizer is quenched or (2) $^1O_2$ is quenched in the case of oxygen-dependent photosensitizers. Protection by carotenoids also suggests that the membrane is an important target for lethal events. The absence of protection by carotenoids against inactivation by a photosensitizer can result from the fact that the photosensitizer either (1) targets sites other than the membrane for lethal damage or (2) generates superoxide leading to H_2O_2 and, ultimately, hydroxyl radical in the case of oxygen-dependent photosensitizers. It is probable in the case of most oxygen-dependent photosensitizers that the partitioning of the sensitizer into the aqueous and lipid portion of the cell determines whether it generates chiefly $^1O_2$ or superoxide and, ultimately, hydroxy radical (electron transfer reaction).

Acknowledgments

This research was supported in part by grants from the U.S. Department of Agriculture (AG 87 CRCR 1-2374; AG 89-372804-897) and U.S. Public Health Service (1 R01 ES14397-01). Sincere appreciation is expressed to R. E. Lloyd for the preparation of the figures.

[29] Biosynthesis of Carotenoids: An Overview

By T. W. GOODWIN

Introduction

It was in the early 1930s that the three basic carotenoid structures lycopene (**I**), α-carotene (**II**), and β-carotene (**III**) were elucidated.[1] All known naturally occurring carotenoids (about 400) are variations on these

[1] O. Isler, ed., "Carotenoids." Birkhauser, Basel, 1971.

structures. Carotenoids are tetraterpenoids; that is, they are composed of eight branched C_5 isoprenoid *(ip)* units made up into two C_{20} units formed by tail to head condensation *(ipipip)* joined head to head at the center of the molecule, thus *ipipipippipipipi*. This satisfies Ruzicka's generalization (biogenetic isoprene rule)[2] that all terpenoids can be considered to be derived by cyclization or other rearrangements from an aliphatic precursor made up of isoprene units.

Early Investigations

Ruzicka's rule had a profound effect on structural organic chemistry but gave no hint of how the basic isoprene unit was formed in living tissues. Progress was only made after the key role of acetyl-CoA was demonstrated in the early 1950s when it was found to be an important intermediate in metabolic processes in animals, plants, and microorganisms and the key precursor of fatty acid and cholesterol biosynthesis.[3] The first unequivocal demonstrations that acetyl-CoA was also a carotenoid precursor were made by Grob et al.[4] in fungi, by Braithwaite and Goodwin[5] in tomatoes, and by Steele and Gurin[6] in photosynthetic tissues.

A major breakthrough in terpenoid biosynthesis was the discovery that mevalonic acid (MVA), a growth factor for certain lactobacilli,[3] was a very active precursor of cholesterol and that it was formed from three acetyl-

[2] L. Ruzicka, *Experientia* **9**, 357 (1957).
[3] J. W. Porter and S. L. Spurgeon, "Biosynthesis of Isoprenoid Compounds," Vol. 1. Wiley, New York, 1981.
[4] E. C. Grob, G. G. Poretti, A. von Muralt, and W. H. Schopfer, *Experientia* **7**, 218 (1951).
[5] G. D. Braithwaite and T. W. Goodwin, *Biochem. J.* **76**, 1 (1960).
[6] W. J. Steele and S. Gurin, *J. Biol. Chem.* **235**, 2778 (1960).

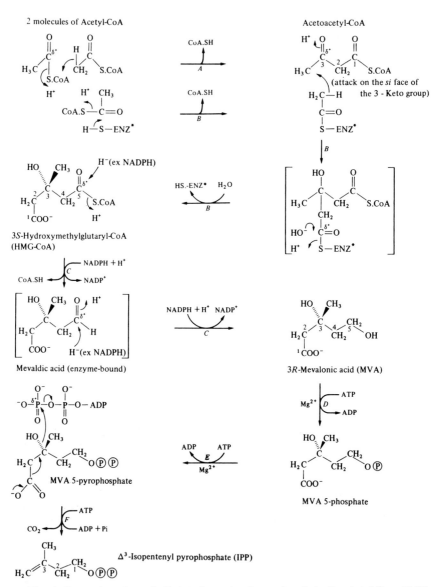

FIG. 1. Conversion of acetyl–CoA to isopentenyl pyrophosphate. Reprinted from T. W. Goodwin and E. I. Mercer, "Introduction to Plant Biochemistry," 2nd ed., Copyright 1983, with permission from Pergamon Press Ltd, Headington Hill Hall, Oxford OX3 0BX, UK.

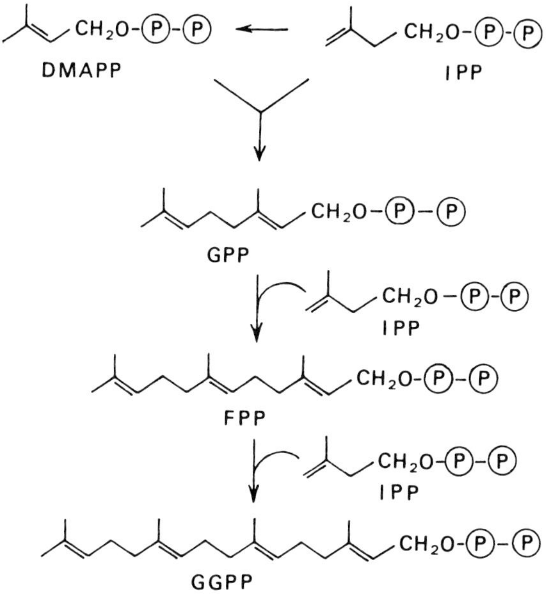

FIG. 2. Conversion of IPP to GGPP.[9]

CoA molecules with the loss of one carboxyl group (pathways A, B, Fig. 1).[7] Its activity in carotenogenesis was soon demonstrated, and the labeling pattern was the same as that in sterol synthesis.[8] The next development was the elucidation of the pathway from the first committed C_6 precursor (MVA) to the first committed C_5 precursor, isopentenyl pyrophosphate (pathways C, D, E, F, Fig. 1).

Formation of Phytoene: The First C_{40} Precursor

The first step involves the isomerization of IPP to dimethylallyl pyrophosphate (DMAPP) and the sequential addition of three further IPP molecules to DMAPP. These reactions, catalyzed by prenyl transferases, yield the C_{20} compound geranygeranyl pyrophosphate (GGPP) (Fig. 2).[9] This pathway is shared with the sterol pathway up to the C_{15} stage (farnesyl

[7] T. W. Goodwin and E. I. Mercer, "Introduction to Plant Biochemistry," 2nd Ed. Pergamon, Oxford, 1983.
[8] G. Britton, in "Plant Pigments" (T. W. Goodwin, ed.), p. 133. Academic Press, London, 1988.
[9] T. W. Goodwin, "The Biochemistry of Carotenoids," p. 33. Vol. 1. Chapman & Hall, London, 1980.

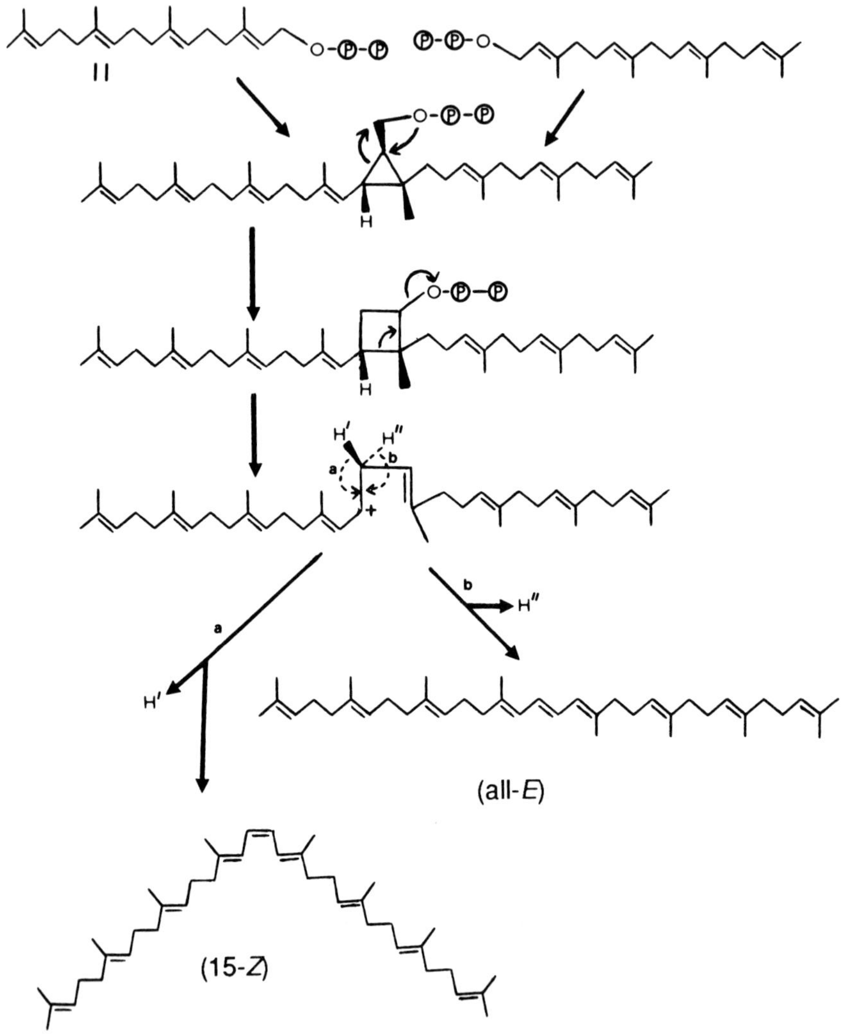

FIG. 3. Formation of (all-*E*) and (15*Z*)-phytoene from GGPP via prephytoene pyrophosphate.[8]

pyrophosphate, FPP), when FPP branches off to form squalene, the C_{30} acyclic precursor of sterols. In carotenogeneic systems two GGPP molecules condense head to head to form, via prephytoene pyrophosphate, phytoene. The stereochemistry of the reaction has been fully described (Fig. 3) for the formation of both (all-E)- or *trans*-phytoene and (15Z)- or *cis*-phytoene.[8] The latter is the predominant form in higher plants, whereas the former is found in some bacteria, for example, *Mycobacterium* and *Halobacterium* species.

Desaturation of Phytoene

If the stereochemical problems associated with the existence of phytoene isomers are ignored for the moment, structural, genetic, and enzymatic studies have shown that phytoene is stepwise desaturated to lycopene[3,9] (Fig. 4). The stereochemistry of the removal of the hydrogens in the

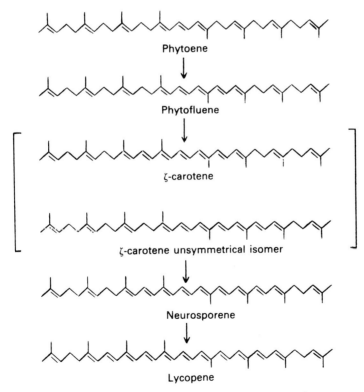

FIG. 4. Sequential conversion of phytoene to lycopene.[8]

FIG. 5. Stereochemistry of double-bond formation in the formation of lycopene from phytoene.[8]

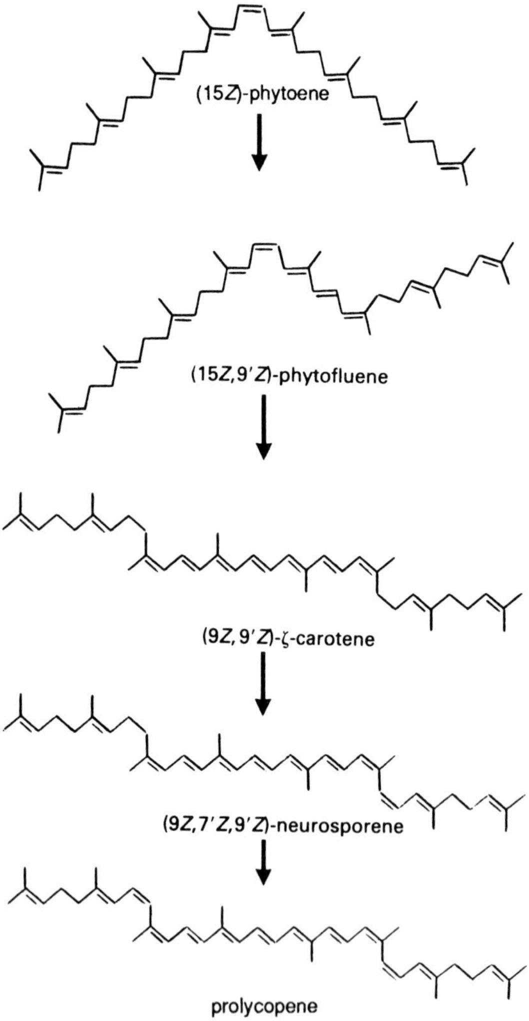

FIG. 6. Conversion of (15Z)-phytoene to prolycopene.[8]

production of each double bond in lycopene formation has been clearly demonstrated (Fig. 5).[8]

As lycopene is an all-trans compound, the isomerization of the 15Z double bond of phytoene must occur at some stage in the desaturation process. At the present time almost all options seem to be open: isomerization of (15Z)-phytoene occurs in *Flavobacterium,* of (15Z)-phytofluene in fruit chromoplasts, and of (15Z)-ζ-carotene in a *Scenedesmus* mutant.[8] However (15Z)-neurosporene and (15Z)-lycopene do not yet appear to have been implicated.

A unique reaction occurs in the tangerine mutant of *Lycopersicum esculentum* in which (15Z)-phytoene is converted into the main pigment prolycopene (7Z, 9Z, 7'Z, 9'Z)-lycopene. The pathway (Fig. 6) is based on the isolation of the proposed intermediates. The full stereochemistry of the hydrogens lost in producing the Z double bonds in prolycopene is not known, but hydrogen removal from C-7 and C-10 has the same stereospecificity as in the production of (all-E)-lycopene.[8]

Cyclization and Other Reactions of Lycopene

Inhibitors (e.g., nicotine) which allowed the accumulation of lycopene in organisms normally synthesizing cyclic carotenoids[8,9] were used in experiments which showed that lycopene was the precursor of cyclic carotenoids and that cyclization was initiated by H$^+$ attack in the formation of the β and ε rings (Fig. 7). The mechanism for the formation of the γ ring is almost certainly the same, but it has not been proved experimentally.

NMR studies with ^{2}H$_2$O have clearly defined the stereochemistry of the H removal at C-4 during the cyclization leading to the formation of

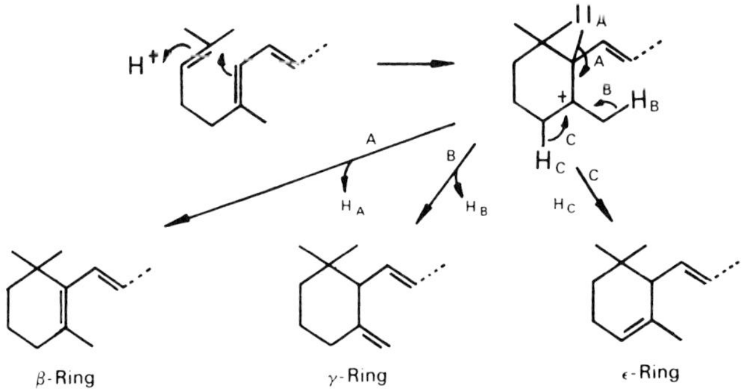

FIG. 7. Mechanism of formation of β- and ε-carotenoid rings from a common precursor (lycopene).[9]

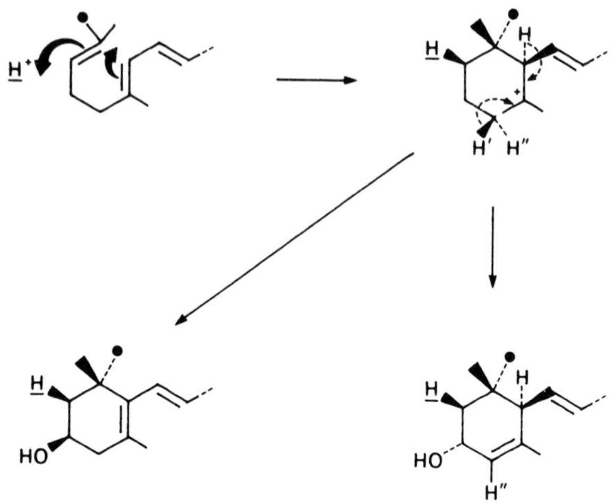

FIG. 8. Stereochemistry of cyclization leading to the formation of the β rings in zeaxanthin and lutein and the ε ring in lutein.[8]

zeaxanthin (3,3'-dihydroxy-β-carotene) and lutein (3,3'-dihydroxy-α-carotene) (Fig. 8).[8,9] ^{13}C NMR studies with 2-[^{13}C]MVA have also demonstrated that the prochiral methyl groups at C-1 arising from MVA maintain their stereochemical individuality on cyclization (Fig. 8).[8,9]

Lycopene can also undergo reactions other than cyclization, such as hydration and hydrogenation across the 1,2 double bond. These reactions are particularly characteristic of photosynthetic bacteria, and the arylcarotenoids, with trimethylphenyl rings, occur in the green Chlorobiaceae and the purple Chromatiaceae.[9] In *Chloropseudomonas ethylicum* the methyl

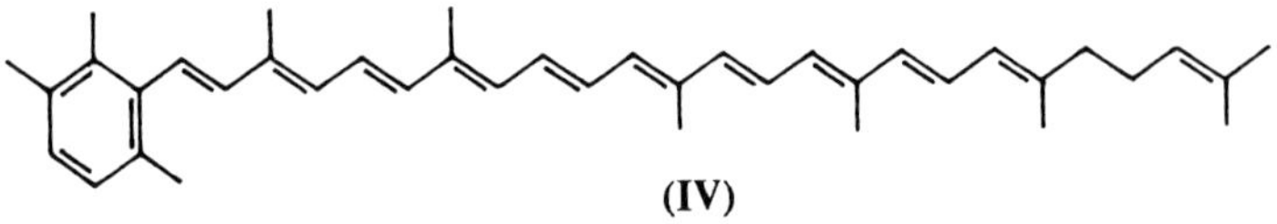

(IV)

group migrating from C-1 to C-2 in chlorobactene (**IV**) arises stereospecifically from the C-3' methyl group of MVA (Fig. 9).[8]

The hydroxyl at C-3 in cyclic carotenoids is inserted after ring closure and occurs by direct stereospecific replacement of the 3-*pro-R* hydrogen by OH. Many other later reactions of cyclic carotenoids lead to, for example, cyclopentanoid derivatives (capsorubin, **V**), acetylenic derivatives (diadinoxanthin, **VI**), allenic carotenoids (neoxanthin, **VII**), and C_{37} norcaroten-

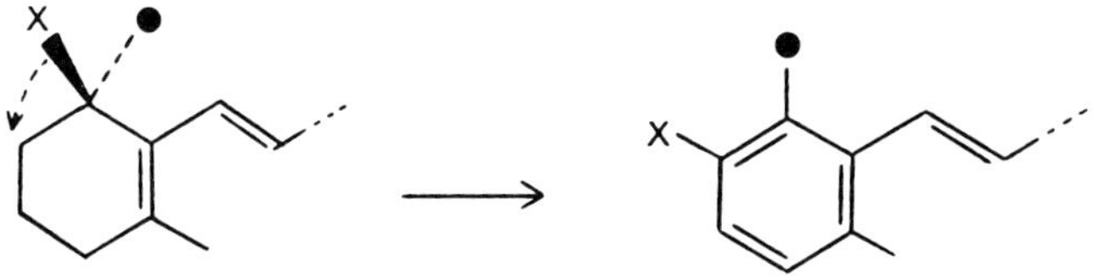

FIG. 9. Stereochemistry of rearrangement to form the 1,2,5-trimethylphenyl end group of chlorobactene.

oids (peridinin, **VIII**). Reasonable mechanisms for the formation of these pigments have been proposed, but detailed proof for the pathways is still awaited.[8,9] Other structural variations are fully described by Goodwin[9] and Britton.[8]

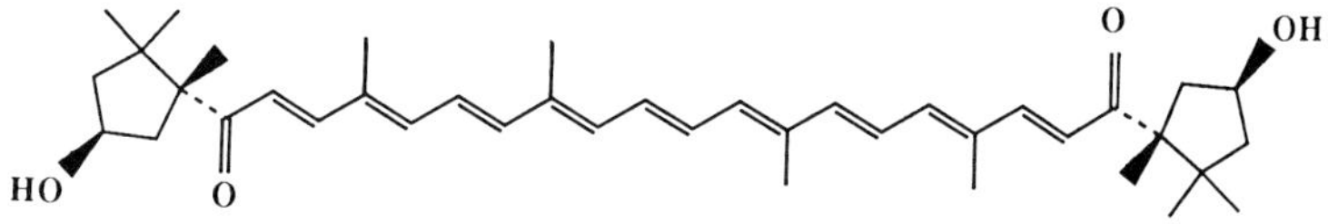

(V)

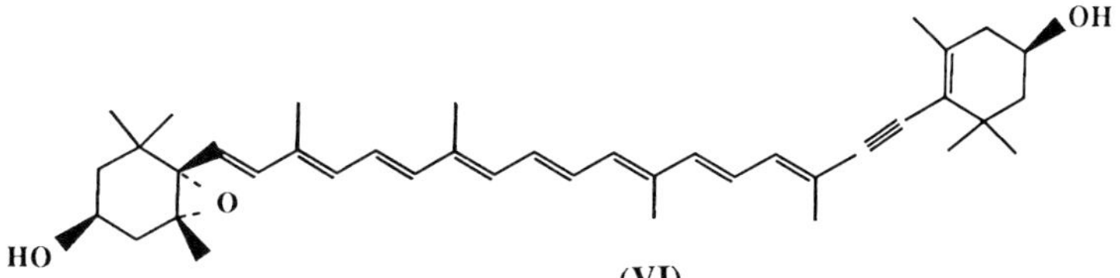

(VI)

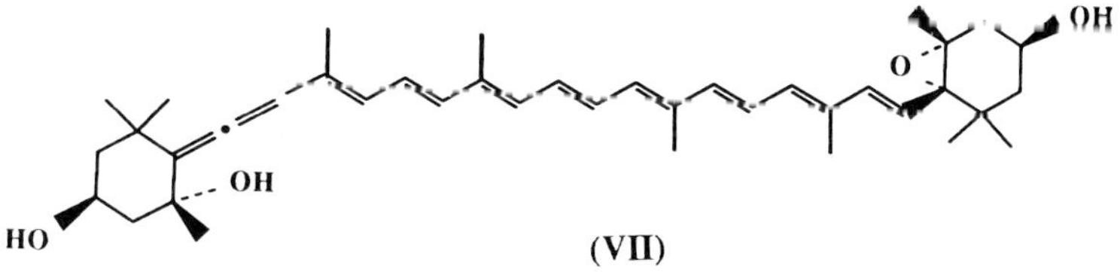

(VII)

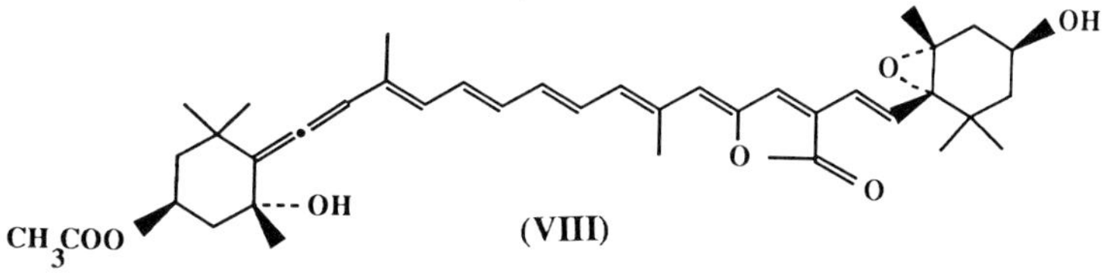

(VIII)

Enzymology

The preparation of cell-free systems active in forming early precursors is well documented,[3] but those catalyzing the steps from phytoene onward are less clearly defined mainly because they are intimately associated with membranes and rapidly lose activity during laboratory manipulation[8,10]; phytoene synthase is an exception because although it is loosely combined with membranes it is not an integral part thereof.

Active but still impure preparations from plastids have frequently been reported since Porter's pioneering work with tomatoes.[3,11] The most active plastid preparations so far recorded are those from *Capsicum*[12] and *Narcissus* species.[10] An active bacterial system is found in a *Flavobacterium* sp.,[8] and a preparation solubilized from the fungus *Phycomyces* which contains the original tight assembly line will convert (15Z)-phytoene into β-carotene.[13]

Regulation of Synthesis in Plastid Development in Higher Plants

There is little doubt that chloroplast carotenoids are synthesized as part of an integrated system which brings about plastid development. The question of whether the plastids can synthesize the pigments from IPP or whether IPP has to be imported is still controversial, but it does appear that the site of synthesis of the early precursors depends on the developmental stage of the chloroplast.

The marked quantitative and qualitative changes which carotenoids undergo as chloroplasts are converted to chromoplasts (e.g., rapid and massive synthesis of lycopene and capsanthin in ripening tomatoes and peppers, respectively) are currently the object of important genetic and molecular biological studies. It has been suggested that the change in pigment synthesis, characterized by nuclear-encoded enzymes, may be the triggering event in the chloroplast to plastid transformation in fruit. How this regulation takes place is still doubtful, but transcriptional regulation of the biosynthetic genes is a possibility. For example, mutations in pigment synthesis have no effect on plastid synthesis, and two temperature-dependent pathways exist. This area is ripe for exploration by the techniques now available to molecular biologists.

---

[10] P. Beyer and H. Kleinig, in "Carotenoids: Chemistry and Biology" (N. Krinsky, M. M. Mathews-Roth, and R. F. Taylor, eds.). Plenum, New York, 1989.

[11] J. W. Porter and S. L. Spurgeon, "Biosynthesis of Isoprenoid Compounds," Vol. 1. Wiley, New York, 1981.

[12] B. Camara, *Pure Appl. Chem.* **57**, 675 (1985).

[13] P. M. Bramley, *Pure Appl. Chem.* **57**, 671 (1985).

[30] Carotenoid Analysis in Mutants from *Escherichia coli* Transformed with Carotenogenic Gene Cluster and *Scenedesmus obliquus* Mutant C-6D

By GERHARD SANDMANN

Introduction

Carotenoid biosynthesis in plants is well regulated, and acyclic carotene precursors of α- and β-carotene and of corresponding oxygenated xanthophylls are below detection limits. Therefore, the search for mutants in which those precursors accumulate has been undertaken for many years.[1] With new developments in the field of molecular genetics, carotenoid mutants have become available by direct manipulation of appropriate genes, for example, by transposon mutagenesis[2] or by deletion restriction of gene fragments.[3] Nevertheless, conventional mutagenesis is still very helpful and provides specific mutants which have not yet been obtained in other ways. Two examples will be presented in this chapter. One is *Escherichia coli* transformed with a carotenogenic gene cluster from *Erwinia herbicola*.[4] Mutants have been generated by partial digestion of the genes involved in carotenogenesis.[3] Analysis by HPLC showed formation of zeaxanthin diglycoside in the transformant and various intermediates in these mutants. The other organism is the green alga *Scenedesmus obliquus*. Ultraviolet mutagenesis resulted in the mutant C-6D[5] which under heterotrophic conditions forms untypical poly-*cis*-carotenes.[6]

Characterization of Mutants

Cultivation of Mutants

From the genomic library of *Erwinia herbicola* a 12.4-kilobase (kb) fragment containing all the genes necessary for carotenoid formation has been cloned.[4,7] Growth of the transformed *E. coli* LE392/pPL376 is carried

---

[1] P. M. Bramley and A. Mackenzie, *Curr. Top. Cell. Regul.* **29**, 291 (1988).
[2] K. M. Zsebo and J. E. Hearst, *Cell (Cambridge, Mass.)* **37**, 937 (1984).
[3] G. Schnurr, A. Schmidt, and G. Sandmann, *FEMS Microb. Lett.* **78**, 157 (1991).
[4] K. L. Perry, T. A. Simonitch, K. J. Harrison-Lavoie, and S. T. Liu, *J. Bacteriol.* **168**, 607 (1986).
[5] N. I. Bishop, in "Methods in Chloroplast Molecular Biology" (M. Edelmann *et al.*, eds.), p. 51. Elsevier Biomedical, Amsterdam, 1982.
[6] S. Ernst and G. Sandmann, *Arch. Microbiol.* **150**, 590 (1988).
[7] G. Sandmann, W. S. Woods, and R. W. Tuveson, *FEMS Microb. Lett.* **71**, 77 (1990).

out in Luria-Bertani (LB) liquid medium containing 10 g/liter Bactotryptone (Difco), 5 g/liter yeast extract, and 10 g/liter NaCl, adjusted to pH 7.5. As the plasmid confers ampicillin resistance, 75 µg/ml of this antibiotic is added. After inoculation, the bacterial cultures are incubated at 30–35° with continuous shaking.

The C-6D mutant of *Scenedesmus obliquus* was provided by Prof. N. Bishop (University of Corvallis, OR). It is grown as described[8] in complete darkness under heterotrophic conditions. Therefore, the liquid cultures are supplemented with 5 g/liter glucose and 2.5 g/liter yeast extract and grown at 20° on a shaker, with a stream of air being bubbled through. Cells are harvested after 7 days.

Extraction of Carotenoids

All extraction steps should be carried out under dim light to avoid photodegradation and isomerization. Lyophilized cells (20 mg) are suspended in 20 ml of methanol, and 2 ml of 60% KOH (w/v) is added. Carotenoids are extracted at 65° in darkness for 20 min. Then the extract is poured into 15 ml of diethyl ether in a separatory funnel, and 10 ml of saturated NaCl solution is added. After partitioning, the upper diethyl ether layer is collected and the partitioning repeated with the lower phase. The combined diethyl ether extract is dried over anhydrous Na_2SO_4, and the solvent is evaporated under reduced pressure. The residue can be stored under nitrogen in a sealed vessel in a freezer or directly dissolved in 100 µl of methanol for further use.

Chromatographic Separation, Photoisomerization, and Reference Compounds

For analysis of carotenoids, the isocratic HPLC system described by Märki-Fischer *et al.*[9] is employed. It is suitable for separation of various carotenes, including their cis/trans isomers, and of different xanthophylls. The column (25 cm) is filled with Spherisorb ODS-1 5 µm, the eluent consists of acetonitrile/methanol/2-propanol (85:10:5, v/v), and the flow rate is 1 ml/min. Absorbances at different wavelengths are measured with a Waters 994 programmable photodiode array detector, and spectra are recorded from the eluent peaks.

Poly-*cis*-carotenes can be converted to their all-*trans* forms by iodine isomerization. The carotenes are dissolved in 2 ml of hexane to give an absorbance at the wavelength of the central spectral maximum of 1.5. This corresponds to a total amount of 10 to 15 µg. Then 10 µl of an iodine

[8] G. Sandmann and P. Böger, *Photosynth. Res.* **2**, 281 (1981).
[9] E. Märki-Fischer, U. Marti, R. Buchecker, and C. H. Eugster, *Helv. Chim. Acta* **66**, 494 (1983).

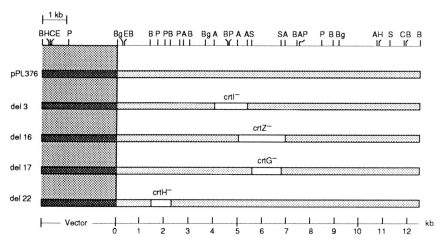

FIG. 1. Deletion mutants in a carotenogenic gene cluster of transformant *E. coli* LE392/pPL376.

solution in hexane (30 μg/ml) is added, and the test tube is illuminated for 30 min with an Osram L40/W25 fluorescence lamp at a distance of 20 cm.

The following reference carotenoids are employed for HPLC: all-*trans*-carotene, neurosporene, and lycopene are isolated from fungal mutants,[6] and pro-ζ-carotene, proneurosporene, and prolycopene from the tomato variety Tangella.[10] 15-*cis*-Phytoene is extracted from cress seedlings germinated for 3 days on filter paper in the dark in water containing 0.5 μM of the bleaching herbicide norflurazon (trade names: Zorial, Evital, or Solican). Zeaxanthin and β-cryptoxanthin are isolated from the cyanobacterium *Aphanocapsa* PCC 6714.[11] Isolation procedures are similar to the one used for *Scenedesmus* and *E. coli*, and purification is performed by two subsequent TLC steps. Carotenes are first separated on silica gel plates with 15% toluene in hexane and then on activated Al_2O_3 plates with 0.7% acetone in hexane. The two subsequent TLC systems for xanthophyll purification are silica plates with toluene/ethyl acetate/methanol (75:20:5, v/v) and then Al_2O_3 plates with 35% acetone in hexane.

Carotenoids of Deletion Mutants from Escherichia coli LE392/pPL376

Deletions in pPL376 are obtained by partial digestion of plasmid DNA with different restriction enzymes and directional cloning of restriction fragments. The details are given elsewhere.[3] Figure 1 shows the 12.4-kb

---

[10] J. M. Clough and G. Pattenden, *J. Chem. Soc. Chem. Commun.*, 616 (1979).
[11] P. M. Bramley and G. Sandmann, *Phytochemistry* **24**, 2919 (1985).

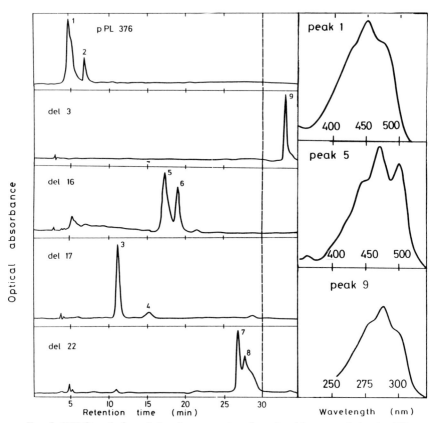

FIG. 2. HPLC analysis and absorbance spectra of carotenoids accumulating in different deletion mutants of *E. coli* LE392/pPL376.

gene cluster from *Erwinia herbicola* which contains six genes encoding the carotenoid pathway to zeaxanthin diglycoside.[3] Deletions in the genes *crtI*, *crtZ*, *crtH*, and *crtG* have been generated. The resulting mutants *del 3, 16, 17,* and *22* accumulate different carotenoid intermediates.

Depending on the growth conditions, *E. coli* LE392/pPL376 accumulates large amounts of colored carotenoids. The dominant pigments (>90%) in *Erwinia herbicola* are zeaxanthin mono- and diglycoside,[7] which are the same carotenogenic end products as in *Erwinia uredevora*.[12] These highly polar glycosides chromatograph at 5 (diglycoside) and 7 min

[12] N. Misawa, M. Nakagawa, K. Kobayashi, S. Yamano, Y. Izawa, K. Nakamura, and K. Harashima, *J. Bacteriol.* **172,** 6704 (1990).

(monoglycoside) in the reversed-phase HPLC system used in Fig. 2. All deletion mutants have been analyzed for carotenoid content by HPLC with corresponding standards for all the carotenoids found. The deletion mutant *del 3* did not accumulate any colored carotenoid. Instead, 15-*cis*-phytoene was found at R_t 33.1 min (peak 9, Fig. 2). It showed its typical UV spectrum with absorbance maxima at 276, 287, and 298 nm. Two carotenes with a lycopene-like spectrum were found in strain *del 16* at R_t 16.8 (absorbance maxima at 446, 471, and 502 nm) and 18.5 min (peaks 5 and 6, Fig. 2). The fine structure in the region between 330 and 380 nm is typical for an all-trans isomer of lycopene in case of peak 5 (Fig. 2) and for a mono-cis configuration of one of the central double bonds in peak 6 (Fig. 2). In *del 17* the major peak (peak 3, Fig. 2) consisted of zeaxanthin (R_t 10.8 min). In addition, small amounts of its monohydroxylated precursor β-cryptoxanthin (peak 4, Fig. 2; R_t 15.5 min) were present; *del 22* accumulated only nonhydroxylated all-*trans*-β-carotene (peak 7, Fig. 2; R_t 26.9 min) and a cis isomer (peak 8, Fig. 2; at 27.7 min).

The carotenoid pattern in these deletion mutants indicates that in *del 3* pytoene desaturase encoded by the gene *crtI* is inactivated. In *del 16*

TABLE I
ABSORBANCE MAXIMA AND RETENTION TIMES OF all-*trans* AND POLY-*cis*-CAROTENES FROM *Scenedesmus* C-6D

Carotenes	Absorbance maximum (nm) in eluent for HPLC			R_t (min)
A. Reference compounds for all-*trans* pathways				
Lycopene (all-*trans*)	440	471	498	14.67
Neurosporene (all-*trans*)	416	439	468	15.33
γ-Carotene (all-*trans*)	440	460	490	16.97
ζ-Carotene (all-*trans*)	382	402	427	18.40
α-Carotene (all-*trans*)	421	445	472	21.92
β-Carotene (all-*trans*)	425	452	477	22.82
Phytofluene (all-*trans*)	332	349	368	25.00
Phytoene (15-*cis*)	277	286	298	27.83
B. Carotenes from *Scenedesmus* C-6D				
Prolycopene (7,9,7′,9′-poly-*cis*)	418	438	460	16.85
Proneurosporene (9,7′,9′-poly-*cis*)	408	432	459	18.65
cis-ζ-Carotene II (?-*cis*)	382	402	427	19.00
Pro-ζ-carotene (9,9′-di-*cis*)	382	402	427	20.55
Prophytofluene (9,15-di-*cis*)	332	349	368	25.00
Phytofluene II (?-*cis*)	332	349	368	25.00
Phytoene (15-*cis*)	277	286	298	27.83

lycopene cyclization (gene *crtZ*) is impaired, and *del 17* with a modified *crtH* gene lacks expression of β-carotene hydroxylase. Finally, in mutant *del 22* the glycosylation step (gene *crtH*) is missing.

Poly-cis-carotenes from Scenedesmus C-6D

In *Scenedesmus* C-6D only acyclic carotenes are present. None of them exhibits the usual all-trans configuration which is typical for colored carotenoids. The HPLC retention times of all-*trans*-carotenes which normally participate in the biosynthetic pathway are shown in Table I together with their absorbance maxima. The only carotene from C-6D also formed by wild-type *Scenedesmus* or other photosynthetic organisms is 15-*cis*-phytoene.

Aside from phytofluene II and ζ-carotene II, the isomeric configurations could be assigned as the ones known from the series of procarotenes found in the tomato mutant Tangella.[11] By cochromatography with carotenes from this fruit, prophytofluene, pro-ζ-carotene, proneurosporene, and prolycopene can be identified. In addition, proneurosporene and prolycopene show absorbance spectra which are different from those of the corresponding all-trans isomers. For the separation of phytofluene isomers the HPLC solvent system is modified to acetonitrile/methanol/2-propanol (95:3:2, v/v) at a flow rate of 0.6 ml/min.[10]

An additional possibility to identify *cis*-carotenoids is by iodine-catalyzed photoisomerization to all-trans and other cis isomers. Then, changes in their spectral properties and separation characteristics are observed. The isomerization conditions employed cause the complete conversion of poly-*cis*-carotenes to several other cis isomers or into all-trans forms. These isomers are formed from pro-ζ-carotene (Fig. 3A). ζ-Carotene I runs with the same retention time on HPLC as pro-ζ-carotene. However, in contrast to pro-ζ-carotene, the spectrum of ζ-carotene I exhibits a pronounced cis peak at 297 nm and a different ratio of the two main absorbance maxima. Both maxima show a shift of 3 nm to lower wavelengths compared to pro-ζ-carotene. ζ-Carotene II is the same cis isomer already specified in Table I. At the shortest retention time of all-ζ-carotene isomers, all-*trans*-ζ-carotene ($t - \zeta C$) can be identified with a standard and by its spectrum. In case of proneurosporenes this isomer and all the others obtained after isomerization can be well separated from each other by HPLC (Fig. 3B). They also differ by the position of the main absorbance maximum. The major isomerization products are two different cis isomers, whereas all-*trans*-neurosporene is found to a lesser extent. The spectrum of prolycopene shows the main absorbance peak at the lowest wavelengths of all lycopene isomers. After isomerization no significant

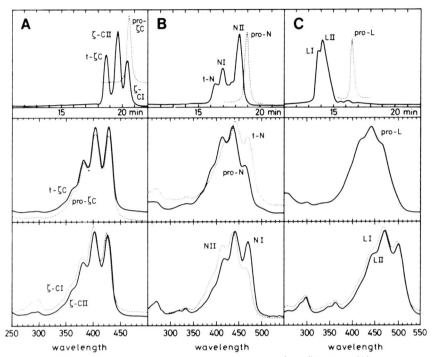

FIG. 3. HPLC separation of photoisomerization products of pro-ζ-carotene (A), proneurosporene (B), and prolycopene (C).

amounts of all-*trans*-lycopene could be detected by cochromatography (Fig. 3C). The spectra of the two lycopene isomers found are those of cis isomers as indicated by their cis peaks at 362 nm and a wavelength shift of the main absorbance maximum

Acknowledgments

This work was supported by the Deutsche Forschungsgemeinschaft.

[31] Effects of δ-Aminolevulinic Acid and 3-Amino-1,2,4-Triazole on Carotenoid Accumulation

By GEORGES T. DODDS, SYAMALA S. ASHTAKALA, and STEPHEN W. LAMOUREUX

Introduction

Although a committed precursor of tetrapyrroles, δ-aminolevulinic acid (ALA) stimulates carotenoid accumulation when applied exogenously to cucumber cotyledons[1,2] and also to *Euglena gracilis*.[3] We showed[1] that this stimulation is reversed by the chlorosis-inducing herbicide 3-amino-1,2,4-triazole (AT), whose mode of action has been reviewed elsewhere.[4] Our experience has been with cotyledons of the cucumber *(Cucumis sativus)* cv. Wisconsin SMR 18 K44,[1] which have also been used in other studies including the effect of cytokinins on chlorophyll accumulation.[5,6] In using other cucumber cultivars it may be necessary to redetermine optimal conditions. Similar systems were developed with various monocot leaves and with radish seedlings (see Ref. 1).

Methods

An outline of the procedure is given in Table I. It is impractical to generate the cucumber cotyledons in a completely aseptic manner, but the use of initially sterile solutions, glassware, and instruments significantly reduces variability and minimizes losses due to contamination. Exogenously applied ALA has been shown to stimulate carotenoid accumulation in *Euglena gracilis*[3] and cucumber cotyledons.[1,2] The mode of action of the chlorosis-inducing herbicide AT has been reviewed.[4]

Cucumber seeds are generally coated with pink fungicide, which is removed by washing in a sieve with cool tap water for about 1 min, followed by surface sterilization in 5% commercial bleach for 5 min. Subsequent washing with sterile distilled water improves germination but is not absolutely necessary. Seeds, in 50-ml batches, are transferred to 100 ml of 2 mM sodium phosphate buffer with or without AT (pH 6.2)

---

[1] S. S. Ashtakala, G. T. Dodds, and S. W. Lamoureux, *J. Plant Physiol.* **135**, 86, (1989).
[2] L. Huang and N. E. Hoffman, *Plant Physiol.* **94**, 375 (1990).
[3] J. T. O. Kirk, *Planta* **78**, 200 (1968).
[4] M. C. Carter, *in* "Herbicides: Chemistry, Degradation, and Mode of Action" (P. C. Kearney and D. D. Kaufman, eds.), p. 377. Dekker, New York, 1975.
[5] R. A. Fletcher, V. Kallidumbil, and P. Steele, *Plant Physiol.* **69**, 675 (1982).
[6] R. A. Fletcher and D. McCullagh, *Can. J. Bot.* **49**, 2197 (1971).

TABLE I
PROCEDURE FOR EXPERIMENTS ON ACCUMULATION OF CAROTENOIDS IN CUCUMBER COTYLEDONS

1. Seeds are washed with water then sterilized 5 min in 5% commercial bleach
2. Seeds (50 ml) are placed in 100 ml of 2 mM sodium phosphate buffer ± AT[a] (pH 6.2) in a 250-ml flask and placed in a low-speed shaker for 24 hr in the dark (imbibition)
3. In the dark,[b] imbibed seeds are spread on a bed of vermiculite in a plastic tray (30 × 60 cm), covered with 1 cm of vermiculite, wetted with about 1.5 liters water, and placed in the dark
4. After 4 days of growth, cotyledon pairs are harvested in the dark by cutting the seedling hypocotyl 0.5 mm below the cotyledon–hypocotyl junction
5. In the dark, replicates of 10 cotyledon pairs are weighted to within 0.1 mg and each group placed in a 6-cm petri dish containing 3 ml of 2 mM sodium phosphate buffer ± AT (1 mM), ±ALA (10 mM) (pH 6.2)
6. In the dark, tissue in covered petri dishes is vacuum infiltrated 3 times for 1 min, then placed for 24 hr in the dark (preincubation)
7. Tissue is placed in the light (32 μmol m^{-2} sec^{-1}, from both top and bottom) for 6 hr when ALA is absent or 30 min if ALA is present (incubation)
8. Carotenoids are measured[c] and expressed in mg/g fresh weight

[a] Note that AT is considered a possible carcinogen and should be handled appropriately.
[b] For all work in the dark, a green safelight was used.
[c] For methods, see Volume 213, this series.

(enough to just cover the seeds) in sterile 250-ml Erlenmeyer flasks. The addition of AT at this stage leads to much higher levels of inhibition of pigment accumulation than simple preincubation (Table I) alone.[1,7] The highest concentration of AT to inhibit pigment accumulation without substantially inhibiting or altering radicle development is generally 1 mM, but AT concentrations ranging from 10 to 0.1 mM should be tried when other cucumber cultivars are used. The sealed flasks are placed on a low-speed (just enough to keep seeds moving) rotary shaker for 24 hr, in the dark. If the shaker is left at high speed the tiny emerging radicle tends to get damaged.

Imbibed seeds are spread in a single layer on a bed of vermiculite in plastic trays (30 × 60 cm). Rinsing of seeds decreases the AT effect.[1,7] Seeds are covered with a 1-cm layer of vermiculite and wet with 1.5 liters of water. Treatment groups should be planted in different trays since some diffusion of AT occurs through the vermiculite. The effect of watering with a solution containing AT has not been investigated.[1] There is no apparent effect on seedling growth whether trays are placed in complete darkness in

[7] W. Rudiger and J. Benz, Z. Naturforsch. **34c**, 1055 (1979).

a sealed cupboard at room temperature or in an incubator (22°, 80-90% relative humidity).

The first seedlings begin to emerge at 4 days (including 1 day of preimbibition), but the majority of seedlings are obtained only at or after 5 days. Cotyledons which optimized subsequent chlorophyll accumulation in the light were 5 days old,[1] but this should be determined before proceeding to formal experiments.

Under the illumination of a green safelight, hypocotyls are cut off at the surface of the vermiculite and the cotyledon pairs excised on a sterile glass plate with a sharp sterile razor blade or scalpel. Having these initially sterile is good practice but not essential. Excess hypocotyl is removed by slicing (chopping tends to crush the hypocotyl) as close as possible to the cotyledons without detaching them (~0.5 mm), which minimizes variability.[6] Any damaged cotyledons are discarded. Cut cotyledons are kept in a covered 10-cm petri dish with moistened filter paper until they are weighed, since they tend to dry very quickly when left in the open. Weights (within 0.1 mg) of replicates consisting of 10 cotyledon pairs are recorded. Cutting and weighing are best done by two or more people. Each replicate is placed in 3 ml of solution in a sterile, acid-washed, 6-cm glass petri dish, which permits better light transmission (less prone to scratching) than plastic dishes, and subjected to baking to remove contaminant traces. In the absence of an oven, dishes are washed first with soap (Alconox, New York, NY), rinsed, soaked 24 hr in acid (30 g CrO_3 in 100 ml water, adjusted to 1 liter with concentrated HNO_3), rinsed in running tap water for 30 min, then rinsed twice with distilled water for successful cleaning.[1] Larger petri dishes (10 cm) may be used with approximately 50 cotyledon pairs; however, this involves considerable cutting if the same number of replicates are used.

Cotyledon pairs were bathed but not submerged in the sterile preincubation solution,[6] thus only 3 ml of a solution containing the appropriate additions was needed for a 6 cm petri dish with 10 cotyledon pairs. A pH between 5.8 and 6.0 was suggested as optimal,[6] however pH 6.2 has also been used[1]; therefore, an optimal pH between 5.5 and 6.5 should be determined. The substitution of 40mM KCl for sodium phosphate buffer enhanced a response to cytokinins[5] but was not buffered. The substitution of 100mM sodium phosphate buffer (pH 6.2), 2mM or 100mM Tris (pH 6.3) for 2mM sodium phosphate buffer (pH 6.2) had no significant effect on subsequent chlorophyll accumulation. However, substitutions of 2mM or 100mM HEPES (pH 6.3) for 2mM sodium phosphate buffer (pH 6.2) were both highly inhibitory (30% and 50% respectively). The addition of 10% dimethyl sulfoxide (DMSO) to the preincubation medium also severely inhibited subsequent chlorophyll accumulation,[1] although it was used successfully in such a system before.[8] The addition of AT to the buffer has little effect on pH, but the addition of ALA requires NaOH to readjust the pH. Furthermore, solutions containing AT can be autoclaved, but ALA must be added to cold sterile buffer by filter sterilization.

The covered petri dishes are then placed in a vacuum desiccator, with a vacuum (from flowing water) applied for 1 min and then released, three times in succession before preincubation in the dark for 24 hr[1] or 18 hr[6] at room temperature. The preincubation time should be optimized prior to formal experiments; with too short a preincubation absorption of exogenous compounds is poor, whereas contamination with microorganisms may become a problem with long preincubations.

With incubation in the light it is essential that equal light intensities are obtained from both above and below the cotyledon pairs. A light rack with banks of fluorescent lights above and below a glass shelf on which the dishes are placed is easily constructed with a slotted angle framework. Light intensity can be varied by changing the distance between the shelf and lights, or by interposing filters of cheesecloth or fiberglass. It can be very convenient to have the lights turned on by a timer. An incubator (Sherer Model RT 24B-SE, Sherer-Gillett Co., Marshall, MI) equipped with a full-size glass shelf and two pairs of 40-W Sylvania cool white bulbs (one above, one below) gave a light intensity[9] of 32 μmol m^{-2} sec^{-1}. This is higher than intensities reported by others [22 μmol m^{-2} sec^{-1} (Ref. 6); 24 μmol m^{-2} sec^{-1} (Ref. 8); 24–26 μmol m^{-2} sec^{-1} (Ref. 10)], but with this system and at this light intensity chlorophyll accumulation showed greater differences between 0, 0.1, 1.0, and 10 mM AT concentrations than at 18–20 or 50–52 μmol m^{-2} sec^{-1}. The light intensity should be measured in μmol m^{-2} sec^{-1} (=μE m^{-2} sec^{-1}) or W m^{-2}. The use of an incubator allows for constant temperature and relative humidity throughout dark preincubation and light incubation.

Some condensation tends to form on the lid of the petri dish in an incubator even with the relative humidity set high (80–90%), thus somewhat dispersing the incoming light. To avoid this no lid should be used and small amounts of the appropriate solution added at intervals to maintain the level of solution (e.g., after 2 and 4 hr of a 6-hr light incubation). The total duration of incubation in the light will depend on the nature of the experiment. For experiments not involving ALA, 6 hr of light maximized pigment accumulation while maintaining a reasonably low level of variability, but a 4-hr light incubation has also been used.[5,6] When ALA is

---

[8] S. I. Beale and P. A. Castelfranco, *Biochem. Biophys. Res. Commun.* **52**, 143 (1973).

[9] Light intensities reported here were measured in foot-candles; they have been converted to μmol m^{-2} sec^{-1} using the approximation of 2000 fc ≈ 200 μmol m^{-2} sec^{-1} for fluorescent bulbs.

present during preincubation a light incubation of 30 min is sufficient, because tissue breakdown begins after 30 min and after 6 hr the tissue is brown and necrotic.[1,10] An ALA concentration of 10 mM during preincubation[1,10] results in total carotenoid levels that are 2-3 times of those of control cotyledons either when dark incubated for 6 hr or light incubated for 30 min. In the dark, ALA stimulation of carotenoid accumulation was unaffected by AT (1 mM), but in the light (30 min) AT (1 mM) halved the ALA stimulation. Although lower ALA concentrations may be effective, higher ones tend to be toxic. Solutions containing ALA are also particularly prone to bacterial proliferation, so extra care should be taken to minimize contamination.

Carotenoids are measured after an appropriate incubation time.[11] Tissue can be frozen in the dish with liquid N_2 to stop the light incubation and placed in a freezer until needed. Alternatively, samples can be stored for a short time in the dark, prior to pigment extraction.

This system lends itself well to the addition of many exogenous compounds besides AT and ALA. It is simple and reliable (coefficient of variation of generally between 5 and 10%) and relatively inexpensive.

[10] P. A. Castelfranco, P. M. Rich, and S. I. Beale, *Plant Physiol.* **53**, 615 (1974).
[11] This series, Volume 213.

[32] Plant Phytoene Synthase Complex: Component Enzymes, Immunology, and Biogenesis

By BILAL CAMARA

Introduction

In plants, the synthesis of phytoene, the first C_{40} carotenoid, requires the coordinated activity of the different component enzymes of the phytoene synthase complex which catalyze the sequence of reactions presented in Fig. 1. The entire pathway is readily demonstrated *in vitro* using isolated plastids.[1-4] However, despite the conservation of this pathway, isolated

[1] B. Camara, F. Bardat, and R. Monéger, *Eur. J. Biochem.* **127**, 255 (1982).
[2] . Dogbo, F. Bardat, A. Laferrière, J. Quennemet, J. Brangeon, and B. Camara, *Plant Sci. (Limerick, Irel.)* **49**, 89 (1987).
[3] B. Camara, this series, Vol. 110A, p. 244.
[4] H. Kleinig and P. Beyer, this series, Vol. 110A, p. 267.

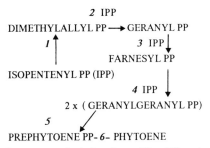

FIG. 1. Pathway of phytoene synthesis in plants. The different steps are catalyzed by operationally soluble enzymes: isopentenyl PP isomerase *(1)*, geranylgeranyl PP synthase *(2, 3, 4)*, and phytoene synthase *(5, 6)*, collectively termed the phytoene synthase complex.

nongreen plastids are by far more active than chloroplasts. This chapter reports general considerations concerning the isolation of chloroplasts and chromoplasts from *Capsicum* fruits and their use for three specific purposes, namely, the characterization of the phytoene synthase component enzymes, their immunochemistry, and their biogenesis.

Isolation of Plastids and Subfractions

The plastids are purified according to the procedures outlined in Fig. 2. The purity of these fractions is illustrated in Figs. 3 and 4. At the completion, the purified plastid suspensions are diluted to 0.4 M sucrose and pelleted after centrifugation for 10 min at 4000 g. The pellet is homogenized in a Potter homogenizer with 50 mM Tris-HCl (pH 7.6) containing 5 mM dithiothreitol (DTT). The plastic stroma obtained after centrifugation for 1 hr at 100,000 g is saved and used for the purification of phytoene synthase component enzymes. This preparation is stable for up to 1 year at $-20°$, especially if the protein concentration is high (≥ 10 mg/ml).

Component Enzymes of *Capsicum* Phytoene Synthase Complex

Initially, the individual component enzymes of the phytoene synthase complex [isopentenyl pyrophosphate (PP) isomerase (EC 5.3.32), geranylgeranyl PP synthase (EC 2.5.1.1), and phytoene synthase (EC 2.5.1.32)] were separated after concentrating the stromal extract with polyethylene glycol 6000 followed by DEAE-Sephacel (Pharmacia, Piscataway, NJ) and aminophenethyl chromatography.[5,6] Owing to the time consumed in mak-

---

[5] O. Dogbo and B. Camara, *Biochim. Biophys. Acta* **920**, 140 (1987).
[6] O. Dogbo and B. Camara, *Proc. Natl. Acad. Sci. U.S.A.* **85**, 7054 (1988).

HOMOGENIZATION
  Pericarp tissue of firm green or red fruit (2 kg fresh weight) is chopped into 2 liters of chilled isolation medium (1 mM 2-mercaptoethanol, 1 mM EDTA, 0.4 M sucrose, Tris-HC1, pH 8)
↓
DISRUPTION (2 times for 1 sec each in a Waring blender); pH readjusted to 8 with concentrated Tris
↓
FILTRATION (through 4 layers of Blutex, 50 μm apertures)
↓
CENTRIFUGATION
  Sorvall GSA rotor, 150 g for 5 min

Plastid supernatant	Pellet discarded

Sorvall GSA rotor (2200 g for 30 sec for chloroplasts and 10 min for chromoplasts)

Pellet	Supernatant discarded

↓ (crude plastids washed twice with the isolation medium)

SUCROSE STEP GRADIENT
  Chloroplast suspension (10 ml) is layered on top of a sucrose gradient (0.75, 1, 1.5 M, 10 ml per fraction) containing 1 mM 2-mercaptoethanol and 50 mM Tris-HC1, pH 7.6
  Centrifugation: Sorvall Hb4 rotor, 1000 g for 15 min
  Interface 0.75–1 M (broken membranes)
↓ Interface 1–1.5 M Intact chloroplasts

  Chromoplast suspension (2 ml) is layered on top of a sucrose gradient (0.45, 0.84, 1.45 M, 9 ml per fraction) containing 1 mM 2-mercaptoethanol and 50 mM Tris-HC1, pH 7.6
  Centrifugation: Beckman SW 27 rotor, 62,000 g (R_{max}) for 1 hr
  Interface 0.45–0.84 M (broken membranes)
↓ Interface 0.84–1.45 M Intact chromoplasts

FIG. 2. Method for the isolation of *Capsicum* chloroplasts and chromoplasts.

ing the column and its instability after several runs, we have devised a more expeditious procedure. Accordingly, solid NaCl is added to the stromal fraction to bring the solution to 1 M NaCl. The resulting solution is loaded onto a Phenyl-Sepharose (Pharmacia) column (2.5 × 25 cm) equilibrated and eluted with buffer A (50 mM Tris-HCl, pH 7.6, containing 20% glycerol and 5 mM 2-mercaptoethanol) supplemented with 1 M NaCl. The flow-through and the following 1 M NaCl washing remove up to 90% of the isopentenyl PP isomerase and the endogenous phosphatase activities (fraction I). The fraction enriched with geranylgeranyl PP synthase activity is eluted with a decreasing gradient of 1 to 0 M NaCl (fraction II). Indeed most of the geranylgeranyl PP synthase activity elutes between 0.5 and 0 M NaCl. Finally the phytoene synthase activity *stricto sensu* is eluted with buffer A containing 40% dimethyl sulfoxide (DMSO). The presence of dimethyl sulfoxide inhibits the phytoene synthase activity (80%). Full activity is restored by applying the 40% dimethyl sulfoxide solution directly to a column of Q-Sepharose (Pharmacia) equilibrated with buffer A. After washing with buffer A (4 to 5 bed volumes), the phytoene synthase activity is eluted with 0.3 M NaCl in buffer A (fraction III).

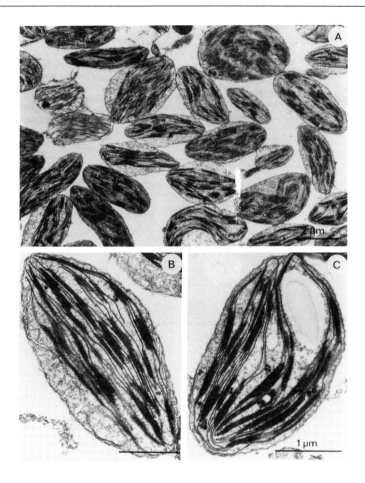

FIG. 3. Electron micrographs showing the integrity and purity of chloroplasts isolated from *Capsicum* pericarp. (A) Appearance of a typical purified chloroplast fraction; (B, C) detailed structure of purified chloroplasts depicted in (A).

Characterization of Isopentenyl Pyrophosphate Isomerase

Enzyme Assay Procedure. Isopentenyl PP isomerase is routinely assayed at 30° by dimethylallyl PP formation in an Eppendorf tube containing the following reaction mixture (0.1 ml final volume): 50 mM Tris-HCl (pH 7.6), 5 mM MgCl$_2$, 1 mM MnCl$_2$, 2 mM dithiothreitol, 10 μM [1-^{14}C]isopentenyl PP, and enzyme extract. Alternatively, 10 mM KF is in-

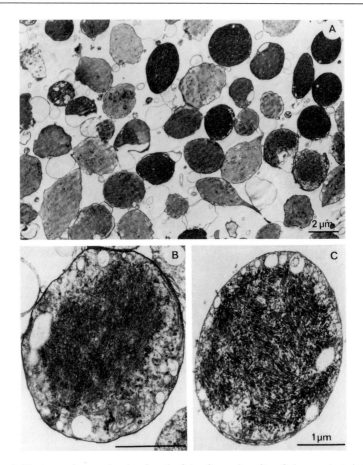

Fig. 4. Electron micrographs showing the integrity and purity of chromoplasts isolated from *Capsicum* pericarp. (A) Appearance of a typical purified chromoplast fraction; (B, C) detailed structure of purified chromoplasts depicted in (A). Note that the high osmolarity of the gradient and the high speed of centrifugation induce the formation of condensed stroma beneath the increased sucrose permeable space.

cluded to test for the presence of phosphatases. The reaction is stopped after 15 min of incubation by adding 0.5 ml of concentrated HCl/methanol (1:4, v/v) followed by incubation at 37° for 15 min. The reaction mixture is extracted with 0.5 ml of chloroform, and after centrifugation an aliquot of the lower organic phase is used for liquid scintillation counting.

Purification. Fraction I containing isopentenyl PP isomerase is diluted to 0.25 M NaCl with buffer A and subjected to hydroxyapatite chromatography (column 2.5 × 15 cm), prepared by mixing dropwise a 4.4% (w/v) solution of $CaCl_2$ and a 7.1% (w/v) solution of Na_2HPO_4 (Ref. 7) followed by repeated washing with water. The column is equilibrated with buffer A, and elution is done with a 10 mM to 0.2 M Na_2HPO_4 linear gradient in buffer A. Peak fractions are pooled, and after dilution (5 times) the enzyme solution is applied to a Q-Sepharose column (2.5 × 15 cm) eluted with a 0 to 0.4 M NaCl linear gradient in buffer A. Pooled peak fractions of activity are concentrated by dialysis against buffer A saturated with polyethylene glycol 20,000 and applied to a Sephacryl S-200 (Pharmacia) column (2.5 × 45 cm) equilibrated with buffer A containing 0.15 M NaCl. Active fractions are adjusted to 10 mM $MgCl_2$ and eluted through a column (1 × 5 cm) of Reactive Red 120 agarose (Sigma, St. Louis, MO) equilibrated with buffer A containing 0.15 M NaCl and 10 mM M_gCl_2. Unadsorbed isopentenyl PP isomerase is adjusted to 50% (v/v) glycerol and stored at $-20°$.

Properties. Homogeneous isopentenyl PP isomerase is a 33-kDa monomer. The pH optimum is about 7.5.(5) The K_m value for isopentenyl PP is 6 μM. The enzyme requires Mn^{2+} or Mg^{2+} for maximal activity. The activity as measured by the conversion of isopentenyl PP to dimethylallyl PP is more than 90% inhibited by 20 μM dimethylallyl PP. The enzyme is sensitive to thiol directed reagents.

Characterization of Geranylgeranyl Pyrophosphate Synthase

Enzyme Assay Procedure. The assay mixture (0.1 ml) contains 50 mM Tris-HCl, pH 7.6, 5 mM $MgCl_2$, 1 mM MnCl, 2 mM dithiothreitol, 5 μM [1-^{14}C]isopentenyl PP, 20 μM dimethylallyl PP, geranyl PP, or farnesyl PP, and the enzyme extract. After incubation for 20 min the reaction is stopped and the mixture processed as described above in the isomerase section.

Purification. Solid NaCl is added (60 g/ml) to fraction II enriched with geranylgeranyl PP synthase activity. The resulting solution is applied to a column of Phenyl-Sepharose (2.5 × 15 cm) equilibrated with buffer A containing 1M NaCl. The column is washed by a 1 and 0.75 M NaCl step gradient in buffer A followed by a 0.5–0 M NaCl linear gradient in buffer A. Active fractions are diluted three times and adsorbed to a hydroxyapatite column (2.5 × 15 cm) (prepared as described above in the isomerase section). Geranylgeranyl PP synthase is eluted with a 10 mM to 0.3 M Na_2HPO_4 linear gradient in buffer A. The pooled peak fractions are diluted 10 times, and the solution is adsorbed onto a Q-Sepharose column (2.5 × 15 cm) preequilibrated with buffer A. Active fractions are eluted

---
[7] C. K. Mathews, F. Brown, and S. S. Cohen, *J. Biol. Chem.* **239**, 2957 (1964).

with a 0-0.4 M NaCl linear gradient in buffer A. After dilution (5 times), the pooled peak fractions previously adjusted to 10 mM MgCl$_2$ are adsorbed onto a Reactive Red 120 agarose column (2.5 × 15 cm) preequilibrated with buffer A containing 10 mM MgCl$_2$. The column is eluted with a 0.1-0.5 M NaCl linear gradient in the same buffer. If impurities remain, the active fractions are diluted (4 times) and adsorbed onto a small Q-Sepharose column (1.5 × 5 cm). The column is washed with buffer A devoid of 2-mercaptoethanol before eluting geranylgeranyl PP using the same buffer adjusted to 0.3 M NaCl. Finally, the enzyme solution is applied onto a small (1.5 × 3 cm) column of Affi-Gel 501 (Bio-Rad, Richmond, CA) washed with buffer A containing 0.3 M NaCl but devoid of 2-mercaptoethanol. Geranylgeranyl PP synthase is eluted specifically with buffer A containing 10 mM dithiothreitol. Pooled fractions are concentrated against dry polyehtylene glycol 20,000 and dialyzed against buffer A containing 0.5 M NaCl before storing at $-20°$.

Properties. Geranylgeranyl PP synthases from all preparations thus far examined are dimeric proteins composed of apparently identical subunits of approximately 36-37 kDa and have a pH optimum of about 7-7.6. The K_m values are as follows: 3 μM for isopentenyl PP; 0.95 μM for dimethylally pp; 1 μM for geranyl PP; and 1.2 μM for farnesyl PP.[5] The enzyme is sensitive to thiol-directed reagents.

Characterization of Phytoene Synthase

Enzyme Assay Procedure. Two methods are used according to the presence or the absence of isopentenyl PP isomerase and geranylgeranyl PP synthase. In the presence of these companion enzymes, the reaction mixture (0.2 ml final volume) contains 50 mM Tris-HCl buffer, pH 7.6, 5 mM MgCl$_2$, 2.5 mM MnCl$_2$, 2 mM dithiothreitol, 10 μM [1-^{14}C]isopentenyl PP, and enzyme extract. After 20 min to 1 hr of incubation at 30° the reaction is stopped by adding 0.5 ml of chloroform/methanol (2:1, v/v). After addition of unlabeled standards (geranylgeraniol and phytoene), labeled phytoene in the chloroform phase is purified by thin-layer chromatography on silica gel plates developed with benzene/ethyl acetate (90:10, v/v) or by HPLC using a C$_{18}$ μBondapak column (Waters, Milford, MA) and methanol/cyclohexane/hexane (80:10:10, v/v) as the mobile phase.[5] The incorporated radioactivity is determined by liquid scintillation counting. In the second method, the same incubation mixture is adopted except that [1-^{14}C]isopentenyl PP is replaced by 2 μM [1,5,9-^{14}C]geranylgeranyl PP enzymatically synthesized from dimethylallyl PP and [1-^{14}C]isopentenyl PP using the geranylgeranyl PP synthase purified above.

Purification. Fraction III containing the bulk of phytoene synthase is treated as described for the purification of geranylgeranyl PP synthase except that during the Reactive Red 120 agarose chromatography a 0.1–1.5 M NaCl linear gradient in buffer A containing 10 mM MgCl$_2$ is used to elute the enzyme. If impurities remain, the active fractions are diluted (10 times) and chromatographd rapidly onto a small column of Q-Sepharose and Affi-Gel 501 as described for geranylgeranyl PP synthase.

Properties. Purified phytoene synthases from all preparations thus far examined are monomeric proteins of approximately. The K_m values for geranylgeranyl PP and prephytoene PP are 0.30 and 0.27 μM, respectively. The activity of the enzyme is strictly dependent on Mn^{2+}. The enzyme is sensitive to arginine-directed reagents.[6]

Immunochemistry of Phytoene Synthase Complex Component Enzymes

The different component enzymes of the phytoene synthase complex are used to elicit immune responses in rabbits according to standard protocols.[8] The different polyclonal antibodies obtained under these conditions are not active site directed, since they fail to inhibit directly *in vitro* the enzymatic activity. Their specificity is evaluated as described below.

Immunoinhibition Assay Procedure

Samples of soluble fractions (200 μl) are incubated with antibodies against isopentenyl PP isomerase, geranylgeranyl PP synthase, or phytoene synthase (5–20 μl) for 30 min at room temperature, then for 2 hr to overnight at 4°. Thereafter two kinds of immunoprecipitations can be used. In the first method, protein A-Sepharose beads (Sigma; twice the volume of preimmune serum or antibodies) are added to the protein solution, and the suspension is incubated at 4° for 1 hr while swirling to disperse any clumps that may form. The protein A-Sepharose–antibody–antigen complex is pelleted by centrifugation at 17,000 rpm for 20 min in JA 18.1 rotor (Beckman). The resulting supernatant is used for enzyme assay. In the second method, the antibody–antigen complex is precipitated without the addition of protein A-Sepharose beads. Depending on the concentration of antibodies used, preparations obtained in these manners usually have less than 50 to 90% activity compared to controls.

---

[8] N. Harboe and A. Ingild, *in* "Quantitative Immunoelectrophoresis" (N. H. Axelsen, J. Kroll, and B. Weeks, eds.), p. 161. Blackwell, Oxford, 1977.

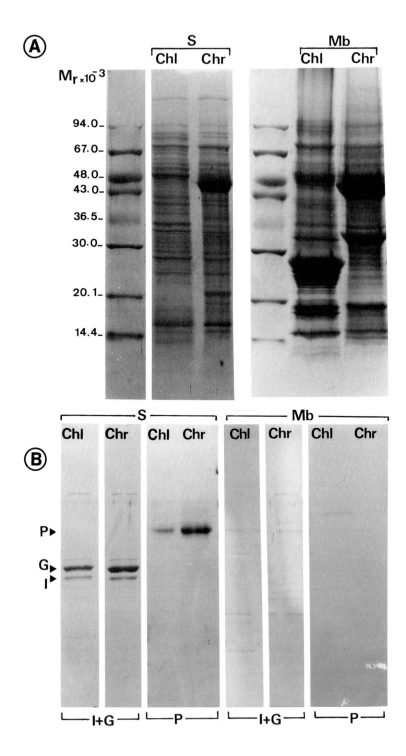

Western Blotting

Proteins are separated by one-dimensional sodium dodecyl sulfate (SDS)-10% acrylamide gel electrophoresis.[9] The transfer of proteins from the gel to nitrocellulose or Immobilon-P is performed in a cold room at 4° using a Transphor (Hoefer Scientific Instruments) and buffered (pH 8.3) transfer medium containing 0.3% (w/v) Tris, 0.15% (w/v) glycine, and 20% (v/v) methanol. Alternatively a Novablot (LKB) apparatus is used with a transfer medium containing 0.3% (w/v) glycine, 0.6% (w/v) Tris, 0.05% (w/v) sodium dodecyl sulfate, and 20% (v/v) ethanol. The blotting is carried out for 2 to 4 hr. The proteins are visualized using specific antibodies (1/1000 dilution) against the individual enzymes according to the procedure described.[10] A typical result is presented in Fig. 5, showing that isopentenyl PP isomerase, geranylgeranyl PP synthase, and phytoene synthase are quantitatively recovered in the soluble stromal fraction of chloroplasts and chromoplasts. Furthermore, the data presented demonstrate that these enzymes are more highly expressed in chromoplasts compared to chloroplasts.

Immunocytochemistry

Pericarp tissue from green and red fruits are fixed at 4° (2 times, 15 min each) in the presence of 2% paraformaldehyde and 1% glutaraldehyde dissolved in a medium containing 25 mM KH_2PO_4-Na_2HPO_4 buffer (pH 7.2) supplemented with 0.1 M sucrose. The tissues are embedded in Epon.[11] Polymerization is carried out at 60° for 2 to 3 days. Sections (1 to 1.5 μm thick) are cut and mounted on glass slides. The sections are overlaid with phosphate-buffered saline (PBS) containing 1% (w/v) bovine serum albumin and 0.1% (w/v) Triton X-100 (BT-PBS) and containing preimmune serum or diluted (1/100 for green tissues and 1/1000 for red tissues) antigeranylgeranyl PP synthase or antiphytoene synthase antibod-

---

[9] B. D. Hames, in "Gel Electrophoresis of Proteins: A Practical Approach" (B. D. Hames and D. Rickwood, eds.), p. 1. IRL Press, Oxford, 1984.
[10] M. S. Blake, K. H. Russel-Jones, and E. C. Gotslich, *Anal. Biochem.* **136**, 175 (1984).
[11] J. H. Luft, *J. Biochem. Cytol.* **9**, 409 (1961).

FIG. 5. Electrophoretic analysis followed by immunoblotting of the different plastid subfractions. (**A**) Stromal (S) and membrane proteins (Mb) from chloroplasts (Chl) and chromoplasts (Chr) were separated by SDS-polyacrylamide gel electrophoresis. Size standards are indicated at left. (**B**) After electrophoretic transfer the polypeptides were visualized with antiisopentenyl PP isomerase antibodies mixed with antigeranylgeranyl PP synthase antibodies (I + G) or with antiphytoene synthase antibodies (P). Arrowheads indicate positions of isopentenyl PP isomerase (I), geranylgeranyl PP synthase (G), and phytoene synthase (P) subunits.

ies. Following incubation in a moist chamber and washing 3 times (over a 40-min interval) with BT-PBS, the slides are drained and overlaid with 1% (w/v) fluorescein isothiocyanate-coupled goat anti-rabbit immunoglobulin G (Sigma), for up to 2 hr in the dark. Then the slides are washed with BT-PBS followed by phosphate buffer and finally treated with 0.02% Evan's Blue (Merck, Darmstadt, Germany). The slides are subsequently

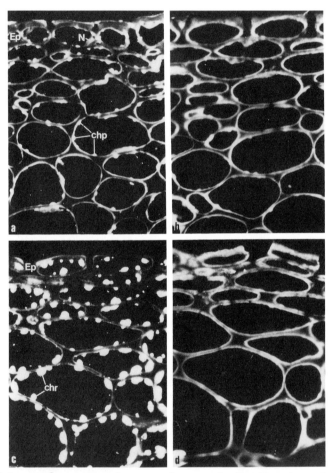

FIG. 6. Immunofluorescence microscopy of green and red pericarp tissue of *Capsicum* immunolabeled with polyclonal antibodies to geranylgeranyl PP synthase or preimmune serum (control). The fluorescene micrographs were taken with a Zeiss photomicroscope. (a, b) Green tissues treated with antigeranylgeranyl PP synthase antibodies and preimmune serum. (c, d) Red tissues treated with antigeranylgeranyl PP synthase antibodies and preimmune serum. Ep, Epidermia; N, nucleus; chp, chloroplast; chr, chromoplast.

mounted with Citifluor (Agar Aids, Stansted, England). The tissues are observed and photographed with a Zeiss photomicroscope equipped with epifluorescence illumination. Photomicrographs are recorded on Agfachrome 200 RS film. The results obtained at completion are shown in Figs. 6 and 7, which clearly indicate that geranylgeranyl PP synthase and phytoene synthase are exclusively localized in the plastid compartment.

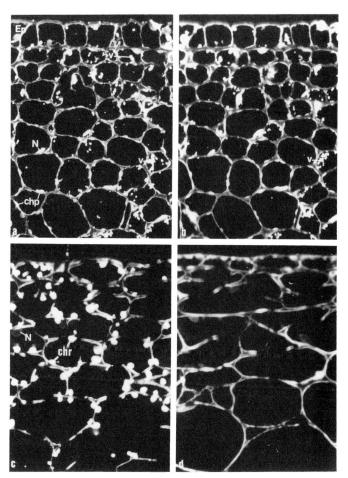

FIG. 7. Immunofluorescence microscopy of green and red pericarp tissue of *Capsicum* immunolabeled with polyclonal antibodies to phytoene synthase or preimmune serum (control). (a, b) Green tissues treated with antiphytoene synthase antibodis and preimmune serum. (c, d) Red tissues treated with antiphytoene synthase antibodies and preimmune serum. Ep, Epidermis; N, nucleus; chp, chloroplast; chr, chromoplast; v, vacuole.

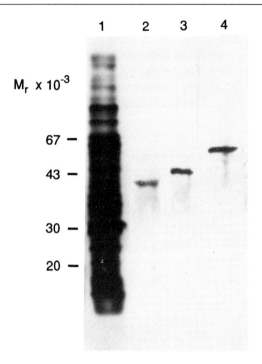

FIG. 8. Fluorogram after electrophoresis of translation products directed by poly(A)+ RNA isolated from red pericarp tissue of *Capsicum*. Lane 1, total translation products; lanes 2-4, polypeptides immunoprecipitated with antiisopentenyl PP isomerase, antigeranylgeranyl PP synthase, and antiphytoene synthase antibodies.

Biogenesis of Phytoene Synthase Component Enzymes

RNAs are isolated from pericarp tissues frozen in liquid nitrogen according to a previously described method.[12] Poly(A)+ RNAs are isolated by affinity chromatography on oligo(dT)-cellulose.[13] Poly(A)+ RNAs are translated *in vitro* in a rabbit reticulocyte lysate kit (Amersham) in the presence of 20 mCi/ml [^{35}S]methionine, as recommended by the manufacturer. After 90 min at 30°, translation is terminated by the addition of sodium dodecyl sulfate to 1% (w/v) followed by boiling for 1 min. An aliquot (25 µl) of the boiled translation mixture is diluted with 250 µl of the immunoprecipitation buffer (IB) (50 mM Tris-HCl buffer, pH 7.6, 0.15 M

---

[12] M. Kuntz, J. L. Evrard, A. d'Harlingue, J. H. Weil, and B. Camara, *Mol. Gen. Genet.* **216**, 156 (1989).
[13] H. Aviv and P. Leder, *Proc. Natl. Acad. Sci. U.S.A.* **69**, 1408 (1972).

NaCl, 1 mM EDTA, 10 mM methionine, 1 mg/ml bovine serum albumin, and 1% Triton X-100). After centrifuging at full speed in a microcentrifuge at room temperature, antibodies (5 µl) are mixed with the supernatant and incubated for 1 hr at room temperature, followed by overnight incubation at 4°. Then 40 µl of a suspension of protein A-Sepharose washed twice in IB buffer is added. The mixture is rocked for 30 min at room temperature. The resulting immune complexes are washed twice with IB buffer and once with 50 mM Tris-HCl buffer, pH 7.6, by centrifugation and resuspension. The pellet is finally boiled in electrophoresis sample buffer containing 2% sodium dodecyl sulfate. Radioactive polypeptides are visualized after electrophoresis followed by fluorography. A typical result observed at the completion of the above steps is presented in Fig. 8, indicating that the component enzymes of the phytoene synthase complex are synthesized as high molecular weight precursors (3 to 5 kDa higher than the mature form).

[33] Solubilization and Purification Procedures for Phytoene Desaturase from *Phycomyces blakesleeanus*

By PAUL D. FRASER and PETER M. BRAMLEY

Introduction

The name phytoene desaturase, or phytoene dehydrogenase, is given to the enzyme or enzymes which catalyzes the sequence of four didehydrogenations which convert phytoene (usually the 15-*cis* isomer) to phytofluene, ζ-carotene, neurosporene, and lycopene. As well as dehydrogenation, an isomerization, probably at the phytofluene stage, from 15-*cis* to the all-*trans* configuration also occurs.[1] The number of enzymes required for this sequence is unknown and probably varies among organisms. Genetic studies with higher plants indicate that two enzymes are necessary,[2] but complementation analyses of mutants of *Phycomyces blakesleeanus* have shown that a single gene, *carB*, encodes phytoene desaturase. This protein may form a tetrameric aggregate *in vivo*.[3]

[1] P. M. Bramley, *Adv. Lipid Res.* **21**, 243 (1985).
[2] J. T. O. Kirk and R. Tilney-Basset, "The Plastids," 2nd Ed. Freeman, San Francisco, 1978.
[3] C. M. G. Aragon, F. J. Murillo, M. D. De la Guardia, and E. Cerdá-Olmedo, *Eur. J. Biochem.* **63**, 71 (1976).

With the exception of halophilic bacteria, where the enzyme is located in the cytoplasm,[4] phytoene desaturase is a membrane-bound protein, associated with the plastid envelope of higher plants.[5] Therefore, a prerequisite of purification is the solubilization of the protein from its membrane environment, with a minimal loss of catalytic activity.

Many crude extracts, from a wide range of tissues, show phytoene desaturase activity[6]; however, relatively few solubilization procedures have been reported, and its purification has proved to be very difficult. In this chapter, protocols for the solubilization and purification of phytoene desaturase from *Phycomyces blakesleeanus* are described.

Preparation of Crude, Carotenogenic Cell Extract

Mycelia from the C9*carR21*(−) (lycopene-accumulating) strain of *Phycomyces*, grown for 60 hr at 24° in light (60 W/m^2)[7] are lyophilized to complete dryness. The mycelial ball should be "biscuit dry," not soft. The dry mycelia can be stored, desiccated at −70°, for several weeks.

The cell extract is prepared[8] by rubbing the mycelia through a fine (180 μm) mesh sieve and adding ice-cold 0.4 M Tris-HCl, pH 8.0, containing 5 mM dithiothreitol (DTT) (1:6, w/v) to the powder. The paste is thoroughly stirred and centrifuged at 10,000 g for 20 min at 4°. The resulting supernatant (S_{10}) is used as the cell extract. Although it can be frozen at −70° for several weeks, it is usually freshly prepared.

Assay of Phytoene Desaturase

Enzyme activity is estimated by the incorporation of radioactivity (typically ^{14}C) into phytofluene, ζ-carotene, neurosporene, and lycopene. With crude extracts, which contain all the enzymes of the isoprenoid pathway, water-soluble substrates such as mevalonic acid (MVA) or isopentenyl diphosphate (IDP) can be used.[7,9] However, none is a specific substrate for the enzyme, and it is preferable and more accurate to use [^{14}C]phytoene. This can be prepared biosynthetically from [2-^{14}C]MVA using cytosolic

---

[4] S. C. Kushwaha, M. Kates, and J. W. Porter, *Can. J. Biochem.* **54**, 816 (1976).
[5] F. Lütke-Brinkhaus, B. Liedvogel, K. Kreuz, and H. Kleinig, *Planta* **156**, 176 (1983).
[6] P. M. Bramley, in "Carotenoids: Chemistry and Biology" (N. I. Krinsky, M. M. Mathews-Roth, and R. F. Taylor, eds.), p. 185. Plenum, New York, 1990.
[7] A. Than, P. M. Bramley, B. H. Davies, and A. F. Rees, *Phytochemistry* **11**, 3187 (1972).
[8] P. M. Bramley and B. H. Davies, *Phytochemistry* **14**, 463 (1975).
[9] A. De la Concha, F. J. Murillo, E. J. Stone, and P. M. Bramley, *Phytochemistry* **22**, 441 (1983).

preparations of a *carB* mutant of *Phycomyces*,[10] or from cell extracts of several other tissues.[1]

Once synthesized, the [^{14}C]phytoene must be purified and added to an aqueous incubation mixture. Several approaches have been used to overcome the immiscibility of phytoene in the incubation. Early studies added the substrate in a detergent micelle, typically using Tween 80,[11] but more reproducible assays are achieved using phytoene-containing liposomes.[12]

We have adopted an alternative strategy, using a coupled assay with [2-^{14}C]MVA and a cytosolic preparation of the C5*carB10*(−) strain of *P. blakesleeanus*.[10] Once the [^{14}C]phytoene has been synthesized *in situ*, the extract to be assayed for phytoene desaturase is added to the incubation. Thus, no purification of the [^{14}C]phytoene is required, nor is there any need to solubilize the lipophilic substrate in the incubation mixture. This approach has been used with cell extracts of the C9*carR21*(−) strain of *Phycomyces*,[13] *Aspergillus giganteus*,[14] and *Aphanocapsa*.[15]

Incubation System

The coupled assay incubation[10,14,15] mixture (initial volume 350 µl) contains 0.4 M Tris-HCl, pH 8.0, 5 mM DTT, 6 mM ATP, 2 mM NAD, 2 mM NADP, cell extract of C5*carB10*(−) (∼2.0 mg protein), and 18 µM (0.5 µCi) sodium (3R)-[2-^{14}C]mevalonate. After 2 hr at 35°, in darkness, the cell extract (150 µl) to be assayed for phytoene desaturase is added and the incubation continued for 1 hr. The reactions are stopped with methanol (1.5 ml), "carrier" carotenes (25 µg each of phytofluene, ζ-carotene, neurosporene, lycopene, and β-carotene) are added, and the lipids are extracted (3 times) with petroleum ether (bp 40–60°), in dim light and at 4°. Typically, the incorporation of (3R)-[2-^{14}C]MVA into phytoene by the S_{105} extract of C5 is 41.0% (S.E. 0.92, $n = 5$), whereas conversion of [^{14}C]phytoene to unsaturated carotenes of a Tween 60-solubilized C9 extracts is 22.3% (S.E. 0.91, $n = 11$).[10]

Purification and Radioassay of [^{14}C]Carotenes

Different laboratories have their own, favored protocols for the purification of ^{14}C-labeled carotenes using combinations of TLC and HPLC,

---

[10] P. D. Fraser, J. De la Rivas, A. Mackenzie, and P. M. Bramley, *Phytochemistry* **30**, 3971 (1991).
[11] P. M. Bramley and B. H. Davies, *Phytochemistry* **15**, 1913 (1976).
[12] P. Beyer, this series, Vol. 148, p. 392.
[13] P. D. Fraser, Ph.D. Thesis, University of London, 1991.
[14] M. El-Jack, A. Mackenzie, and P. M. Bramley, *Planta* **174**, 59 (1988).
[15] G. Sandmann and P. M. Bramley, *Planta* **164**, 259 (1985).

such as those described in Refs. 12 and 16–18. Whatever system is used, it is essential to ensure that the radiolabeled carotenes are radiochemically pure and to follow the routine precautions to minimize degradation and isomerization of carotenes.[18,19]

We use two TLC steps.[10,14,15] First, silica gel GF_{254} (Merck 5735) thin layers developed with 15% (v/v) toluene in light petroleum (bp 80–100°) separates phytoene (R_f 0.80) and lycopene (R_f 0.40) from other isoprenoids which comigrate (R_f 0.50–0.70). The band containing phytofluene, ζ-carotene, β-carotene, and neurosporene is quickly scraped off and rechromatographed on activated aluminium oxide F_{254} (Merck No. 5550, Germany) developed with 3% (v/v) toluene in light petroleum (bp 80–100°), which separates phytofluene (R_f 0.60), β-carotene (R_f 0.45), neurosporene (R_f 0.40), and ζ-carotene (R_f 0.30). Bands are located by color or by staining with I_2 vapor. The gels are scraped off and radioassayed by liquid scintillation counting using automatic color quench compensation.[20]

Solubilization of Phytoene Desaturase

Despite the wealth of literature on the solubilization of membrane-bound proteins, the complexity of the interactions between lipids and proteins within a membrane dictates that an empirical approach has to be used to find the optimal surfactant for a particular protein.[13,21,22] The solubilization must take place under nondenaturing conditions to preserve enzymatic activity and other native properties of the protein. A vast range of ionic and nonionic surfactants are commercially available. It is important, however, to use the purest detergents available, since contaminants such as peroxides and carbonyls can disrupt the conformation of a protein. Essentially, the variables to be considered are the type and concentration of detergent, the time and temperature of treatment, and the detergent:protein ratio employed. Typically, the treated membrane preparation is centrifuged at 105,000 g for at least 1 hr at 4° and the supernatant (S_{105}) assayed for enzyme activity and protein content.

We have found that the Tween's are the most suitable detergents for the solubilization of carotenogenic enzymes in *Phycomyces*. Carotenogenic

[16] W. Rau and U. Mitzka-Schnabel, this series, Vol. 110, p. 253.
[17] B. L. Jones and J. W. Porter, this series, Vol. 110, p. 209.
[18] B. Camara, this series, Vol. 110, p. 224.
[19] G. Britton, this series, Vol. 111, p. 112.
[20] P. M. Bramley, B. H. Davies, and A. F. Rees, in "Liquid Scintillation Counting" (M. A. Crook and P. Johnson, eds.), p. 76. Heyden, London, 1974.
[21] P. M. Bramley and R. F. Taylor, *Biochim. Biophys. Acta* **839**, 155 (1985).
[22] A. Helenius and K. Simons, *Biochim. Biophys. Acta* **415**, 29 (1975).

activity in S_{105} fractions after treatment of crude extracts of the C115 (β-carotene-accumulating) mutant of *P. blakesleeanus* with nine nonionic detergents was highest in Tween 40- and 60-treated preparations (Fig. 1). A similar screen with ionic surfactants showed that only the zwitterionic detergent Zwittergent 3-08 (an N-alkyl-N,N-dimethyl-3-ammoniopropane sulfonate) was effective in solubilizing phytoene desaturase in an active form.[21] The Tween-treated extracts are more stable to storage than the Zwittergent 3-08 preparation, retaining activity for over 4 weeks at $-70°$.[21]

Similar results have been found with the C9 strain of *Phycomyces*. Maximal phytoene desaturase activity was found with 1% (w/v) Tween 60, at a detergent:protein ratio of about 5:1, when incubated for 1 hr at 4°.[13] After solubilization, the Tween 60 concentration can be reduced to 0.5% (w/v) without loss of enzymatic activity. Lower concentrations, however, result in a significant loss of phytoene desaturase activity.

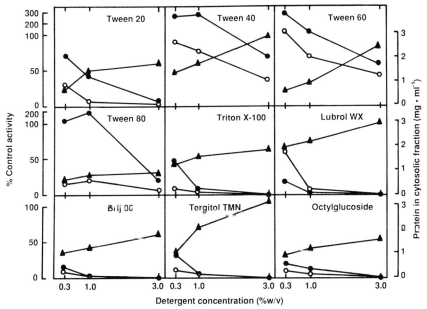

FIG. 1. Carotenogenic activities and protein release from *Phycomyces* C115 cell extract after preincubation with nonionic detergents for 30 min at 4°. Enzyme activities were estimated by the incorporation of 3(RS)-[2-^{14}C]MVA (1 μCi) into phytoene (●) and β-carotene (○), and are expressed as a percentage of control values (16,283 and 56,578 dpm for phytoene and β-carotene, respectively). Protein concentrations (▲) were estimated on S_{105} fractions after detergent treatment. The control S_{105} contained 0.32 mg protein/ml. [From P. M. Bramley and R. F. Taylor, *Biochim. Biophys. Acta* **839**, 155 (1985); reproduced with permission from Elsevier Science Publishers.]

TABLE I
PRODUCTS OF PHYTOENE DESATURASE IN CRUDE AND TWEEN 60-SOLUBILIZED
PREPARATIONS OF C9carR21(−)[a]

Preparation	Total incorporation (dpm/mg protein)[b]	% Total[c]		
		Phytofluene	ζ-Carotene	Lycopene
Crude extract (S_{10})	2246	11	14	75
S_{10} + 1% Tween 60	1962	19	20	61
S_{105} after Tween 60 treatment	2059	43	28	29

[a] P. D. Fraser and P. M. Bramley, unpublished (1992).
[b] From 41,322 dpm [^{14}C]phytoene.
[c] No radioactivity was detected in neurosporene.

Two other effects of treatment with Tween are apparent. There is a large stimulation in phytoene synthesis from [2-^{14}C]MVA (Fig. 1), and the pattern of phytoene desaturase products changes from that found with a crude (S_{10}) extract (Table I). The S_{10} fraction produces predominantly lycopene, but the solubilized enzyme yields phytofluene as its major product. There is, however, only a small loss of total phytoene desaturase activity on solubilization with Tween 60. Solubilization by such mild, nonionic detergents suggests that phytoene desaturase (and lycopene cyclase) is a peripheral, rather than integral, membrane protein as such detergents cannot dissolve the lamellar structure.[22]

Tween 40 is the most suitable detergent for the solubilization of carotenogenic enzymes from *Aphanocapsa*,[23] but sodium cholate (1 mg/10 mg protein) is the surfactant of choice with *Neurospora crassa*.[16] CHAPS {3-[(3-cholamidopropyl)dimethylammonio]-1-propanesulfonate} is effective in solubilizing the membrane-bound carotenogenic enzymes of daffodil chromoplasts. The highest yields of activity were found with CHAPS concentrations ranging from 1.25 to 10 mM.[24] In both of the latter two examples, it was necessary to add microsomal lipids to the incubations in order to observe maximum enzymatic activity. Presumably, a close association of certain membrane lipids with the native protein is necessary for enzymatic activity. The fatty acid residues on Tween 40 and 60 may compensate for the removal, during solubilization, of fatty acids from the microenvironment of the enzyme.

[23] P. M. Bramley and G. Sandmann, *Phytochemistry* **26**, 1935 (1987).
[24] P. Beyer, G. Weiss, and H. Kleinig, *Eur. J. Biochem.* **153**, 341 (1985).

TABLE II
PURIFICATION OF PHYTOENE DESATURASE FROM *P. blakesleeanus* C9 *car21*(−)[a]

Purification step	Total protein (mg)	Specific activity[b]	Recovery (%)	Purification (-fold)
Crude extract	114	957	100	1
1% Tween 60 S_{105}	96	1098	55	1.1
4% PEG precipitate	24	6950	86	7.2
CM-Sepharose	2.0	25,475	29	29
Gel filtration	0.3	86,167	13	90
Isoelectric focusing	0.12	237,917	4	249

[a] P. D. Fraser and P. M. Bramley, unpublished (1991).
[b] Expressed as dpm [^{14}C]phytoene incorporated into desaturase products per mg protein per hr. Each assay contained 300,000 dpm [^{14}C]phytoene.

Purification Procedures

The following separation techniques have been found to be the most applicable for the purification of phytoene desaturase from *Phycomyces* (Table II). They may not, however, be appropriate for all detergent-solubilized preparations of carotenogenic enzymes. Whenever possible, successive purification procedures are performed without interruption. General methodology for protein purification can be found in several excellent texts such as Scopes,[25] Deutscher,[26] and Jackoby.[27]

Polyethylene Glycol Precipitation

Phytoene desaturase from the C9 strain is precipitated with 4% (w/v) polyethylene glycol (PEG) 8000. The requisite PEG concentration is determined by obtaining an analytical precipitation curve, using the Tween 60-solubilized extract with increasing amounts of 50% (w/v) PEG solution.[28] Unlike procedures involving ammonium sulfate, it is not necessary to remove the PEG prior to enzyme assay, as it does not inhibit phytoene desaturase at this concentration. PEG has the additional advantage over ammonium sulfate that the protein precipitate will pellet, rather than float, after centrifugation of the detergent-solubilized preparation. The 4% PEG

[25] R. K. Scopes, "Protein Purification, Principles and Practice," 2nd Ed. Springer-Verlag, Berlin and New York, 1987.
[26] M. P. Deutscher, ed., this series, Vol. 182.
[27] W. B. Jackoby, ed., this series, Vol. 104.
[28] K. C. Ingham, this series, Vol. 104, p. 351.

precipitate can be stored for at least 6 months at $-70°$ with a minimal loss of enzyme activity in the presence of 20% ethylene glycol or 50% glycerol. It is redissolved in 20 mM Tris HCl, pH 8.0, containing 5 mM DTT and 0.5% Tween 60 prior to ion-exchange chromatography.

Ion-Exchange Chromatography

The ion-exchange step is only applicable to phytoene desaturase preparations solubilized with a nonionic detergent. Chromatography of a 4% PEG fraction on DEAE-Sepharose showed that phytoene desaturase activity was recovered principally in the wash fraction. However, it does bind significantly to CM-Sepharose and can be eluted with a stepwise gradient of NaCl (150, 300, 450 mM). Phytoene desaturase activity is found in both the 150 and 300 mM eluates. It was also noticeable that the lycopene present in the 4% PEG fraction is not eluted with the enzyme. The may account for the apparent stimulation of activity observed in the eluted fractions, as endogenous carotene may cause feedback inhibition of phytoene desaturase activity.[11]

Although column chromatography, with stepwise elution, can be used to elute phytoene desaturase, we find that it is more convenient, and far quicker, to use a bath method of adding CM-Sepharose directly to the reconstituted PEG fraction, stirring for 30 min at 4°, removing the supernatant, and then eluting phytoene desaturase from the gel with 300 mM NaCl. This fraction is then subjected to gel filtration.

Chromatographic Gel Filtration

TSK gel SW type columns are used in for HPLC gel filtration, either the G2000 SW (fractionation range 5–60 kDa), G3000 SW, or G3000 SW$_{xl}$ (10–300 kDa). Calibration of the gel filtration column must be carried out using standard proteins (Sigma, St. Louis, MO, MW-GF-200), which are dissolved in a 0.5% Tween 60 buffer. Using 20 mM Tris-HCl buffer, containing 5 mM DTT, 20% glycerol, 0.5% Tween 60, and 100 mM NaCl, the G2000 SW column separates proteins of 12.5 to 60 kDa, at a flow rate of 0.2 ml/min. Preparative G2000 SW columns can also be used and are run at 2 ml/min.

Using the 300 mM NaCl fraction from CM-Sepharose, multiple peaks of phytoene desaturase activity elute from TSK columns, with apparent M_r values of 15K, 25K, 50K, and 95K. Rechromatography of the low-M_r fractions, under identical conditions, yields higher M_r aggregates. It is not possible to know whether these aggregates are artifacts or reflect the situation *in vivo*. The latter possibility has been suggested by genetic evidence.[3]

The gel filtration fractions can be stored at $-70°$, with 20% glycerol or 20% ethylene glycol, for at least 2 weeks.

Isoelectric Focusing

Precast isoelectric focusing (IEF) gels (Phamacia-LKB, dimensions 245 × 110 × 1 mm) of pH range 3.5–9.5 are used for the final purification step. Electrode solutions are 0.1 M glutamic acid (anode) and 0.1 M β-alanine (cathode). Samples (50 μl per track) from the gel filtration fraction (M_r 15K) after desalting are run for 2 hr at 1500 V, 50 mA, and 4°. After focusing, slices of gel (80 × 10 mm) are assayed for phytoene desaturase activity by adding the macerated gel slices directly to the incubations. The pH gradient of the IEF gel can be verified with a standard IEF calibration kit (Merck, Electran range). IEF gels can be stained with Coomassie or silver staining reagents, but only after removal of ampholytes by washing gels in 20% (w/v) trichloroacetic acid for 1 hr.

Purification of the phytoene desaturase protein is monitored by SDS-PAGE[29] or two-dimensional IEF–SDS-PAGE,[30] and verified by silver staining of IEF gels (pI 4.7). Protein samples for SDS-PAGE are concentrated and dilipidated with a mixture of chloroform/methanol/water (1:4:3, v/v) according to the method of Wessel and Flügge.[31] Proteins precipitate at the interface and are carefully removed, dried under N_2, and redissolved in sample buffer prior to electrophoresis.

Acknowledgments

P. D. Fraser was in receipt of a Science and Engineering Research Council Cooperative Award for Science and Engineering with Schering Agrochemicals, to whom we express our thanks. We also thank Dr. A. Mackenzie for useful discussions during part of the work.

---

[29] U. R. Laemmli, *Nature (London)* **227**, 680 (1970).
[30] P. H. O'Farrell, *J. Biol. Chem.* **250**, 4007 (1975).
[31] D. Wessel and U. I. Flügge, *Anal. Biochem.* **138**, 141 (1984).

[34] Functional Analysis and Purification of Enzymes for Carotenoid Biosynthesis Expressed in Photosynthetic Bacteria

By GLENN E. BARTLEY, ANETTE KUMLE, PETER BEYER, and PABLO A. SCOLNIK

Introduction

The bright yellow, orange, and red colors of carotenoids are due to absorption of light by the chromophore, a central array of conjugated double bonds. These double bonds are synthesized on the C_{40} skeleton of phytoene, a colorless precursor of the carotenoid biosynthesis pathway in both prokaryotes and eukaryotes which has a chromophore of only three conjugated double bonds. Further addition of double bonds to this chromophore results in increases of the absorption maxima that range from 7 to 35 nm.[1] Our laboratories are currently using combined genetic and biochemical approaches to study the structural and functional properties of carotenoid desaturases. The genetic approach is aimed at analyzing structure–function correlations through the complementation of microbial carotenoid mutants with homologous and heterologous genes that code for carotenoid desaturases. The biochemical approach consists of purifying carotenoid desaturases to homogeneity and, since carotenoids are highly nonpolar molecules, perfecting enzyme assays in nonaqueous phases.

Carotenoid desaturases are membrane-bound enzymes in both bacteria and plants.[2,3] To date, no *in vitro* system for testing enzymatic activity with purified protein has been described. However, the existence of enzymatically active reconstitution systems with partially purified enzymes[4] suggests that enzyme assays with purified carotenoid desaturases may be feasible. To this end, we have devised a procedure for the purification of phytoene desaturase from photosynthetic bacteria.

We have achieved functional expression of *Neurospora crassa*[5] and

---

[1] W. Vetter, G. Englert, N. Rigassi, and U. Schwieter, *in* "Carotenoids" (O. Isler, ed.), p. 189. Birkhaüser, Basel, 1971.
[2] G. E. Bartley and P. A. Scolnik, *J. Biol. Chem.* **264**, 13109 (1989).
[3] F. Lütke-Brinklhaus, B. Liedvogel, K. Kreuz, and H. Kleining, *Planta* **156**, 176 (1982).
[4] P. Beyer and H. Kleining, this series, Vol. 213 [8].
[5] G. E. Bartley, T. J. Schmidhauser, C. Yanofsky, and P. A. Scolnik, *J. Biol. Chem.* **265**, 16020 (1990).

Glycine max (soybean)[6] cDNAs coding for carotenoid desaturases in mutants of the photosynthetic bacterium *Rhodobacter capsulatus*. Thus, it is logical to expect functional expression in *R. capsulatus* of carotenoid desaturase cDNAs from other species. Also, it is possible that cDNAs coding for carotenoid enzymes other than desaturases could be expressed in *R. capsulatus* carotenoid mutants.

Phytoene desaturases from various organisms catalyze a different number of desaturations. *Rhodobacter capsulatus* phytoene desaturase (CrtI) converts phytoene to neurosporene, a three-step desaturation reaction.[7] *Neurospora crassa* Al-1 catalyzes six desaturations.[5] Soybean Pds1 converts phytoene to ζ-carotene, a two-step desaturation reaction.[6] Expression of *al-1* in a phytoene-accumulating strain of *R. capsulatus* leads to accumulations of lycopene and 3,4-dehydrolycopene.[5] Expression of *pds* in this bacterial strain results in the accumulation of ζ-carotene.[6] Lycopene, 3,4-dehydrolycopene, and ζ-carotene are not normally accumulated in *R. capsulatus*. Thus, expression of heterologous genes for phytoene desaturases in *R. capsulatus* mutants leads to the accumulation of carotenoids not normally found in this bacterium, illustrating the usefulness of this genetic complementation system in the characterization of phytoene desaturases from a variety of species.

In this chapter we present methods for the inducible expression of carotenoid biosynthesis genes in *R. capsulatus* and *Escherichia coli* along with novel procedures for the purification of phytoene desaturases.

Expression in Photosynthetic Bacteria

Plasmid Vectors

The expression vectors used are based on high-level transcription provided by the promoter of the operon for nitrogen fixation structural genes *(nifHDK)*. This promoter is induced by low levels of fixed nitrogen in the medium, and it can be completely silenced by oxygen and ammonia. Intermediate levels of expression can be obtained by supplementing with amino acids. We currently work primarily with pNF3, a plasmid based on a ColE1 origin of replication and containing the *nif* promoter, leader sequence, and initiator codon (Fig. 1). Other plasmids based on this *nif* promoter have been described.[8] Prokaryotic genes or eukaryotic cDNAs

---

[6] G. E. Bartlet, P. V. Viitanen, Iris Pecker, Daniel Chamouitz, Joseph Hirschberg, and P. A. Scolnik, *Proc. Natl. Acad. Sci. U.S.A.* **88**, 6532 (1991).
[7] G. E. Bartley and P. A. Scolnik, *J. Biol. Chem.* **264**, 13109 (1989).
[8] D. Pollock, C. E. Bauer, and P. A. Scolnik, *Gene* **65**, 269 (1988).

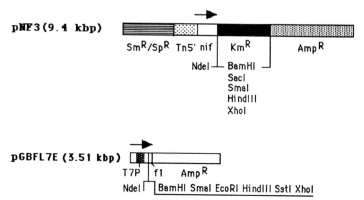

FIG. 1. *nif* and T7 expression vectors. The *nif* expression vector pNF3[8] is based on a pBR322 skeleton and carries resistance markers for streptomycin (Sm)/spectinomycin (Sp), and ampicillin (Amp). The *nif* segment carries the *R. capsulatus* nifHDK promoter and translation initiation region. The *Nde*I restriction site can be used for in-frame translational fusions. The remaining restriction sites in the polylinker region can be used for transcriptional fusion.

are cloned, preferably in-frame, into pNF3 in the orientation that places the coding region under *nif* control (Fig. 1). If an antibody is also desired, the gene or cDNA can also be cloned into an expression vector for overproduction of proteins in *E. coli*.[9] Our highest rate of success in this procedure has been achieved with vectors containing the T7 RNA polymerase promoter driving T7 gene 10 protein fusions.[10] We constructed a vector (pGBFL7E, Fig. 1) which contains an *Nde*I site at the T7 ϕ10 start codon. We use site-specific mutagenesis to obtain one or more segments of DNA with an in-frame *Nde*I site at the N-terminal initiator codon. These DNA fragments are simultaneously cloned in pGBFL7E and pNF3 for use in expression in both *E. coli* and *R. capsulatus*.

Growth and Characterization of Rhodobacter capsulatus Carotenoid Mutant Strains

To date, no efficient transformation system for *R. capsulatus* has been developed. Thus, we rely on conjugation for introducing plasmids into *R. capsulatus* carotenoid mutants. Efficient conjugations depend on using well-characterized donor and recipient cultures.

---

[9] E. Harlow and D. Lane, "Antibodies," pp. 88–91. Cold Spring Harbor Laboratory, Cold Spring Harbor, New York, 1988.
[10] F. W. Studier, A. H. Rosenberg, J. J. Dunn, and J. W. Dubendorff, this series, Vol. 185, p. 60.

We use the minimal media RCV and RCVNF for growing *R. capsulatus* carotenoid mutants (modified from Ref. 11).

Stock Solutions

10% $(NH_4)_2SO_4$	50 g/500 ml
10% DL-Malic acid	50 g/500 ml
NaOH	30 g
Adjust pH to 6.8	
1% EDTA	1.00 g/100 ml
20% $MgSO_4 \cdot 7H_2O$	20.00 g/100 ml
7.5% $CaCl_2 \cdot 2H_2O$	7.50 g/100 ml
Thiamin hydrochloride	0.10 g/100 ml
0.5% $FeSO_4 \cdot 7H_2O$	1.25 g/250 ml

Trace Elements

$MnSO_4 \cdot H_2O$	0.3975 g
H_3BO_3	0.7000 g
$Cu(NO_3)_2 \cdot 3H_2O$	0.0100 g
$ZnSO_4 \cdot 7H_2O$	0.0600 g
$NaMOO_4 \cdot 2H_2O$	0.1875 g
Distilled H_2O	to 250 ml

Super Salts

1% EDTA	20 ml
20% $MgSO_4 \cdot 7H_2O$	10 ml
7.5% $CaCl_2 \cdot 2H_2O$	10 ml
Trace elements	10 ml
0.5% $FeSO_4 \cdot 7H_2O$	24 ml
Thiamin	10 ml
Distilled H_2O	500 ml

RCV Medium

10% $(NH_4)_2SO_4$	10 ml
10% DL-malic acid	40 ml
Super salts	50 ml
0.64 M KPO_4	15 ml
Distilled H_2O	to 1000 ml
Adjust pH to 6.8	

RCVNF Medium. For RCVNF (nitrogen-free RCV), omit $(NH_4)_2SO_4$.

---

[11] B. C. Johansson and H. Gest, *J. Bacteriol.* **156**, 251 (1983).

Conjugation

Introduce the plasmid containing the carotenoid gene or cDNA into *E. coli* TEC5 cells either by transformation or electroporation.[12] For antibiotic selection in *E. coli* we use trimethoprim at 100 μg/ml or spectinomycin at 100 μg/ml. Grow recipient cells either phototrophically or heterotrophically. We keep our *R. capsulatus* strains as 1 ml 25% glycerol stocks frozen in cryovials at $-70°$. Add an entire cell stock to 25 ml RCV in a 50-ml flask. Grow at 35° with gentle (~50 rpm) shaking for 24–36 hr (some mutant strains may take up to 48 hr). For phototrophic growth, start a heterotrophic culture. Next day, transfer 1–2 ml to a 16.5 ml screw-capped tube (Pyrex No. 9826) and fill to the top with RCV. Place under illumination at 35° for 2–3 days.

A simple setup for growing cells phototrophically can be constructed with a 20-gallon home aquarium fitted with a heating and circulation pump and set at 35° in a cold room. Illumination is provided by 16 incandescent tubes (General Electric 40T8/F).

Grow the *E. coli* donor cells overnight at 37° in Luria broth (LB) containing either trimethoprim or spectinomycin. We generally start from a single colony or a loopful of frozen stock cells. In a 1.5-ml microcentrifuge tube mix 1 ml *R. capsulatus* and 0.4 ml *E. coli* cells. Pellet at 6500 rpm for 4 min at room temperature. Remove the supernatant and resuspend the pellet in 0.4 ml RCV. Centrifuge again, discard the supernatant, and resuspend in 0.2 ml RCV. With a micropipette, spot 10–20 drops of about 20 μl each onto the surface of a 10-day-old plate containing 25 ml RCV agar. Allow to dry partially at room temperature and then incubate at 35° overnight.

Remove plates from the incubator and overlay with 3 ml top RCV agar containing spectrinomycin, to give a final concentration of 10 μg/ml. Alternatively, cells can be resuspended in 0.5 ml RCV, concentrated by centrifugation, and spread directly on RCV plates containing 10 μg/ml spectinomycin. After the top agar hardens, place the plates in anaerobic jars (BBL Gas Pack 100 anaerobic system: No. 60626, Becton Dickinson and Co., Cockeysville, MD). Place the Gas Pack hydrogen and carbon dioxide generator envelope (No. 70304) and anaerobic indicator (No. 70504) in the jar and place the jar in the aquarium tank at 35° for 5–7 days.

Purify exconjugants from colonies by streaking in RCV plates contain-

---

[12] F. M. Ausubel, R. Brent, R. E. Kingston, D. D. Moore, J. G. Seidman, J. A. Smith, and K. Struhl, "Current Protocols in Molecular Biology," Vol. 1, p. 181. Wiley (Interscience), New York, 1990.

ing spectinomycin. Prepare frozen stocks from single-colony overnight cultures.

Induction of nif Expression

Grow *R. capsulatus* cells containing the appropriate plasmid under heterotrophic conditions. Generally, we grow two 25-ml cultures under the conditions described above. When cultures reach stationary phase, harvest by centrifugation in sterile tubes, resuspend in 1.5 ml RCVNF, and add 1 ml each to two screw-capped tubes. Top off one tube with RCV and the other with RCVNF, both containing spectinomycin at 10 μg/ml. Incubate in the aquarium tank at 35° for 3–4 days. Pressure buildup, owing to hydrogen production by the nitrogenase reaction, should be evident at this point in the tube containing RCVNF. Harvest cells by centrifugation and process for carotenoid extraction (see below) or freeze at −70°.

Carotenoid Extraction and Analysis

Resuspend cells in 2–5 ml of distilled water, save 50 μl for protein assay, and pellet the rest by centrifugation for 10 min at room temperature in glass screw-cup test tube. Extract the pellet with diethyl ether/methanol (50:50, v/v) until all pigment has been extracted (usually 3 times). Transfer the solvent to screw-capped glass tubes and dry under a stream of nitrogen. Resuspend the pigment in the solvent of choice for thin-layer chromatography (TLC), high-performance liquid chromatography (HPLC), or saponification (see below).

Saponification removes bacteriochlorophyll from the sample. However, if one desires to obtain a rapid qualitative characterization of the main carotenoids accumulated, TLC can be used without saponification because bacteriochlorophyll is retained at the origin of the chromatogram. Also, methods are being developed for analyzing both bacteriochlorophyll and carotenoids in the same sample using HPLC. At the present time, however, we prefer to use saponification for quantitative analysis of carotenoids.

To saponify a sample, resuspend the pigments in 2 ml of 100% ethanol. Add 0.2 ml of 60% KOH (w/v). Mix and incubate overnight at room temperature under nitrogen. Add 2 ml diethyl ether and 0.3 ml distilled water and shake under N_2 for a few seconds. Allow for full separation of phases. Transfer the top phase to a glass screw-capped tube. Repeat the extraction steps until no more pigment can be seen in the upper phase. If working with a colorless compound such as phytoene, repeat the extraction steps 5 times. Dry the combined top phases under nitrogen purge and resuspend pigments in the solvent of choice.

Overexpression in *Escherichia coli* and Generation of Antibodies

The following protocol is currently used in our laboratory for expression of foreign genes in *E. coli* using the bacteriophage T7 RNA polymerase. A complete description of the system, including other bacterial strains and plasmids, can be found in Ref. 10.

Prepare competent cells of the *E. coli* strain BL21(DE3) by standard procedures.[12] This strain contains a gene encoding the phage T7 DNA polymerase gene under control of a *lacUV5* promoter.[10] Transform *E. coli* BL21(DE3) cells with the plasmid containing the carotenoid gene under control of the T7 promoter. Dilutions of the plasmid DNA must be made to ensure a reasonable number of colonies. Expression of foreign proteins can be toxic to *E. coli*, and this toxicity can result in instability of the plasmid carrying the foreign gene. Stability of the plasmid should be checked by plating cells onto LB, LB containing 100 μg/ml ampicillin (Amp), LB plus 1 mM of the *lacUV5* inducer isopropyl-β-D-thiogalactopyranoside (IPTG), and LB plus Amp and IPTG. Cells containing an active T7 promoter will not grow in the presence of the inducer IPTG. Thus, if the plasmid expressing the target DNA is not lost at a high frequency, very few colonies (fewer than 0.1–1% of the LB controls) will form on plates containing both Amp and IPTG.[10]

Grow overnight liquid cultures from a single colony and prepare stocks by adding glycerol to 25% and freezing at $-70°$. Inoculate two tubes containing 5 ml of LB plus Amp (100 μg/ml) and grow overnight at 37°, shaking at 250 rpm. Use the entire 5-ml overnight cultures to inoculate two 100 LB plus Amp cultures. Grow with shaking (250 rpm) at 37° until the OD$_{600}$ reaches 1.2. Pellet one 1-ml aliquot from each culture at 16,000 g, 4°, for 1 min in microcentrifuge tubes. Discard supernatants and save pellets, which are uninduced controls. Add 400 μl of 100 mM IPTG to one of the cultures and continue growing both cultures for 3–4 hr. Harvest cells by centrifugation at 10,000 g at 4° for 10 min. Discard the supernatant and freeze the pellet at $-70°$ until ready for preparation of protein.

When ready to extract protein, thoroughly resuspend cells in resuspension buffer [50 mM Tris-HCl pH 7.8, 2 mM dithiothreitol (DTT), 5 mM EDTA, 2 mM benzamidine hydrochloride, 1 mM phenylmethylsulfonyl fluoride (PMSF), and 10% (v/v glycerol]. Transfer cells to microcentrifuge tubes. Add 0.25 ml lysis buffer (resuspension buffer plus 0.6% Triton X-100 and 0.2 mg/ml lysozyme) and incubate at 4° for 30 min. Sonicate on ice with seven 30-sec pulses spaced by 15-sec pauses. Centrifuge tubes in a microcentrifuge for 15 min at 4°. Save the supernatant. Resuspend the pellet in 1× protein sample buffer [50 mM Tris-HCl, pH 7.00, 2 mM EDTA, 1% (w/v) sodium dodecyl sulfate (SDS), 1% (w/v) mercaptoethanol, 10% glycerol, and 0.025% bromphenol blue].

Run aliquots of both pellet and supernatant in SDS-PAGE gels. Most enzymes for carotenoid biosynthesis are expected to be present in the pellet. For antibody production, sections of SDS-polyacrylamide gels containing the expressed protein can be cut, ground, and, after mixing with Freund's adjuvant, inoculated into New Zealand White rabbits.[9]

Purification of CrtI

Most of the enzymes of the carotenoid biosynthesis pathway are bound to membranes. This, plus the relatively low abundance of some of these proteins, makes purification in an active form rather difficult. Overexpression in photosynthetic bacteria could greatly facilitate this process. We are currently developing general methods for the purification of phytoene desaturases from overproducing strains of photosynthetic bacteria. The following section summarizes our current purification protocols.

Plasmid Construction

To obtain an *R. capsulatus* strain overproducing phytoene desaturase, we used site-directed mutagenesis to place an *NdeI* site at the initiator ATG codon of *crtI*. The mutagenized gene was used to construct an in-frame translation fusion into pNF3. The resulting plasmid (pGBN4.2) was transferred from *E. coli* into *R. capsulatus* BPY69 by conjugation. This *R. capsulatus* strain contains a point mutation in *crtI*, the gene encoding phytoene desaturase.[7]

Purification

All separation steps are carried out at 4°. A 200-ml bacterial culture [*R. capsulatus* strain BPY69 (pGBN4.2)] induced for *nif* is harvested by centrifugation at 17,000 g for 30 min. After resuspension in buffer A (100 mM Tris-HCl, pH 7.2, 10 mM MgCl$_2$, and 2 mM dithioerythritol), the cells are disrupted by use of a French pressure cell at 84,000 MPa (12,000 psi). The protein content at this stage is usually 5.5 mg/ml. Nondisrupted material and crude debris are removed by centrifugation at 62,000 g for 15 min, and a membrane pellet is obtained by subsequent centrifugation at 187,000 g for 60 min. The membrane pellet is sonicated on ice at low energy in 5 ml of buffer A. The membrane suspension (4.1 mg/ml protein) is then adjusted to a detergent concentration {CHAPS, 3-[(3-

cholamidopropyl)dimethylammonio]-1-propanesulfonate} of 300 mM, corresponding to about 30-fold the critical micellar concentration (CMC). The suspension is flushed with nitrogen and stirred for 40 min on ice. Nonsolubilized material is removed by centrifugation for 60 min at 187,000 g. The clear red supernatant (2.6 mg/ml) is then used for further purification.

A 5-ml aliquot of the supernatant is subjected to size-exclusion chromatography using Ultrogel AcA 22 (gel bed 3 × 45 cm). The column is equilibrated with CHAPS starting buffer (buffer A containing CHAPS at a 20 mM concentration, 2 times the CMC). During development (flow rate 0.5 ml/min), some red-colored material elutes in the void volume and two additional red bands are resolved. The corresponding spectra indicate the separation of native light-harvesting complex (B800–850) at higher molecular mass from the respective denatured complex and/or solubilized bacteriochlorophyll, pheophytin, and carotenoids (Fig. 2). The CrtI protein is detected in dot blots using the anti-CrtI antibodies (see above). As can be seen in Fig. 2, the desaturase elutes between these two red bands and thus could be easily localized. The two positive fractions (11 and 12, 10 ml each) show only faint color and are pooled (protein content 0.04 mg/ml).

Any residual amounts of colored protein complexes and lipid/pigment-containing mixed detergent micelles are removed by a batch procedure employing hydroxyapatite (HAP type III, Sigma). The material is equilibrated to 2 times the CMC, washing 3 times (5 ml each) with CHAPS starting buffer. After adsorbing the positive fractions from size-exclusion chromatography with HAP, nonbinding material is removed by two additional washings with CHAPS starting buffer; resuspensions are carried out by sonication at low energy. The bound protein is eluted with 5 ml CHAPS starting buffer containing 0.2 M Na$_3$PO$_4$ (pH 7.5). This step is repeated once.

The phosphate concentration in the 10-ml eluate from HAP is lowered to 20 mM by 10-fold dilution with CHAPS starting buffer. An ultrafiltration Diaflow Cell equipped with a hydrophilic filter, YM 30 (Amicon, Danvers, MA), is used to concentrated mixed micelles without concentrating free detergent micelles and phosphate. Ultrafiltration starting with 100 ml proceeds to a final volume of 10 ml. This concentrate is then applied to an anion-exchange fast protein liquid chromatography (FPLC) column, Mono Q HR 5/5 (Pharmacia Piscataway, NJ). The nonbinding material is eluted with CHAPS starting buffer.

As can be seen in Fig. 3, the CrtI protein elutes essentially quantitatively with the nonbinding fraction. Developing with a linear NaCl gradient (0–1 M NaCl in CHAPS starting buffer, same flow rate) elutes the

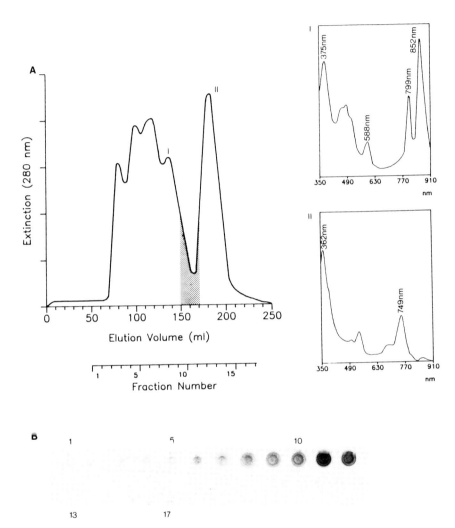

FIG. 2. Size-exclusion chromatography of solubilized membrane proteins. (A) Elution profile. The shaded area indicates the position of the CrtI protein. Insets show the spectra of peaks I and II. (B) Detection of the CrtI protein by immunoblotting, employing an alkaline phosphatase label.[7]

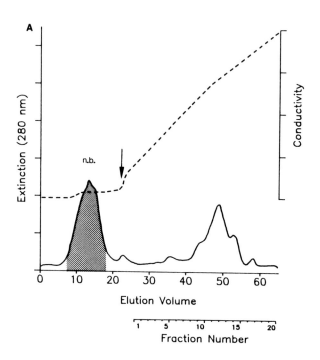

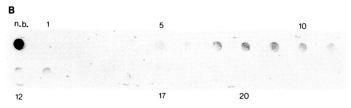

FIG. 3. Final anion-exchange chromatography of the CrtI protein. (A) Elution profile. The shaded area indicates the position of the protein with the nonbinding fraction (n.b.). The conductivity detector response indicates the course of the gradient. The arrow marks the start of the NaCl gradient (0–1 M). (B) Detection of the CrtI gene product in immunoblots, stained as in Fig. 2.

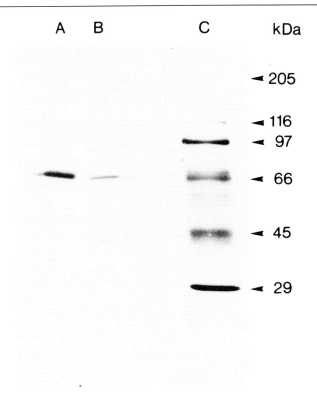

FIG. 4. SDS–polyacrylamide gel electrophoresis of the purified CrtI protein. Lanes A and B, purified CrtI protein at two different concentrations; lane C, marker proteins. The protein bands were visualized by silver staining.

binding proteins; small amounts of bound CrtI protein are also detected in fractions 5–13 in dot blots. The purity of the protein in the nonbinding fraction is judged using SDS–polyacrylamide gel electrophoresis according to Neville.[13] Figure 4 shows the purified desaturase. In some cases copurification of some faint bands of lower molecular weight proteins may occur. These bands also give a positive signal with the anti-CrtI antibody in immunoblots and are probably the result of either premature termination during overexpression or some degree of proteolytic cleavage. The purified protein seems to be homodimerically associated as judged by apparent molecular mass on gel filtration; its enzymatic characterization is currently underway.

[13] D. M. Neville, Jr., *J. Biol. Chem.* **246**, 6328 (1971).

[35] β-Carotene Synthesis in *Rhodotorula*

By HEBE MARTELLI and IRACEMA M. DA SILVA

Introduction

Few yeasts are able to form carotenoid pigments. However, all eight species of the genus *Rhodotorula* synthezize β-carotene and other carotenoids.[1] The overall pathways for carotenogenesis in yeasts have been reviewed,[2] and it has been shown that β-carotene formation in *Rhodotorula* follows the pathway of carotenogenesis in other carotene-forming systems.

Although β-carotene has attracted increasing interest as a provitamin A and food colorant, few studies have been done to improve the techniques for producing β-carotene using *Rhodotorula* cells. Research has been done in the Soviet Union[3,4] and Bangladesh.[5] In this chapter the preparation of β-carotene from *Rhodotorula* cells is presented. All the techniques described have been developed, or adapted, and used in our laboratory.

Production of β-Carotene

Medium and Growth of Rhodotorula

The strain *Rhodotorula glutinis* EQ, from the author's collection, seems to be a good producer of β-carotene. Cells are routinely grown aerobically in a chemically defined medium containing the following (in grams per liter of distilled water): $(NH_4)_2SO_4$, 5.5; KH_2PO_4, 5.3; Na_2HPO_4, 3.5; $MgSO_4 \cdot 7H_2O$, 0.5; $MnSO_4$, 0.1; yeast extract, 1.0. Sucrose (2.0%) is added as the carbon source; the pH is adjusted to 6.8 before sterilization at 120° for 15 min.

The inocula are prepared in 500-ml conical flasks containing 100 ml of medium, at 27°, for 18 hr. The experiments are carried out in a bench

---

[1] J. W. Fell, A. Statzell-Tallman, and D. G. Ahearn, in "The Yeasts: A Taxonomic Study" (N. J. W. Krege van Rij, ed.), p. 893. Academic Press, New York, 1984.

[2] K. L. Simpson, C. O. Chichester, and H. J. Phaff, in "The Yeasts" (A. H. Rose and J. S. Harrison, eds.), Vol. 2, p. 493. Academic Press, New York, 1971.

[3] I. F. Koroleva, M. V. Zalasko, V. D. Andreevskaya, and S. G. Demina, *(Inst. Mikrobiol. Minsk, USSR), Vestsi Akad. Navuk BSSR, Ser. Biyak Navuk*, **2**, 37 (1984) [in Belorussian]; *Chem. Abstr.* **101**, 51432d (1984).

[4] I. F. Koroleva and M. V. Zallasko, *(Inst. Mikrobiol. Minsk, USSR). Vestsi Akad. Navuk BSSR, Ser. Biyak Navuk* **4**, 49 (1988) [in Belorussian]; *Chem. Abstr.* **109**, 187018w (1988).

[5] J. U. Bhuyan, M. Hahman, and K. Ahmad, *(Dep. Biochem. Univ. Dhaka, Dahka-2, Bangladesh) Bangladesh Acad. Sci.* **91**, 65 (1985); *Chem. Abstr.* **82**, 26470 (1985).

fermentor with 2 liters working volume of the growing medium, at 27°, an agitator speed of 400 rpm, and an aeration rate of 1.0 vvm. During growth the pH decreases and is maintained at 5.0. After 20 hr the maximum biomass yield is attained, and no more sugar is found in the medium. The cells are slightly colored. Aeration of the culture proceeds for a further 24 hr (total time 44 hr).

The strongly colored orange cells are centrifuged and washed with distilled water. A sample is taken and oven-dried at 60° to constant weight, for the determination of the cell dry weight. Cold acetone (10 ml) is added to the biomass, and the cells are broken with 10 g of glass beads in a Mickles apparatus for 15 min. The temperature is maintained at 4° in an ice bath. The extraction is repeated until no more pigment remains in the cell debris. For the isolation of β-carotene the extracts are combined, and an adaptation of the technique of Peterson et al.[6] is followed.

Extraction and Isolation Procedures

Direct light must be avoided during the following procedures. The acetone extracts are carefully mixed with 20 ml of petroleum ether in a separatory funnel. If the layers do not separate well, 10 ml of distilled water improves the separation. The acetone is drawn off and extracted with 10 ml of petroleum ether to assure the total removal of β-carotene in the petroleum ether extract. This extract is washed 3 times with an equal volume of distilled water. After careful removal of the last wash, the petroleum ether extract is transferred to a volumetric flask, the separatory funnel is washed with petroleum ether, and the combined solutions are brought to volume. The pigmented extract is again transferred to a separatory funnel and mixed with 50 ml of methanolic potash (90 ml methanol and 10 ml of a 0.1 N KOH solution) for extracting pink pigments (torularhodine). Two or three extractions are required. The ether petroleum layer is then washed with distilled water until the last wash presents no color when assayed with phenolphthalein. The petroleum ether solution is drawn under vacuum through an adsorption column (15 cm high and 3 cm in diameter) made up with a mixture of MnO Supercell, previously washed with 20 ml petroleum ether β-carotene is eluted from the column with absolute ethanol, and the ethanol extract is brought to volume. An adequate volume of this extract is taken for determination its β-carotene concentration.

[6] T. A. Peterson, J. L. Bell, W. G. W. Etchells, and J. R. Smart, *J. Bacteriol.* **67,** 708 (1958).

Quantification

The sample is transferred to a separatory funnel and equal volumes of distilled water and petroleum ether added; after mixing the pigment passes to the petroleum ether layer. After the extraction of the ethanol by washing with distilled water, this layer is brought to volume.

For colorimetric assessment of β-carotene the absorbance of the petroleum ether layer is measured at 450 nm and the β-carotene concentration calculated by comparison with a standard curve prepared with pure commercial β-carotene. From this value the β-carotene concentration by weight of dry biomass is calculated. Statistical analysis of the data[7] from 10 independent batches under identical conditions shows that the standard error for a single determination is less than 2%.

Purification and Conservation

The solvent from the remaining ethanol solution is fully evaporated under vacuum. The β-carotene is dissolved in a small amount of petroleum ether, and two volumes of methanol are added to induce crystallization. the crystals are separated and the residual solvent evaporated in a vacuum. One more crystallization is needed. Addition of a small volume of a highly refined edible oil assures good conservation of the product if the solution is kept under vacuum and stored in a freezer.

β-Carotene Synthesis in *Rhodotorula* Cultures

Rhodotorula glutinis EQ cells are cultivated as described above. During the 44-hr cultivation period, samples (10 ml) are taken at intervals, centrifuged, and the cells washed. From the combined washes and the volume the residual sugar is determined, as well as the dry weight of the cells. Analytical methods have been described elsewhere.[8] At the same intervals other 200 to 50-ml samples (according to the cell concentration of the culture) are taken, centrifuged, and the cells extracted with acetone and the β-carotene concentration measured. If samples are saved for further extraction, they have to be kept under a few milliliters of distilled water in the refrigerator, away from air and light. Storage under acetone must be avoided because the pigments are partially destroyed. Figure 1 shows the kinetics of β-carotene synthesis under the conditions assayed. Synthesis of β-carotene in *Rhodotorula* is not associated with culture growth, increases

---

[7] J. P. Holman, "Experimental Methods for Engineers," p. 40. McGraw-Hill, New York, 1966.
[8] I. Costa, H. L. Martelli, I. M. da Silva, and D. Pomeroy, *Biotechnol. Lett.* **9**(5), 373 (1987).

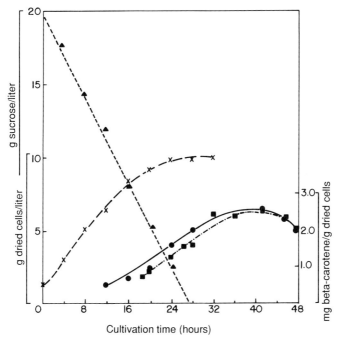

FIG. 1. Kinetics of β-carotene formation in *Rhodotorula*. ▲, Sucrose consumption; X, growth curve; ●, kinetics of β-carotene production in cultures, ■, kinetics of β-carotene production in cells suspended in distilled water. Conditions are described in the text.

after all the sugar is consumed and attains a maximum at 42 hr of cultivation.

β-Carotene Synthesis in *Rhodotorula* Cells Suspended in Distilled water

Rhodotorula glutinis EQ cells are cultivated as usual for 20 hr. At this time the cells, not yet colored, are centrifuged, washed twice in distilled water, and suspended in an adequate volume of distilled water. A sample is taken to determine the cell concentration in the suspension, which is constant throughout the experiments. The cell suspension is aerated and agitated as during growth (1.0 vvm, 400 rpm, 27°, 2.0 ml of Nalco is added to prevent foaming). For 44 hr, 50-ml samples are taken at intervals, centrifuged, and the cells extracted with acetone and the β-carotene concentration per gram of dried cells measured. The pH of the supernatants does not change, and no excreted substance is detected. Results can also be seen in Fig. 1.

TABLE I
Carbon Sources for β-Carotene Production[a]

Compound	Concentration (g/liter)	Biomass (g/liter)	β-Carotene (mg/g dry cells)
Sucrose[b]	20.0	10.0	1.7
Sucrose[c]	20.0	12.0	1.7
Xylose	22.5	10.0	1.0
Glycerol[d]	21.5	11.0	2.7
Acetate[d]	28.0	9.5	2.4

[a] Conditions are described in the text.
[b] From Ref. 8.
[c] Added as sugar cane juice. [H. L. Martelli, I. M. da Silva, N. O. Souza, and D. Pomeroy, *Biotechnol. Lett.* **12**(3), 207 (1990)].
[d] Adaptation is needed for growth.

Carbon Sources

Many carbon sources can be added to the basic medium for β-carotene production (Table I). Similar experiments with *R. lactosa* and *R. rubra* reveal the same pattern for growth and β-carotene production.

[36] Characteristics of Membrane-Associated Carotenoid-Binding Proteins in Cyanobacteria and Prochlorophytes

By K. J. Reddy, George S. Bullerjahn, and Louis A. Sherman

Introduction

Carotenoids are ubiquitous components of photosynthetic organisms, and these pigments have several important functions: (1) carotenoids can act as an accessory light-harvesting antenna; (2) they serve to quench toxic singlet oxygen and triplet chlorophyll that arise due to endogenous photosensitization[1]; and (3) the accumulation of carotenoid pigments can dissipate energy at photoinhibiting light intensities.[2] While much work has

---

[1] D. Siefermann-Harms, *Physiol. Plant.* **69**, 561 (1987).
[2] A. Ben-Amotz, A. Shaish, and M. Avron, *Plant Physiol.* **91**, 1040 (1989).

focused both on the synthesis and on the mechanisms of carotenoid function, less is known about the organization of carotenoids within photosynthetic systems.

More recently, several papers have described the existence in cyanobacteria of protein-containing complexes which exclusively bind carotenoids.[3-7] Sherman and collaborators have identified two distinct intrinsic membrane carotenoproteins which are components of the cell envelope in *Synechocystis* sp. PCC6714[4] and *Synechococcus* sp. PCC7942.[6,8] The *Synechocystis* sp. apoprotein is a hydrophobic xanthophyll-binding polypeptide of 45 kDa, and it accumulates to high levels under elevated oxygen concentration.[4] Antibodies to this protein have also identified the complex in the cell envelope of *Synechococcus* sp. PCC7942.[5] The other intrinsic membrane apoprotein has been identified in *Synechococcus* sp. PCC7942; it is a 42-kDa cytoplasmic membrane polypeptide which accumulates during growth at high light intensity.[6] The gene encoding this apoprotein has been sequenced, and RNA blots have shown that the gene is regulated by light at the transcriptional level.[8] More recently, a water-soluble cell surface carotenoid–protein complex has been purified from the prochlorophyte *Prochlorothrix hollandica*.[9] The complex is composed of two structurally similar polypeptides of 58 and 56 kDa which participate with the lipopolysaccharide to bind most, if not all, of the cellular zeaxanthin. Adaptation of *P. hollandica* to supersaturating light results in the accumulation of both the apoproteins and the xanthophyll pigment.[9] The characteristics of all of these carotenoid-binding proteins suggest strongly that their primary function is photoprotection and not to serve as photosynthetic antennas.

This chapter discusses the purification of the hydrophobic carotenoproteins from *Synechococcus* sp. PCC7942, and the regulation of the gene encoding the 42-kDa carotenoprotein. Additionally, we describe the properties of the *P. hollandica* surface-associated soluble zeaxanthin–protein complex.

---

[3] T. K. Holt and D. W. Krogmann, *Biochim. Biophys. Acta* **637**, 408 (1981).
[4] G. S. Bullerjahn and L. A. Sherman, *J. Bacteriol.* **167**, 369 (1986).
[5] G. S. Bullerjahn, H. C. Riethman, and L. A. Sherman, in "Progress in Photosynthesis Research" (J. Biggins, ed.), Vol. 2, p. 145. Martinus Nijhoff, Dordrecht. The Netherlands, 1987.
[6] K. Masamoto, H. C. Riethman, and L. A. Sherman, *Plant Physiol.* **84**, 633 (1987).
[7] M. Diverse-Pierluissi and D. W. Krogmann, *Biochim. Biophys. Acta* **933**, 372 (1989).
[8] K. J. Reddy, K. Masamoto, D. M. Sherman, and L. A. Sherman, *J. Bacteriol.* **171**, 3486 (1989).
[9] J. M. Engle, W. Burkhart, D. M. Sherman, and G. S. Bullerjahn, *Arch. Microbiol.* **155**, 453 (1991).

Procedure

Cell Growth. Synechococcus sp. PCC7942 and *Prochlorothrix hollandica* are grown in BG-11 medium[10] with constant illumination (50 μmol quanta/m^2/sec) and aeration at 25° in 5 to 15 liter carboys. Growth under high and low intensity white light employ photon flux densities of 250 and 30 μmol quanta/m^2/sec, respectively. For expression studies on the carotenoprotein gene *cpbA*, stationary phase cultures of *Synechococcus* sp. PCC7942 are harvested by centrifugation and resuspended in 25 ml of BG-1. This concentrated cell suspension is used to inoculate 500 ml of BG-11 medium to give an absorbance of 0.20 at 750 nm. The cultures are next incubated for 24 hr at 5 μmol quanta/m^2/sec prior to shifting to the appropriate experimental light intensities. The cultures are allowed to grow for an additional 22 hr before cells are harvested for RNA isolation.

Characterization of Carotenoid-Binding Proteins

Preparation of Synechococcus sp. Cytoplasmic Membranes and Cell Walls. Membrane fractions enriched for envelope components are prepared by sucrose gradient centrifugation essentially as described by Murata and Omata,[11] except that 50 mM morpholinoethanesulfonic acid (MES), pH 6.5, is used to buffer all solutions. In this procedure, the cytoplasmic membrane fraction is recovered as a yellow band at the top of the sucrose gradient, whereas the pellet is enriched in cell wall components. Henceforth, buffered solutions in all procedures contain the protease inhibitors benzamidine, 6-aminohexanoic acid, and phenylmethylsulfonyl fluoride (Sigma, St. Louis, MO), each at 1 mM final concentration.

Isolation of Synechococcus sp. 45-kDa Outer Membrane-Associated Carotenoid-Binding Protein. The carotenoid-binding protein is isolated by first pelleting cell walls prepared from 5 liters of autotrophically grown culture (3 hr at 115,000 g in a Beckman 60 Ti rotor) and then suspending the pellet in 1.5 ml of 50 mM MES (pH 6.5) containing 0.2% Triton X-100 and 0.2% dodecyl β-D-maltoside (Calbiochem-Behring, San Diego, CA). The membrane suspension is passed through an Affi-Gel 501 organomercury column (Bio-Rad Laboratories, Richmond, CA; 1 × 15 cm bed) equilibrated in the same buffer; this step removes all contaminating chlorophyll and phycobilin-containing components. The orange material which elutes in the void volume is collected, pelleted, and suspended as described above. The Affi-Gel 501 eluate is next applied to a DEAE-Sephacel column (1 × 25 cm bed, equilibrated in MES–Triton X-100–

[10] M. M. Allen, *J. Phycol.* **4**, 1 (1968).
[11] N. Murata and T. Omata, this series, Vol. 167, p. 245.

dodecyl β-D-maltoside) and washed with 4 volumes of the same buffer, containing 10 mM NaCl, followed by washing with 2 volumes of buffer containing 100 mM NaCl, prior to elution of the carotenoid-binding protein with 350 mM NaCl. The carotenoid-binding protein elutes from the column as a sharply defined orange band.

Isolation of Synechococcus sp. Cytoplasmic Membrane Carotenoprotein. Cytoplasmic membranes are collected by ultracentrifugation (115,000 g, 12 hr) and resuspended in 1.5 ml of 50 mM MES (pH 6.5) containing 0.05% dodecyl β-D-maltoside.[6] This sample is passed through a DEAE-Sephacel column equilibrated in the suspension buffer. Almost all yellow material elutes in the void volume. The yellow eluate is applied to an Affi-Gel 501 organomercury column (Bio-Rad) equilibrated in the same buffer. The yellow eluate in the void volume from the Affi-Gel column is centrifuged at 130,000 g for 24 hr in the Ti 60 rotor, and the upper, clear portion of the centrifuged solution is discarded. The yellow fraction, containing the carotenoid-associated protein, is resuspended in a small volume of 50 mM MES (pH 6.5) containing 0.05% (w/v) dodecyl β-D-maltoside. The properties of the carotenoid on the columns and in ultracentrifugation suggest that the pigment is specifically bound to the protein.

Isolation of Prochlorothrix hollandica Soluble Protein–Zeaxanthin Complex. Prochlorothrix hollandica filaments from 3 liters of stationary phase cultures are harvested by low-speed centrifugation (3000 g, 5 min) washed once in 50 mM MES (pH 6.5), and resuspended in 30 ml of the same buffer. Resuspended filaments are disrupted by two passages through a chilled (4°) French pressure cell (70 MPa), and unbroken cells are removed by centrifugation at 3000 g for 5 min. Following high-speed centrifugation of the supernatant (34,000 g for 30 min), the resulting yellow supernatant, containing approximately 5 mg protein, is collected. The material is dialyzed extensively against deionized water, then added to 1/10 volume of Bio-Rad 3-10 ampholytes. This mixture is next subjected to preparative isoelectric focusing; we routinely employ the Bio-Rad Rotofor preparative cell running for 2 hr at 12 W constant power. The eluted yellow material banding toward the acidic pole (pH ~3.5) of the cell is diluted to 30 ml in deionized water and refocused as described above. The pigmented material (3 ml) retrieved from the cell is dialyzed against 50 mM MES, pH 6.5, to remove the ampholytes, then centrifuged for 12 hr at 120,000 g to pellet the complex, and the resulting orange pellet is resuspended in 500 μl of 50 mM MES buffer.

Alternatively, *P. hollandica* cultures left undisturbed for 10 weeks or more will accumulate the yellow complex in the culture fluid; thus, spent culture medium can be used as starting material for purification of the complex. Stationary phase cells are removed either by filtration or by low-speed centrifugation; the supernatant is then lyophilized and dissolved in 5–10 ml of deionized water. After dialysis against deionized water and low-speed centrifugation to remove insoluble material, the mixture is suitable for isoelectric focusing and all subsequent steps as described above. Chromatographic analysis of the carotenoids extracted from the material will reveal that zeaxanthin is the sole pigment present, at a pigment/protein ratio of approximately 0.2.

Thin-Layer Chromatography and Protein Assay. Carotenoid concentrations are determined from the molar extinction coefficients described previously[12]; separation of carotenoid pigments on silica gel plates is according to Liaaen-Jensen and Jensen,[12] and the developing solvent is light petroleum benzene/2-propanol/water (100:11:0.5, v/v).[13] Under these conditions, zeaxanthin migrates with an R_f value of 14. Protein concentrations are based on Lowry-type assays employing bovine serum albumin as the standard.[14]

Cloning and Expression of Synechococcus sp. cpbA Gene

Cloning and Sequencing of cbpA Gene Encoding 42-kDa Carotenoid-Binding Protein. The *cbpA* gene from *Synecococcus* sp. PCC7942 is identified and cloned using an antibody probe. A λgt11 expression library of *Synechococcus* sp. PCC7942 DNA is constructed by cloning randomly sheared chromosomal fragments as previously described.[15] Screening of this library has yielded an immunopositive clone, λgtAN42, bearing a 0.9-kb *Eco*RI insert.

Following isolation of immunopositive clones from λgt11 expression libraries, it is essential to confirm the clone identity by some other means. This is especially important when polyclonal antibodies exhibit reactivity to several proteins in whole cell extracts. Confirmation of the identity of a clone is achieved by affinity purification of the polyclonal antibody by adsorption to the fusion protein produced by the immunopositive clone.[16] Testing such purified antibodies on immunoblots will reveal whether the fusion protein has selected for antibodies recognizing the polypeptide of interest. Characterization of the λgtAN42–LacZ fusion protein confirmed that it shares immunological identity with the 42-kDa carotenoprotein.

---

[12] S. Liaaen-Jensen and A. Jensen, this series, Vol. 23, p. 586.
[13] U. J. Jurgens and J. Weckesser, *J. Bacteriol.* **164**, 384 (1985).
[14] M. A. K. Markwell, S. M. Haas, L. L. Bieber, and N. E. Tolbert, *Anal. Biochem.* **87**, 206 (1978).
[15] R. A. Young and R. W. Davis, *Proc. Natl. Acad. Sci. U.S.A.* **80**, 1194 (1983).
[16] R. Webb, K. J. Reddy, and L. A. Sherman, *DNA* **8**, 69 (1989).

Screening of the EMBL-3 library with λgAN42 resulted in the isolation of a clone, λEM109, which contains an 18-kb chromosomal insert. The *cbpA* gene is located near the center of this insert with 7–8 kb flanking DNA on either side. The λEM109 clone is further mapped with various restriction enzymes to identify the smallest restriction fragment that contains the entire *cbpA* gene. This fragment is subcloned into plasmid and M13 vectors for chain-termination DNA sequencing.[17]

RNA Isolation. To 500 ml of cyanobacterial culture, 25 ml of 20× stop buffer [200 mM Tris-HCl (pH 8.0), 20 mM EDTA, 20 mM sodium azide] and 5 ml of 20 mM aurin tricarboxylic acid are added and kept on ice.[18] Cells are pelleted by centrifugation and the pellet resuspended in 2 ml of STET buffer [8% sucrose, 50 mM Tris-HCl (pH 7.00), 5 mM EDTA, 5% Triton X-100] containing 20 mM vanadyl ribonucleoside complex (VRC). The cells are extracted twice with phenol–chloroform (1:1, v/v) and the supernatant precipitated at room temperature by the addition of 1/10 volume of 3 M sodium acetate and 2 volumes of ice-cold ethanol. The nucleic acid pellet is next resuspended in 2 ml of diethyl pyrocarbonate (DEPC)-treated water, extracted twice with phenol–chloroform, and ethanol precipitated. The pellet is redissolved in 2 ml DEPC-treated water, then 1 g solid CsCl is added. This solution is layered onto 0.75 ml of 5.7 M CsCl in a TLA 100.3 polycarbonate centrifuge tube. After centrifugation at 70,000 rpm (200,000 g) in a TL-100 Beckman tabletop ultracentrifuge, the upper CsCl solution is carefully removed to avoid disturbing the RNA pellet. The RNA pellet is dissolved in DEPC-treated water and ethanol precipitated. The RNA is then separated under denaturing conditions on MOPS (morpholinepropanesulfonic acid)–formaldehyde gels and transferred to Nytran charge-modified nylon membranes (Schleicher and Schuell, Keene, NH) for Northern blot hybridizations.

Comments on Procedures

Purification of Carotenoproteins. The isolation procedures above yield complexes which exhibit the visible absorption spectra shown in Fig. 1. With respect to the membrane-bound *Synechococcus* sp. complexes, the spectra may yield bandshifts due to interactions of the pigment with the solubilizing detergent.[7] Nonetheless, the hydrophobic nature of these proteins requires the presence of at least 0.03% dodecyl maltoside.

The cellular location of the polypeptides has been established only through immunocytochemical techniques (Fig. 2). In earlier papers, the 45-kDa protein was reported to be a component of the cytoplasmic mem-

---

[17] F. Sanger, S. Nicklen, and A. R. Coulson, *Proc. Natl. Acad. Sci. U.S.A.* **74,** 5463 (1977).
[18] K. J. Reddy, R. Webb, and L. A. Sherman, *BioTechniques* **8,** 251 (1990).

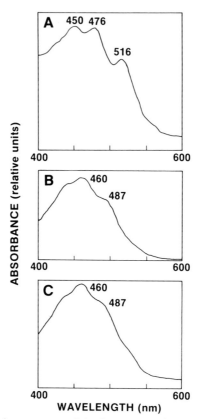

FIG. 1. Visible absorbance spectra of the three carotenoid–protein complexes. (A) *Synechococcus* sp. PCC7942 45-kDa polypeptide complex; (B) *Synechococcus* sp. 42-kDa polypeptide complex; (C) *P. hollandica* cell surface zeaxanthin–protein complex.

brane,[4,19] whereas the 42-kDa polypeptide had been identified as a either a thylakoid protein[6] or a cytoplasmic membrane protein.[20] Immunodecoration of Lowicryl HM20-embedded thin sections with polyclonal antibodies to these apoproteins reveals unequivocally that the 45-kDa protein is found in the outer membrane and the 42-kDa polypeptide is localized to the cytoplasmic membrane (Fig. 2A,C). The *P. hollandica* component is localized exclusively to the outer membrane (Fig. 2B); this is consistent

---

[19] T. Omata, M. Ohmori, N. Arai, and T. Ogawa, *Proc. Natl. Acad. Sci. U.S.A.* **86**, 6612 (1989).
[20] T. Omata and T. Ogawa, *Plant Physiol.* **80**, 525 (1986).

with the observation that the complex is released from growing cells into the medium.[9]

Cloning and Expression of cbpA Gene. The carotenoid-binding protein gene sequence yielded an open reading frame of 450 codons capable of producing a protein with a calculated molecular weight of 49,113.[8] The hydropathic analysis of this protein according to Kyte and Doolittle[21] revealed a 49-residue stretch at the amino terminus sharing the structural features of a signal sequence. This is in general agreement with other studies indicating that the 42-kDa carotenoprotein is associated with the cytoplasmic membrane.

Owing to the fact that the majority of prokaryotic RNAs are known to have short half-lives, we have developed a rapid method that yields undegraded transcripts.[18] An essential feature of this method is the addition of the ribonuclease inhibitors VRC and aurin tricarboxylic acid. With this procedure, we were able to detect a 6.2-kb transcript which likely represents a polycistronic *cbpA* operon.

The accumulation of the 42-kDa carotenoprotein occurs under conditions of high light stress.[6] As described in Ref. 6, the protein is abundant under high-light growth conditions but undetectable in low-light-grown cells. Transcription of the *cbpA* gene is also induced by growth under high light intensities (Fig. 3); additional studies indicate that transcription is repressed during growth under iron-deficient conditions. Furthermore, the expression of *cbpA* is affected by DNA supercoiling. This is clearly evident in high-light-grown cells treated with novobiocin, a DNA gyrase inhibitor; such treatment abolishes transcription (Fig. 3). Our contention that the carotenoid-binding protein is involved in protecting cells from photodamage is further supported by the high level of expression of *cbpA* under conditions which induce oxygen radical formation. Recent data indicate that levels of the *cbpA* transcripts are 4 to 5-fold higher in cells treated with agents known to generate oxygen radicals.

Concluding Remarks

Further perspectives on the function of carotenes and xanthophylls in photosynthetic systems stem from the work of Demmig-Adams and collaborators.[22,23] Their data indicate that zeaxanthin is an important component of chloroplast photosynthetic membranes, serving to contribute to

---

[21] J. Kyte and R. F. Doolittle, *J. Mol. Biol.* **157**, 105 (1982).
[22] B. Demmig-Adams, W. W. Adams, F.-C. Czygan, U. Schreiber, and O. L. Lange, *Planta* **190**, 582 (1990).
[23] B. Demmig-Adams, *Biochim. Biophys. Acta* **1020**, 1 (1990).

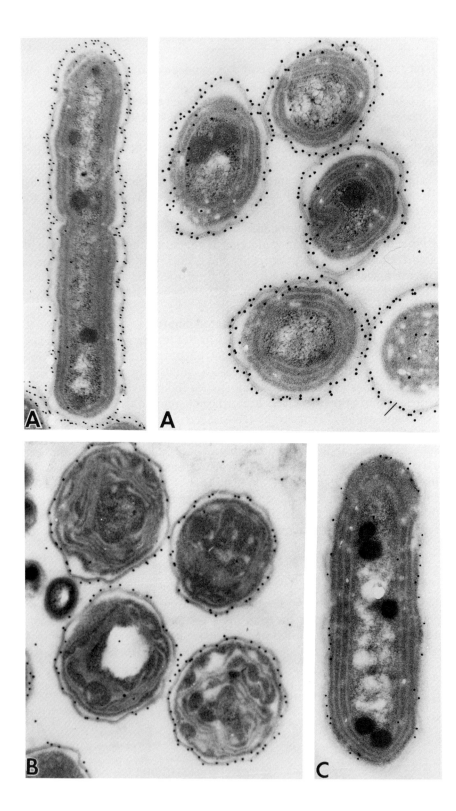

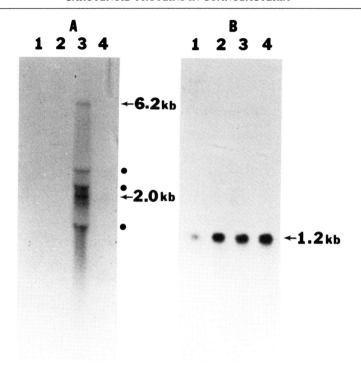

FIG. 3. Effect of light intensity and novobiocin on *cbpA* and *psbO* transcription. Lanes 1, low light (5 μmol quanta/m²/sec); lanes 2, medium light (50 μmol quanta/m²/sec); lanes 3, high light (200 μmol quanta/m²/sec); and lanes 4, same as lane 3 but treated with 50 μg/ml novobiocin. (A) Probed with the *cbpA* gene. (B) Results of reprobing the same blot with a *psbO* gene fragment. The *psbO* gene was chosen as a control as the levels of the PsbO protein (the Mn^{2+} stabilizing protein of Photosystem II) remain fairly constant under these experimental conditions. The arrows indicate *cbpA*-specific transcripts. All other bands are artifactual bands of rRNA arising from degraded transcripts.

photoprotection by conversion to violaxanthin in the xanthophyll cycle. It is evident from studies on cyanobacteria and prochlorophytes that such a cycle does not occur in these organisms; the cellular location of the pigment proteins in the envelope suggests that they may act primarily to dissipate excess excitation energy at high photon flux densities.[6,9] Nevertheless, analysis of cyanobacterial envelope membranes and the pigment proteins described here indicates that xanthophylls are the predominant

FIG. 2. Immunocytochemical localization of the carotenoid-binding proteins. (A) Immunodecoration of *Synechococcus* sp. thin sections with antibody to the 45kDa protein; (B) immunodecoration of *P. hollandica* with antibody raised against the 58- and 56-kDa surface layer apoproteins; (C) immunodecoration of *Synechococcus* sp. with antibody to the 42-kDa protein.

MNEFQPVNRRQFLFTLGATAASAILLKGCG--NPPSSSGGGTSSTTQPTAAGAS-52
:: :::.: :.:::.:. : .:: : . . :.:. .:.
MSQFSRRKFLLTAGGTAAAALWLNACGSNNSSTDTTGSTSTPAPSGTSGGD-51

DLEVKTIKLGYIPIFEAAPLIIGREKGFFAKYGL-DVEVSKQASWAAARDNVIL-105
::: . :: : . .:::.::. :::.:::::: : : :: ::: :::. :
APEVKGVTLGFIALTDAAPVIIALEKGLFAKYGLPDTKVVKQTSWAVTRDNLEL-105

GSAGGGIDGGQWQMPMPALLTEGAISNGQK-VPMYVLACLSTQGNGIAVSNQLK-158
:: :::::. ::: ::: : :...:: .:::.:: : :::.:..::..
GSDRGGIDGAHILSPMPYLLTAGTITKSQKPLPMYILARLNTQGQGISLSNEFL-159

AQNLGLKLAPNRDFILNYPQTSGRKFKASYTFPNANQDFWIRYWFAAGGIDPDK-212
:. . .: : : ::. .:: ::: .. :.:.:::.:: .:::.
AEKVQIK-DPKLKAIADQKKASGKLLKAAVTFPGGTHDLWMRYWLAANGIDPNN-212

DIELLTVPSAETLQNMRNGTIDCFSTGDPWPSRIAKDDIGYQAALTGQMWPYHP-266
: .:. .: . . :: :::.: : :.:: ..: .:: ::.:::.: .::
DADLVVIPPPQMVANMQTGTMDTFCVGEPWNARLVNKKLGYTAAVTGELWKFHP-266

EEFLALRADWVDKHPKATLALLMGLMEAQQWCDQKANRAEMAKILSGRNFFNVP-320
: :..:::: ::.:::.:: ::.. .. :: :. . ..
EKALTIRADWADKNPKATMALLKAVQEAQIWCEDPANLDELCQITAQDKYFKTS-320

VSILQPILEGQIKVGADGKDLNNFDAGPLFWKSPRGSVSYPYKGLTLWFLVESI-374
: . : :.: : : ::. . : :: :..:::. :::: : :
VEDIKPRLQGDIDYG-DGRSVKNSDLRMRFW---SENASFPYKSHDLWFLTEDI-370

RWGFNKQVLPDIAAAQKLNDRVTREDLWQEAAKKLGVPAADIPTGSTRGTETFF-428
:::. :: : ..: ::: :::: .: . ::: .:: .:: ::::
RWGY----LPASTDTKALIEKVNRSDLWREAAKAIG-REQDIPASDSRGVETFF-419

DGITYNPDSPQAYLQSLKIKRA-450   42 kDa Protein
::.:...:. ::::: ..:.:
DGVTFDPENPQAYLDGLKFKAIKA-443   45 kDa Protein

FIG. 4. Sequence comparison of the 45-kDa (bottom row) and 42-kDa (top row) polypeptides deduced from the nucleotide sequences of the cloned *Synechococcus* sp. genes [T. Omata, T. Ogawa, T. J. Carlson, and J. Pierce, in "Current Research in Photosynthesis" (M. Baltscheffsky, ed.), Vol. 3, p. 525. Kluwer, Dordrecht, The Netherlands, 1990; K. J. Reddy, K. Masamoto, D. M. Sherman, and L. A. Sherman, *J. Bacteriol.* **171**, 3486 (1989)]. The alignment indicates both amino acid identities (colons) and conservative replacements (periods).

carotenoids, whereas β-carotene is found almost exclusively in the thylakoid compartment.[12,24] These findings underscore the overall importance of xanthophylls in photoprotective mechanisms irrespective of a functional xanthophyll cycle.

Recently, the deduced amino acid sequences encoded by both *cbpA* and the gene encoding the 45-kDa carotenoprotein were compared and shown to be homologous (Fig. 4). Another study also demonstrated that mutants lacking the 45-kDa polypeptide were defective in nitrate transport.[25] Owing to the location of the protein in the cell wall, it is possible that the 45-kDa polypeptide acts either to form or to stabilize pores in the outer membrane which could be involved in ion uptake. The structural similarity of the two membrane-associated proteins also suggests that they share common function(s), despite their location in different membranes of the envelope. It is certainly possible that these structurally similar features represent common xanthophyll-binding domains.

The *P. hollandica* carotenoid–protein complex represents a very different type of structure. The complex is an acidic, soluble component which contains two structurally related proteins, forms oligomers *in vitro*, associates with lipopolysaccharide, and is located at the cell surface.[9] These are characteristics which are shared among a class of prokaryotic protein complexes which are termed S- (surface) layers.[26] S-layers are common features seen in many natural isolates of bacteria, but this complex is the first which has been shown to be associated with xanthophyll pigments. The accumulation of this complex under high photon fluxes[9] suggests that the synthesis or turnover of this complex is somehow regulated by light.

Whereas a great deal of structural information is available on the carotenoproteins known to be components of the invertebrate carapace,[27] much work remains to clarify the structure and assembly of these pigment proteins in photosynthetic systems. Furthermore, it is largely unclear why the pigment molecules are arranged onto protein scaffolds in the first place, as the physical and chemical properties of free carotenoids alone could possibly be effective in photoprotection. These questions will lead to additional studies aimed at examining in detail the physiological properties of mutants which are defective in the synthesis of these apoproteins. The light regulation of the *cbpA* gene also points to future experiments which seek to identify light-activated transcriptional components in cyanobacteria.

[24] T. Omata and N. Murata, *Arch. Microbiol.* **139**, 113 (1984).
[25] T. Omata, T. Ogawa, T. J. Carlson, and J. Pierce, in "Current Research in Photosynthesis" (M. Baltscheffsky, ed.), Vol. 3, p. 525. Kluwer, Dordrecht, The Netherlands, 1990.
[26] U. B. Sleytr and P. Messner, *Annu. Rev. Microbiol.* **37**, 311 (1983).
[27] P. F. Zagalsky, E. E. Eliopoulos, and J. B. C. Findlay, *Comp. Biochem. Physiol.* **97B**, 1 (1990).

[37] Xanthosomes: Supramolecular Assemblies of Xanthophyll–Chlorophyll a/c Protein Complexes

By TETZUYA KATOH, AYUMI TANAKA, and MAMORU MIMURO

Introduction

In the light-harvesting systems of brown algae and diatoms, fucoxanthin is confined, together with chlorophyll (Chl) a and Chl c, in a protein to form fucoxanthin–Chl a/c pigment protein complexes.[1-5] Like phycobilisomes in cyanobacteria and red algae, the fucoxanthin–Chl a/c protein complexes are assembled with each other to form a supramolecular assembly, which was originally named the fucoxanthin–chlorophyll a/c protein assembly (FCPA).[4] If attention is paid to the habitat of brown algae in the sea, it seems most plausible that these algae have large-sized assemblies for light harvesting under a dim light environment in the depths. As peridinin–Chl a/c pigment proteins were also isolated in an assembled form from *Glenodinium*, a dinoflagellate,[6] the supramolecular assemblies of fucoxanthin–chlorophyll pigment proteins and those of peridinin–chlorophyll proteins have been named xanthosomes.

Xanthosomes of *Petalonia*, a brown alga, have a molecular weight of 7.0×10^5, contain 128 Chl a, 27 Chl c, and 69 fucoxanthin units, and absorb light over much of the visible range.[5] Energy absorbed by any of fucoxanthin or Chl c molecules is efficiently transferred to Chl a, although the mechanism of energy migration is currently unknown.[7,8] The energy transferred to Chl a is funneled into photosystems in the thylakoids. Here we describe the procedures in use for the isolation of xanthosomes as well as xanthosome–photosystem I complexes from brown algae. Purification of monomeric pigment proteins, the component unit of xanthosomes, is also described.

---

[1] J. Barrett and J. M. Anderson, *Biochim. Biophys. Acta* **590**, 309 (1980).
[2] R. S. Alberte, A. L. Friedman, D. L. Gustavson, M. S. Rudnick, and H. Lyman, *Biochim. Biophys. Acta* **635**, 304 (1981).
[3] T. G. Owens and E. R. Wold, *Plant Physiol.* **80**, 732 (1986).
[4] T. Katoh, M. Mimuro, and S. Takaichi, *Biochim. Biophys. Acta* **976**, 233 (1989).
[5] T. Katoh and T. Ehara, *Plant Cell Physiol.* **31**, 439 (1990).
[6] T. Katoh, unpublished data.
[7] M. Mimuro, T. Katoh, and H. Kawai, *Biochim. Biophys. Acta* **1015**, 450 (1990).
[8] M. Mimuro and T. Katoh, *Pure Appl. Chem* **63**, 123 (1991).

Problems Encountered in Isolation of Xanthosomes

Two major problems associated with the isolation and purification of xanthosomes are their instability to detergents and, in spite of this, the necessity to use detergent to separate them from thylakoid vesicles. Xanthosomes of brown algae, which have fucoxanthin as the sole carotenoid, are orange-brown in color and show an efficient excitation energy transfer from fucoxanthin to Chl a and from Chl c to Chl a.[4] They are, however, fairly unstable at room temperature and rather readily uncoupled with an accompanying change of color from brown to green, reflecting the blue shift of fucoxanthin. In an energetically coupled state, the configurations of fucoxanthin molecules are distorted in such a way that the energy absorbed by them migrates to Chl a efficiently, and the long-wavelength form of fucoxanthin would correspond to this configuration.[7,8] Thus, one can judge the intactness of xanthosome preparations from their color. In fact, the color of xanthosome preparations changes to green instantaneously if brought into contact with the conventional detergents, like Triton X-100, lauryldimethylamine N-oxide (LDAO), and Zwittergent 3-14, and the energy migration to Chl a is no longer observed in the preparations which turned to green.

On the other hand, the separation of xanthosomes from thylakoid membranes involves solubilization with detergent. Although Triton X-100,[1,3] sodium dodecyl sulfate (SDS),[2] or CHAPS[9] have previously been used in the isolation of fucoxanthin–chlorophyll a/c protein complexes, these detergents caused a rapid dissociation of xanthosomes to component units and sometimes a concomitant blue shift of fucoxanthin. Milder detergents, such as decanoyl sucrose,[10] give a fairly good separation of xanthosomes from brown algae as well as from other chromophytic algae without causing dissociation of xanthosomes and uncoupling of excitation energy transfer.

Isolation of Xanthosomes from Brown Algae

Selection, Harvesting, and Transportation of Seaweeds

Choice of Algae. Generally, the isolation of xanthosomes from brown algae is not easy. The seaweeds, like most of the *Sargassum* family, have extensive, tough cell walls which resist cell disruption. Seaweeds like *Fucus*

---

[9] A. L. Friedman and R. S. Alberte, *Plant Physiol.* **76**, 483 (1984).
[10] Lauryl sucrose (SM1200), decanoyl sucrose (SM1000), and octanoyl sucrose (SM800) can be obtained from Mitsubishi-Kasei Foods, Co., 13-3, Ginza 5 chome, Chuo-ku, Tokyo 104, Japan; Fax 81-3-3545-4860, Telex BISICH J24901 ATT.MFC.

or *Undaria*, though appearing more tender, secrete copious mucilage when injured, which interferes extensively with the separation and collection of bands in density gradient centrifugations. Selecting seaweeds with tender thalli and with little slime is crucial in this work. *Dictyota* and *Petalonia* seem to be suitable for the purpose of xanthosome isolation, although the growth of these seaweeds is limited to colder seasons.

Collection and Transport of Seaweeds. The thalli of *Dictyota* (or *Petalonia*) collected from sea have to be freed from the red algae which attach at their basement in clean seawater quickly and carefully. Thalli with calcareous red algae on them are also excluded. Slight contamination of phycobilin pigments makes it impossible to measure fluorescence characteristics of xanthosome preparations. The cleaning must be performed in running seawater, otherwise the thalli turn greenish very soon owing to the limited oxygen supply. Xanthosomes with efficient energy coupling can no longer be isolated from the thalli which turned green. Care must be also taken not to contaminate samples with seaweeds of the *Desmarestia* group, which often grow at the same place where *Dictyota* does, because these algae, when brought into air or when dying, secrete a lot of sulfuric acid which makes the entire preparation green.

When transportation takes long time, washed thalli are drained and frozen rapidly with dry ice or with liquid nitrogen, for transport and for storage until use. Freezing and thawing, if warming to room temperature is avoided, not induce the dissociation of xanthosomes nor serious uncoupling of energy transfer, but rather facilitate their separation from thylakoid membranes.

Procedure of Xanthosome Isolation

1. Thawed thalli of *Dictyota* are minced and rinsed in an icecold solution containing 0.3 M sorbitol and 0.2 M Tris-HCl (pH 7.6). Rinsed fragments of thalli are suspended in twice the volume of the same solution, then disrupted by passing through a chilled French press 3 times.

2. Pastelike disintegrates of thalli are mixed with 7 to 10 volumes of an ice-cold solution of 0.3 M sorbitol and 50 mM Tris-HCl (pH 7.6), then centrifuged at 12,000 g for 30 min at 4° to remove soluble materials and mucilage. The pellets are homogenized with twice the volume of solution containing 0.3 M sorbitol and 50 mM Na–HEPES (pH 7.0).

3. Decanoyl sucrose (n-decyl-β-D-fructofuranosyl-α-D-glucopyranoside; SM1000[10]) is added to 0.8% (w/v) and stirred at 0° for 3 hr. The disrupted thalli contain, in addition to thylakoid lamellae, the bulk of the fibrous and amorphous materials which adsorb the detergent, so it is useless to note the precise amount of detergent in terms of a detergent to Chl *a* ratio. How-

ever, the concentration of detergent should not be higher than 0.8%, otherwise various Chl-containing vesicles appear in the supernatant. Furthermore, the xanthosomes are destablized by this detergent when it is present in excess. By centrifugation at 22,000 g for 60 min, the decanoyl sucrose-treated thalli give an orange-brown supernatant. When the extract is not dense enough, another treatment with the same concentration of decanoyl sucrose will yield dense brown extracts. Care should be taken to maintain the pH during the treatment, not exceeding pH 8.0, otherwise sugar esters might be hydrolyzed to yield free fatty acids.

4. The brown color of the supernatant is practically entirely accounted for by the isolated xanthosomes. They are purified by sucrose density gradient centrifugation. Sucrose gradients (0.3 to 1.2 M sucrose) of 20 ml volume are prepared in a solution containing 50 mM Na–HEPES (pH 7.0) and 0.2% decanoyl sucrose, on which 2.0 ml of supernatant is overlaid. The gradients are centrifuged in a Hitachi RP30 fixed-angle rotor at 28,000 rpm (~76,000 g) for 15 hr. At the end of the centrifugation, four bands are separated: a pale green band at the top, a pale brown band closely below the top band, a sharp and dense brown band in the middle of the gradients (at ~1.0 M sucrose), and a greenish precipitate. Xanthosomes are recovered in the third brown band, which is collected with a syringe.

Isolation of Xanthosomes, Photosystem I–Xanthosome Vesicles, and Monomeric Forms from *Ectocarpus*

Ectocarpus is one of the brown algae which can be easily cultured under laboratory conditions. This alga serves as an excellent starting material for the study of the photosynthetic systems of brown algae without worrying about contamination with other algae. In this alga, however, the xanthosomes bind to the thylakoid lamellae very tightly. Treatment of disrupted cells with decanoyl sucrose does not solubilize xanthosomes in a free state but rather in the form of vesicles in which xanthosomes and photosystem I are combined. Other means are needed to deal with the xanthosomes of this alga.

Algal Source, Culture Medium, and Growth Conditions

Ectocarpus siliculosus B63.81 is obtained from Sammlung von Algenkulturen, Göttingen University, Germany. The algae are cultured in 10-liter flasks containing 8 liters of ES-enriched brackish medium,[11] which is

[11] J. McLachlan, in "Handbook of Phycological Methods, Culture Methods and Growth Measurements" (J. R. Stein, ed.), p. 25. Cambridge Univ. Press Cambridge, 1973.

modified as follows: 100 mg $NaNO_3$, 10 mg disodium β-glycerophosphate, 1.8 mg $FeSO_4 \cdot 7H_2O$, 1.0 mg $MnCl_2 \cdot 4H_2O$, 0.66 mg $CuSO_4 \cdot 5H_2O$, 0.4 mg $ZnSO_4 \cdot 7H_2O$, 1.6 μg cyanocobalamin, 0.8 μg biotin, 100 μg thiamin hydrochloride, and 80 mg Tris are dissolved in a mixture of 500 ml filtered seawater and 500 ml of distilled water, and the pH is adjusted to 7.6 with 0.2 N H_2SO_4. The culture is gassed with air containing 5% (v/v) CO_2. The culture bottles are incubated at 20° under illumination from four 40-W fluorescent tubes, and cultures are harvested in the late exponential phase of growth using nylon cloth.

Isolation of Xanthosome–Photosystem I Vesicles

The procedures involve the extraction of thylakoid vesicles, followed by the separation of xanthosomes from vesicles. Decanoyl sucrose (SM1000) is not so efficient in isolating xanthosomes in the free form.

1. The collected algae are rinsed rapidly by suspension in an ice-cold solution of 0.3 M sorbitol and 0.1 M Na–HEPES, pH 7.0 (HS buffer), collection on nylon cloth, and resuspension in 2 volumes of the same solution. The following steps are carried out at 0°.

2. The algal suspension is passed through a French pressure cell at 200 kg/cm^2 3 times. Disrupted material is homogenized with 10 volumes of ice-cold HS buffer, centrifuged for 5 min at 100 g to remove unbroken cells, and then centrifuged at 22,000 g for 60 min to pellet thylakoid membranes. The brownish supernatant containing soluble proteins is discarded.

3. The thylakoid membranes are homogenized with HS buffer to a concentration of 2 to 3 mg Chl a/ml, to which 1/6 volume of 10% (w/v) octanoyl sucrose (n-octyl-β-D-fructofuranosyl-α-D-glucoside; SM800[10]) is added, and stirred for 4 hr at 0°, followed by centrifugation at 22,000 g for 60 min. The concentration of octanoyl sucrose should not exceed 2%.

4. Five milliliters each of the resultant brown supernatant is overlaid in 25-ml centrifuge tubes containing 3.0 ml of 1.5 M sucrose, 4.0 ml of 1.0 M sucrose, and 10.0 ml of 0.7 M sucrose, all containing 50 mM Na–HEPES, and centrifuged at 84,000 g for 3 hr. The dense brown band at the 1.0–1.5 M sucrose interface is collected by a syringe with a flat-tipped canula. This fraction contains vesicles in which xanthosomes are bound to PSI (Fig. 5).

Isolation of Xanthosomes

Separation of xanthosomes from the photosynthetic vesicles of *Ectocarpus* is not so easy as from *Petalonia* or *Dictyota*, probably due to the tighter binding of xanthosomes to the thylakoid lamellae in the former. The procedure of extraction with decanoyl sucrose (SM1000), as applied to

Petalonia, did not yield a good preparation of xanthosomes free from chlorophyll-containing vesicles. Treatment of the vesicles with lauryl sucrose (*n*-dodecyl-β-D-fructofuranosyl-α-D-glucoside; SM1200[10]) seems to result in a better separation, though this detergent induces a partial degradation of energetic coupling between the pigments.

5. The xanthosome–PSI complexes are diluted with 3 volumes of 50 mM Na–HEPES (pH 7.0) and spun down at 144,000 g for 2 hr. The pellets are homogenized with 0.1% (w/v) lauryl sucrose (SM1200) in 50 mM Na–HEPES, pH 7.0 (0.8–1.5 mg Chl a/ml), stirred for 1 hr at 0°, and then subjected to density gradient centrifugation. The composition of gradients and the conditions of centrifugation are described above. A brown band in the middle of the gradients contains xanthosomes, which are collected with a syringe.

Isolation of Monomeric Form of Fucoxanthin–Chlorophyll a/c Protein Complexes

6. Vesicles containing xanthosome–PSI complexes, recovered at the 1.0–1.5 M sucrose interface, are mixed with an equal volume of 0.6 M Tris-HCl buffer (pH 8.8) containing 20% glycerol, and subjected to polyacrylamide gel electrophoresis (PAGE). The nondenaturing buffer system is a modification of the Anderson system.[12] The resolving gel (6 cm long, 0.5 cm diameter) consists (w/v) of 40 mM borate–41 mM Tris (pH 8.64), 10% glycerol, 8% acrylamide, 0.2% bisacrylamide, 0.003% ammonium persulfate, and 0.1% N,N,N',N'-tetramethylethylenediamine (TEMED). The upper reservoir buffer contains 50 mM Tris–borate (pH 9.5) and 0.005% SDS. After 2 hr of electrophoresis at an electric current of 0.5 mA per tube in the cold room, two bands are resolved, of which the faster migrating with orange brown color is the dissociated monomer of fucoxanthin–chlorophyll a/c protein, and the slower migrating green band of P700-chlorophyll a protein complex (CP I) to which xanthosomes have bound. To obtain a preparation of the monomeric form which shows highly efficient energy transfer from fucoxanthin to Chl a, the samples are mixed with SDS-free and glycerin-containing buffer just prior to the electrophoresis. The electrophoresis should not be prolonged, otherwise bands of free Chl a and of free fucoxanthin appear at the top.

Characterization of Xanthosomes

The preservation of xanthosome structure and function can be assessed by three types of analyses. Absorption spectra provide a rapid measure of

[12] J. M. Anderson, J. C. Waldron, and S. W. Thorne, *FEBS Lett.* **122**, 149 (1978).

the degree to which fucoxanthin molecules in xanthosomes remain in the native, long-wavelength form. Fluorescence excitation spectra detected at the emission of Chl *a* provide a sensitive measure of the functional integrity of a preparation. Transmission electron microscopy provides a measure of the preservation of structure at the level of gross morphology and reveals the degree of dissociation and/or aggregation of the particles.

Absorption Spectrum. Fresh xanthosomes of brown algae show a considerable, broad absorption between 490 and 550 nm owing to the long-wavelength form of fucoxanthin. The absorbance at 510 nm is 5.5 times as high as that at 600 nm in the intact xanthosomes (Fig. 1A, solid line),

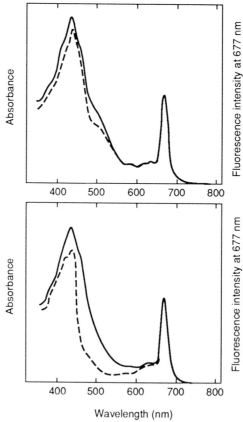

FIG. 1. Absorbance and fluorescence excitation spectra of freshly prepared (A) and denatured (B) *Petalonia* xanthosomes, suspended in 50 mM Na–HEPES (pH 7.2). Solid lines show the absorbance and dashed lines the fluorescence excitation spectra as detected at 677 nm. For denaturation, the fresh sample used in (A) is treated at 40° for 5 min.

whereas the A_{510}/A_{600} ratio is 3.2 in the denatured preparation, in which fucoxanthin is blue shifted (Fig. 1B, solid line).

Fluorescence Emission and Excitation Spectra. At room temperature, the intact xanthosomes of brown algae have a fluorescence emission maximum at 677 nm, independent of excitation wavelength (Fig. 2). In addition to this, excitation by 460 nm light yields a very weak emission around 640 nm, originating from Chl *c*. Although the emission at 640 nm is nearly indiscernible in energetically well-coupled xanthosomes, it increases to a separate peak with the uncoupling of energy transfer (Fig. 2, dashed line). Increase in the 640 nm emission is diagnostic of the denaturation of xanthosomes.

In the fluorescence excitation spectrum as detected at 677 nm, the fresh, intact xanthosomes of brown algae (Fig. 1A, dashed line) show a close conformity with the absorption spectrum except for a slight deviation around 490 nm. The shoulder around 500 to 540 nm indicates an excitation energy transfer from fucoxanthin to Chl *a* with an efficiency practically the same as from Chl *a* itself. This efficiency lowers nearly to zero when xanthosomes are denatured (Fig. 1B, dashed line).

Electron Microscopic Features. When stained negatively with uranyl acetate, freshly isolated xanthosomes from *Petalonia* are observed as rod-

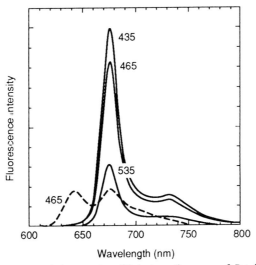

FIG. 2. Fluorescence emission spectra of intact xanthosomes of *Petalonia* (solid lines), excited by light absorbed preferentially by Chl *a* (435 nm), Chl *c* (465 nm), and fucoxanthin (535 nm). Dashed line indicates fluorescence emission of degraded xanthosomes, excited by 465 nm light.

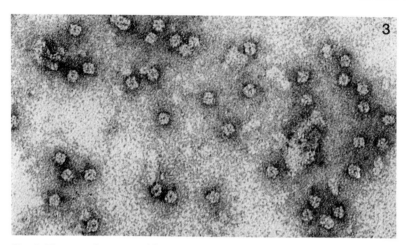

FIG. 3. Electron micrograph of freshly prepared xanthosomes from *Petalonia fascia*. Most of the particles have a dot in the center, while the remainder stain linearly, suggesting the shape of particles to be discoidal with dimensions of 11.2 nm diameter and 10.2 nm height, with a small hole piercing through. Uranyl acetate negative stain. Magnification: ×342,000.

shaped particles with a high uniformity in size and morphology (Fig. 3). Most of the particles have a dot at the center while others show a linear stain, indicating a cylindrical morphology with a hole piercing through it. Xanthosomes isolated from *Petalonia* are cylindrical in shape with a diameter of 11.2 nm and a height 10.2 nm, which gives a volume of 1005 nm^3 or a molecular mass of 744×10^3 if the specific volume of xanthosomes is assumed as 0.81. This figure is consistent with the figure (697×10^3) obtained from sedimentation velocity experiments ($s_{20,w}$ 21.6).[5]

When aged and the color changed to green, the particles no longer have uniform morphology but show distorted shapes and many aggregates (Fig. 4).

Polypeptide Composition. Xanthosomes consist of only one species of polypeptide, with a molecular size around 20 kDa. Pigment analysis of *Petalonia* xanthosomes, the polypeptide of which is 19.6 kDa, indicates that 3 fucoxanthin, 1 Chl *c*, and 4 Chl *a* molecules are confined in this polypeptide.

Properties of Photosystem I–Xanthosome Vesicles

The orange-brown band collected from the octanoyl sucrose extracts of *Ectocarpus* shows an oxidation–reduction difference around 702 nm,

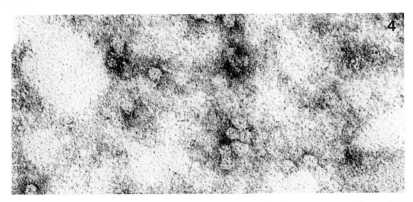

FIG. 4. Electron micrograph of aged xanthosomes, namely, the same preparation as in Fig. 3 but after 4 days of storage at 4°. The fine structure of the xanthosomes is no longer retained. Magnification: ×342,000.

characteristic of P700. Under the electron microscope (Fig. 5), it is observed as vesicles with dimensions ranging from 40 to 100 nm. Higher magnification reveals many particles with the dimensions of xanthosomes on the surface of vesicls. On SDS-PAGE, this preparation contains, in addition to the main band of 20.5 kDa, two faint bands of 67 and 71 kDa. These two polypeptides show a potent reactivity to the antisera prepared against the P700 apoprotein of spinach.

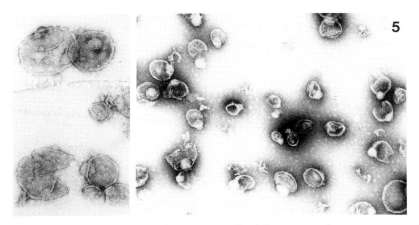

FIG. 5. Electron micrograph of orange-brown band from octanoyl sucrose extracts of *Ectocarpus siliculosus,* which consists of vesicles with diameters ranging from 40 to 100 nm. Magnification: ×75,600. With higher magnification (left side: ×163,000), particles with a size similar to that of xanthosomes (nearly 10 nm diameter) are seen on the surface of the vesicles.

Comments

The above methods can be applied to isolate xanthosomes as energetically coupled assemblies from most of the chromophytic algae, including brown algae *(Sphacelaria, Halopteris, Cutleria, Cladosiphon, Analipus, Ecklonia)*, diatoms *(Pheodactylum, Cyclotella)*, chrysophytes *(Chromulina, Ruttnera)*, raphidophytes *(Heterosigma)*, prymnesiophytes *(Prymnesium, Isochrysis)*, and dinoflagellates *(Amphidinium, Glenodinium, Prorocentrum)*. However, it should be borne in mind that chromophytic algae vary extensively in the stability of xanthosomes and in the readiness with which xanthosomes are separated from the thylakoid membranes. For instance, in xanthosomes obtained from the brown algae *Scytosiphon* or *Undaria*, a significant part of the fucoxanthin if blue shifted even if prepared carefully.

Generally the saccharose esters are effective in separating xanthosomes from thylakoid lamellae, and the samples obtained with these detergents show better signs of intactness in absorption spectra, in fine structures, and in energetic coupling than those obtained with n-decylmaltoside (Sigma, St. Louis, MO) or n-octylthioglucoside (Sigma), though the latter detergents may yield better results depending on the materials. It should be noted that other classes of detergents, including Triton X-100, Nonidet P-40, LDAO, Deriphat 160, and Zwittergent 3-14, act in such a manner as to disintegrate the energetic coupling of xanthosomes and cause the accompanying blue shift of fucoxanthin. As we did not make an extensive survey of detergents, these methods should be considered as only a starting point.

[38] Photoregulated Carotenoid Biosynthetic Genes of *Neurospora crassa*

By GIORGIO MORELLI, MARY ANNE NELSON, PAOLA BALLARIO, and GIUSEPPE MACINO

Introduction

Plants and algae, as well as many fungi and bacteria, are able to direct the *de novo* synthesis of carotenoids, and in many cases this synthesis has been demonstrated to be under light control.[1] In the filamentous fungus

[1] T. W. Goodwin, "The Biochemistry of the Carotenoids," 2nd Ed., Vol. 1. Chapman & Hall, London, 1980.

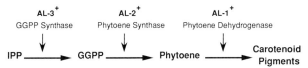

FIG. 1. Model for carotenoid biosynthesis in *Neurospora crassa*. AL-3$^+$, AL-2$^+$, and AL-1$^+$ refer to the wild-type albino genes. IPP, Isopentenyl pyrophosphate; GGPP, geranylgeranyl pyrophosphate.

Neurospora crassa, carotenoid biosynthesis is under blue light control in the vegetative mycelium, but it proceeds even in the absence of light induction in the asexual spores or conidia.[2] When grown in the light, wild-type *Neurospora* is a uniform bright orange, while in the dark it produces white mycelia and orange conidia.

Since carotenoids are not required for growth of *Neurospora*, the isolation of strains harboring mutations in regulatory and structural genes of the carotenoid biosynthetic pathway proved to be straightforward. Two regulatory mutants, known as *white collar (wc-1* and *wc-2)* because they produce carotenoids normally in the conidia but fail to produce them in the underlying mycelium, have been isolated.[3,4] The *wc* mutations exert a pleiotropic effect on all the known light-induced phenomena in *N. crassa,* which include stimulation of carotenoid biosynthesis,[2] shifting and photosuppression of the circadian rhythm,[5] production of the female reproductive structures or protoperithecia,[6] and phototropism of the perithecial beaks.[7] The sweeping control exerted by the *wc* genes leads one to conclude that the products of these two loci are involved in the initial steps of photoinduction. Indeed, one of the *wc* genes might encode the *Neurospora* photoreceptor, which has not yet been identified.

Carotenoids are very hydrophobic and thus difficult to use as substrates in the cell-free extracts that are utilized for the characterization of enzymatic activities. However, such experiments, coupled with the analysis of mutant strains, have led to the generally accepted scheme of carotenoid biosynthesis in *Neurospora* that is diagrammed in Fig. 1.[3] Three *albino* mutants are known in *Neurospora crassa (al-1, al-2,* and *al-3)*, each affected in a structural gene of the carotenogenic pathway. The *al-1* mutant

[2] R. W. Harding and W. Shropshire, *Annu. Rev. Plant Physiol.* **31,** 217 (1980).
[3] R. W. Harding and R. V. Turner, *Plant Physiol.* **68,** 745 (1981).
[4] F. Degli Innocenti and V. E. A. Russo, in "Blue Light Effects in Biological Systems" (H. Senger, ed.), p. 213. Springer-Verlag, Berlin, 1984.
[5] J. F. Feldman, *Annu. Rev. Plant Physiol.* **33,** 583 (1982).
[6] F. Degli Innocenti and V. E. A. Russo, *Photochem. Photobiol.* **37,** 49 (1983).
[7] R. W. Harding and S. Melles, *Plant Physiol.* **72,** 996 (1983).

strain accumulates phytoene and is thought to be defective in phytoene dehydrogenase.[8] The *al-2* mutant has been shown to be defective in phytoene synthase, whereas *al-3* mutants are almost completely lacking the enzymatic activity required for geranylgeranyl pyrophosphate (GGPP) biosynthesis.[3]

In our investigation of the light regulation of carotenogenesis, we chose to clone one of the structural genes involved in the pathway, namely, the GGPP synthase or *al-3$^+$* gene.[9] Another group has cloned the structural gene for phytoene dehydrogenase, *al-1$^+$*.[10] The transcription of the *al-1$^+$* and *al-3$^+$* genes is induced by light treatment, and this induction requires functional *white collar* gene products.[9,10] Analyses of the structural *albino* and regulatory *white collar* genes should yield much information about this photoinduction pathway.

Walking to the Albino-1 Gene

When one wishes to isolate a gene encoding a nonselectable phenotype (like the *albino* genes of *Neurospora*), it is often efficacious to isolate a nearby gene whose function is selectable and then use that gene as the starting point in a "walk" to the gene of interest. That is the approach which was used to isolate the *al-1$^+$* gene of *Neurospora;* a cosmid carrying a nearby gene of selectable phenotype (*hom,* homoserine-requiring) also contained the *al-1$^+$* gene.[10] When that cosmid was transformed into a *hom al-1* strain, many of the *hom$^+$* transformants produced carotenoids, demonstrating the presence on the cosmid of the intact *al-1$^+$* gene.

Walk to Albino-3 Failed

We attempted to walk to the *al-3$^+$* gene from the nearby *inl* (inositol-requiring) gene, using a cosmid library prepared by Vollmer and Yanofsky.[11] We now know that the *al-3$^+$* gene is not present in that library, and so our attempts had no chance of success. Not wanting to relinquish our search for *al-3$^+$*, we undertook a more laborious (but ultimately successful) scheme for its isolation.

---

[8] A. H. Goldie and R. E. Subden, *Biochem. Genet.* **10**, 275 (1973).

[9] M. A. Nelson, G. Morelli, A. Carattoli, N. Romano, and G. Macino, *Mol. Cell. Biol.* **9**, 1271 (1989).

[10] T. J. Schmidhauser, F. R. Lauter, V. E. A. Russo, and C. Yanofsky, *Mol. Cell. Biol.* **10**, 5064 (1990).

[11] S. J. Vollmer and C. Yanofsky, *Proc. Natl. Acad. Sci. U.S.A.* **83**, 4869 (1986).

Identification of *al-3*⁺ Transformants

Strains and Vectors

The *Neurospora* strains were obtained from the Fungal Genetics Stock Center (FGSC; University of Kansas, Kansas City, KS): *al-3* (RP100; FGSC No. 2082) and *qa-2 aro-9* (M246 Y325M6; FGSC No. 3958). Double *qa-2* (quinate utilization) *aro-9* (aromatic cluster) mutants require an aromatic amino acids supplement for growth. The *al-3 qa-2 aro-9* triple mutant was constructed using standard procedures.[12] The *al-3* mutant that was used has a leaky phenotype, characterized by the production of extremely light orange conidia. Methods for culturing *N. crassa* were standard.[12]

The plasmid library that was used is based on the pRAL1 vector, which contains the *qa-2*⁺ gene for selection in *Neurospora* and the chloramphenicol resistance gene for selection in *Escherichia coli*.[13] The plasmid clone bank consists of 10 pools of about 1800 different plasmids each, and it has proved quite useful in the sib selection of various genes.

Spheroplast Preparation

The method used for preparation of *Neurospora* spheroplasts is based on that of Schweizer *et al.*[14] The *al-3 qa-2 aro-9* strain is grown with aromatic amino acid supplements in *Neurospora* growth flasks, and the conidia are harvested with sterile water. The conidia are inoculated into 0.5X Vogel minimal medium,[15] 2% sucrose plus supplements at 10^9 conidia per 100 ml, and germinated by shaking at 30° for 4 to 6 hr. Germinated conidia are harvested by centrifuging for 20 min at 8000 rpm in a Sorvall SA rotor (in sterile centrifuge bottles), then washed 3 times with 50 ml sterile 1 M sorbitol (centrifuging for a few minutes each time at 4000 rpm in 50-ml conical polypropylene tubes). The final pellet is resuspended in 10 ml sterile 1 M sorbitol. Novozym 234 (BioLabs, Novo Allo', Denmark) is dissolved in sterile 1 M sorbitol and added to the germinated conidia at a final concentration of 0.5 mg/ml. Digestion is carried out at 30° with gentle shaking and allowed to continue until the efficiency of spheroplast formation is about 80 to 90% (as visualized with a phase-contrast microscope). The spheroplasts are diluted with cold sterile 1 M sorbitol to 50 ml and washed gently 3 times with the same (centrifuging for

---

[12] R. H. Davis and F. J. de Serres, this series, Vol. 17, p. 79 (1970).
[13] R. A. Akins and A. M. Lambowitz, *Mol. Cell. Biol.* **5**, 2272 (1985).
[14] M. Schweizer, M. E. Case, C. C. Dykstra, N. H. Giles, and S. R. Kushner, *Proc. Natl. Acad. Sci. U.S.A.* **78**, 5086 (1981).
[15] H. J. Vogel, *Am. Nat.* **98**, 435 (1964).

a few minutes each time at 2000 rpm in 50-ml conical polypropylene tubes). At this point the spheroplasts are counted and washed a final time with cold sterile 1 M sorbitol, 50 mM Tris-HCl, pH 8, 50 mM CaCl$_2$. To the pellet is added, per 10^8 spheroplasts, the following: 0.7 ml of 1 M sorbitol, 50 mM Tris-HCl, pH 8, 50 mM CaCl$_2$; 0.2 ml of 40% polyethylene glycol (PEG) 4000, 50 mM Tris-HCl, pH 8, 50 mM CaCl$_2$; and 10 μl dimethyl sulfoxide (DMSO). The spheroplasts are either used immediately or stored in small aliquots at $-70°$.

Transformation Protocol

DNA that is to be used for transformation of *Neurospora* must be free of contaminating RNA. We carry out a CsCl purification of the plasmids, followed by precipitation with PEG. Specifically, we adjust the DNA solutions to final concentrations of 10% PEG 4000 and 0.5 M NaCl and incubate them on ice for at least 1 hr, followed by centrifugation and two 70% ethanol washes. The transformation protocol that we use is based on that of Vollmer and Yanofsky.[11] Transformations with each of the 10 plasmid banks[13] are carried out separately. Five microliters (5 μg) of DNA is mixed with 2 μl of 50 mM spermidine and 5 μl heparin solution (5 mg/ml heparin, 1 M sorbitol, 50 mM Tris-HCl, pH 8, 50 mM CaCl$_2$) and incubated on ice for 20 min. Then 100 μl of *al-3 qa-2 aro-9* spheroplasts are added, mixed in gently, and stored on ice for 30 min. One milliliter of 40% PEG 4000, 50 mM Tris-HCl, pH 8, 50 mM CaCl$_2$ is added and mixed in gently, after which the mixture is incubated at room temperature for 20 min. The entire transformation mixture is transferred to 5 ml of regeneration mix (0.5 M MgSO$_4$, 0.5X Vogel's, 2% sucrose, 1% glucose) and shaken gently at room temperature overnight. Aliquots of the regenerated spheroplasts are spread onto selective plates (lacking the aromatic amino acids supplement that is required for the growth of *qa-2 aro-9* double mutants) and incubated at 30° for 4 to 5 days. The transformant *(qa-2+)* colonies are allowed to conidiate and then are screened for the production of carotenoids.

Results

We identified a population of plasmids containing the GGPP synthase *(al-3+)* gene by the ability of this population to complement an *al-3* mutation and cause the synthesis of orange rather than white mycelia and conidia.[9] About 10,000 stable *qa-2+* transformants were screened for each of the 10 plasmid pools. Primary transformants producing orange conidia (clearly more pigmented than the conidia produced by the *al-3* mutant strain) were observed when plasmid pools 7 and 9 were used in the transformations. However, stable *al-3⁺* transformants were quite rare, occurring with frequencies of 1 in every 5000 to 10,000 *qa-2⁺* transformants obtained with the 7 and 9 plasmid pools.

Cloning of Geranylgeranyl Pyrophosphate Synthase (*al-3⁺*) Gene

When complementation of a particular mutation has been detected after transformation in *Neurospora,* two general methods have been used to isolate the vector carrying the gene of interest. Sib selection (the repeated subdivision of a genomic library) can be used to enrich progressively for and ultimately isolate the gene of interest.[11,13] Alternatively, the complementing sequences can be recovered from the *Neurospora* transformant strain. Transforming DNA is always integrated into the host genome, preventing the recovery of the plasmid by direct retransformation of *E. coli*. However, in many cases the integrated plasmid remains associated with the bacterial sequences necessary for selection and propagation in *E. coli,* so that it is possible to rescue the transforming plasmid from the recipient *N. crassa* strain. This is accomplished by digestion of recipient chromosomal DNA with an appropriate restriction enzyme and ligation to obtain a circular vector for transformation into *E. coli*. In some cases, the integrated sequences are heavily methylated, and plasmids can be recovered only on transformation into *E. coli* strains with defective methylcytosine restriction systems.[16]

Attempts were made to use the sib selection procedure for the isolation of the *al-3⁺* containing plasmid from plasmid pools 7 and 9, but those attempts failed (see discussion below). Only with the selection of plasmids in *E. coli* were we able to isolate the *al-3⁺* gene. The gene was isolated after selection in a strain with functional methylcytosine restriction systems, suggesting that in this case the transforming DNA had not been extensively methylated.

Preparation of al-3⁺ Transformant DNA

Homokaryotic derivatives of the stable *qa-2⁺ al-3⁺* transformants were isolated by successive single colony isolations under restrictive conditions, and a transformant producing apparently normal levels of carotenoids (named AL3-1) was chosen for further study. We attempted to isolate some or all of the original transforming vector from this transformant as follows. The mycelia produced after overnight growth of AL3-1 in 100 ml liquid culture are filtered and frozen in liquid nitrogen; about 1 g of dry frozen

[16] M. J. Orbach, W. P. Schneider, and C. Yanofsky, *Mol. Cell. Biol.* **8**, 2211 (1988).

mycelia is recovered. The frozen mycelia are ground to a powder in liquid nitrogen, resuspended in 15 ml of 50 mM Tris-HCl, pH 8, 20 mM EDTA, 1% sodium dodecyl sulfate (SDS), and incubated at 65° for 30 min. After the addition of 4.5 ml of 5 M potassium acetate, the mixture is put on ice for 1 hr, followed by centrifugation for 10 min at 12,000 rpm. The DNA in the supernatant is precipitated with 2 volumes of ethanol. After 1 hr at room temperature, the DNA is recovered by centrifugation at 12,000 rpm for 20 min. The pellet is washed with 70% ethanol and resuspended in 2 ml of TE (10 mM Tris-HCl, pH 8, 1 mM EDTA). Three hundred micrograms of DNase-free RNase is added, and the solution is incubated at 37° for 30 min. The solution is brought to 0.3 M sodium acetate, and the DNA is precipitated with 1 volume of 2-propanol.

Isolation of Chloramphenicol-Resistant Plasmids from AL3-1 Transformant

Southern analysis of AL3-1 DNA suggested that the transforming vector sequence was probably integrated at a single site (and in only one copy).[9] Also, the results of restriction analysis led us to believe that the vector sequence had not been rearranged. Since pRAL1 contains unique HindIII and SalI sites that lie outside the essential regions of the vector,[13] we used these two restriction enzymes in an attempt to rescue plasmids in E. coli containing both vector and flanking sequences.

Ten micrograms of AL3-1 DNA is digested with HindIII or SalI, diluted in ligation buffer to 1 μg/ml (the dilute solution is used to avoid intermolecular annealing), and ligated at 4° overnight with T4 DNA ligase (1 U/ml). The ligated DNAs are concentrated to 200 μl with 2-butanol and then ethanol precipitated. A 1-μg sample of ligated DNA is used to transform 5×10^{10} competent E. coli HB101 cells to chloramphenicol resistance.

A 5.2-kbp plasmid (pNC6) was isolated from SalI-digested AL3-1 DNA, and a 10.8-kbp plasmid (pNC3) was isolated from the HindIII-digested sample. The pNC3 and pNC6 plasmids (both of which contain functional Neurospora qa-2$^+$ genes) were used to transform al-3 qa-2 aro-9 spheroplasts to prototrophy (qa-2$^+$); no al-3$^+$ transformants were obtained, indicating that an intact al-3$^+$ gene is not contained in either pNC3 or pNC6. The structures of these plasmids are shown in Fig. 2.

Isolation of Plasmid Containing Intact al-3$^+$ Gene

Restriction fragments from pNC3 and pNC6 were used as probes in the isolation of the original transforming plasmid (pNC39) from pool 9.[9] pNC39 transformed al-3 qa-2 aro-9 spheroplasts to qa-2$^+$al-3$^+$ at high frequency; about 20% of the primary heterokaryotic transformants pro-

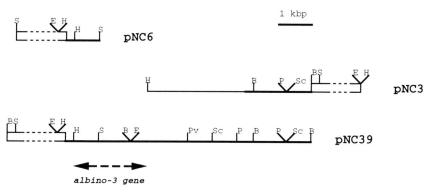

FIG. 2. Structures of the pNC3, pNC6, and pNC39 plasmids. *Neurospora* insert sequences are indicated by lines; the thick lines denote sequences that are present in the original pNC39 insert, whereas light lines denote chromosomal sequences flanking the inserted vector. pNC6 contains only sequences that are also present in the transforming plasmid pNC39, and pNC3 contains also chromosomal flanking sequences. The *N. crassa* sequences are drawn approximately to scale (see bar). Vector (pRAL1) sequences are indicated by open boxes and have been compressed. The position of the *al-3+* gene is shown. Restriction sites: S, *Sal*I; E, *Eco*RI; H, *Hin*dIII; B, *Bam*HI; Pv, *Pvu*II; Sc, *Sac*I; P, *Pst*I.

duced dark orange conidia. The color of the remaining transformants ranged from light orange to white; analysis of homokaryotic derivatives of the transformants showed that about 80% contained functional *al-3+* genes. The pNC39 plasmid was present at low frequencies (10^{-4}) in plasmid pool 9; its low representation might partially explain why our attempts at sib selection of the *al-3+* gene (above) failed. Also, poor expression of the *al-3+* gene in the primary heterokaryotic transformants reduced the likelihood of a successful sib selection.

Comparative restriction analysis of the pNC3, pNC6, and pNC39 plasmids (Fig. 2) has confirmed that pNC39 was the vector used to generate the AL3-1 transformant, and that the vector sequences were not rearranged during the transforming event. Conventional subcloning techniques, followed by transformation of *al-3 qa-2 aro-9* spheroplasts to *qa-2+* and screening for orange (*al-3+*) colonies, were used to further localize the *al-3+* gene within pNC39[9,17]; its position is indicated in Fig. 2.

Mapping to Confirm Identity of *al-3+* Gene

To rule out the possibility that we had cloned an unlinked suppressor of the *al-3* mutation rather than the *al-3+* gene itself, we determined the map position of the pNC39 insert using RFLP (restriction fragment length

---

[17] A. Carattoli, N. Romano, P. Ballario, G. Morelli, and G. Macino, *J. Biol. Chem.* **266**, 5854 (1991).

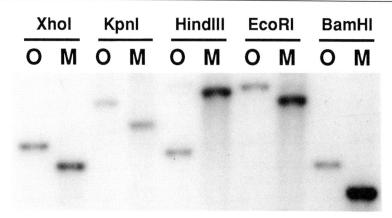

FIG. 3. Restriction fragment length polymorphisms (RFLPs) used to map the al-3^+ gene. A 4-kbp EcoRI–BamHI fragment of pNC39 was used to probe digests of genomic $Neurospora$ DNA from two strains, Mauriceville-1c A (M) and multicent-2-a (Oak Ridge, O). The restriction enzymes used to digest DNA from the two mapping strains are indicated above the lanes.

polymorphism) mapping. The technique used was that devised by Metzenberg et al.[18] This technique takes advantage of the large number of restriction site polymorphisms in two strains of $N.$ $crassa$. Segregation of both conventional genetic markers and RFLPs is analyzed, and cosegregation of RFLPs within the cloned DNA and these markers is evidence of genetic linkage. A 4-kbp EcoRI–BamHI fragment of pNC39 was used in this analysis; sites for five restriction enzymes were tested, and all were polymorphic in the two $Neurospora$ mapping strains, Mauriceville and Oak Ridge (Fig. 3). The segregation of these sites among the progeny of a Mauriceville × Oak Ridge cross was analyzed, and the pNC39 fragment was mapped to linkage group V, between the cyh-2 and inl genes, as expected based on conventional mapping of al-3 mutants.[9]

Photoinduction Experiments

Culture Conditions and Light Treatment

$Neurospora$ $crassa$ strains are cultured on vegetative growth medium (Vogel minimal medium, 2% sucrose, 1.5% agar)[15] for 8 days at 30°. The conidia are harvested with sterile distilled water and filtered through cheesecloth. Conidial stocks (4 × 10^8/ml) are stored at −20° in distilled

[18] R. L. Metzenberg, J. N. Stevens, E. U. Selker, and E. Morzycka-Wroblewska, $Proc.$ $Natl.$ $Acad.$ $Sci.$ $U.S.A.$ **82**, 2067 (1985).

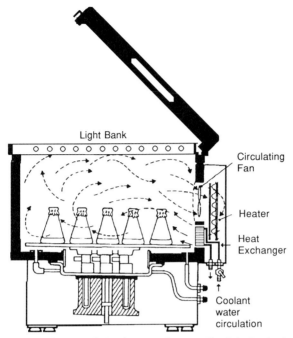

FIG. 4. System used for photoinduction experiments. The light bank also serves as the chamber lid, as there is a plexiglass sheet between the lights and the chamber.

water. For the photoinduction experiments, conidia are inoculated at 4×10^5/ml in 100 ml Vogel minimal medium plus 2% sucrose (in 250-ml Erlenmeyer flasks). The submerged mycelia are grown in the dark with agitation (200 rpm) in a New Brunswick Scientific G25 incubator shaker at 30° for 20 hr at which time the cells were at the end of log phase; the mycelia consist of a fairly homogeneous network of short hyphae at this stage.

The cultures are illuminated directly in the growth flasks with a light bank located at the top of the shaker (Fig. 4). A heat exchanger is used to maintain constant temperature during the period of illumination. The light bank consists of 12 Sylvania GRO-LUX F 18W-GRO lamps; these lamps generate an energy fluence rate of 8 W/m² in the blue region. The light-induced cultures and the dark-grown controls are collected by filtration on filter paper disks and immediately frozen in liquid nitrogen.

RNA Extraction

Total RNA is extracted from frozen mycelia that had been powdered in a Waring blendor with liquid nitrogen. Frozen powdered mycelia

(0.4 g) are dissolved in 8 ml GT buffer (5 M guanidinium thiocyanate, 25 mM sodium citrate, pH 7, 0.5% sarcosyl, 2 mM EDTA, 5% 2-mercaptoethanol)[19] and incubated at 50° for 15 min. Cell debris is removed by centrifugation at 8000 rpm for 15 min in a Sorvall SS34 rotor. The supernatant is placed in a polyallomer ultracentrifuge tube on a CsCl cushion (4.95 M CsCl, 0.1 M sodium acetate, pH 5, 5 mM EDTA),[20] and centrifuged (34,000 rpm for 18 hr at 20°) in a Beckman SW41T rotor. The supernatant is carefully aspirated, and the RNA pellet is gently resuspended in 400 μl of 7 M urea, 2% sarcosyl. A half-volume of phenol is added, and the solution is mixed by vortexing; the urea prevents the formation of phases. Then a half-volume of chloroform is added, and the phases are separated by microcentrifugation. The RNA is precipitated with 0.3 M sodium acetate, pH 5.2, and 3 volumes of ethanol and pelleted by microcentrifugation; finally, the RNA is resuspended in 200 μl distilled water.

Northern Hybridizations

For Northern blot analysis, total RNA is denatured and electrophoresed on 1.2% agarose – 1.9% formaldehyde gels[21] with 20 mM morpholinopropanesulfonic acid, 5 mM sodium acetate, 1 mM EDTA, pH7, running buffer. The separated RNAs are transferred to nylon membranes or nitrocellulose.[22] Samples are loaded in duplicate and hybridized with either an *al-3*-specific oligonucleotide probe (27 nucleotides long) or a control probe from the *N. crassa* β-tubulin gene.[23] The ^{32}P-labeled oligonucleotide probe is prepared by 5'-end-labeling with T4 polynucleotide kinase,[24] and the ^{32}P-labeled β-tubulin probe is made with the random oligomer-primer method.[25] When oligonucleotide probes are used, filters are hybridized with 5 × 10^6 counts/min (cpm)/ml of probe at 50° for 18 hr in 5× SSC, 5× Denhardt's solution, 50 mM sodium phosphate, pH 7, 0.5% SDS, 50 μg/ml denatured herring sperm DNA (1× SSC is 0.15 M NaCl, 15 mM sodium citrate, pH 7; 1× Denhardt's is 0.02% bovine serum albumin, 0.02% Ficoll, 0.02% polyvinylpyrrolidone). Unhybridized oligonucleotide probe is removed by washing in 1× SSC, 0.1% SDS at 50° for 30 min. In

---

[19] J. M. Chirgwin, A. E. Przybyla, R. J. MacDonald, and W. J. Rutter, *Biochemistry* **18**, 5294 (1979).
[20] V. Glisin, R. Crkvenjakov, and C. Byus, *Biochemistry* **13**, 2633 (1974).
[21] H. Lehrach, D. Diamond, J. M. Wozney, and H. Boedtker, *Biochemistry* **16**, 4743 (1977).
[22] P. S. Thomas, this series, Vol. 100, p. 255 (1983).
[23] M. J. Orbach, E. B. Porro, and C. Yanofsky, *Mol. Cell. Biol.* **6**, 2452 (1986).
[24] T. Maniatis, E. F. Fritsch, and J. Sambrook, "Molecular Cloning: A Laboratory Manual." Cold Spring Harbor Laboratory, Cold Spring Harbor, New York, 1982.
[25] A. P. Feinberg and B. Vogelstein, *Anal. Biochem.* **132**, 6 (1983).

hybridizations with random primer-labeled probes, filers are hybridized with 2×10^6 cpm/ml of probe at 42° for 18 hr in 6× SSPE, 1× Denhardt's solution, 50% formamide, 5% dextran sulfate, 0.1% SDS, 200 μg/ml denatured herring sperm DNA (1× SSPE is 0.18 M NaCl, 10 mM sodium phosphate, pH 7.7, 1 mM EDTA); the filters are washed in 0.1× SSC, 0.1% SDS at 48° for 30 min. The filters are autoradiographed with preflashed Hyperfilm-MP film (Amersham) at −70°; signals are increased with intensifying screens (Cronex Quanta III, Du Pont Corp.).

Results of Transcript Analyses

The expression of the *al-3* gene in light-induced and dark-grown mycelial cultures has been examined.[9,17] Since the *white collar* genes are thought to encode regulatory products necessary for the expression of the *albino* genes,[3,4] *al-3* expression was analyzed in *wc-1* and *wc-2* mutants as well as in a wild-type strain. The results of those analyses demonstrated that photoinduction causes a dramatic increase in the transcription of the *al-3* gene, and that functional *white collar* products are required for this photoinduced response.[9,17] Similar regulation of the expression of the *al-1* gene has also been shown.[10]

Conserved Domains in Neurospora crassa Geranylgeranyl Pyrophosphate Synthase and Other Prenyltransferases

GGPP synthase is a member of the prenyltransferase family. Prenyltransferases carry out the major synthetic steps in isoprenoid metabolism and produce various prenyl compounds that are the precursors of a variety of products including carotenoids, steroids, chlorophylls, heme *a*, prenylated tRNAs, glycosyl carrier lipids, plant hormones, and side chains of quinones.[26] We have analyzed *N. crassa* GGPP synthase and other prenyltransferases for conserved regions,[17,21] and a similar comparison (of yeast hexaprenyltransferase and farnesyl diphosphate synthases) has been made by Ashby and Edwards[28]; three conserved domains have been identified in these analyses. Domain I (LXXDDXXDXSXXRRGXP) contains many conserved highly charged amino acids (where X is any amino acid and the other amino acids are indicated by single-letter code). The second domain is more variable in the different prenyltransferases.[17] Domain III (GXXFQIXDDYLD) also contains many highly charged residues. Do-

---

[26] C. D. Poulter and H. C. Rilling, *in* "Biosynthesis of Isoprenoid Compounds" (J. W. Porter and S. L. Spurgeon, eds.), p. 162. Wiley, New York, 1981.
[27] V. Ilardi, F. Luti, G. Macino, and G. Morelli, manuscript in preparation (1992).
[28] M. N. Ashby and P. A. Edwards, *J. Biol. Chem.* **265**, 13157 (1990).

mains I and III both contain "DDXXD" motifs; these aspartate residues could serve in the binding of Mg^{2+} or Mn^{2+} cations, which is required for the activities of some prenyltransferases.[29] The positively charged amino acids in domains I-III may be involved in the enzyme activity[17]; site-directed mutagenesis could be used to test the functions of these conserved residues. The genes encoding prenyltransferases in other organisms could be isolated using oligonucleotides corresponding to the amino acid sequences of the conserved domains.

Acknowledgments

We are grateful to Robert L. Metzenberg for assistance with RFLP mapping. We wish to thank Maurizio DiFelice for skilled technical assistance, and we are indebted to A. Carattoli, N. Romano, and S. Baima for helpful discussions. This work was supported by grants from the Piano Nazionale Tecnologie Avanzate Applicate alle Piante, Ministero Agricoltura e Foreste.

---

[29] O. Dogbo, A. Laferriere, A. d'Harlingue, and B. Camara, *Proc. Natl. Acad. Sci. U.S.A.* **85,** 7054 (1988).

Author Index

Numbers in parentheses are footnote reference numbers and indicate that an author's work is referred to although the name is not cited in the text.

A

Abrams, P., 236, 239(46)
Acuff, K. J., 44, 45(27)
Adams, L. L., 33
Adams, W. W., 397
Adriaens, P., 127
Agalidis, I., 209
Ahearn, D. G., 386
Ahmad, K., 386
Akins, R. A., 415, 416(13), 417(13), 418(13)
Alam, B. S., 227
Alam, S. Q., 227
Albers, J. J., 43–44
Alberte, R. S., 402–403
Alberti, M., 297, 298(4)–298(5), 299(3)–299(5), 301(4), 307(3)–307(5), 309(4), 310(3)–310(4), 324
Alberts, D. S., 97
Al-Hasni, S. M., 69
Allemand, B. H., 82
Allen, M. M., 392
Allen, R. D., 47
Alwine, J. C., 321
Ameduc-Manoome, O., 86
American Institute of Nutrition, 126
American Psychiatric Association, 117
An, G.-H., 283
Anderson, B., 12, 14(34)
Anderson, I. C., 312, 313(15)
Anderson, J. M., 402, 403(1), 407
Anderson, S. M., 56
Ando, S., 75
Andreevskaya, V. D., 386
Andreoni, L., 124
Aponte, G. W., 22, 23(15)
Aragón, C.M.G., 284, 365, 372(3)
Arai, N., 396
Armstrong, G. A., 297–298, 299(3)–299(5), 301(4), 307(3)–307(5), 307(10), 309(4), 310(3)–310(4), 324
Arnaboldi, A., 124
Arnold, W., 202–203
Arnrich, L., 169
Arroyave, G., 136
Ashby, M. N., 303, 305(23), 306(23), 423
Ashtakala, S. S., 348, 349(1), 350(1), 351, 352(1)
Association of Official Analytical Chemists, 126
Aswad, A., 137
Attlesey, M., 21, 32(8), 74, 169, 200, 258, 266(15), 267(15)
Ausubel, F. M., 378, 380(12)
Avalos, J., 284, 285(16), 287(16), 289(16), 290, 293, 293(16), 293(21)
Aviv, H., 364
Avron, M., 390

B

Bachmann, M. D., 312, 313(15)
Bachorik, P. S., 43, 44(24)
Bader, F., 141
Ball, E. G., 82, 85(31)
Ballario, P., 419, 423(17), 424(17)
Balsalobre, J. M., 284, 293
Balun, J. E., 257
Banji, S., 122
Bardat, F., 352
Barkan, A., 311, 315(11), 321
Baron, J. A., 94, 136
Barrett, J., 402, 403(1)
Bartley, G. E., 298, 307, 310, 374–375, 381(7), 383(7)
Barua, A. B., 31, 89, 104, 106, 143

Bates, C. J., 169, 173
Bates, M., 33
Batra, P. P., 183, 269, 273, 284, 286, 289(10), 291(7), 292(7)
Batres, R. O., 104, 106(18), 143
Battey, J. F., 320
Bauer, C. E., 375
Bauernfeind, J. C., 67, 86
Beale, S. I., 351–352
Beaudry, J., 97
Beecher, G. R., 3, 4(1), 5–7, 11, 12(1), 34, 46(7a), 56, 102, 104–105, 106(16), 107, 119, 122(10), 130
Behrens, W. A., 42
Beijersbergen van Henegouwen, G.M.J., 124–125
Bejarano, E. R., 284, 285(16), 287(16), 289(16), 290(16), 291, 293
Bell, J. L., 387
Ben-Amotz, A., 390
Bendich, A., 74, 102, 137, 228, 230, 234–235
Benjamin, I. J., 235
Bensasson, R. V., 124, 192
Benton, W. D., 318
Bentzen, C. L., 44, 45(27)
Benya, R., 3, 14(8)
Benz, J., 349
Bergman, K., 284, 288(15)
Berlin, E., 33, 34(7), 35
Berman, M., 146
Bermond, P., 124
Bernhard, K., 167
Berry, R. A., 284
Bertram, I., 311
Bertram, J. S., 21, 23, 27, 31(4), 55(3), 56(4), 57(8), 58, 59(9), 61(9), 62(4), 62(9), 63, 64(9), 65(9), 66(8), 66(13), 67–68, 86, 230, 236
Bertsch, W., 203
Betz, E. P., 94
Beyer, P., 297–298, 299(5), 307, 310, 340(10), 352, 367, 368(12), 370, 374(4)
Bhagavan, H. N., 32
Bhawan, J., 136
Bhuyan, J. U., 386
Bianchi, A., 124
Bianchi, L., 124
Bickers, D. R., 127

Bieber, L. L., 394
Bierer, T. L., 4, 102, 107, 109(6), 109(24), 114(24)
Bieri, J. G., 3, 4(1), 5–7, 12(1), 33, 34(7), 35, 46, 118–119, 122(10), 130, 133(31)
Bindl, E., 274
Bird, C., 301
Bischof, S., 149, 157(8), 163(8)
Bishop, N. I., 178, 341
Björn, L. O., 284
Bjornson, L. K., 7, 33, 34(6), 36, 37(6)
Blake, M. S., 361
Blakeley, C. F., 286
Blakely, C. F., 273
Blanc, P. L., 324, 327(9)
Blaner, W. S., 63
Block, G., 9
Bobior, R. M., 235
Bobowski, C., 275
Boedtker, H., 422
Boerman, M.H.E.M., 193
Böger, P., 342
Bohnert, H. J., 303
Bo-Linn, G. W., 3, 14
Boone, C. W., 3, 7(3), 94
Bordin, F., 124
Borland, C. F., 193
Borowitzka, L. J., 209
Borowitzka, M. A., 209
Bossuyt, E., 209
Boston, R. C., 147
Bowen, P. E., 3, 7(2), 7(4), 8, 12–14, 131, 136(34)
Boxer, S. G., 226
Boyer, V., 125
Braithwaite, G. D., 331
Bramley, P. M., 284, 292(17), 293, 297, 340–341, 343, 346(11), 365–368, 369(21), 370, 372(11)
Brandt, R. B., 257
Brangeon, J., 352
Brembilla, M., 311
Brent, R., 378, 380(12)
Britton, G., 178, 284, 297, 333, 334(8), 335(8), 336(8), 337(8), 338(8), 339, 340(8), 368
Brody, S. S., 203
Broich, C. R., 102, 109(10)
Brookmeyer, R., 94

Brown, E. D., 3, 4(1), 5(1), 6-7, 9-10, 12, 38, 46, 118-119, 122, 130, 133(31)
Brown, F., 357
Brown, J. C., 236
Browne, M. L., 117, 120(8), 122(8)
Brubacher, G. B., 22, 69, 94
Bruggeman, E., 209
Bryant, J. D., 84
Buchecker, R., 342
Buckner, B., 311, 312(14), 314(14)
Buddrus, D. J., 3, 14
Buess, E., 94
Bullerjahn, G. S., 75, 83(17), 391, 396(4), 397(9), 399(9), 401(9)
Burchard, R. P., 183, 284, 324
Burgess, M., 7, 131, 136(34)
Burke, D., 297, 298(5), 299(5), 307(5)
Burkhart, W., 391, 397(9), 399(9), 401(9)
Burns, R. D., 193
Burr, B., 311
Burr, F. A., 311
Burr, J., 99
Burstein, M., 43-44
Burton, G. W., 74, 86, 137
Busch, U., 127
Bustos, P., 125, 136(18)
Buten, B., 82, 85(27)
Byus, C., 422

C

Cama, H. R., 75, 84(23), 168, 169(7), 261, 268(18)
Camara, B., 301, 340, 352-353, 358(5), 359(6), 364(12), 368, 424
Canfield, L. M., 9, 99
Cantilena, L. R., 97, 98(14), 101(14)
Capozzi, A., 124
Carattoli, A., 264, 414, 416(9), 418(9), 419, 420(9), 423(9), 423(17), 424(17)
Carcamo, G., 125
Carlson, K., 33
Carlson, T. J., 400-401
Carson, C. B., 311
Carter, M. C., 348
Case, M. E., 415
Casper, R. C., 116, 122(4), 123
Castelfranco, P. A., 351-352

Catignani, G. L., 208
Causse, M. B., 257
Ceccaldi, H. J., 82, 85(29)
Ceccanti, M., 3, 7
Cerdá-Olmedo, E., 178, 284, 285(16), 287(16), 289(16), 290(16), 291-293, 365, 372(3)
Chambers, J.A.A., 264
Chambon, P., 67
Chamouitz, D., 375
Chansang, H., 69, 168, 256, 267(6)
Cheesman, D. F., 75, 81(21), 82, 85(24), 85(29), 224, 225(27)
Chen, J., 311
Chenoweth, W., 3, 7(3)
Chew, B. P., 101
Chichester, C. O., 102, 103(11), 104(11), 109(11), 136, 214, 386
Chirgwin, J. M., 422
Choi, Y.-L., 303
Chou, S.-C., 47, 55(15)
Churley, M., 23, 58, 59(9), 61(9), 62(9), 63(9), 64(9), 65(9), 67(9), 230
Chytil, F., 257
Claes, H., 176, 178(7), 182(7), 183
Clark, C., 99
Clark, J. S., 148
Clark, L. C., 100
Clark, R. E., 147
Clevidence, B. A., 3, 33, 34(7), 35, 45
Clough, J. M., 343, 346(10)
Cogdell, R. J., 191, 193, 209
Cohen, I. R., 235
Cohen, S. S., 357
Cohen-Bazire, G., 83, 323
Cole, R. S., 124
Combs, G. F., 100
Conconi, M. T., 124
Cone, K. C., 311
Connor, M. J., 137
Cooney, R. V., 23, 58, 59(9), 61(9), 62(9), 63(9), 64(9), 65(9), 67(9), 68
Cornstock, G. W., 94
Cornwell, D. G., 33, 34(1), 37, 38(1), 43(1), 70, 75
Costa, I., 388, 390(8)
Coulson, A. R., 395
Craft, N. E., 3, 4(1), 5-6, 7(1), 9-10, 12(1), 38, 46(11), 111, 119, 122(10)

Crain, F. D., 74
Crean, C., 126
Criel, G. R., 209, 210(11), 211(11)
Crkvenjakov, R., 422
Curran-Celentano, J., 116, 123
Czygan, F.-C., 397

D

Daenens, P., 127
Dagadu, J. M., 86
Dain, B. J., 136
Dally, P. J., 116, 122
D'Antonio, J. A., 33
Das, R., 69
Da Silva, I. M., 388, 390(8)
Davenport, R., 327
Davies, B. H., 103, 214, 366–368, 372(11)
Davis, G. R., 3, 14
Davis, J. M., 123
Davis, L. G., 320
Davis, R. H., 415
Davis, R. W., 318, 394
Davis, T. P., 97
de Chaffoy de Courcelles, D., 213
De Fabo, E. C., 183, 284, 285(19)
Degli Innocenti, F., 178, 293, 413, 423(4)
De la Concha, A., 366
De la Guardia, M. D., 365, 372
De la Rivas, J., 367, 368(10)
Delbrück, M., 175, 285, 289(31), 292
De Leenheer, A. P., 209–210, 211(11), 214(11), 217, 224(13)
Dellaporta, S. L., 311
DeLorenzo, A. A., 3, 7(4)
Demina, S. G., 386
Demmig-Adams, B., 397
de Mol, N. J., 124–125
Denhart, D. T., 316
Dennis, E., 311
De Ritter, E., 131
de Serres, F. J., 415
Deshmukh, D. S., 169
Desquilbet, N., 257
Deuel, H. J., Jr., 74
Deutscher, M. P., 371
Devereux, J., 299
de Wolff, F. A., 127
d'Harlingue, A., 364, 424

Diamond, D., 422
Dibner, M. D., 320
Dieber-Rotheneder, M., 36
Di Fonzo, N., 311
Dimitrov, N. V., 3, 7(3), 94
Dimitrovskii, A. A., 69
Diverse-Pierluissi, M., 391, 395(7)
Dodds, G. T., 348, 349(1), 350(1), 351(1), 352(1)
Dogbo, O., 352–353, 358(5), 359(6), 424
Doolittle, R. F., 299, 301(16), 303(16), 397
Douglas, K. T., 309
Downum, K. R., 328
Dratz, E. A., 73, 259
Driskell, W. J., 208
Dubendorff, J. W., 376, 380(10)
Dujacquier, R., 3
Dunkel, V. P., 236
Dunn, B. P., 94, 229
Dunn, J. J., 376, 380(10)
Dworkin, M., 324
Dykstra, C. C., 415

E

Eder, H. A., 44
Edwards, B. K., 3, 4(1), 5–7, 12(1)
Edwards, E. K., 119, 122(10)
Edwards, P. A., 303, 305(23), 306(23), 423
Eggink, G., 309, 310(28)
Ehara, T., 402, 410(5)
Eisenstark, A., 327
El-Gorab, M. I., 32
Elías, M., 293
Eliopoulos, E. E., 75, 81(22), 82(22), 85(22), 401
El-Jack, M., 284, 292(17), 293, 367, 368(14)
Endo, H., 228, 229(19), 236(19)
Engel, H., 309, 310(28)
Engle, J. M., 391, 397(9), 399(9), 401(9)
Englert, G., 167, 298, 374
Enright, W. J., 193
Epler, C. S., 111
Erdman, J. W., Jr., 4, 9, 84, 87, 102, 107(6), 107(9), 107(24), 109, 116(6), 123(6)
Ernst, S., 341, 343(6)
Eslava, A. P., 284, 288(15)
Esselbaugh, N. C., 74
Esterbauer, H., 36

Etchells, W.G.W., 387
Eugster, C. H., 342
Evans, M., 12, 14(34)
Evrard, J. L., 364

F

Fahey, G. C., Jr., 4, 9, 102, 107(6), 109(6)
Farr, A. L., 258
Farrow, P. W., 106
Faure, 78
Fazakerley, S., 69
Federoff, N. V., 311
Feinberg, A. P., 422
Feldman, J. F., 413
Fell, J. W., 386
Felts, J. M., 41, 42(20)
Ferro-Luzzi, A., 3, 7(4)
Fidge, N. H., 168, 169(2), 170(2)
Findlay, J.B.C., 75, 81(22), 82(22), 85(22), 401
Fink, G., 3, 7(3)
Fischli, A., 24, 30
Fishwick, M. J., 69
Fitzpatrick, T. B., 124, 137
Fletcher, R. A., 348, 350(5)–350(6), 351(5)–351(6)
Flores, H., 136
Flügge, U. I., 373
Flynn, E., 227, 231
Fong, F., 312
Fong, L. G., 34
Fontes, M., 284, 293(5)
Forbes, P. D., 129, 136(30)
Ford, C., 3, 14(8)
Fordtran, J. S., 3, 14
Foster, D. M., 147
Fox, J. G., 4, 70, 71(18), 73(18), 74(12), 75(12), 102, 168, 257
Fraikin, G. Y., 292
Frank, H. A., 186, 191
Fraser, P. D., 297, 367, 368(10), 368(13), 369(13)
Freeling, M., 311, 315(11)
Friedman, A. L., 402–403
Friedman, H., 3, 7(4), 12, 14(34)
Frisoli, S. K., 228
Fritsch, E. F., 317, 422
Fuchs, J., 125

Fujisaki, S., 303
Fukuda, M., 127
Furr, H. C., 86, 89, 104, 106(18), 143
Furrow, G., 56, 104, 106(16), 130
Furtek, D. B., 311
Furuya, M., 280

G

Gale, M., 94
Galland, P., 287
Ganguly, J., 69, 74, 169
Garmyn, M., 136
Gawienowski, A. M., 168
Gelson, W. A., 227
Gerber, L. E., 70, 102, 109(10)
Gest, H., 377
Gettner, S., 3, 7(4)
Gidez, L. I., 44
Gilchrest, B. A., 136
Gilchrist, B. M., 224
Giles, A., 124
Giles, N. H., 415
Gillbro, T., 191
Giuliano, A. R., 9
Giuliano, G., 298
Gleason, R. M., 284, 291(7), 292(7)
Glickman, B. W., 125
Glinz, E., 149, 157(8), 159, 183(8)
Glisin, V., 422
Gloor, U., 69
Glover, J., 69, 136
Goldie, A. H., 414
Goli, M. B., 11, 34, 46(7a), 107
Gollnick, P., 303, 305(20)
Goodman, D. S., 21–22, 70, 74, 168–170, 172(8), 256, 264
Goodwin, P., 68
Goodwin, T. W., 55, 56(1), 74, 148, 226, 298, 324, 331–332, 333(7), 333(9), 335(9), 337(9), 338(9), 339, 412(1)
Gordon, S. A., 324
Gotslich, E. C., 361
Gottfried, D. S., 226
Govind, N. S., 291
Grabowski, S. M., 47
Grater, S. J., 116, 123(6)
Green, J. B., 147
Green, M. H., 147

Greenberg, E. R., 94, 136
Greenblatt, I., 311
Greene, J. R., 320
Gregory, R. A., 74
Grierson, D., 301
Griffiths, M., 83, 323
Grob, E. C., 331
Gronowska-Senger, A., 169, 172(12)
Grossweiner, L. I., 124
Grumbach, K. H., 175
Grundman, H. P., 94
Guarente, L., 320
Gubler, M. L., 63
Gugger, E. T., 4, 87, 107, 109(24), 114(24)
Gulbrandsen, C. L., 33, 34(5), 36
Gunter, E. W., 208
Gurin, S., 331
Gustavson, D. L., 402, 403(2)
Gutmann, H., 139
Guzhova, N. V., 292

H

Haas, S. M., 394
Habig, W. H., 227
Haeberli, P., 299
Hahman, M., 386
Hake, S., 311, 320
Hallberg, D., 33
Hallden, G., 22, 23(15)
Hamano, M., 127
Hames, B. D., 361
Hamilton, R. H., 292
Hancock, R.W.E., 328
Handelman, G. J., 11, 73, 103, 106(14), 259
Hansen, S., 21, 169, 173
Hara, H., 303
Harashima, K., 297, 298(6), 299(6), 301(6), 307(6), 344
Harboe, N., 359
Harding, R. W., 178, 183, 264, 269, 274, 284–285, 287, 292, 413, 414(3), 423(3)
Harlow, E., 376, 381(9)
Harris, J. C., 106
Harrison-Lavoie, K. J., 324, 327(7), 341
Hart, J. W., 176
Hartmann, K. M., 180, 287, 290(37)
Hasegawa, K., 278
Hata, M., 214

Hatano, M., 75
Hatch, F. T., 39, 41(16)
Hay, M. B., 94
Hayaishi, O., 21, 74, 200, 256, 265
Hayden, J., 47
Hearst, J. E., 297–298, 299(3)–299(5), 301(4), 307(3)–307(5), 307(10), 309(4), 310(3)–310(4), 324, 341
Heilbrun, L. K., 94
Helene, C., 125
Helenius, A., 368, 370(22)
Helland, D. E., 125
Helsing, K. J., 94
Henderson, C. T., 3, 7, 7(2)
Henderson, W. R., 67
Hendricks, S. B., 183, 284
Hengartner, U., 149
Hennekens, C. H., 21, 94
Henner, D. J., 303, 305(20)
Henze, T. M., 4, 107, 109(24), 114(24)
Herfst, M. J., 127
Hertzberg, S., 166
Herve, F., 257
Hervo, G., 209
Hess, R., 22, 23(15)
Hicks, J. B., 311
Hill, R., 203
Himeno, K., 228, 229(19), 236(19)
Hinkelammert, K., 264
Hirohata, T., 228, 229(19), 236(19)
Hirschberg, J., 375
Hoehner, S. C., 117, 120(8), 122(8)
Hoffman, N. E., 348
Höfle, G., 156
Hol, W.G.J., 309, 310(27)
Holbrook, J., 9
Holden, J. M., 8
Hollander, D., 22
Holman, J. P., 388
Holmes, M. G., 287
Holmgren, S. C., 4, 74(12), 75(12), 102
Holt, T. K., 75, 83(16), 391
Honigsman, H., 124
Hopkinson, J. M., 99
Horiuchi, K., 303
Hornby, A. P., 94
Hossain, M. Z., 86
Howes, C. D., 183, 273, 284, 286, 289(10)
Hsiao, K. C., 284
Hsu, W.-J., 214–215

Huang, D. S., 231, 241(29)
Huang, H. S., 21-22, 74, 170, 256, 264(1)
Huang, L., 348
Huang, P. C., 269, 287
Huff, D. L., 208
Huflejt, M. E., 125
Hug, B. A., 324
Humbeck, K., 176, 179(8), 182(8), 183
Hummel, D., 126
Hundle, B. S., 297-298, 307
Hyden, S., 16

I

Idei, T., 275
Ilardi, V., 423
Imanishi, I., 56
Ingham, K. C., 371
Ingild, A., 359
Ingold, K. U., 74
Inoue, Y., 182-183, 280
Isler, O., 103, 139, 141, 267, 330
Ito, Y., 148
Itri, L. M., 55
Iwahashi, I., 280
Iwai, K., 278
Iwasaki, R., 56
Iwashima, A., 56
Izawa, Y., 297, 298(6), 299(6), 301(6), 307(6), 344

J

Jackoby, W. B., 371
Jacob, R. A., 123
Jacquez, J. A., 147
Jagger, J., 280
Jayaram, M., 175, 285, 289(31), 292
Jencks, W. P., 82, 85(27)
Jenkins, J., 284, 291(7), 292(7)
Jensen, A., 394, 401
Jensen, H. B., 136
Jensen, L. C., 39
Jhaveri, A., 257
Jimura, H., 148
Jochum, P., 285
Johansson, B. C., 377
Johnson, E. A., 283

Johnson, F. L., 41, 43(19)
Johnson, J. H., 264
Jokoby, W. B., 227
Jonas, R. B., 326
Jonckheere, J. A., 209, 210(11), 211(11)
Jones, B. L., 368, 370(17)
Joshi, P. C., 124
Jurgens, U. J., 394

K

Kafatos, F., 307
Kagan, J., 325, 326(16), 327(16), 328, 329(16)
Kahlweit, M., 61
Kalkwarf, H. J., 125, 136(18)
Kallidumbil, V., 348, 350(5), 351(5)
Kamath, S. K., 169
Kanai, M., 170
Kanof, M. E., 234, 238(32)
Kaplan, L. A., 7, 33-34, 86, 112, 123, 132
Kappock, T. J. IV, 23, 58, 59(9), 61(9), 62(9), 63(9), 64(9), 65(9), 67(9)
Karrer, P., 69
Kastner, P., 67
Katagiri, K., 148-149
Katayama, T., 214
Kates, M., 366
Kato, F., 284
Katoh, T., 402, 403(4), 403(7)-403(8), 410(5)
Katsuyama, M., 148
Kawai, H., 402, 403(7)
Kawamukai, M., 303
Kayden, H. J., 7, 33, 34(6), 36(6), 37(6)
Keddie, A., 22, 23(15)
Kelloff, G. J., 94
Kelson, T. L., 311, 312(14), 314(14)
Kemman, E., 116
Kemp, D. J., 321
Kermicle, J. L., 311
Kern, D. G., 47
Key, J. L., 175
Khachik, F., 7, 11, 34, 46(7a), 56, 102, 104-105, 106(16), 107, 130
Kikendall, J. W., 7, 131, 136(34)
Kim, C., 125, 136(18)
Kimura, O., 56, 230
Kimura, Y., 278, 292

King, M., 128
Kingston, R. E., 378, 380(12)
Kioni, R., 3, 7(2)
Kirk, J.T.O., 348, 365
Kirschner, B., 123
Kitajima, M., 182-183
Kittler, L., 125
Klaui, H., 86
Kleinig, H., 297-298, 299(5), 306(8), 307(5), 310, 340, 352, 366, 370, 374
Kobayashi, K., 297, 298(6), 299(6), 301(6), 307(6), 344
Koehler, D. E., 312
Kohmura, E., 56
Kolonel, L. N., 55
Komano, T., 303
Kondo, M., 213
Kornhauser, A., 124
Koroleva, I. F., 386
Koshino, Y., 149
Koss, F. W., 127
Kottko, B. R., 13, 14(35)
Koyama, Y., 194, 284
Kramer, D. M., 124
Krause, R. F., 74
Kreuz, K., 366, 374
Krinsky, N. I., 4, 11, 33(1), 34, 38, 43(1), 56, 70, 71(18)-71(19), 73(18)-73(19), 74(20), 75, 83, 102-103, 106(14), 124, 137, 170, 200, 214, 257, 323, 326(1)
Kroger, B., 235
Krogmann, D. W., 75, 83(16), 391, 395(7)
Kröncke, U., 310
Kruger, F. A., 37
Krust, A., 67
Kuller, L. H., 33
Kumle, A., 307, 310(26)
Kuntz, M., 364
Kushner, S. R., 415
Kushwaha, S. C., 366
Kyte, J., 397

L

Laemmli, U. K., 53, 78, 373
Laferrière, A., 352, 424
Laird, A., 311
Lakshman, M. R., 21, 32, 69-70, 74, 168-169, 200, 256, 258, 266, 267(5)-267(6), 267(15), 268(4)
Lalemand, D., 257
Lambowitz, A. M., 415, 416(13), 417(13), 418(13)
Lamden, M. P., 133
Lamoureux, S. W., 348, 349(1), 350(1), 351(1), 352(1)
Land, E. J., 124, 193
Landard, E. J., 192
Lane, D., 376, 381(9)
Lang, W., 274
Lange, O. L., 397
Langenberg, P., 3, 7
Lang-Feulner, J., 292
Lanza, E., 9
LaPorte, R. E., 33
Larson, R. A., 325, 326(16), 327(16), 328(16), 329(16)
Latscha, T., 148
Lau, J. M., 33, 86, 112
Lausen, C. G., 102
Lausen, N.C.G., 4, 170, 200
Lauter, F. R., 264, 297, 298(7), 325, 414, 423(10)
Lavens, P., 209, 210(11), 211(11), 217, 224(13)
Leach, F., 297, 299(3), 307(3), 310(3)
Lechelt, C., 311
Leder, P., 364
Le Doan, T., 125
Lee, J. A., 41, 42(20)
Lee, M.-J., 97, 98(13)
Lee, W. L., 75, 81(21), 82(21), 85(20), 224, 225(27)
Lees, R. S., 39, 41(16)
Léger, P., 209
Lehmann, J., 45
Lehrach, H., 422
Leipala, J., 125
Leuenberger, F. J., 156
Levenson, S. M., 68, 227-228
Lewis, K. C., 147
Lewis, S., 9
Liaaen-Jensen, S., 104, 166, 394, 401
Liao, Y., 3, 14(8)
Liedvogel, B., 366, 374
Light, G., 227
Lima, A. F., 99
Lindemann, I., 292

Lindgren, F. T., 39
Lindlar, H., 139
Lindler, H., 141
Link, P., 22, 23(15)
Lipid Research Clinics Program, 39, 43
Lipson, E. D., 284, 285(16), 287, 289(16), 290, 293(16)
Liu, S.-T., 324, 327(7), 341
Lloyd, R. E., 324
Lo, S. J., 47, 52(4), 54(11), 55(10)
Loerch, J. D., 32
Loewenstein, W., 86
Löffelhardt, W., 303
Longley, R., 168
Lopez, M. C., 231, 241(29)
López-Díaz, I., 293
Lotan, R., 193
Lotspeich, F. J., 74
Lowe, N., 137
Lowry, L., 257
Lowry, O. H., 258
Lucchese, D., 12, 14(34)
Lucchesi, D., 3, 7(2)
Luft, J. H., 361
Lusby, W. R., 11, 34, 46(7a), 102, 107
Luti, F., 423
Lütke-Brinkhaus, F., 366
Lütke-Brinklhaus, F., 374
Lutz, M., 209
Lyman, H., 402, 403(2)
Lynch, T. J., 47, 52(18), 53(7), 54(11), 54(18), 55(10), 55(18)

M

MacDonald, R. J., 422
Machlin, L. J., 22, 102, 126, 132(21), 133(21)
Macino, G., 264, 414, 416(9), 418(9), 419, 420(9), 423, 424(17)
Mackenzie, A., 284, 292(17), 293, 341, 367, 368(10), 368(14)
Maddaloni, M., 311
Madere, R., 42
Maes, D., 125
Mahley, R. W., 44, 46(26)
Maiani, G., 3, 7(4)
Malone, W., 3, 7(3), 94
Mandel, J. S., 94

Mangels, A. R., 8
Maniatis, T., 317, 422
Maoka, T., 148–149
Marbet, R., 139
Marechal, B., 44, 45(27)
Marenchic, I. G., 106
Marenus, K., 125
Maret, W., 21, 169, 173
Märki-Fischer, E., 342
Markwell, M.A.K., 394
Marsh, A. C., 208
Martelli, H. L., 388, 390(8)
Marti, U., 342
Martienssen, R. A., 311, 315
Martin, G. S., 99
Martínez-Laborda, A., 284, 293
Martta, R., 311
Marzetta, C. A., 41, 43(19)
Masamoto, K., 75, 83(18)–83(19), 391, 393(6), 396(6), 397(6), 397(8), 399(6), 400
Mason, P. J., 318
Matgner, M. B., 9
Mathews, M. M., 83
Mathews, C. K., 357
Mathews-Roth, M. M., 4, 33, 34(5), 36, 74, 75(10), 102, 126, 170, 200, 323
Mathur, S. N., 69
Matsumoto, J., 47
Matsumoto, K., 280
Matsuno, T., 102, 148–149
Matsuo, K., 278, 292
Mattingly, D., 122
Maudinas, B., 192
Maunders, M., 301
May, W. E., 111
Mayer, H., 24, 30(17)
Mayer, M., 310
Mayne, S. T., 74, 75(13), 78(13)
Mayrent, S. L., 21
McCarty, D. R., 311, 321
McCord, J. D., 84
McCullagh, D., 348, 350(6), 351(6)
McGee, D. L., 99
McKie, J. H., 309
McLachlan, J., 405
McLaughlin, M., 311
McNiven, M. A., 47
McShane, L. M., 100
McVie, J., 191

Meduna, V., 159
Mehl, J. W., 74
Mehta, P. P., 86
Meister, W., 167
Mejia, L. A., 136
Melles, S., 413
Menasveta, P., 148
Menkes, M. S., 94
Mercer, E. I., 332, 333(7)
Merchen, N. R., 4, 102, 107(6), 109(6)
Merriman, R. L., 63, 66(13)
Messner, P., 401
Metzenberg, R. L., 420
Meyer, C., 3, 7(3)
Meyer, J. C., 94
Meyskens, F. L., Jr., 55
Michalowski, C. B., 303
Michelakis, A., 3, 7(3)
Michell, H. K., 274
Micozzi, M. S., 3, 4(1), 5–7, 12(1), 119, 122(10)
Miles, D., 321
Miller, E., 7, 33, 34(6), 36(6), 37(6)
Miller, G. J., 44
Miller, J. A., 132
Miller, J. H., 325, 326(18)
Mimuro, M., 402, 403(4), 403(7)–403(8)
Misawa, N., 297, 298(6), 299(6), 301(6), 307(6), 344
Mitchell, H. K., 269, 287, 292
Mittal, P. C., 169, 172(9)
Mitzka-Schnabel, U., 293, 368
Miyoshi, Y., 280
Mobarhan, S., 3, 7, 12, 14
Moens, L., 209, 210(11), 211(11), 224(13)
Mohn, G. R., 125
Mohr, H., 183–184
Moir, A., 303, 305(20)
Monger, T. G., 191
Montavon, M., 139, 141
Montreewasuwat, N., 257
Moody, S., 7
Moon, R. C., 21, 55
Moore, D. D., 378, 380(12)
Moore, T., 21, 257
Morawski, S. G., 3, 14
Morelli, G., 264, 414, 416(9), 418(9), 419, 420(9), 423, 424(17)
Morf, R., 69
Morikawa, F., 127
Morris, M. D., 41, 42(20)
Morzychka-Wroblewska, E., 420
Moshell, A. N., 7, 33, 34(6), 36(6), 37(6)
Moss, S. H., 327
Mott, D. J., 22, 102, 126, 132(21), 133(21)
Motto, M., 311
Moustacchi, E., 125
Mueller, D. G., 257
Muller, P., 141
Müller, R. K., 155
Mummery, R. S., 183, 274
Mummery, S., 284
Muni, I. A., 128
Murakoshi, M., 56
Murata, N., 392, 401
Murillo, F. J., 284, 291, 293, 365–366, 372(3)
Murphy, M. R., 102, 107(6), 109(6)
Murskoshi, M., 230
Muto, S., 284
Mychkovsky, I., 21, 32(8), 74, 169, 200, 258, 266(15), 267(15)

N

Naito, H. K., 38
Nakagawa, M., 297, 298(6), 299(6), 301(6), 307(6), 344
Nakamura, K., 297, 298(6), 299(6), 301(6), 307(6), 344
Nakayama, Y., 127
Napoli, J. L., 21–22, 32, 74, 86, 169–170, 193–194, 197(6), 198(6), 199(8)–199(9), 201(6), 261
Naqvi, K. R., 226
Nathanson, N., 82
National Research Council, 8, 9(24)
Navarro, J., 257
Nelis, H.J.C.F., 209–210, 211(11), 217, 224(13)
Nelki, D., 293
Nelson, M. A., 264, 414, 416(9), 418(9), 419(9), 420(9), 423(9)
Nelson, O. E., Jr., 311
Nelson, R. A., 116, 123(6)
Neville, D. M., Jr., 385
Newton, D., 67
Nicar, M. J., 14
Nicklen, S., 395

Nierenberg, D. W., 97-98, 101(14), 104, 112(15), 136
Niesbach-Kloesgen, U., 311
Nievelstein, V., 297, 298(5), 299(5), 307(5), 310
Ninnemann, H., 226
Nishida, T., 303
Nishimura, M., 84, 303
Nishino, H., 56
Noble, R. P., 38
Nomoto, K., 228, 229(19), 236(19)
Nomura, A. M., 94
Norden, D. A., 82

O

Oelmüller, R., 183-184
O'Fallen, J. V., 101
O'Farrell, P. H., 53, 373
Ogawa, T., 182-183, 396, 400-401
Ohmori, M., 396
Okamatsu, H., 275
Okazumi, J., 56
Olson, J. A., 21, 31-32, 68-70, 74, 86, 102, 104, 106(18), 136, 137(5), 143(9), 168-170, 172(8), 200(11), 256-257, 265, 267, 268(4), 268(22)
Olsson, T. A., 128
Omata, T., 392, 396, 400-401
Oncley, J. L., 33, 34(1), 38(1), 43(1), 70, 75
Ong, D. E., 4
Ookubo, M., 148
Orbach, M. J., 417, 422
O'Reilly, C., 311
Osman, M., 183-184
Ostrea, E. M., Jr., 257
Otto, M. K., 292
Owens, T. G., 402, 403(3)

P

Pabst, M. J., 227
Packer, L., 125
Padawer, J., 68, 227-228
Palazzo, R. E., 47, 54(11), 55
Parker, B. A., 321

Parker, R. S., 3, 74, 75(13), 78(13), 86, 102, 107, 112(1), 114(1), 114(22)
Parra, F., 291
Parrish, D. B., 69
Parrish, J. A., 137
Parson, W. W., 191
Partali, V., 166
Parthasarathy, S., 34
Pasquale, S. A., 116
Pathak, M. A., 124, 127, 137
Pattenden, G., 343, 346(10)
Peacock, J. W., 311
Pearson, W. R., 299
Pecker, I., 375
Pelham, H.R.B., 235
Peng, G., 47, 55(10)
Peng, Y.-M., 97
Pereira, A., 311
Perry, K. L., 324, 327, 341
Petersen, L. A., 206
Peterson, G. L., 212
Peterson, P. A., 311
Peterson, T. A., 311, 387
Petkovich, M., 67
Phaff, H. J., 386
Pierce, A. W., 86
Pierce, J., 400-401
Pins, M., 94
Pizzala, R., 124
Pollock, D., 298, 375
Pomeroy, D., 388, 390(8)
Ponziani, G., 311
Poor, C. L., 4, 87, 102, 107, 109
Pope, J. L., 69, 256, 268(4)
Poppe, W., 124
Pops, M. A., 116, 122(2)
Poretti, G. G., 331
Porro, E. B., 422
Porter, J. W., 331, 335(3), 340, 366, 368, 370(17)
Porter, T., 257
Posch, K. C., 193
Poulter, C. D., 206, 423
Powls, R., 178
Presti, D., 175, 285, 289(31)
Prince, M. R., 228
Przybyla, A. E., 422
Pung, A., 21, 23, 27(4), 31(4), 56-58, 59(9), 61(9), 62(4), 62(9), 63, 64(9), 65(9), 66(8), 67(4), 67(9), 68, 230, 236

Puppione, D. L., 38-39
Purcell, A. E., 131
Pustell, J., 307

Q

Quackenbush, F. W., 208
Quennemet, J., 352
Quick, T. C., 4
Quinn, M. T., 34

R

Race, K. R., 21-22, 32, 74, 86, 169-170, 193, 198(6), 201(6), 261
Raica, N., Jr., 257
Ranaldi, L., 3, 7(4)
Randall, R. J., 258
Rao, B.S.N., 3
Rao, C. N., 3
Rau, W., 175, 176(4), 177(4), 183, 264, 269, 273-275, 283-284, 286, 289, 291-293, 368
Rau-Hund, A., 289, 292
Ray, J., 301
Raynor, W., 94
Read, C. M., 21, 27(4), 31(4), 56, 58(4), 62(4), 67(4), 68(4), 236
Reay, C. C., 73
Reddy, K. J., 75, 83(19), 391, 394-395, 397(8), 397(18), 400
Reddy, P. P., 33, 34(7), 35
Redfearn, E. R., 69
Redgrave, T. G., 41
Reed, B. C., 264
Rees, A. F., 366, 368
Reiser, J., 321
Reiss-Husson, F., 209
Reithman, H. C., 75, 83(18)
Renart, J., 321
Rettura, G., 68, 227-228
Reynolds, R. D., 117, 120(8), 122(8)
Rhodes, J., 229, 236
Ribaya-Mercado, J. D., 4, 74, 75(12), 102, 136
Rich, P. M., 352
Riethman, H. C., 391, 393(6), 396(6), 397(6), 399(6)
Rigassi, N., 139, 374

Rilling, H. C., 183, 264, 269, 272, 284, 292(11), 423
Rinkenberger, J. L., 324
Robboy, M. S., 116, 122(3)
Roberts, D.C.K., 41
Robertson, D. S., 311-313, 314(14)
Robinson, H. B., 37
Rock, C. L., 7, 9, 122
Roe, D. A., 125, 136(18)
Roelandts, R., 127
Roels, O. A., 3
Romano, N., 264, 414, 416(9), 418(9), 419, 420(9), 423(9), 423(17), 424(17)
Rose, A., 3, 4(1), 5-6, 7(1), 9, 12(1), 119, 122(10)
Rosebrough, N. J., 258
Rosenberg, A. H., 376, 380(10)
Rosenthal, M. A., 44, 45(27)
Rosin, M. P., 94
Rossner, S., 33
Rothfuss, L. M., 125
Rubin, L. B., 292
Ruble, P. E., 22
Ruddat, M., 284
Rudel, L. L., 41-43
Rudiger, W., 349
Rudnick, M. S., 402, 403(2)
Ruegg, R., 139, 141
Ruiz-Vázquez, R. M., 293
Rundhaug, J. E., 21, 27, 31, 56-57, 58(4), 62(4), 63, 66(8), 67(4), 68(4), 68(8), 230, 236
Rushin, W. G., 208
Russel-Jones, K. H., 361
Russell, R. M., 4, 70, 71(18)-71(19), 73(18)-73(19), 74, 75(12), 79(20), 102, 136, 168, 257
Russett, M. D., 4, 11, 102, 103(14), 106, 170, 200
Russo, V.E.A., 178, 264, 293, 297, 298(7), 325, 413-414, 423(4), 423(10)
Rutter, W. J., 422
Ruzicka, L., 331
Ryser, G., 139, 141

S

Saedler, H., 311
Sage, E., 125

Sakai, H., 303
Sakai, T., 56
Salamini, F., 311
Saleh, F. K., 284
Salet, C., 124
Salgado, L. M., 293
Salkeld, R. M., 94
Salomon, K., 82, 85(30)
Samaille, J., 43
Sambrook, J., 317, 422
Sameshina, M., 148
Sammartano, L. J., 324, 327
Sampson, L., 117, 120(8), 122(8)
Sander, L. C., 111
Sandmann, G., 298, 307(10), 324, 326(13), 329(13), 341–343, 344(3), 344(7), 346(11), 367, 368(15), 370
Sandstead, H. H., 123
Sanger, F., 395
Santa Ana, C. A., 3, 14
Santagati, G., 124
Santamaria, L., 124
Sapuntzakis, M., 7
Sargent, M. L., 324, 327(9)–327(10)
Sastry, P. S., 69
Sato, A. S., 116, 122(3)
Sauberlich, H. E., 257
Saucy, G., 139, 141
Savage, D. D., 33
Schaeren, S. F., 139, 141
Scheele, T., 311
Schiedt, K., 148–149, 156, 157(8), 159, 162, 163(6), 163(8), 166
Schiller, L. R., 14
Schmid, J., 127
Schmidhauser, T. J., 264, 297–298, 307(12), 310(12), 325, 374, 375(5), 414, 423(10)
Schmidt, A., 298, 307(10), 341, 343(3), 344(3)
Schmidt, R. J., 311
Schmitz, H. H., 102, 107(9), 109(9)
Schneider, F. H., 128
Schneider, W. P., 417
Schnurr, G., 341, 343(3), 344(3)
Schopfer, W. H., 331
Schöpp, K., 69
Schreiber, U., 397
Schroeder, J. R., 257
Schrott, E. L., 175, 176(4), 177(4), 264, 283–285, 289
Schuch, W., 301

Schumaker, V. N., 38–39
Schwabe, A. D., 116, 122(2)–122(3)
Schwartz, J. L., 227–228, 231, 234–235, 240, 241(48)
Schwartz, S. J., 208
Schwarz-Sommer, Z., 311
Schweizer, M., 415
Schwieter, U., 139, 141, 374
Schynder, U. W., 94
Scolnik, P. A., 298, 307, 310(12), 310(26), 374–375, 381(7), 383(7)
Scopes, R. K., 371
Scott, J., 257
Searle, G.F.W., 226
Searle, S. R., 96
Seeger, B., 94
Sehgal, P. K., 4, 102, 170, 200
Seidel, K. E., 9
Seidman, J. G., 378, 380(12)
Seifter, E., 68, 227–228
Selker, E. U., 420
Senger, H., 178, 182
Shaish, A., 390
Shapiro, S. S., 22, 74, 75(11), 102, 126, 132(21), 133, 228, 234
Sharma, R. V., 69
Sheikh, M. S., 14
Shekelle, R. B., 94
Shenai, J. P., 257
Shepherd, N. S., 311
Sherman, D. M., 391, 397(8)–397(9), 399(9), 400, 401(9)
Sherman, L. A., 75, 83(17)–83(19), 391, 393(6), 394–395, 396(4), 396(6), 397(6), 397(8), 397(18), 399(6), 400
Sherman, M. I., 63
Sherman, M. M., 75, 83(19), 206
Shiau, A., 3, 14(8)
Shibata, K., 182–183
Shipley, R. A., 147
Shiratori, T., 21, 170, 256
Shiroishi, M., 264, 270–271, 273(8), 274–275, 278, 284, 286, 289(35)
Shklar, G., 227–228, 231, 234, 235(14), 240, 241(48)
Shriver, D. F., 58
Shropshire, W., Jr., 178, 183, 280, 282(26), 284, 285(19), 413
Siefermann-Harms, D., 83, 185, 226, 390
Simms, P., 12, 14(34)
Simon, W., 24, 30(17)

Simonitch, T. A., 324, 327(7), 341
Simons, K., 368, 370(22)
Simpson, K. L., 70, 102, 103(11), 104(11), 109(11), 136, 208, 214, 386
Sinclair, R. S., 191
Singh, H., 168, 169(7), 261, 268(18)
Singh, R. P., 240
Sistrom, M. R., 83, 323
Sistrom, W. R., 83
Skaf, R., 116
Sklan, D., 168, 169(3)
Slagle, S., 44
Sleytr, U. B., 401
Sloane, D., 227, 235(14)
Smart, J. R., 387
Smith, F. R., 168, 169(2), 170(2)
Smith, J. A., 378, 380(12)
Smith, J. C., Jr., 3, 4(1), 5-7, 9-10, 12(1), 33, 34(7), 35, 38, 46, 118-119, 122(10), 130, 133(31)
Smith, J. D., 312
Smith, K. C., 327
Smith, P., 234, 238(32)
Smith, W. P., 125
Smithies, O., 299
Snodderly, D. M., 103, 106(14)
Snodderly, M. D., 11
Soave, C., 311
Sober, A., 7, 123
Soejima, T., 214
Solmssen, U., 69
Sorgeloos, P., 209, 210(11), 211(11), 214, 218, 224(13)
Southern, E. M., 315
Sowell, A. L., 208
Spencer, S. K., 94
Sporn, M. B., 67
Spurgeon, S. L., 331, 335(3), 340
Stacewicz-Sapontzakis, M., 3, 7, 12, 14(8), 14(34), 131, 136, 168
Stahelin, H. B., 94
Stahlman, M. T., 257
Stain, E. A., 7
Stampfer, M. J., 7, 94, 123
Stapp, H., 298
Stark, G. R., 321
Starlinger, P., 311
Statzell-Tallman, A., 386
Steele, P., 348, 350(5), 351(5)
Steele, W. J., 331

Steffen, M. A., 226
Steglich, W., 156
Stein, E. A., 33, 86, 112, 123, 132
Steinberg, D., 34
Steiner, A., 124
Stemmerman, G. N., 94
Stern, K. G., 82, 85(30)
Stern, R. S., 137
Stevens, J. N., 420
Stevens, M. M., 94
Stich, H. F., 94, 229
Stinard, P. S., 311
Stokes, P., 236, 239(46)
Stoller, H. J., 24, 30(17)
Stone, E. J., 366
Strassberger, G., 178
Stratford, F., 68, 227
Strehler, B. L., 202-203
Stricker, K., 141
Striegl, G., 36
Stringham, J. M., 240
Struhl, K., 378, 380(12)
Stryker, W. S., 7, 123
Studier, F. W., 376, 380(10)
Stukel, T. A., 94, 136
Subbarayan, C., 75, 84(23)
Subbiatr, M.T.R., 13, 14(35)
Subden, R. E., 275, 414
Suda, D., 227-228
Sugarman, S. B., 3, 7, 12, 14(34)
Sugimoto, T., 56
Sugino, M., 275
Suglimoto, K., 99
Sullivan, J. M., 7
Sundstrum, V., 191
Sutter, R. P., 283
Suzuki, T., 278
Sweeney, J. P., 208
Swendseid, M. E., 7
Szabolcs, J., 208

T

Tada, M., 264, 270-271, 273-275, 278, 280, 284, 286, 289(35), 292
Tada, Y., 280, 284
Takagi, S., 278, 292
Takaichi, S., 402, 403(4)
Takamatsu, K., 84

Takayasu, J., 56, 230
Takii, T., 194
Takimoto, H., 278, 292
Takruri, H.R.H., 170
Tamm, R., 141
Tanaka, Y., 148
Tang, G., 70, 71(18)–71(19), 73(18)–73(19), 74(20), 168, 257
Tangney, C. C., 94
Tanumihardjo, S. A., 86
Tappel, A. L., 133
Tarr, E., 82
Taylor, J. D., 47, 52(4), 52(16)–52(18), 53(7), 53(16), 54(11), 54(13), 54(16), 54(18), 55(10), 55(13)–55(15), 55(17)–55(18)
Taylor, J. O., 94
Taylor, P. R., 3, 4(1), 5–7, 12(1), 33, 34(7), 35, 119, 122(10)
Taylor, R. F., 103, 106, 368, 369(21)
Taylor, S. L., 133
Taylor, W. C., 311, 315(11), 321
Tchen, T. T., 47, 52(4), 52(16)–52(18), 53(7), 53(16), 54(11), 54(13), 54(16), 54(18), 55(10), 55(13)–55(15), 55(17)–55(18)
Teicher, B., 240, 248
Tengerdy, R. P., 236
Terpstra, P., 309, 310(27)–310(28)
Than, A., 366
Theimer, R. R., 291–293
Theres, N., 311
Thomas, P. S., 321, 422
Thomas, S. A., 324, 327(10)
Thompson, J., 203
Thompson, J. N., 42
Thompson, R., 311
Thorne, S. W., 407
Thurnham, D. I., 170
Tiede, D. M., 191
Tilney-Basset, R., 365
Toda, K., 127
Tolbert, N. E., 394
Tomita, Y., 228, 229(19), 236
Torres-Martínez, S., 284
Toth, G., 208
Towers, G.H.N., 328
Travis, R. L., 175
Trickler, D., 231
Trout, M., 3

Truscott, T. G., 191, 193, 228
Tsou, S.C.S., 102, 103(11), 104(11), 109(11)
Tsubouchi, M., 278, 292
Tsukida, K., 194
Turnbull, B. W., 100
Turner, R. V., 178, 264, 413, 414(3), 423(3)
Tuveson, R. W., 324–328, 329(13), 329(16), 341, 344(7)

U

Ullrey, D. E., 3, 7(3)
Underwood, B. A., 32, 94, 105, 136
Urbach, F., 129, 136(30)
U.S. Department of Health and Human Services, 126
Utsumi, R., 303

V

Vahlquist, A., 33
Valadon, L.R.G., 175, 183–184, 274, 284
Van Boven, M., 127
Vanderslice, J. T., 56, 104, 106(16), 130
Vanhaecke, P., 209
van Kleef, P. M., 125
Van Kuijk, F.J.G.M., 73, 259
Van Steenberge, M.M.Z., 209, 217
Varik, O. Y., 292
Vecchi, M., 155–156, 159, 167
Versichele, D., 209
Vetter, W., 374
Viitanen, P. V., 375
Vile, G. E., 231
Villard-Mackintosh, L., 169, 173
Villijara, M. O., 94
Viulleumier, J. P., 94
Vogel, H. J., 415, 420(15)
Vogelstein, B., 422
Volk, M. E., 44, 45(27)
Vollbrecht, E., 311
Vollmer, S. J., 414, 416–417
von Muralt, A., 331
Vorbrüggen, H., 156
Vriend, G., 309, 310(28)
Vuilleumier, J. P., 94

W

Wada, M., 38
Waeg, G., 36
Wahl, G. M., 321
Walbot, V., 311, 320(25)
Wald, G., 82
Waldron, J. C., 407
Walker, G. R., 47, 54(13), 55(10), 55(13)
Warner, W., 124
Wang, G. R., 325, 328
Wang, T. P., 325, 328
Wang, X.-D., 70, 71(18)–71(19), 73(18)–73(19), 74(20)
Ward, J. B., 47
Warmer, W. G., 236
Warnick, G. R., 43–44
Wasielewski, M. R., 191
Watanabe, M., 280, 284
Watkins, L. O., 33
Watson, R. R., 231, 241
Webb, R., 394–395, 397(18)
Weber, G. F., 235
Weckesser, J., 394
Weedon, B.C.L., 208
Weeks, O. B., 284
Wei, R. R., 236
Weil, J. H., 364
Weir, J.C.J., 227
Weiser, H., 22
Weisgraber, K. H., 44, 46(26)
Weiss, G., 370
Weiss, M. F., 146
Welankiwar, S., 4, 102, 170, 200
Wellman, R. B., 102, 107(9), 109(9)
Wessel, D., 373
Wessels, J.S.C., 226
Wessler, S. R., 311
West, C. E., 41
West, J., 203
Weydemann, U., 311
Wheeler, L. A., 137
Whetter, P., 94
White, C. B., 9
White, W. S., 4, 107, 109(24), 114(24), 125, 136(18)
Wienand, U., 311
Wierenga, R. K., 309, 310(27)
Wilkens, L. R., 23, 58, 59(9), 61(9), 62(9), 63(9), 64(9), 65(9), 67(9), 86
Will, O. H., 284
Willett, W. C., 7, 21, 94, 117, 120(8), 122(8), 123
Williams, J. G., 318
Williams, R. S., 235
Wilson, D. S., 125
Winkler, R., 257
Winterbourn, C. C., 231
Wirahadikusumah, M., 284
Wise, S. A., 111
Wiss, O., 69
With, T. K., 136
Witholt, B., 309, 310(28)
Witt, E., 125
Wold, E. R., 402, 403(3)
Wolf, C., 124
Wolf, G., 169, 172(12)
Wong, X.-D., 257
Woods, W. S., 324, 326(13), 329(13), 341, 344(7)
Worawattanamateekul, W., 148
Woytkiw, L., 74
Wozney, J. M., 422
Wu, B.-Y., 47, 52(18), 54(18), 55(10), 55(14), 55(18)

Y

Yager, J., 122
Yamada, S., 148
Yamagishi, S., 284
Yamano, S., 297, 298(6), 299(6), 301(6), 307(6), 344
Yang, C.-F., 47, 55(15)
Yang, C. S., 97, 98(13)
Yang, S. Y., 203
Yanofsky, C., 264, 297–298, 307(12), 310(12), 325, 374, 375(5), 414, 416–417, 422–423
Yazdi, M. A., 303, 305(20)
Yelle, L. M., 106
Yoshizawa, C. N., 21, 57, 63(8), 66(8), 68(8), 230, 236
Young, R. A., 394
Yu, F.-X., 47, 52(17)–52(18), 54(18), 55(10), 55(14)–55(15), 55(17)–55(18)

Z

Zagalsky, P. F., 75, 81(21)-81(22), 82, 85(20), 85(22), 85(29), 224, 401
Zalasko, M. V., 386
Zalokar, M., 269, 284, 285(18), 289(18)
Zech, L. A., 147
Zelent, A., 67
Zeller, P., 141
Zeng, Z.-C., 47, 52(16), 53(16), 54(16), 55(15)
Zhang, L. X., 68
Ziegler, R. G., 137
Zimmer, A., 127
Zipp, H., 127
Zsebo, K. M., 341
Zurcher, P., 141

Subject Index

A

Absorbance spectroscopy
  of carotenoids in deletion mutants from
    E. coli LE392/pPL376, 344
  of carotenoprotein complexes, 396
  of xanthosomes, 408
Acetyl-CoA
  conversion to isopentenyl pyrophosphate, 332
  role in carotenoid biosynthesis, 331
4'-(9-Acridinylamino)methanesulfon-m-anisidide. See m-AMSA
Action spectroscopy, of photoinduced carotenogenesis, 180–182, 280–283, 290–291
Adipose tissue, carotenoids in, measurement of, 87–93
Affinity chromatography, heparin-sepharose, for lipoprotein separation, 44–46
ALA. See δ-Aminolevulinic acid
Albino-1 (al-1) gene
  N. crassa, 298
  nucleotide sequencing of, 297
  walking to, 414
Albino-3 (al-3) gene
  photoinduction experiments, 421–423
  transcript analyses of, results of, 423
  walking to, 414
Albino-3$^+$ (al-3^+) gene
  cloning of, 417–419
    isolation of chloramphenicol-resistant AL3-transformant plasmid, 418
    isolation of plasmid containing, 418–419
    preparation of al-3^+ transformant DNA for, 417–418
    intact, plasmid containing, isolation of, 418–419
    mapping of, 419–420
Albino-1 (al-1) gene product. See Al-1 protein(s)

Albino-3$^+$ (al-3^+) gene transformants, N. crassa, identification of
  results, 416–417
  spheroplast preparation for, 415–416
  strains and vectors for, 415
  transformation protocol for, 416
Albino hairless mice (Skh/HR-1), 125–126
  effects of psoralen photosensitization on plasma and cutaneous β-carotene levels in, 124–137
  photosensitized, irradiation-induced edema responses for, 128
Algae, brown, isolation of xanthosomes from, 403–405
Alleles, transposon-tagged, analysis of, 322
Al-1 proteins, 298
  FAD/NAD(P)-binding domains in, 310
  sequence alignment of, 307–309
AL3-1 transformant, chloramphenicol-resistant plasmids from, isolation of, 418
Amino acid sequences
  of carotenoid biosynthesis enzymes
    alignment of, procedures for, 299
    comparison and analysis of, methods and applications, 299–311
    relationships of, 300–301
    for phytoene dehydrogenase, 324–325
δ-Aminolevulinic acid, effects on carotenoid accumulation, 348–352
3-Amino-1,2,4-triazole, effects on carotenoid accumulation, 348–352
m-AMSA, inhibition of β-carotene or retinal conversion to retinol and retinoic acid in rat intestinal cytosol by, 202
Anacystis nidulans R2, carotenoid-binding protein of, 75, 83
Animal lymphocytes and macrophages, determination of carotenoid activity in, 234–236
Animal models, of carotenoid absorption, 4
Animal neoplastic cells, determination of carotenoid activity in, 241
Animal tissues, carotenoids in, 102–116

Anion-exchange chromatography, of CrtI protein, 384
Anorexia nervosa, plasma carotenoid levels in, 116–123
Antibodies, generation of, 380–381
Apo-β-8′-carotenal, from human plasma, 91–93
β-Apo-8′-carotenal, 138
  structure of, 138
  UV/visible absorption spectra of, 139
10′-β-apocarotenal, cleavage of, 268
Apocarotenoids, metabolism in humans, 137–147
β-Apocarotenoids
  chromatography of, 71
  formation of, from incubation of β-carotene with crude homogenates, 73
  identification of, 71–73
  production of, 70
β-Apo-8′-carotenol, UV/visible absorption spectra of, 139
10′-β-apocarotenol, cleavage of, 268
Apolipoproteins, 33
Ar5 cells, proliferation of, 237
Artemia
  carotenoids in, 222
    organ and tissue distribution of, 223
    subcellular localization of, 224
  development of
    all-*trans*-, *cis*-, and total canthaxanthin in, concentration of, 220
    all-*trans*- and *cis*-canthaxanthin in, fate of, 217–225
    carotenoid isomers in, 208–225
    carotenoid profiles of, 216–217
    later stages, analysis of, 213
  dissected tissues from, analysis of, 213
  geographical strains, 218
  life cycle of, levels of β-carotene, echinenone, and all-*trans*-canthaxanthin during, 221
  samples of, 209
Artemia cysts
  analysis of, 212
  cis- and all-*trans*-canthaxanthin in
    distribution of
      effect of hydration time on, 220
      effect of repeated dehydration and hydration on, 219
    levels of, 217, 219

cis-canthaxanthins in, demonstration of, 214–216
  extracts, chromatography of, 214–215
    normal-phase, 214–215
  SFB 288/2606 and GSL Z-627, subcellular fractions, quantitative analysis of, 224
  subcellular fractions, analysis of, 213–214
Artemia nauplii
  analysis of, 212–213
  development of
    early, *cis*- and all-*trans*-canthaxanthin in, fate of, 218–220
    later, *cis*- and all-*trans*-canthaxanthin in, fate of, 220–221
Aspergillus gignateus, photoinduction of carotenoids in, 291
Astacene, R_f values of, on thin-layer chromatography, 160
Astacene diacetate, R_f values of, on thin-layer chromatography, 160
Astaxanthin
  additive, 153
  chromatography of
    column, 156–158
    high-performance liquid, 158–162
    preparative thin-layer, 159
    systems for, 156–161
    thin-layer, 159
    R_f values on, 160
  in crustaceans, 82
  (−)-dicamphanates of, preparation of, 155–156
  diesters, of lipid extract, 157
  extraction of, 153
  isolation procedure, ratio, and specific radioactivity of, 164
  of lipid extract, 157
  monoesters, of lipid extract, 157
  relative abundance of, in carapace of *Penaeus japonicus*, 166
  saponification of
    anaerobic, 153–155
    conventional, 155
  stereoisomers of, ratio and specific radioactivity of, via diastereomeric (−)-dicamphanate derivatives, 165
  in vivo racemization of, 163
Astaxanthin dicamphanate, R_f values of, on thin-layer chromatography, 160

Astaxanthin dipalmitate, R_f values of, on thin-layer chromatography, 160
Astaxanthin monopalmitate, R_f values of, on thin-layer chromatography, 160
AT. See 3-Amino-1,2,4-triazole

B

Bacteria
　photoinduction of carotenoids in, 291
　photosynthetic, expression of carotenoid biosynthesis genes in, 375-380
　functional analysis and purification of enzymes for, 374-385
Bacterial antenna complexes, isolation of, 186-189
Bacteriochlorophyll
　singlet-singlet energy transfer from carotenoids to, measurement of, 189-191
　triplet-triplet energy transfer from carotenoids to, measurement of, 191-193
Bacteriochlorophyll a, singlet and triplet energies of, 186
BC. See β-Carotene
BCC enzyme. See β-Carotene cleavage enzyme
Bilirubin, from human plasma, 91-93
Biogenetic isoprene rule, 331
Blood sampling, 142
Breast carcinoma, human, MCF-7, effects of carotenoids on, 242-244
Brown algae, isolation of xanthosomes from, 403-405

C

Calf liver, carotenoids in, Vydac separation of, 112
Calf tissues
　β-carotene concentrations in, 115
　carotenoids in, analysis of, experimental design for, 107
Calves, 4
Canthaxanthin
　beadlet preparation of, 56
　concentrations
　　in *Artemia* development, 220
　　in *Artemia* subcellular fractions, 224
　　in crustaceans, 82, 148

effects of
　on human breast carcinoma MCF-7, 242-244
　on MCA-induced neoplastic transformation, 63, 65, 67
　as provitamin A source, 67
　uptake and half-life of, in cell cultures, 62-63
　UV/visible absorption spectra of, 140
all-*trans*-Canthaxanthin
　in *Artemia*, organ and tissue distribution of, 223
　in *Artemia* cysts
　　distribution of
　　　effect of hydration time on, 220
　　　effect of repeated dehydration and hydration on, 219
　　levels, 217, 219
　in *Artemia* development, concentration of, 220
　in *Artemia* life cycle, levels, 221
　in mature *Artemia*, levels, 222
　quantitation of, 212
cis-Canthaxanthin, 210
　in *Artemia*
　　identification of, 211
　　organ and tissue distribution of, 223
　in *Artemia* cysts
　　demonstration of, 214-216
　　distribution of
　　　effect of hydration time on, 220
　　　effect of repeated dehydration and hydration on, 219
　　levels, 217, 219
　during *Artemia* development, concentration of, 220
　in mature *Artemia*, levels, 222
　quantitation of, 212
Capsicum
　chloroplasts
　　electron microscopy of, 355
　　integrity and purity of, 355
　　isolation of, method for, 354
　chromoplasts
　　electron microscopy of, 356
　　integrity and purity of, 356
　　isolation of, method for, 354
　phytoene synthase complex, component enzymes of, 353-359
　plastid preparations, 340

Capsorubin, 338
  structure of, 339
Carbon sources, for β-carotene synthesis, 390
Carotene
  chlorophyll and, energized state migration between, 207–208
  ^{14}C-labeled, purification and radioassay of, 367–368
  energized state of, life-time of, 206
  illumination of, relation between intensities of delayed light emission and duration of dark intervals after, 206
α-Carotene
  association with lipoproteins, 35
  certified reference material (SRM) for, informational values from, 11
  delivered in tetrahydrofuran, activity of, 63–64
  effects on MCA-induced neoplastic transformation, 63, 65, 67
  from human plasma, 91–93
  from human thigh adipose tissue, 91
  structure of, 330–331
  uptake and half-life of, in cell cultures, 62–63
β,ϵ-Carotene, cleavage of, 268
β,β-Carotene
  in crustaceans, 148
  R_f values of, on thin-layer chromatography, 160
β-Carotene
  absorption
    in animals, 74–75
    estimation of, by total gastrointestinal lavage, 14–17
  as anticarcinogen, 21
  aqueous solution of, physical state, 59–62
  in *Artemia*, organ and tissue distribution of, 223
  association with lipoproteins, 35
  beadlet preparations, 56, 60–61
  certified reference material (SRM) for, in human plasma, 11
  chromatographic analysis of, 96
  ^{14}C-labeled, incubation with BCC enzyme, [^{14}C]retinal (*O*-ethyl) oxime formed from, crystallization of, 259–260
  cleavage, 256–269
    assay method, 257–260
      animals for, 258
      chromatographic analysis, 259
      neonatal tissues for, 258
      principle of, 257–258
    in cell culture, 21–32
    in cultured cells, 27–32
    enzymatic activity for, 73
    excentric, 69–70
      biological significance of, 73–74
      and central, differentiation between, 69–74
    major findings, 261–267
    precautions, 267
  cleavage products
    identification of, 70
    quantitation of, 70
    separation of, 70
  contacting with Chl*a* aggregates, light emission of, 207
  delivered in tetrahydrofuran
    activity of, 63–64
    versus β-carotene in beadlet form, effect on MCA-induced transformation, 63–64
  dietary, cancer-preventive properties of, 55–56
  E. coli, 324
  effects of
    on [^{3}H]thymidine incorporation, 239
    on human breast carcinoma MCF-7, 242–244
    on MCA-induced neoplastic transformation, 63, 65, 67
    photoprotective, against psoralen photosensitization, 124–125
    on SCC-25 cell growth, 250
    ultrastructural, on SCC-25 cells, 246–247
  extraction and isolation of, from *Rhodotorula*, 387
  half-life, in cell cultures, 62–63
  handling, 194
  illuminated, and aggregated chlorophyll, determinations of energy transfer between, 204–205
  incubation of, with tissue homogenates, 71
  ingestion of
    long-term, plasma carotenoid response to, 20

single, plasma carotenoid response to, 20
inhibition of HUT 78 by, 239-240
levels. *See also* β-Carotene, plasma levels
during *Artemia* life cycle, 221
in calf tissue, 115
cutaneous, psoralen photosensitization and, in hairless mice, 124-137
in epidermal/dermal homogenates for two skin separation methods, 129
in ferret tissue, 115
in tissue extracts
standardization, 131
statistical analysis of, 134
metabolism, 69
and dietary fiber, 9
pH optimum for, 73
modulation of delayed light emission by, 202-208
photoinduction of, in bacteria and fungi, 291
plasma, photosensitization and, 124-137
plasma extracts, after 8-MOP photosensitization and irradiation, 135
plasma levels
after 8-MOP photosensitization and irradiation, 134
mean
before and after ingesting test dose of β-carotene, 121
error rates for, 99
psoralen photosensitization and, in hairless mice, 124-137
standardization, 130
statistical analysis of, 96, 134
variability in quantitation, 94-95, 97-101
methods for determining, 95-96
random effects analysis of, 97
plasma standards, 96
postprandial incorporation into lipoproteins, 36-37
preilluminated, chlorophyll *a* aggregates induced by, delayed light emission from, 205
purification of, 70, 194
quantitation of, 212
quenching of [^{3}H]thymidine by, 232-233
retinoic acid synthesis from
pathway of, 199-202

by rat intestinal cytosol, 200
inhibition by *m*-AMSA, 202
inhibition by 4-methylpyrazole, 201
by rat liver cytosol
effects of cofactors on, 199
effects of pH on, 198
in vitro, 193-202
retinol synthesis from, by rat intestinal cytosol
inhibition by *m*-AMSA, 202
inhibition by 4-methylpyrazole, 201
from *Rhodotorula*
purification and conservation of, 388
quantification of, 388
sera/plasma response to, with single ingestion, 6
singlet and triplet energies of, 186
solubilized in THF, bioavailability of, 62-63
storage in liver, 75
structure of, 138, 330-331
synthesis of
carbon sources for, 390
in *Rhodotorula*, 386-401
kinetics of, 389
medium for, 386-387
in *Rhodotorula* cells suspended in distilled water, 389
in *Rhodotorula* cultures, 388-389
uptake in cell cultures, 21-27, 62-63
all-*trans*-Carotene, from *Scenedesmus* C-6D, absorbance maxima and retention times of, 345
all-*trans*-β-Carotene, 204
enzymatic conversion to retinal, 256-269
from human plasma, 91-93
from human thigh adipose tissue, 91
cis-β-Carotene
from human plasma, 91-93
from human thigh adipose tissue, 91
poly-*cis*-Carotene, from *Scenedesmus* C-6D, 346-347
absorbance maxima and retention times of, 345
pro-ζ-Carotene, photoisomerization products of, HPLC of, 347
β-Carotene cleavage enzyme
assay, 258-259
β-[^{14}C]carotene incubation with

O-ethyl oxime derivative formed from, mass spectrum of, 259–260
[^{14}C]retinal oxime formed from, crystallization of, 259–260
isolation of, 258
properties, 267–268
rabbit
assay, lipid extracts from, HPLC of, 261–262
boiled, assay, lipid extracts from, HPLC of, 261–262, 264
β,β-Carotene-3,3′-diol. *See* Zeaxanthin
β,β-Carotene-4,4′-dione. *See* Canthaxanthin
Carotenodermia, 20–21
Carotenogenesis
under continuous illumination, 270–271
in dark reaction process, 272–273
effect of cycloheximide on, 274–275
effect of illumination time in light process on, 272
effect of photoinducible enzymes on, 277–280
effect of temperature on, 271–272
photoinduced, 175–184, 283–294
action spectroscopy of, 180–182, 184, 280–283, 290–291
in bacteria and fungi, 291
biochemistry of, 293–294
chemical modification of, 291–292
culture and illumination conditions for, 286–287
difference spectroscopy of, 178–180
effect of cycloheximide on, 279–280
effect of mevinolin on, 278–279
effect of oxygen on, 273
fluence-response curves for, 182, 281–282
kinetics of, 181
light sources for, 287–288
methods for investigating, 269–283
stimulus for, 286–288
and response, relation between, 288–291
stimulus–response curve for, 289–290
time course of, 288–289
photoreceptors involved in, 182–184, 269–283
Carotenoid biosynthesis enzymes
amino acid sequences of

alignment of, procedures for, 299
comparison and analysis of, methods and applications, 299–311
relationships of, 300–301
protein sequences of, sequence alignment of, 299–303
Carotenoid biosynthesis gene products, 297–311
Carotenoid biosynthesis genes, 297
cloning of, from maize, 311–323
expression of, in photosynthetic bacteria, functional analysis and purification of enzymes for, 374–385
of *N. crassa*, 412–424
nucleotide sequencing of, 297
overexpression of, in *E. coli*, 380–381
Carotenoid cyclases, FAD/NAD(P)-binding domains in, 309–311
Carotenoid dehydrogenases
FAD/NAD(P)-binding domains in, 309–311
sequence alignment of, 307–309
β-Carotenoid 15,15′-dioxygenase, 168
Carotenoid dioxygenase assay, 168–174
blanks, 171
chromatographic separation, 171
components of, 171
extraction, 171
incubation, 171
preparation of enzyme extract for, 170–171
procedure, 170–173
recovery, 171
scintillation counting, 171
substrate, 171
unit of activity, 173
Carotenoid droplet protein p57
labeling of
by cell-free systems, 53–54
in intact xanthophores, 52–53
phosphorylation of, 52–54
Carotenoid droplets
isolation of, 53
light and heavy, 53–54
translocation of
in goldfish xanthophores, 47–55
in squirrel fish erythrophores, 47
Carotenoid–protein complexes, 74–86
β-Carotenoid rings, formation of
from lycopene, mechanism of, 337

in zeaxanthin and lutein, stereochemistry
of cyclization leading to, 338
ε-Carotenoid rings, formation of
in lutein, stereochemistry of cyclization
leading to, 338
from lycopene, mechanism of, 337
Carotenoids
absorption of
factors affecting, 7
in humans, 3-17
basal diet for, 8
blood collection in, 9-10
metabolic balance method, 11-14
prestudy equilibrium period, 8
serum analysis, 10-11
subject selection, 7
measurement of, 3
relative, plasma/sera response technique, 4-11
in shrimp, 163
accumulation of
in cucumber cotyledons, experiments on
methods for, 348-352
procedure for, 349
effects of δ-aminolevulinic acid and 3-amino-1,2,4-triazole on, 348-352
activity of, determination of
in animal lymphocytes and macrophages, 234-236
in animal neoplastic cells, 241
with beadlets, 230
by conjugation to membranes, 231
with emulsions, 230
in human lymphocytes and monocytes, 236-238
in human neoplastic cells, 241-247
in human nonneoplastic cells, 239-241
with liposomes, 231
in neoplastic cells, 241-247
in nonneoplastic cells, 234-241
by serial dilution of stock solution with or without alcohol, 229-230
in vitro, 226-256
analogs
metabolism in humans, 137-147
preparation of, 139-142
selection of, 138
structures of, 138
analysis of

chromatographic separation, photoisomerization, and reference compounds for, 342-343
by HPLC, 34, 58, 87
in mutants, 341-347
in animal tissues, analysis of, general considerations for, 103-107
in *Artemia*
levels of, 222
organ and tissue distribution of, 223
subcellular localization of, 224
association with human plasma lipoproteins, 33-46
beadlets of, 230
bioactivity of, in cells in culture, 63-67
biosynthesis of
early investigations, 331-333
enzymology of, 340
in *N. crassa*, model for, 413
overview, 330-340
in plastid development in higher plants, regulation of, 340
blood sampling, 142
cell content
biomass absorbance as estimate of, 285-286
dependence on light intensity, 271
method for determining, convenient, 270
cells, undissolved, Ficoll-Hypaque gradient centrifugation technique for, 232-234
chloroplast, synthesis of, regulation of, 340
clearance of, factors affecting, 7
in combination with chemotherapy, suppression of tumor growth with, 245-247
composition and metabolites of, in shrimp, 163-164
conjugation to membranes, 231
crystals, undissolved, Ficoll-Hypaque gradient centrifugation technique for, 232-234
of deletion mutants from *E. coli* LE392/pPL376, 343-346
HPLC and absorbance spectra of, 344
derivatives, R_f values of, on thin-layer chromatography, 160
dietary, cancer-preventive properties of, 55-56

dissolution of, emulsions for, 230
distribution among lipoproteins, 33–37
effects of
  on activity and synthesis of cellular proteins, 248–255
  antiproliferative, 68
  cell cycle, 247–248
  on MCA-induced neoplastic transformation, 63, 65, 67
  proliferative, 241–243
    on cells in culture, 232–248
  on RNA, DNA, and protein synthesis, 243–244
  on tumor cell viability, 244–245
  ultrastructural, 247
extraction of, 153, 342
  from human liver, effect of extraction method and saponification on, 105
  from plasma, high-performance liquid chromatography for, 46
  from *R. capsulatus*, 379–380
  from tissue, 108
in *F. acquaeductuum*, after transient and continuous illumination, time course for accumulation, 286
fractions, of lipid extract, 157
functions of, in photosynthesis, 185–193
gastrointestinal modification of, 11
genes, cloned, photoprotection by, 323–330
HPLC of, reference compounds for, 343
from human breast tissue, 92
in human plasma, analysis of, 86–93
in humans
  kinetic analysis of, 146–147
  metabolism of, 137–147
  subjects, dosing, and blood sampling, 142
in human tissues
  analysis of, 86–93
  concentrations, 114
ingestion of
  acute single, plasma carotenoid response to, 4–6
  chronic multiple daily, plasma response to, 6–7
  long-term, plasma carotenoid response to, 20
  single, plasma carotenoid response to, 20

intake of
  correlation with blood carotenoid levels, 19–20
  evaluation of, 17–21
  human dietary assessment of, 17–19
isomers, in *Artemia* development, 208–225
  identification of, 211
  liquid chromatographic conditions for, 210–211
  liquid chromatographic instrumentation for, 210
  quantitation of, 212
  sample preparation, 212–214
  standard compounds for, 210
isotopes of, 3–4
liposomes encapsulating, 231
metabolism of
  factors affecting, 7
  in humans, 137–147
  in shrimp, 148–168
    experiment, 152–156
    methods and instruments, 151–152
    objectives and strategy, 149–151
oxidation of, during lipoprotein manipulation, 38
photoprotection by, 124, 326–330
  methods for using *E. coli* to assess, 325–326
plasma levels
  in anorexia nervosa and obesity, 116–123
  determinations, 119
  measurements, 118–119
  statistical analysis, 119
  study of
    characteristics of subjects in, 119
    methods, 117–118
    procedures, 117–119
    results, 119–121
  mean, in anorexia nervosa and obese patients, 120
  variability in quantitation, 94–101
    methods for determining, 95–96
    chromatographic analysis, 96
    extraction procedure, 95–96
    materials for, 95
    results, 97–101
    sample protocol, 95
    standards, 96

statistical analysis, 96
random effects analysis of, 97
polar
  and nonpolar, resolution of, 87
  relative enrichment of adipose tissue in, 92
quenching characteristics of, 232–234
from *R. capsulatus*, analysis of, 379–380
relative abundance of, in carapace of *Penaeus japonicus*, 166
response to light, 284–286
retinoid synthesis from, *in vitro*, 193–202
R_f values of, on thin-layer chromatography, 160
serum samples, extraction and chromatographic analysis of, 142–146
singlet and triplet energies of, 186
singlet–singlet energy transfer to bacteriochlorophyll from, measurement of, 189–191
solubilized in THF, bioavailability of, 62–63
solutions, preparation of, 56–57
stability of
  in isolated lipoprotein fractions, 38
  in whole plasma, 38
structure of, 330–331
thin-layer chromatography of, 394
in tissues, 102–116
  analysis of
    chromatographic, 106–107, 110
      equipment for, 108–109
      procedures, 109
      standards for, 109
    experimental design for, 107–108
    experimental procedures for, 108–109
    results, 109–114
      qualitative, 109–113
      quantitative, 113–114
      variables in, 104–106
      by Vydac separation, 110
  physicochemical characteristics of, 103–104
  variability of, 105–106
transfer of
  among lipoproteins or from lipoproteins to tissues, 37
  from plasma lipoproteins to tissues, 33
triplet formation, induced by Q-switched ruby laser flash
  difference spectrum of, 192
  kinetics of, 192
triplet–triplet energy transfer to bacteriochlorophyll from, measurement of, 191–193
whole cells, methods for treatment of, *in vitro*, 229–231
zones resolved by preparative TLC, desorption of, 159
Carotenoproteins
from crustaceans, 75, 82
in cyanobacteria, 75, 390–401
  absorbance spectroscopy of, 396
  assay of, 394
  characterization of, 392–394
  immunocytochemical localization of, 398–399
  purification of, 395–397
invertebrate, isolation and purification of, 81–82
from *P. hollandica*
  absorbance spectroscopy of, 396
  immunocytochemical localization of, 398–399
  isolation of, 393–394
  structure of, 401
in prochlorophytes, 390–401
from rat liver, 75–86
  absorption spectrum of, 78–80
  assay, 76–78
    antioxidants in, 76
    controlling for ultraviolet light and high temperatures, 76
    detergents for, 76
    preparations of aqueous extracts for, 76–77
    protease inhibitors in, 76
  chromatographic analysis of, 78
  gel electrophoresis of, 78
  isolation of, 77
    precautions for, 80–81
  properties of, 78–79
  protein and carotenoid assay, 77
  stability of, 80–81
  subcellular distribution of, 77–79, 81
reconstitution of, 84–85
storage conditions for, 85
structure of, 401

from *Synechococcus* PCC7942, 75, 83
  assay of, 394
  isolation of, 393
  42-kDa, 391
    absorbance spectroscopy of, 396
    cbpA gene encoding, cloning and sequencing of, 394–395
    immunocytochemical localization of, 398–399
  45-kDa
    absorbance spectroscopy of, 396
    immunocytochemical localization of, 398–399
    isolation of, 392–393
    and 42-kDa, sequence comparison of, 400
  from *Synechocystis* PCC6714, 75, 83, 391
  vertebrate, isolation of, 75–86
Carrot chromoplasts, carotenoprotein from, 84
cbpA gene
  cloning and expression of, 397
  encoding 42-kDA carotenoprotein, cloning and sequencing of, 394–395
  from *Synechococcus*, cloning and expression of, 394–395
  transcription of, effect of light intensity and novobiocin on, 399
Cell cycle, effects of carotenoids on, 247–248
Chemotherapeutic combinations, suppression of tumor growth with, 245–247
Chicks, β-carotene absorption in, 74–75
Chloramphenicol acetyltransferase assay, 253–254
Chloramphenicol-resistant plasmids, from AL3-1 transformant, isolation of, 418
Chlorella, carotenoid synthesis in, 176
Chlorobactone
  structure of, 338
  1,2,5-trimethylphenyl end group of, stereochemistry of rearrangement to form, 339
Chlorophyll
  aggregated
    delayed light emission from, β-carotene modulation of, 202–208
    illuminated β-carotene and, determinations of energy transfer between, 204–205

and carotene, energized state migration between, 207–208
Chlorophyll *a*, 203–204
  aggregates, 203–204
    β-carotene contacting with, light emission of, 207
    induced by preilluminated β-carotene, delayed light emission from, 205
  singlet and triplet energies of, 186
Chloroplast carotenoids, synthesis of, regulation of, 340
Chloroplasts, from *Capsicum*
  electron microscopy of, 355
  integrity and purity of, 355
  isolation of, method for, 354
Chromatography. *See also* Affinity chromatography; High-performance liquid chromatography
  anion-exchange, of CrtI protein, 384
  of β-apocarotenoids, 71
  of *Artemia* cyst extracts, 214–215
  of astaxanthin, systems for, 156–161
  of β-carotene cleavage, 259
  of carotenoids, 342–343
    in tissues, 110
  column, of astaxanthin, 156–158
  gel-filtration
    of lipoproteins, 41–43
    of phytoene desaturase from *Phycomyces* C9, 372–373
  ion-exchange, 188–189
    of phytoene desaturase from *Phycomyces* C9, 372
  normal-phase, of *Artemia* cyst extracts, 214–215
  of retinoids, 71
  size-exclusion, of CrtI protein, 383
  thin-layer
    of astaxanthin, 159
    of carotenoids, 394
Chromoplasts, from *Capsicum*
  electron microscopy of, 356
  integrity and purity of, 356
  isolation of, method for, 354
C3H/10T1/2 cells
  culture, 57–58
  uptake and half-life of selected carotenoids delivered to, 62–63
  retinoid action of, 66–68

treated with THF alone or with THF and
  β-carotene, growth curves of, 58–59
Chylomicronemia, distribution of β-carotene in, 36
Chylomicrons
  β-carotene in, postprandial levels, 36–37
  isolation of, 41
Cloned carotenoid genes, photoprotection
    by, 323–330
Cloned sequence, confirmation of, as gene
    of interest, 321–323
  analysis of revertants in, 322–323
  analysis of transposon-tagged alleles in,
    322
  linkage for, 321
  RNA analysis for, 321
Cloning
  of carotenoid biosynthetic genes, from
    maize, 311–323
  of *Synechococcus cbpA* gene, 394–395
  of transposon-tagged DNA, 317–321
  by transposon tagging, 311–313
    identification of tagged DNA fragments, 313–316
Column chromatography
  of astaxanthin, 156–158
  magnesium oxide, of astaxanthin, 157
  silica, of astaxanthin, 156–157
Colyte, 15–16
Cotyledons, cucumber, carotenoid accumulation in, experiments on
  methods for, 348–352
  procedure for, 349
C_{40} precursor, first, 333–335
crtA, *R. capsulatus*, nucleotide sequencing
    of, 297
crtB
  E. uredovora, nucleotide sequencing of,
    297
  R. capsulatus, 298
    nucleotide sequencing of, 297
CrtB proteins, 298
  domains in prenyltransferases for, 303–307
  E. herbicola, 299
  R. capsulatus, 299
  related proteins
    domains in prenyltransferases for,
      305
    sequence alignment of, 299–303

related tomato gene product, sequence
    alignments of, 304
  sequence alignments of, 299–302
crtC, *R. capsulatus*, nucleotide sequencing
    of, 297
crtD, *R. capsulatus*, nucleotide sequencing
    of, 297
CrtD proteins, 298
  FAD/NAD(P)-binding domains in, 310
  sequence alignment of, 307–309
crtE
  E. uredovora, nucleotide sequencing of,
    297
  R. capsulatus, 298
    nucleotide sequencing of, 297
CrtE proteins, 298
  domains in prenyltransferases for, 303–307
  related gene products, sequence alignments of, 304
  related proteins
    conserved domains in prenyltransferases for, 305
    sequence alignment of, 299–303
  sequence alignment of, 301–303
crtF, *R. capsulatus*, nucleotide sequencing
    of, 297
crtI
  E. uredovora, nucleotide sequencing of,
    297
  R. capsulatus, 298
    nucleotide sequencing of, 297
CrtI/Al-1 proteins, 298
CrtI proteins, 298
  anion-exchange chromatography of,
    384
  FAD/NAD(P)-binding domains in, 310
  purification of, 381–385
  SDS–PAGE of, 388
  sequence alignment of, 307–309
  size-exclusion chromatography of, 383
crtK, *R. capsulatus*, nucleotide sequencing
    of, 297
crtX, *E. uredovora*, nucleotide sequencing
    of, 297
crtY, *E. uredovora*, nucleotide sequencing
    of, 297
CrtY proteins, 298
  Erwinia, FAD/NAD(P)-binding domains
    in, 309–310

*crt*Z, *E. uredovora*, nucleotide sequencing of, 297
Crustaceans. *See also specific species*
  carotenoproteins in, 75, 82
Crustacyanin, 84–85
  carotenoid–protein linkage in, 85
  storage conditions, 85
Crustaxanthin
  relative abundance of, in carapace of *Penaeus japonicus*, 166
  R_f values of, on thin-layer chromatography, 160
  VIS spectra of, 162
Crustaxanthin tetraacetate, R_f values of, on thin-layer chromatography, 160
Cryptoxanthin
  certified reference material (SRM) for, informational values from, 11
  distribution among lipoproteins, 34–35
α-Cryptoxanthin
  from human plasma, 91–93
  from human thigh adipose tissue, 91
β-Cryptoxanthin
  E. coli, 324
  from human plasma, 91–93
  from human thigh adipose tissue, 91
  tetrahydrofuran as solvent for, 65–66
CTX. *See* Canthaxanthin
Cucumber cotyledons, carotenoid accumulation in, experiments on
  methods for, 348–352
  procedure for, 349
Cyanobacteria
  carotenoid-binding proteins in, 83, 390–401
  carotenoproteins in, 75, 83
Cycloheximide
  effect on carotenoid production in dark reaction process, 274–275
  effect on photoinduced carotenoid production, 279–280
Cytosol
  rat intestinal, retinol and retinoic acid synthesis from β-carotene or retinal by, 200
    inhibition by *m*-AMSA, 202
    inhibition by 4-methylpyrazole, 201
  rat liver, retinoid synthesis from β-carotene by
    effects of cofactors on, 199
    effects of pH on, 198

D

Daily dietary recall, 18
Dark reaction process, carotenoid production in, 272–273
  effect of cycloheximide on, 274–275
  effect of illumination time in light process on, 272
3,4-Dehydro-β, ψ-caroten-16-al, cleavage of, 268
Delayed light emission
  β-carotene modulation of, 202–208
    materials and methods, 203–204
  from chlorophyll *a*, apparatus for measuring, 204
  from chlorophyll induced by energized state, characteristics of, 205
  intensities of, and duration of dark intervals after, relation between, 206
Deletion mutants, from *E. coli* LE392/pPL376, 343
  carotenoids of, 343–346
    HPLC and absorbance spectra of, 344
Diacetylenic asterinic acid, isolation procedure, ratio, and specific radioactivity of, 164
Diadinoxanthin, 338
  structure of, 339
7,8-Didehydroastaxanthin
  preparation of, 155–156
  relative abundance of, in carapace of *Penaeus japonicus*, 166
  R_f values of, on thin-layer chromatography, 160
  stereoisomers of, ratio and specific radioactivity of, via diastereomeric (−)-dicamphanate derivatives, 165
2,3-Didehydro-4-oxomytiloxanthin, R_f values of, on thin-layer chromatography, 160
2,3-Didehydro-4-oxomytiloxanthin triacetate, R_f values of, on thin-layer chromatography, 160
Dietary assessment, errors of, 19
Diet history questionnaires, 19
Diet records, 18
Diets
  basal, 8–9
  low-carotenoid, 5–6
Difference spectroscopy, of light-induced carotenoid synthesis, 178–180

optical system for, 180
5,6-Dihydrocrustaxanthin, relative abundance of, in carapace of *Penaeus japonicus*, 166
5,6-Dihydrocrustaxanthin tetraacetate, VIS spectra of, 162
3,3'-Dihydroxy-β,β-carotene-4,4'-dione. *See* Astaxanthin
4,4'-Dimethoxy-β-carotene, 138
  structure of, 138
  UV/visible absorption spectra of, 140
Dimethylallyl pyrophosphate, isomerization of isopentenyl pyrophosphate to, 333
DLE. *See* Delayed light emission
DMAPP. *See* Dimethylallyl pyrophosphate
DNA
  al-3+ transformant, preparation of, 417–418
  effects of carotenoids on, 243–244
  flanking transposon, subcloning of, 320–321
  isolation of, 317–318
  pBR322, 328
DNA fragments, transposon-tagged, identification of, 313–316
DNA gel blots, hybridization of, 316
DNA transfer, 315–316
Dysbetalipoproteinemia, distribution of β-carotene in, 36

E

Echinenone
  in *Artemia*
    levels, 221–222
    organ and tissue distribution of, 223
    quantitation of, 212
Ectocarpus
  algal source, culture medium, and growth conditions for, 405–406
  isolation of xanthosomes, photosystem I–xanthosome vesicles, and monomeric forms from, 405–407
Ectocarpus siliculosus
  B63.81, 405–406
  xanthosomes, electron microscopic features of, 411
Electron microscopy
  of *Capsicum* chloroplasts, 355
  of *Capsicum* chromoplasts, 356
  of xanthosomes, 409–411

Electrophoresis, 315. *See also* SDS–polyacrylamide gel electrophoresis
Enzymology, 340
Epidermal/dermal homogenates, β-carotene concentrations in, for two skin separation methods, 129
Epidermal tissue, β-carotene extraction, 133–134
5,6-Epoxy-β,ϵ-carotene, cleavage of, 268
5,6-Epoxy-β-carotene, cleavage of, 268
Erwinia
  carotenoid biosynthesis genes, 298
  CrtY protein, FAD/NAD(P)-binding domains in, 309–310
Erwinia herbicola, 297
  carotenoid biosynthesis genes, 297
  CrtB protein, 299
Erwinia uredovora, 297
  carotenoid biosynthesis genes in, 297
  nucleotide sequencing of, 297
Escherichia coli
  β-carotene, 324
  carotenoid biosynthesis genes in, overexpression of, 380–381
  carotenoids in
    biosynthesis pathways for, 298
    photoprotection by, 323–330
    assessing, methods for, 325–326
  β-cryptoxanthin, 324
  mutants from, carotenoid analysis in, 341–347
  phytoene, 324
  strain HB101, 325
    carotenoid photoprotection by, 326–328
    effects of α-T plus near-UV on, 329–330
  strain HB101(pde116), effects of α-T plus near-UV on, 329–330
  strain HB101(pHC79)
    carotenoid photoprotection by, 326–328
    effects of α-T plus near-UV on, 329–330
  strain HB101(pPL376)
    carotenoid photoprotection by, 327
    effects of α-T plus near-UV on, 329–330
  strain LE392, 325–326
    effects of α-T plus near-UV on, 329–330

strain LE392(pde116), effects of α-T plus near-UV on, 329–330
strain LE392(pHC79), effects of α-T plus near-UV on, 329–330
strain LE392(pPL376)
  carotenoid photoprotection by, 327
  carotenoid pigments expressed in, 324
  deletion mutants from, 343
  carotenoids of, 343–346
  effects of α-T plus near-UV on, 329–330
  zeaxanthin, 324
Ethyl β-apo-8'-carotenoate, 138
  concentration versus time curve for, in human subject, 145
  structure of, 138
  UV/visible absorption spectra of, 139

F

FAD cofactors, ADP-binding $\beta\alpha\beta$ folds for, 310
FAD/NAD(P)-binding domains, in carotenoid dehydrogenases and cyclases, 309–311
FASTA (computer program), 299
Fatty tissues, carotenoids from
  extraction of, 88
  quantitative extraction and HPLC analysis of, 87–93
Ferret, 4
  β-carotene absorption in, 74–75
Ferret liver, carotenoids in, Vydac separation of, 111
Ferret tissue
  β-carotene concentrations in, 115
  carotenoids in, analysis of, experimental design for, 107–108
Fibronectin
  FACS analysis of, 254
  production of, effect of carotenoids on, 254–255
Ficoll-Hypaque gradient centrifugation, of undissolved carotenoid crystals and cells, technique for, 232–234
Fish. See Squirrel fish erythrophores
Flavobacterium, plastid preparations, 340
Flavobacterium dehydrogenans, photoinduction of carotenoid in, 291

Fluorescence emission spectroscopy, of xanthosomes, 409
Fluorescence excitation spectroscopy, of xanthosomes, 408–409
Food-frequency questionnaires, 18–19
Food lists, 18–19
Foods
  carotenoid content of, 8
  composition data, 17
  nutrient composition analysis of, 17
Fucoxanthin–chlorophyll a/c protein complexes, monomeric form, isolation of, from Ectocarpus, 407
Fungi, photoinduction of carotenoids in, 291
Fusarium, pic mutants, 293
Fusarium aquaeductuum
  carotenoid accumulation in, after transient and continuous illumination, time course of, 286
  carotenoid synthesis in, 175
  photoinduction of, 291
Fusarium moniliforme, photoinduction of carotenoid in, 291

G

Gastrointestinal lavage, 3
  estimation of apparent absorption of β-carotene by, 14–17
  assessment of, 16–17
  experimental design, 14–17
  preparation of effluent for analysis, 15–16
GCG software package, FIND program, 303, 306
Gel blots, DNA, hybridization of, 316
Gel-filtration chromatography
  of lipoproteins, 41–43
  of phytoene desaturase from Phycomyces C9, 372–373
Geranygeranyl pyrophosphate
  formation of, from isopentenyl pyrophosphate, 333
  formation of phytoene from, via prephytoene pyrophosphate, 334
Geranygeranyl pyrophosphate synthase
  characterization of, 357–358
  enzyme assay procedure for, 357
  immunofluorescence microscopy of, 362–363

from *N. crassa*, conserved domains in, 423–424
properties of, 358
purification of, 357–358
Western blotting of, 360–361
Geranygeranyl pyrophosphate synthase (*al-3+*) gene, cloning of, 417–419
GGPP. *See* Geranygeranyl pyrophosphate
Gibberella, pic mutants, 293
Gibberella fujikuroi, photoinduction of carotenoid in, 291
Glutathione, effects on human breast carcinoma MCF-7, 243–244
Goldfish xanthophores
 isolation of, 47–49
 translocation of carotenoid droplets in, 47–55
Guinea pigs, β-carotene absorption in, 74–75

H

Halobacterium, 335
Hamster cells, HCPC-1
 expression of p53 in, effect of β-carotene on, 252–253
 proliferation of, 241
HDL. *See* High-density lipoproteins
Heparin–sepharose affinity chromatography, for lipoprotein separation, 44–46
Hexane, 203
High-density lipoproteins
 association of carotenoids with, 34–35
 β-carotene in, postprandial levels, 36–37
 carotenoid content of, factors influencing, 36
 isolation of, by sequential ultracentrifugal flotation, 40
Higher plants, plastid development in, carotenoid synthesis in, regulation of, 340
High-performance liquid chromatography
 of astaxanthin, 158–162
 of β-carotene, 96
 of carotenoids, 46, 87
  reference compounds for, 343
 of carotenoids in deletion mutants from *E. coli* LE392/pPL376, 344
 of lipid extracts from standard BCC assay, 262

of photoisomerization products of pro-ζ-carotene, proneurosporene, and prolycopene, 347
resolution of MgO fractions by, 161
of retinoids, 194–197
Human breast carcinoma MCF-7, effects of carotenoids on, 242–244
Human embryonic lung fibroblasts (HLF)
 β-carotene cleavage in, 27–32
 culture, 23
Human kidney
 carotenoids in
  chromatographic separation of, 110
  concentrations, 114
 vitamin A concentrations in, 114
Human liver
 carotenoids in
  chromatographic separation of, 110
  concentrations, 114
  recovery of, effect of extraction method and saponification on, 105
 Supelco separation of, 113
 vitamin A concentrations in, 114
Human lung
 carotenoids in
  chromatographic separation of, 110
  concentrations of, 114
 fibroblasts (WI-38)
  β-carotene cleavage in, 27–32
  conversion of β-carotene to retinol and retinoic acid by, time and concentration dependence of, 27–32
  culture, 23
  uptake of β-carotene in
   concentration and time dependence of, 24–27
   mechanism of, 27
 vitamin A concentrations in, 114
Human lymphocytes and monocytes, determination of carotenoid activity in, 236–238
Human metabolism, of carotenoid analogs and apocarotenoids, 137–147
Human neoplastic cells, determination of carotenoid activity in, 241–247
Human nonneoplastic cells, determination of carotenoid activity in, 239–241
Human tissues, carotenoids in, 86–93, 102–116
 analysis of, experimental design for, 107

Hydroxycarotenoids, yellow, acetylation of, 156
3-Hydroxy-3-methylglutaryl-CoA reductase activity
  relationship between light dose and, 277
  in vivo during dark incubation after exposure to light, time course of, 276
  photoinduction of, 275–277
Hypercarotenemia, diagnosis of, methods for, 116
Hyperlipoproteinemia, distribution of β-carotene in, 36
Hypoxia, effects of, on SCC-25 cell growth, 250

I

Illumination
  length of period, 177
  protocols, 177–178
  temperature during, 178
  use of inhibitors during, 177–178
Ion-exchange chromatography, 188–189
  of phytoene desaturase from *Phycomyces* C9, 372
IPP. *See* Isopentenyl pyrophosphate
Isoastaxanthin, 149
  relative abundance of, in carapace of *Penaeus japonicus*, 166
  R_f values of, on thin-layer chromatography, 160
Isoastaxanthin diacetate, R_f values of, on thin-layer chromatography, 160
all-*trans*-Isoastaxanthin diacetate, VIS spectra of, 162
Isoelectric focusing, of phytoene desaturase from *Phycomyces* C9, 372–373
Isopentenyl pyrophosphate
  conversion of acetyl-CoA to, 332
  conversion to geranygeranyl pyrophosphate, 333
  isomerization to dimethylallyl pyrophosphate, 333
Isopentenyl pyrophosphate isomerase
  characterization of, 355–357
  enzyme assay procedure for, 355–356
  properties of, 357
  purification of, 356–357
  Western blotting of, 360–361
Isozeaxanthin, UV/visible absorption spectra of, 140

K

4-Ketoretinoic acid, as inhibitor of squamous metaplasia, 67

L

Laser, Q-switched ruby flash, carotenoid triplet formation induced by, kinetics of, 192
LDL. *See* Low-density lipoproteins
Light
  control of carotenoid synthesis by, 175–184
  direct measurement of, by difference spectroscopy, 178–180
  illumination protocols, 177–178
  use of mutants in, 178
  response of carotenoids to, 284–286
  sources of, 176–177
Light energy, measurement of, 176–177
Light intensity, effect on *cbpA* and *psbO* transcription, 399
Lipid extract, major carotenoid fractions of, 157
Lipoproteins. *See also* High-density lipoproteins; Low-density lipoproteins; Very low-density lipoproteins
  apolipoprotein B-containing, precipitation of, 43–44
  carotenoid content of, factors influencing, 36
  fractionation, methods for, 38–46
  fractions, extraction of carotenoids from, for chemical analysis, 46
  isolation of, by sequential ultracentrifugal flotation, 39–41
  postprandial incorporation of β-carotene into, 36–37
  precipitation of, 43–44
  separation of
    by gel-filtration chromatography, 41–43
    by heparin-sepharose affinity chromatography, 44–46
  techniques for, 38–46
Liposomes, encapsulating carotenoids, 231
Low-density lipoproteins
  antioxidant components of, 34
  β-carotene in, postprandial levels, 36–37

carotenoid content of, factors influencing, 36
isolation of, by sequential ultracentrifugal flotation, 40
oxidative modification of, and atherogenesis, 34–36
plasma carotenoids associated with, 33–36

Lutein
certified reference material (SRM) for, informational values from, 11
delivered in tetrahydrofuran, activity of, 63–64
distribution among lipoproteins, 34–35
effects of, on MCA-induced neoplastic transformation, 63, 65, 67
formation of β ring in, stereochemistry of cyclization leading to, 338
formation of ϵ ring in, stereochemistry of cyclization leading to, 338
from human plasma, 91–93
from human thigh adipose tissue, 91
uptake and half-life of, in cell cultures, 62–63

Lycopene
association with lipoproteins, 35
certified reference material (SRM) for, informational values from, 11
cyclization of, 337–338
stereochemistry of, 338
effects of, on MCA-induced neoplastic transformation, 63, 65, 67
formation of, from phytoene, 335
stereochemistry of double-bond formation in, 336
formation of β- and ϵ-carotenoid rings from, mechanism of, 337
from human plasma, 91–93
from human thigh adipose tissue, 91
hydration of, 338–339
hydrogenation of, 338–339
structure of, 330–331
uptake and half-life of, in cell cultures, 62–63

all-*trans*-Lycopene, absorption spectra of, 179

Lymphocytes
animal, determination of carotenoid activity in, 234–236
human, determination of carotenoid activity in, 236–238

Lymphoma HUT 78, inhibition by β-carotene, 239–240

M

Macrophages, animal, determination of carotenoid activity in, 234–236
Magnesium oxide column chromatography, of astaxanthin, 157
Maize, cloning carotenoid biosynthetic genes from, 311–323
Mangifera indica, carotenoid–protein complex from, 75, 84
Mangos. See *Mangifera indica*
Metabolic balance techniques, 3
estimation of carotenoid absorption by, 11–14
experimental design, 12–14
fecal preparation and analysis, 13–14
systematic error in, 12
8-Methoxypsoralen (8-MOP) irradiation, effects on plasma and cutaneous β-carotene concentrations in hairless mice, 134–135
8-Methoxypsoralen (8-MOP) photosensitization, effects on plasma and cutaneous β-carotene concentrations in hairless mice, 124–137
trans-Methylbixin, 67
tetrahydrofuran as solvent for, 65–66
Methylcholanthrene-induced neoplastic transformation, in C3H/10T1/2 cells, effect of carotenoids on, 63–68
Methylene blue, as photoreceptor, 292
4-Methylpyrazole, inhibition of β-carotene or retinal conversion to retinol and retinoic acid by rat intestinal cytosol by, 201
Mevalonic acid (MVA), 331–333
Mevinolin, effect on photoinduced carotenoid production, 278, 292
Mice
β-carotene absorption in, 74–75
hairless, plasma and cutaneous β-carotene concentrations in, psoralen photosensitization and, 124–137
Monkeys, 4
Monoacetylenic asterinic acid, isolation procedure, ratio, and specific radioactivity of, 164

Monocytes, human, determination of carotenoid activity in, 236–238
4-MP. *See* 4-Methylpyrazole
Mutants
 carotenoid, from *R. capsulatus*
  growth and characterization of, 376–377
  introducing plasmids into, conjugation for, 378–379
 deletion, from *E. coli* LE392/pPL376, 343
  carotenoids in, 343–346
  HPLC and absorbance spectra of, 344
 from *E. coli*
  carotenoids in
   analysis of, 341–347
   extraction of, 342
  characterization of, 341–347
  cultivation of, 341–342
Mutator transposon system, 312–313
Mycobacterium, 335
 photoinduction of carotenoid in, 291
Mycobacterium marinum
 carotenoid formation in, action spectra of, 183
 photoinduction of carotenoid in, 291
Mycobacterium smegmatis, photoinduction of carotenoid in, 291
Mytiloxanthin, R_f values of, on thin-layer chromatography, 160
Mytiloxanthin triacetate, R_f values of, on thin-layer chromatography, 160
Myxobactone, photoinduction of, in bacteria, 291
Myxococcus, carotenoid photoinduction in, genetic analysis of, 293
Myxococcus xanthus, carotenoid synthesis in
 action spectra of, 183
 photoinduction of, 291

N

NAD(P) cofactors, ADP-binding $\beta\alpha\beta$ folds for, 310
Narcissus, plastid preparations, 340
NBRF database, 306
Neoplastic cells
 animal, determination of carotenoid activity in, 241
 determination of carotenoid activity in, 241–247

human, RNA, DNA, and protein synthesis in, effects of carotenoids on, 243–244
Neoplastic transformation
 assays for, cells and culture conditions for, 57–58
 in cultured cells, activity of β-carotene and other carotenoids as inhibitors of, 55–68
Neoxanthin, 338
 structure of, 339
Neurospora, *wc* mutants, 293
Neurospora crassa, 297, 324
 al-1 gene, 298
  nucleotide sequencing of, 297
  walking to, 414
 al-3 gene
  results of transcript analyses, 423
  walking to, 414
 al-3$^+$ gene, cloning of, 417–419
 al-3$^+$ transformant DNA, preparation of, 417–418
 al-3$^+$ transformants, identification of
  results, 416–417
  spheroplast preparation for, 415–416
  strains and vectors for, 415
  transformation protocol for, 416
 carotenoid biosynthesis in
  genes for, 297, 412–424
  model for, 413
  pathways of, 298
 geranygeranyl pyrophosphate synthase from, conserved domains in, 423–424
 photoinduction of carotenoids in, 291, 420–423
  culture conditions and light treatment for, 420–421
  Northern hybridizations, 422–423
  results of transcript analyses, 423
  RNA extraction, 421–422
 prenyltransferases from, conserved domains in, 423–424
Neurosporaxanthin, photoinduction of, in fungi, 291
nif, expression of
 induction of, 379
 vectors for, 376
Nonneoplastic cells

determination of carotenoid activity in, 234–241
human, determination of carotenoid activity in, 239–241
Normal-phase chromatography, of *Artemia* cyst extracts, 214–215
Northern hybridizations, 422–423
Novobiocin, effect on *cbpA* and *psbO* transcription, 399
Nutrient intake, dietary assessment of, 18–19

O

Obesity, plasma carotenoid levels in, 116–123
Oral carcinoma cell SCC-25
  cell cycle analysis of, 249
  growth of, effect of hypoxia and β-carotene on, 250
  stress proteins on, SDS-PAGE of, 253
  ultrastructure of, effect of β-carotene on, 246–247
  viability of, effect of *o*-phenanthroline on, 244–245
Organelle motility, 47
Ovidin, storage conditions, 85
Ovorubin, 84–85
  storage conditions, 85
Ovoverdin, 84–85
4-Oxomytiloxanthin, relative abundance of, in carapace of *Penaeus japonicus*, 166
Oxygen
  effect on photoinduction of carotenoid production, 273
  singlet and triplet energies of, 186

P

Parsley, carotenoid–protein complex of, 84
Penaeus japonicus, 152
  carapace of, carotenoids in, relative abundance of, 166
  metabolism of carotenoids in, 148–168
Peridinin, 339
  structure of, 339
Petalonia xanthosomes, 402
  absorbance and fluorescence excitation spectra of, 408–409
  electron microscopic features of, 410
  fluorescence emission spectra of, 409
Petroleum ether, 203
o-Phenanthroline, effect on SCC-25 viability, 244–245
Photocarotenogenesis, 269–294
  action spectroscopy of, 280–283, 290–291
  in bacteria and fungi, 291
  biochemistry of, 293–294
  chemical modification of, 291–292
  culture and illumination conditions for, 286–287
  effect of cycloheximide on, 279–280
  effect of mevinolin on, 278–279
  effect of oxygen on, 273
  fluence-response curves for, 281–282
  light sources for, 287–288
  methods for investigating, 269–283
  stimulus for, 286–288
    and response, relation between, 288–291
  stimulus–response curves, 289–290
  time course of, 288–289
Photoinduction
  experiments
    with *N. crassa*, 420–423
    system for, 421
  genetic alteration with, 292–293
Photoisomerization, of carotenoids, 342–343
Photoisomerization products, of pro-ζ-carotene, proneurosporene, and prolycopene, HPLC of, 347
Photoprotection
  by carotenoids, methods for using *E. coli* to assess, 325–326
  by cloned carotenoid genes, 323–330
Photoreceptors, in carotenogenesis, 182–184
  characteristics of, 183
Photosensitization
  irradiation-induced edema responses for, 128
  and plasma β-carotene, 124–137
Photosensitizers, protoporphyrin IX, 323
Photosynthesis, functions of carotenoids in, 185–193
Photosynthetic bacteria, expression of carotenoid biosynthesis genes in, 375–380

functional analysis and purification of enzymes for, 374–385
Photosystem I–xanthosome vesicles
　isolation of, from *Ectocarpus*, 405–407
　procedure for, 406
　properties of, 410
Phycomyces
　carotenoid synthesis in, 175
　mad mutants, 293
　pic mutants, 293
　plastid preparations, 340
Phycomyces blakesleeanus
　C9*carR21*(−) strain
　　cell extract from
　　　phytoene desaturase activity in, 369
　　　phytoene desaturase products in, 370
　　　preparation of, 366
　　phytoene desaturase from
　　　chromatographic gel filtration of, 372–373
　　　ion-exchange chromatography of, 372
　　　isoelectric focusing of, 373
　　　polyethylene glycol precipitation of, 371–372
　　　purification of, 371
　C115 strain, cell extract from, phytoene desaturase activity in, 369
　photoinduction of carotenoid in, 291
　phytoene desaturase from, 365–373
　　purification of, procedures for, 371–373
Phytoene
　desaturation of, 335–337
　E. coli, 324
　formation of, 333–335
　　from geranygeranyl pyrophosphate via prephytoene pyrophosphate, 334
　formation of lycopene from, 335
　stereochemistry of double-bond formation in, 336
　formation of prolycopene from, 336
　photoinduction of, 278
　synthesis of, in plants, pathway of, 353
　tetrahydrofuran as solvent for, 65–66
Phytoene dehydrogenase. *See* Phytoene desaturase
Phytoene desaturase, 365
　amino acid sequences for, 324–325
　assay of, 366–368
　　incubation system for, 367
　from *P. blakesleeanus*, 365–373
　　purification of, procedures for, 371–373
　from *P. blakesleeanus* C9*carR21*(−)
　　chromatographic gel filtration of, 372–373
　　ion-exchange chromatography of, 372
　　isoelectric focusing of, 373
　　polyethylene glycol precipitation of, 371–372
　　purification of, 371
　solubilization of, 368–370
Phytoene synthase
　from *Capsicum*, component enzymes of, 353–359
　characterization of, 358–359
　enzyme assay procedure for, 358–359
　enzyme components
　　biogenesis of, 364–365
　　immunochemistry of, 359–363
　　immunocytochemistry of, 361–363
　　immunoinhibition assay procedure for, 359
　　Western blotting of, 360–361
　immunofluorescence microscopy of, 362–363
　plant, 352–365
　properties of, 359
　purification of, 359
　Western blotting of, 360–361
Pigment organelles, translocation of, 47
Plants
　carotenoid–protein complexes of, 84
　higher, plastid development in, carotenoid synthesis in, regulation of, 340
Plaque lifts, 318–319
Plaque purification, 319–320
Plasma
　β-carotene
　　extraction of, 132–133
　　after 8-MOP photosensitization and irradiation, 135
　　levels
　　　after 8-MOP photosensitization and irradiation, 134
　　　effects of psoralen photosensitization on, in hairless mice, 124–137
　　mean
　　　before and after ingesting test dose of β-carotene, 121

error rates for, 99
statistical analysis of, 96
variability in quantitation, 94–95, 97–101
methods for determining, 95–96
random effects analysis of, 97
photosensitization and, 124–137
carotenoids in
extraction of
for chemical analysis, 46
HPLC analysis of, 90–93
procedure, 95–96
levels
in anorexia nervosa and obesity, 116–123
variability in quantitation, 94–101
human
bilirubin pigments in, extraction of, 89–90
carotenoids in, 89–93
extraction of, 89–90
lipoprotein-depleted, 40
Plasmids, 375–376
chloramphenicol-resistant, from AL3-1 transformant, isolation of, 418
construction of, 381
containing intact *al-3+* gene, isolation of, 418–419
introducing into carotenoid mutants from *R. capsulatus*, conjugation for, 378–379
pBR322 DNA, 328
pde116, *E. coli* strains carrying, 329
pUC79, 325
E. coli strains carrying
carotenoid photoprotection by, 326–328
effects of α-T plus near-UV on, 329–330
pHL545, 325
pNC3, structure of, 419
pNC6, structure of, 419
pNC39, structure of, 419
pNF3, 376
pPL376, 325
E. coli strains carrying, 327
carotenoid photoprotection by, 326
carotenoid pigments expressed in, 324
deletion mutants from, 343

carotenoids in, 343–346
HPLC and absorbance spectra of, 344
effects of α-T plus near-UV on, 329
Plastids
active preparations, 340
development in higher plants, carotenoid synthesis in, regulation of, 340
isolation of, 353
Polyethylene glycol precipitation, of phytoene desaturase from *Phycomyces* C9, 371–372
Pregnancy, in rats, interaction of vitamin A deficiency with, 173–174
Prenyltransferases
CrtE, CrtB, and related protein domains in, 303–307
from *N. crassa*, conserved domains in, 423–424
Preparation assay, multistep, method for determining technical variability in, 94–101
Prephytoene pyrophosphate, formation of phytoene from geranygeranyl pyrophosphate via, 334
Prochlorophytes, carotenoid-binding proteins in, 390–401
Prochlorothrix hollandica
carotenoproteins from
absorbance spectroscopy of, 396
immunocytochemical localization of, 398–399
isolation of, 393–394
structure of, 401
cell growth, 392
Prolycopene
absorption spectra of, 179
conversion of phytoene to, 336
photoisomerization products of, HPLC of, 347
Proneurosporene, photoisomerization products of, HPLC of, 347
Proteins
carotenoid. *See* Carotenoprotein(s)
carotenoid biosynthesis enzymes, sequence alignment of, 299–303
cellular
activity and synthesis of, effects of carotenoids on, 248–255

oxidative state of, effects of carotenoids on, 248–250
protooncogene expression of, effects of carotenoids on, 251–254
determination of, 211–212
fucoxanthin–chlorophyll a/c complexes, monomeric form, isolation from *Ectocarpus*, 407
stress
  effects of carotenoids on, 250–251
  SDS-PAGE of, 250, 253
synthesis of, effects of carotenoids on, 243–244
xanthophyll–chlorophyll *a/c* complexes, supramolecular assemblies of, 402–412
Protoporphyrin IX, as photosensitizer, 323
psbO transcription, effect of light intensity and novobiocin on, 399
Psoralen, administration of, 127
Psoralen photosensitization
effects on plasma and cutaneous β-carotene levels in hairless mice, 124–137
effects on plasma and tissue β-carotene levels in hairless mice
  analytical methods, 130–134
  blood and tissue collection, 129–130
  experimental design, 126
  irradiations, 127–129

R

RA. *See* Retinoic acid
Rabbit β-carotene cleavage enzyme
  assay, lipid extracts from, HPLC of, 261–262
  boiled, assay, lipid extracts from HPLC of, 261–262, 264
RAR. *See* Retinoic acid receptors
Rat
  β-carotene absorption in, 74–75
  β-carotene feeding to, 76
Rat intestinal cytosol, retinol and retinoic acid synthesis from β-carotene or retinal by, 200
  inhibition by *m*-AMSA, 202
  inhibition by 4-methylpyrazole, 201
Rat intestine, vitamin A-deficient, 173–174
Rat liver
  carotenoproteins from, 75–86
  isolation of, 76
  storage of β-carotene in, 75–76
Rat liver cytosol, retinoid synthesis from β-carotene by
  effects of cofactors on, 199
  effects of pH on, 198
Renierapurpurin, tetrahydrofuran as solvent for, 65–66
Restriction endonuclease digestion, 314
Restriction endonuclease fragments, transposon-hybridizing, that cosegregate with mutant genotype, identification of, 314
Restriction fragment length polymorphism, to map *al-3+* gene, 420
Retinal
  from β-carotene, enzymatic reduction and esterification of, 260–261
  from all-*trans*-β-Carotene, enzymatic conversion of, 256–269
  ^{14}C-labeled, enzymatically formed, cocrystallization of, as *O*-ethyl oxime, with authentic retinal (*O*-ethyl) oxime, 266
  HPLC of, normal-phase, 196–197
  retinol and retinoic acid synthesis from, 21–22, 69–73
    by rat intestinal cytosol, 200
      inhibition by 4-methylpyrazole, 201
    in rat intestinal cytosol, inhibition by *m*-AMSA, 202
Retinal (*O*-ethyl) oxime
  ^{14}C-labeled, enzymatically formed, crystallization of, 259–260
  crystalline, mass spectrum of, 261, 265
  enzymatically formed
    HPLC of, 261–263
    mass spectrum of, 259–261, 265
Retinoic acid
  chemical identification of, 24, 30–31
  formation of, 73–74
  HPLC of, normal-phase, 195–197
  synthesis from β-carotene or retinal, 21–22, 69–74
    pathway of, 199–202
    by rat intestinal cytosol, 200
      inhibition by 4-methylpyrazole, 201
    by rat intestinal cytosol, inhibition by *m*-AMSA, 202
  in vitro, 193–202

Retinoic acid receptors, nuclear, 67-68
Retinoids
　activity in C3H/10T1/2 cells, 63-68
　chromatography of, 71
　HPLC of, 194-197
　production of, 70
　synthesis from carotenoids *in vitro*, 193-202
　　detergents for, 199
　　incubations for, 197-199
Retinol. *See also* Vitamin A
　chemical identification of, 24, 30-31
　HPLC of, normal-phase, 196-197
　synthesis from β-carotene or retinal, 21-22
　　by rat intestinal cytosol
　　　inhibition by *m*-AMSA, 202
　　　inhibition by 4-methylpyrazole, 201
Revertants, analysis of, 322-323
Rhodobacter capsulatus, 297
　carotenoid biosynthesis genes, 297-298
　　nucleotide sequencing of, 297
　carotenoid biosynthesis pathways, 298
　carotenoid mutant strains
　　growth and characterization of, 376-377
　　introducing plasmids into, conjugation for, 378-379
　CrtB protein, 299
　expression of carotenoid biosynthesis genes in, heterologous, 374-385
Rhodobacter sphaeroides
　strain 2.4.1, B800-850 complex, absorption, fluorescence emission, and excitation spectra of, 189-190
　strain GA
　　carotenoid triplet formation induced by Q-switched ruby laser flash in, kinetics of, 192
　　carotenoid triplet state induced by Q-switched ruby laser flash in, difference spectrum of, 192
Rhodopseudomonas palustris, fractionation of membranes from, by sucrose density gradient centrifugation, 187
Rhodotorula
　β-carotene in
　　extraction and isolation of, 387
　　purification and conservation of, 388
　　quantification of, 388

　　synthesis of, 386-401
　　　kinetics of, 389
　　　medium for, 386-387
　　cells suspended in distilled water, β-carotene synthesis in, 389
　　cultures, β-carotene synthesis in, 388-389
　　growth of, 386-387
Rhodotorula minuta, photoinduction of carotenoid in, 291
Riboflavin, as photoreceptor, 292
RNA
　effects of carotenoids on, 243-244
　extraction of, 421-422
　isolation of, 395
RNA analysis, for confirmation of cloned sequence as gene of interest, 321
Robertson's *Mutator* transposon system, 312-313
Ruby laser flash, Q-switched, carotenoid triplet formation induced by, kinetics of, 192
Ruzicka's rule, 331

S

Saponification, 103
　anaerobic
　　apparatus and equipment for, 154
　　of astaxanthin, 153-155
　conventional, of astaxanthin, 155
　effect on carotenoid recovery from human liver, 105
Scenedesmus, carotenoid synthesis in, 176
Scenedesmus obliquus C-6D
　absorption by, kinetics of light-induced increase in, 181
　all-*trans*-carotenes from, absorbance maxima and retention times of, 345
　poly-*cis*-carotenes from, 346-347
　　absorbance maxima and retention times of, 345
　carotenoid analysis in, 341-347
　cultivation of, 342
Scintillation counting, 172
SDS-polyacrylamide gel electrophoresis
　of CrtI protein, 388
　of stress proteins, 250, 253
Seaweeds, selection, harvesting, and transportation of, 403-404

Semiastacene, R_f values of, on thin-layer chromatography, 160
Sequences. *See also* Amino acid sequences
 cloned, confirmation of, as gene of interest, 321–323
Sequential flotation ultracentrifugation, of lipoproteins, 39–41
Serum samples, extraction and chromatographic analysis of, 142–146
Shrimp, 152. *See also specific species*
 carotenoids in
  metabolism of, 148–168
  quantification of, 151–152
 feeding, 152
 rearing conditions, 152
 seawater, 152
Silica column chromatography, of astaxanthin, 156–157
Size-exclusion chromatography, of CrtI protein, 383
Skin coloration, carotenoid levels reflected in, 20–21
Small intestinal cells (hBRIE 380), 22
 β-carotene cleavage in, 27–32
  analysis of, 23–24
 β-carotene uptake in
  analysis of, 23–24
  concentration and time dependence of, 24–27
  mechanism of, 27
 conversion of β-carotene to retinol and retinoic acid in
  effect of temperature on, 25–26, 31
  time and concentration dependence of, 27–32
 culture, 23
 retinol uptake and esterification in, 32
Small intestines, homogenates, 70–71
Spectroscopy
 absorbance
  of carotenoids in deletion mutants from *E. coli* LE392/pPL376, 344
  of carotenoprotein complexes, 396
  of xanthosomes, 408
 action, of photoinduced carotenogenesis, 180–182, 280–283, 290–291
 difference, of photoinduced carotenogenesis, 178–180
 fluorescence emission, of xanthosomes, 409

fluorescence excitation, of xanthosomes, 408–409
(3S,3'S)-Astaxanthin
 structure of, 150
 in vivo racemization of, in shrimp, 163
(3S,3'S)[15,15'-^{3}H$_2$] Astaxanthin, specific radioactivity of, 165
Squirrel fish erythrophores, translocation of carotenoid droplets in, 47
Stress proteins
 effects of carotenoids on, 250–251
 SDS-PAGE of, 250, 253
Subcloning, of DNA flanking transposon, 320–321
Subfractions, isolation of, 353
Sucrose density gradient centrifugation, 187–188
 fractionation of membranes from *Rhodopseudomonas palustris* by, 187
Supelco separation, of carotenoids in tissues, 113
Superoxide dismutase, percentage change in, related to thymidine proliferation, 251
Synechococcus PCC7942
 carotenoprotein from, 75, 83
  assay of, 394
  isolation of, 393
 42-kDa, 391
  absorbance spectroscopy of, 396
  cbpA gene encoding, cloning and sequencing of, 394–395
  immunocytochemical localization of, 398–399
 45-kDa
  absorbance spectroscopy of, 396
  immunocytochemical localization of, 398–399
  isolation of, 392–393
  and 42-kDa, sequence comparison of, 400
 cbpA gene, cloning and expression of, 394–395
 cell growth, 392
 cytoplasmic membranes and cell walls, preparation of, 392
Synechocystis PCC6714, carotenoprotein from, 75, 83, 391

T

Temperature, effect on carotenoid production, 271–272
Terpenoids
 biosynthesis of, early investigations, 331–333
 structure of, 331
α-Terthienyl
 chemical structure of, 328
 plus near-UV, effects of, 329
Tetraacetoxypirardixanthin
 R_f values of, on thin-layer chromatography, 160
 VIS spectra of, 162
7,8,7',8'-Tetradehydroastaxanthin
 preparation of, 155–156
 relative abundance of, in carapace of *Penaeus japonicus*, 166
 R_f values of, on thin-layer chromatography, 160
 stereoisomers of, ratio and specific radioactivity of, 165
3,4,3',4'-Tetrahydro-β-carotene, cleavage of, 268
Tetrahydrofuran
 as drug-delivery vehicle, characterization of, 58–59
 lack of effect on MCA-induced transformation, 63, 66
 as solvent for carotenoids, 56–57
 as solvent for lipophilic carotenoids, 64–66
Tetrahydroxycarotenoids, R_f values of, on thin-layer chromatography, 160
Tetrahydroxypirardixanthin, relative abundance of, in carapace of *Penaeus japonicus*, 166
Tetraterpenoids, structure of, 331
T7 expression vectors, 376
THF. *See* Tetrahydrofuran
Thin-layer chromatography
 of astaxanthin, 159
 of carotenoids, 394
 preparative, of astaxanthin, 159
Thymidine
 ³H-labeled
  β-carotene quenching of, 232–233
  effects of β-carotene on, 239
 proliferation of, percentage change in superoxide dismutase related to, 251
[³H]Thymidine incorporation assay, with normal human keratinocytes and SCC-25 cells, 245
Tissues. *See also specific tissue*
 β-carotene levels of, effects of psoralen photosensitization on, in hairless mice, 124–137
 carotenoids in, 102–116
 extracts, HPLC analysis of, 90–93
 homogenates, 70–71
 human, carotenoids in, 86–93, 102–116
α-Tocopherol, effects on human breast carcinoma MCF-7, 243–244
Toluidine blue, as photoreceptor, 292
Torulene, photoinduction of
 in fungi, 291
 in yeast, 291
Transposon
 DNA flanking, subcloning of, 320–321
 Mutator systems, 312–313
Transposon-hybridizing restriction endonuclease fragments, that cosegregate with mutant genotype, identification of, 314
Transposon-tagged alleles, analysis of, 322
Transposon-tagged DNA
 cloning of, 317–321
  hybridization and plaque purification, 319–320
  isolation of DNA to be cloned, 317–318
  plaque lifts, 318–319
 fragments, identification of, 313–316
  DNA transfer in, 315
  electrophoresis in, 315
  hybridization of DNA gel blots for, 316
  rationale for, 313–315
  restriction endonuclease digestion in, 315
Transposon tagging, 311–313
1,2,5-Trimethylphenyl, end group of chlorobactone, stereochemistry of rearrangement to form, 339
Tritium radioactivity, counting of, 152
Tumor cell viability, alteration in, with carotenoids, 244–245

U

Ubiquinone, photoinduction of, 278

V

Vectors
nif, 376
plasmid, 375–376. See also Plasmid(s)
T7, 376
Verticillium agaricinum
carotenogenesis in, 175
light-induced, 183
photoinduction of carotenoid in, 291
Very low-density lipoproteins
association of carotenoids with, 34–35
β-carotene in, postprandial levels, 36–37
carotenoid content of, factors influencing, 36
isolation of, by sequential ultracentrifugal flotation, 40
Vitamin A. See also Retinol
concentrations in human tissues, 114
deficiency of
induction of, 173–174
interaction with pregnancy in rats, 173–174
status of, biochemical evidence of, 174
Vitamin E
and carotenoid components of LDL, inhibition of oxidative modification of LDL, 34–36
effects on human breast carcinoma MCF-7, 242–243
Vitamins, fat-soluble, synergism with various carotenoids, 8–9
VLDL. See Very low-density lipoproteins
Vydac separation, of carotenoids in tissues, 110–111

W

Washout period
for plasma carotenoids, 4–5
for plasma carotenoids to return to baseline, 12
Western blotting, of phytoene synthase complex enzyme components, 360–361

X

Xanthophores
with aggregated pigment, preparation of, 50–51
isolation of, 47–49
maintenance of, 49–50
permeabilized
active factor (anterogin) in, 52, 55
for studies on pigment dispersion, 51
pigment dispersion/aggregation in, quantitation of, 51–52
translocation of carotenoids in, 47–55
motor for, 52, 55
unanswered questions about, 54–55
Xanthophyll–chlorophyll a/c protein complexes, supramolecular assemblies of, 402–412
Xanthosomes, 402–412
absorption spectrum of, 408
characterization of, 407–410
electron microscopic features of, 409–411
fluorescence emission and excitation spectra of, 408–409
isolation of
from brown algae, 403–405
choice of algae for, 403–404
collection and transport of seaweeds for, 404
procedure for, 404–405
from Ectocarpus, 405–407
procedure for, 406–407
problems encountered in, 403
monomeric forms, isolation of, from Ectocarpus, 405–407
polypeptide composition of, 410

Y

Yeast, photoinduction of carotenoids in, 291
Yellow esters, of lipid extract, 157
Yellow hydroxycarotenoids, acetylation of, 156
yl gene, cloning of, 312

Z

Zeaxanthin
certified reference material (SRM) for, informational values from, 11
in crustaceans, 148
distribution among lipoproteins, 34–35
E. coli, 324
formation of β ring in, stereochemistry of cyclization leading to, 338
from human plasma, 91–93
from human thigh adipose tissue, 91